Supplements to the 2nd Edition of

RODD'S CHEMISTRY OF CARBON COMPOUNDS

ELSEVIER SCIENCE PUBLISHERS B.V.
Sara Burgerhartstraat 25
P.O. Box 211, 1000 AE Amsterdam, The Netherlands

Distributors for the United States and Canada:

ELSEVIER SCIENCE PUBLISHING COMPANY INC.
655, Avenue of the Americas
New York, NY 10010, U.S.A.

Library of Congress Card Number: 64-4605
ISBN 0-444-88611-7

Printed in The Netherlands

Supplements to the 2nd Edition of

RODD'S CHEMISTRY OF CARBON COMPOUNDS

VOLUME I

ALIPHATIC COMPOUNDS
★

VOLUME II

ALICYCLIC COMPOUNDS
★

VOLUME III

AROMATIC COMPOUNDS
★

VOLUME IV

HETEROCYCLIC COMPOUNDS
★

Supplements to the 2nd Edition (Editor S. Coffey) of

RODD'S CHEMISTRY OF CARBON COMPOUNDS

A modern comprehensive treatise

Edited by
MARTIN F. ANSELL
Ph.D., D.Sc. (London), F.R.S.C., C.Chem.
Reader Emeritus, Department of Chemistry,
Queen Mary College, University of London, Great Britain

Supplement to

VOLUME IV HETEROCYLIC COMPOUNDS

PART E:
Six-membered Monoheterocyclic Compounds Containing Oxygen, Sulphur, Selenium, Tellurium, Silicon, Germanium, Tin, Lead or Iodine as the Hetero-atom

ELSEVIER
Amsterdam — Oxford — New York — Tokyo   1990

CONTRIBUTORS TO THIS VOLUME

Robert Livingstone, B.Sc., Ph.D., C.Chem., F.R.S.C.
Department of Chemical and Physical Sciences,
The Polytechnic, Huddersfield HD1 3DH

Malcolm Sainsbury, Ph.D., D.Sc., C.Chem., F.R.S.C.
Department of Chemistry,
The University of Bath, Bath BA2 7AY
(Index)

PREFACE TO SUPPLEMENT IV E

The publication of this volume continues the supplementation of the second edition of Rodd's Chemistry of Carbon Compounds, thus keeping this major work of reference up to date. This supplement covers chapters 20, 21 and 22 of volume IV E. Chapter 22 of the second edition was a reproduction of chapter IX of the original edition and was published as a tribute to Sir Robert Robinson. Advances in this area have been incorporated into the main text of the supplement. Although the chapters in this book stand on their own, they are intended to be read in conjunction with the parent chapters in the second edition.

At a time when there are many specialist reviews, monographs and reports available, there is still in my view an important place for a book such as "Rodd", which gives a broader coverage of organic chemistry. One aspect of the value of this work is that it allows the specialist in one field to quickly find out what is happening in other fields of organic chemistry. On the other hand a chemist looking for a way into a field of study will find in "Rodd" an outline of the important aspects of that area of chemistry together with leading references to other works to provide more detailed information.

As editor, I wish to express my thanks to Professor Livingstone who agreed to write the whole of this volume, thus supplementing the chapters he wrote in the second edition. He has produced a readable concise survey of the very considerable amount of work published on six-membered monoheterocyclic compounds having as the hetero atom either oxygen, sulphur, selenium, tellurium, silicon, germanium, tin, lead or iodine. I am grateful to Professor Livingstone and to his secretary, Mrs Susan Smith, for all the care and effort they have put into the preparation of the manuscript for direct reproduction.

My thanks are also extended to Dr Malcolm Sainsbury for the preparation of the extensive and detailed index and to the staff at Elsevier for all the help they have given me and for transforming the author's manuscript into the published work.

Martin F. Ansell   October 1989

CONTENTS

VOLUME IV E

Heterocyclic Compounds: Six-membered Monoheterocyclic Compounds Containing Oxygen, Sulphur, Selenium, Tellurium, Silicon, Germanium, Tin, Lead or Iodine as the Hetero-atom

Chapter 20. Six-membered Ring Compounds with One Hetero Atom: Oxygen
by R. LIVINGSTONE

Chapter 21. *Six-membered Ring Compounds with One Hetero Atom: Sulphur,
Selenium, Tellurium, Silicon, Germanium, and Tin*
by R. LIVINGSTONE

OFFICIAL PUBLICATIONS

B.P.	British (United Kingdom) Patent
F.P.	French Patent
G.P.	German Patent
Sw.P.	Swiss Patent
U.S.P.	United States Patent
U.S.S.R.P.	Russian Patent
B.I.O.S.	British Intelligence Objectives Sub-Committee Reports
F.I.A.T.	Field Information Agency, Technical Reports of U.S. Group Control Council for Germany
B.S.	British Standards Specification
A.S.T.M.	American Society for Testing and Materials
A.P.I.	American Petroleum Institute Projects
C.I.	Colour Index Number of Dyestuffs and Pigments

SCIENTIFIC JOURNALS AND PERIODICALS

With few obvious and self-explanatory modifications the abbreviations used in references to journals and periodicals comprising the extensive literature on organic chemistry, are those used in the World List of Scientific Periodicals.

LIST OF COMMON ABBREVIATIONS AND
SYMBOLS USED

A	acid
$\overset{\circ}{A}$	Ångström units
Ac	acetyl
a	axial; antarafacial
as, $asymm.$	asymmetrical
at	atmosphere
B	base
Bu	butyl
b.p.	boiling point
C, mC and μC	curie, millicurie and microcurie
c, C	concentration
C.D.	circular dichroism
conc.	concentrated
crit.	critical
D	Debye unit, 1×10^{-18} e.s.u.
D	dissociation energy
D	dextro-rotatory; dextro configuration
DL	optically inactive (externally compensated)
d	density
dec. or decomp.	with decomposition
deriv.	derivative
E	energy; extinction; electromeric effect; Entgegen (opposite) configuration
E1, E2	uni- and bi-molecular elimination mechanisms
E1cB	unimolecular elimination in conjugate base
e.s.r.	electron spin resonance
Et	ethyl
e	nuclear charge; equatorial
f	oscillator strength
f.p.	freezing point
G	free energy
g.l.c.	gas liquid chromatography
g	spectroscopic splitting factor, 2.0023
H	applied magnetic field; heat content
h	Planck's constant
Hz	hertz
I	spin quantum number; intensity; inductive effect
i.r.	infrared
J	coupling constant in n.m.r. spectra; joule
K	dissociation constant
kJ	kilojoule

LIST OF COMMON ABBREVIATIONS

k	Boltzmann constant; velocity constant
kcal	kilocalories
L	laevorotatory; laevo configuration
M	molecular weight; molar; mesomeric effect
Me	methyl
m	mass; mole; molecule; *meta-*
ml	millilitre
m.p.	melting point
Ms	mesyl (methanesulphonyl)
[M]	molecular rotation
N	Avogadro number; normal
nm	nanometre (10^{-9} metre)
n.m.r.	nuclear magnetic resonance
n	normal; refractive index; principal quantum number
o	*ortho-*
o.r.d.	optical rotatory dispersion
P	polarisation, probability; orbital state
Pr	propyl
Ph	phenyl
p	*para-*; orbital
p.m.r.	proton magnetic resonance
R	clockwise configuration
S	counterclockwise config.; entropy; net spin of incompleted electronic shells; orbital state
S_N1, S_N2	uni- and bi-molecular nucleophilic substitution mechanisms
S_Ni	internal nucleophilic substitution mechanisms
s	symmetrical; orbital; suprafacial
sec	secondary
soln.	solution
symm.	symmetrical
T	absolute temperature
Tosyl	*p*-toluenesulphonyl
Trityl	triphenylmethyl
t	time
temp.	temperature (in degrees centigrade)
tert.	tertiary
U	potential energy
u.v.	ultraviolet
v	velocity
Z	zusammen (together) configuration

LIST OF COMMON ABBREVIATIONS

α	optical rotation (in water unless otherwise stated)
$[\alpha]$	specific optical rotation
αA	atomic susceptibility
αE	electronic susceptibility
ε	dielectric constant; extinction coefficient
μ	microns (10^{-4} cm); dipole moment; magnetic moment
μB	Bohr magneton
μg	microgram (10^{-6}g)
λ	wavelength
ν	frequency; wave number
$\chi, \chi d, \chi \mu$	magnetic, diamagnetic and paramagnetic susceptibilities
$\sim$	about
(+)	dextrorotatory
(-)	laevorotatory
(±)	racemic
$\ominus$	negative charge
$\oplus$	positive charge

Chapter 20

SIX-MEMBERED RING COMPOUNDS WITH ONE HETERO ATOM: OXYGEN

R. LIVINGSTONE

   There has been considerable growth in the chemistry of
six-membered heterocycles containing one oxygen atom. This
has occurred mainly in existing areas, with the synthesis
of derivatives of well-known parent compounds, rather than
in any particular new area. Some new natural products have
been isolated and their structures determined, but in general
they have been modifications of well-known natural products.
A number of already known natural products, for example,
the anthocyanins, have been obtained from new sources. New
derivatives of existing compounds have been modified in
attempts to develop new medicinals for various purposes and
compounds for use as herbicides, fungicides, and insecticides.
Advances have been made in the application of various
physico-chemical methods, including different types of
chromatography and spectral analysis, to the separation and
identification of related natural products and some of
their degradation products.

1. Pyran and its derivatives

(a) Pyrans and pyrylium salts

   (i) 2H- and 4H-Pyrans (2- and 4-pyrans, α- and γ-pyrans)
Preparation. 2*H*-Pyrans (1) have been obtained by the
cyclization of 1,5-diphenylpenta-2,4-diyn-1-ol with amines
(W. Reid and B. Podkowik, Chem. -Ztg., 1982, 106, 296).

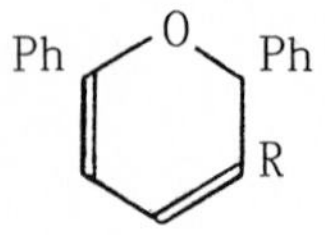

Ph—O—Ph   R = pyrrolidino, morpholino,
                 piperazino,
                 4-(2-methoxyphenyl)piperazino

(1)

The 2-amino-3-ethoxycarbonyl-4H-pyran (2) results from
the Michael addition reaction between PhCH = C(CN)COPh and
ethyl cyanoacetate and its reaction with phenylmagnesium
bromide affords the β-enamino ketone (3). The reaction
between PhCH = C(CN)COPh and malononitrile gives nitrile (4)
(M.R.H. Elmoghayer *et al.*, Monatsh, 1982, **113**, 53).

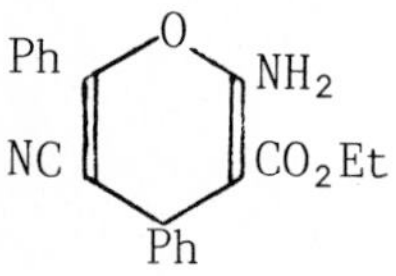

(2) 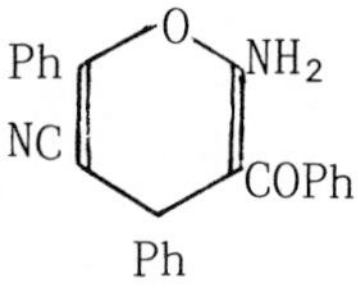   (3) 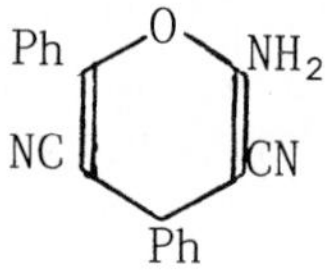   (4)

Treatment of PhCH=CPhCOPh with malononitrile and piperidine
in ethanol at room temperature yields 2-amino derivative (5)
and derivatives (6) are formed from R^{1}CH=CClCHR^{2}COR^2 and
malononitrile and potassium cyanide (J.L. Soto *et al.*,
Heterocycles, 1983, **20**, 803).

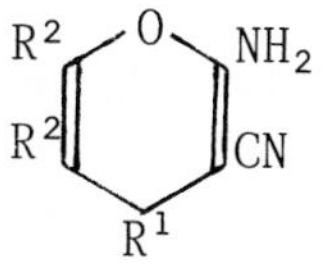

R^2 ―O― NH_2 / R^2 ― CN / R^1

(5) R^1 = R^2 = Ph

(6) R^1 = Ph, $4\text{-}MeC_6H_4$, $4\text{-}MeOC_6H_4$, $4\text{-}ClC_6H_4$

R^2 = Ph, $4\text{-}MeOC_6H_4$

A number of derivatives (7) have been prepared by the cycloaddition reaction between $R^1CH=CR^2COR^3$ and malononitrile, for example, $PhCH=C(CO_2Et)COMe$ with malononitrile and piperidine affords derivative (7; R^1 = Ph, R^2 = CO_2Et, R^3 = Me) (*idem, ibid.*, 1984, **22**, 1).

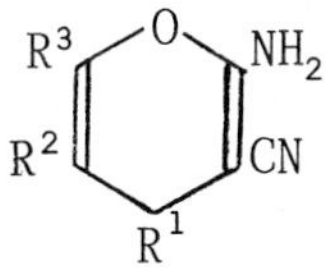

R^3 ―O― NH_2 / R^2 ― CN / R^1

R^1 = Ph, Me, OMe, Cl, NO_2, CN, alkyl, $Cl_2C_6H_3$

R^2 = CO_2Et, Ph, CN, Ac

R^3 = Me, Ph

(7)

2-Amino derivative (8) has been obtained as a minor product from the Knoevenagel reaction between ω-cyanoacetophenone and benzaldehyde (*idem, ibid.*, 1983, **20**, 2393).

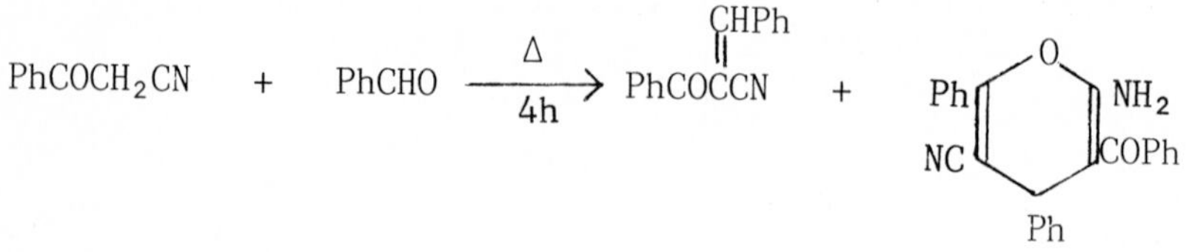

$$PhCOCH_2CN \quad + \quad PhCHO \xrightarrow[\text{4h}]{\Delta} \overset{\overset{\displaystyle CHPh}{\|}}{PhCOCCN} \quad +$$

(8)

The preparation of 2-amino-3,5-dicyano-4-phenyl-6-(2-
-phenylethen-1-yl)-4H-pyran (G.A.M. Nawwar *et al.*,
Heterocycles, 1985, **23**, 2983) has been described.
Bis(4H-pyrans) (9) separated by three, four, and six
ethanediylidene groups have been synthesized and a study made
of their electrochemical properties (C.H. Chen *et al.*, J. org.
Chem., 1983, **48**, 2757).

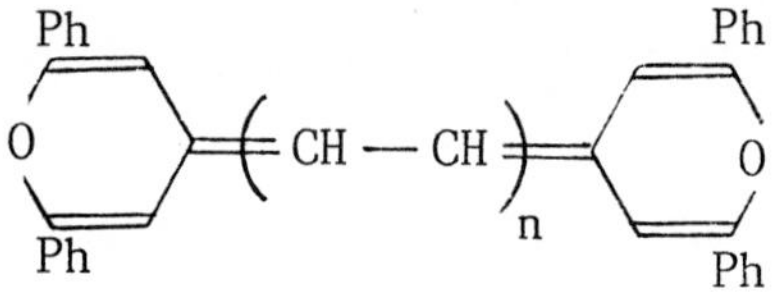

$$n = 3, 4, 6$$

(9)

Two hemispherands (10; R = H and Me) containing a central
4H-pyran unit have been synthesized *via* appropriately
substituted 2,6-diphenylpyrylium salts (P.J. Dijkstra *et al.*,
Tetrahedron Letters, 1986, **27**, 3183).

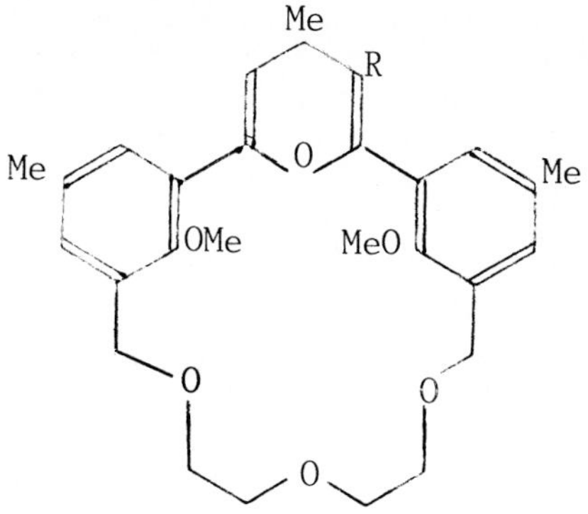

(10)

Reactions. The cycloelimination of ketones or aldehydes from the adduct obtained by cycloaddition of 2H-pyrans with acetylenic dienophiles, results in the formation of substituted benzenes. Methyl propiolate and 2,2,4,6-
-tetramethyl-2H-pyran (11) give methyl 2,4-dimethylbenzoate
(12) (R.G. Salomon, J.R. Burns and W.J. Dominic, *ibid.*, 1976, **41**, 2918).

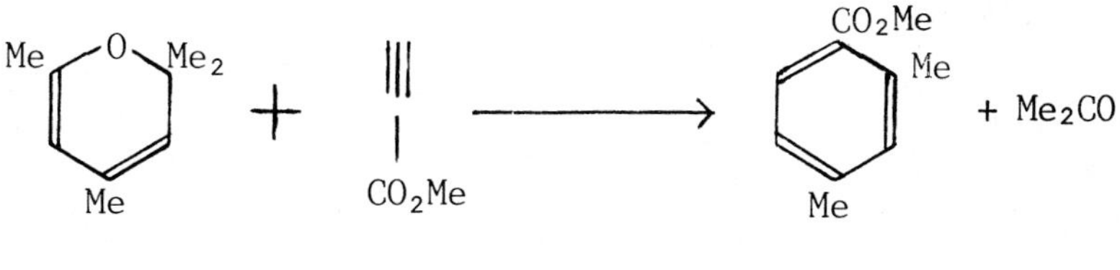

(11)   (12)

The aminopyran (4) on treatment with acetic anhydride affords a pyranopyrimidine (13); with formamide, a pyridine (14); and with 3-methyl-4-phenylpyrazol-5-amine, a pyranopyrazolopyrimidine (15) (E.M. Zayed *et al*., Rev. Port. Quim., 1982, <u>24</u>, 133).

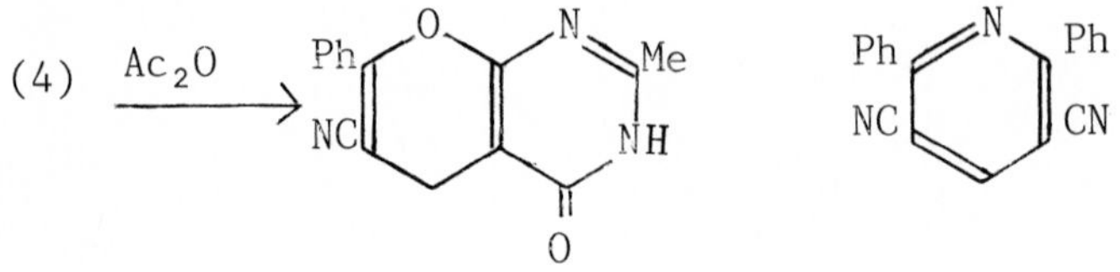

$$(4) \xrightarrow{\text{Ac}_2\text{O}}$$

(13)   (14)

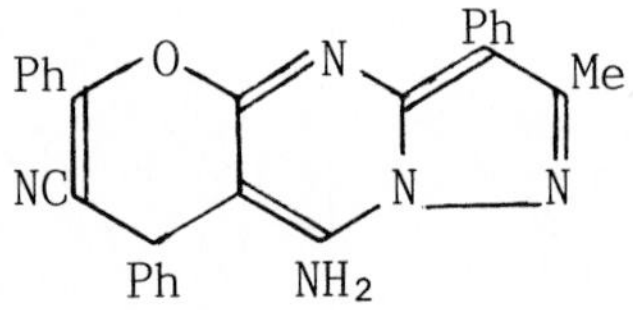

(15)

4-Dicyanomethylene-2,6-dimethyl-4*H*-pyran (16) on irradiation in methanol gives the dimers (17) and (18). The structure of the former has been determined by X-ray analysis (N.W. Alcock and C.J. Samuel, Chem. Comm., 1982, 603).

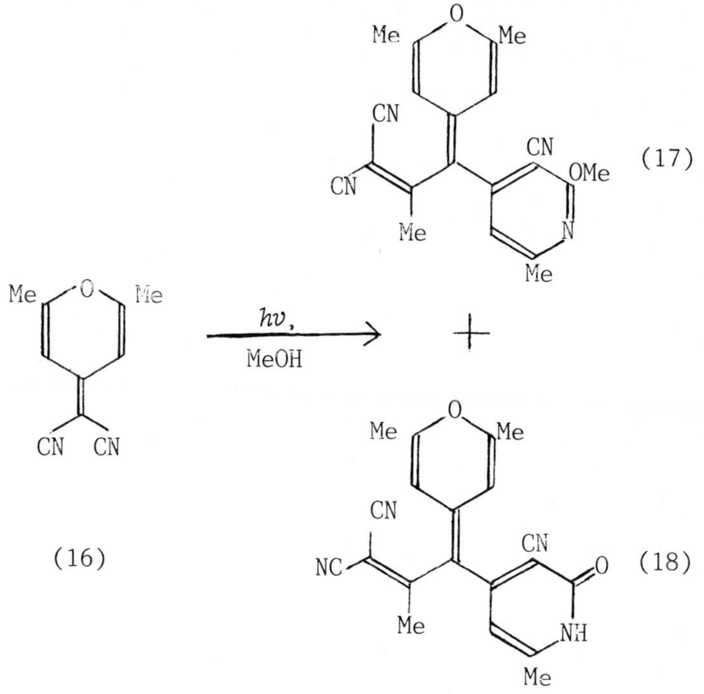

(16) → hv, MeOH → (17) + (18)

When equimolar amounts of 2-methoxy-6-methyl-4H-pyran (19) and dicyanostyrene react together in chloroform at room temperature, the cyclobutane derivative (20; 95%) is formed after $ca.$ 0.5h, but with longer reaction times it gradually disappears, until ultimately only the heptadienoic ester (21) remains. Similar treatment of 3,6-dimethyl-2-methoxy-4H- -pyran affords the pyrone derivative (22). Reactions with related compounds have also been studied (C.G. Bakkar et $al.$, J. org. Chem., 1983, 48, 2736). Both conversions are believed to occur via the abstraction of a hydride ion from C-4 in a rate-determing step leading to a pyrylium ion.

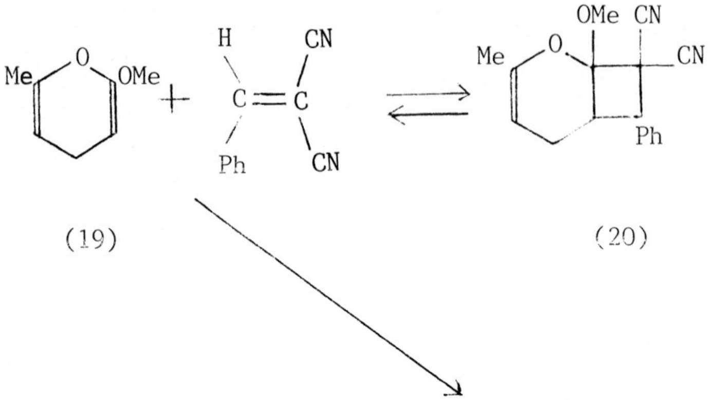

(19)　(20)

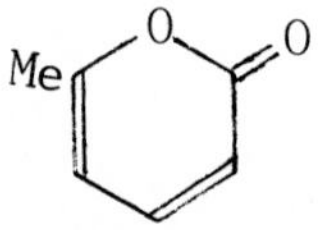

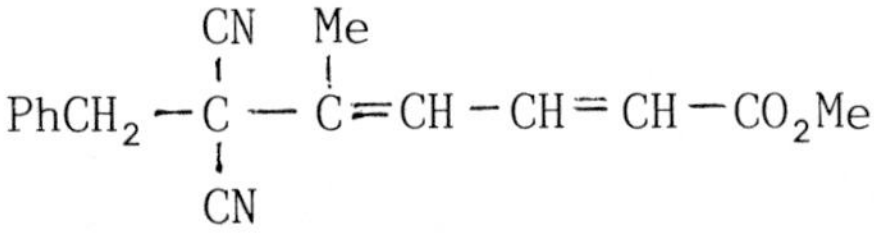

(22)　(21)

The major synthetic routes to the 4*H*-pyran ring system
(Soto and M. Quinterio, Heterocycles, 1980, **14**, 337) and
naturally occurring compounds containing the 4*H*-pyran ring
have been reviewed (I.E. Games, Aromat. heteroaromatic Chem.,
1983 **1**, 369).

(ii) Pyrylium salts
Preparation. The addition of *isobutene* or *tert*-butanol to
acetic anhydride-sulphuric acid gives 2,4,6-trimethylpyrylium
sulphoacetate (1), which can be stored and used to prepare
other heterocyclic or carbocyclic compounds (A. Dinculescu
and A.T. Balaban, Org. prep. Proced. Int., 1982, **14**, 39).

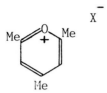

(1) X = O₃SCH₂CO₂H
(2) X = ClO₄
(3) X = BF₄

The treatment of 3,5-di-*tert*-butyl-1,2-benzoquinone with acetic anhydride in perchloric acid (70%) yields 2,4,6--trimethylpyrylium perchlorate (2) and an acetylated benzene derivative; salt (3) is obtained when the reaction is carried out in the presence of boron trifluoride etherate. Similarly propionic anhydride affords 2,6-diethyl-4--methylpyrylium perchlorate. Acetic anhydride and perchloric acid with 2,6-di-*tert*-butyl-1,4-benzoquinone gives the pyrylium salt (2) (L.Yu. Ukhin *et al*., Khim. Geterotsikl. Soedin., 1983, 454). A number of pyrylium salts (e.g. 4 and 5) with furyl substituents have been prepared (V.G. Kharchenko, E.V. Burov and V.A. Sedavkina, *ibid.*, 1981, 1604) and known and previously unavailable pyrylium salts have been obtained by the acid catalyzed cyclization of unsaturated 1,5-diketones (K.T. Potts, U.S.Pat. 4,451,659, 1984).

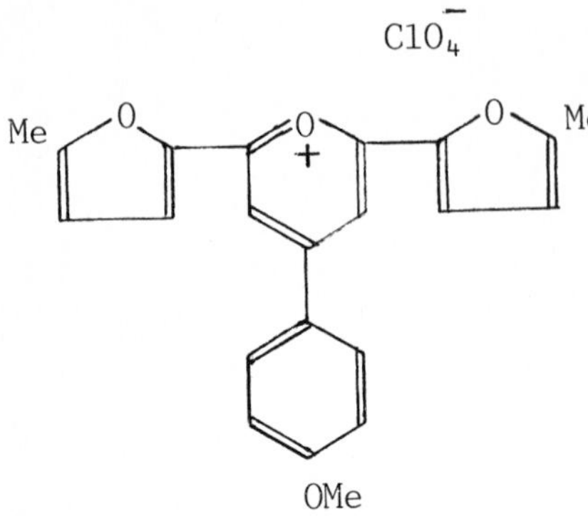

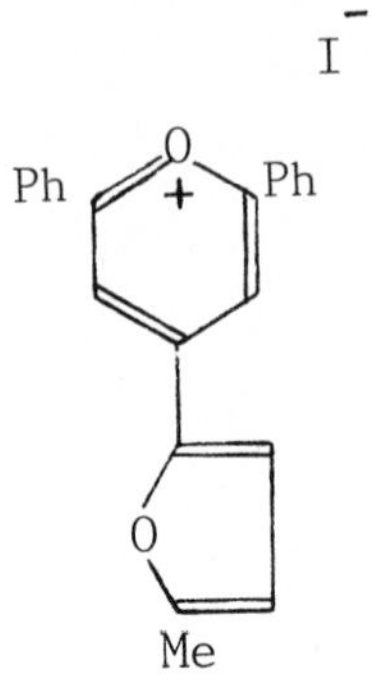

(4)   (5)

Reactions. The rate constants for 2*H* and 4*H* adduct formation
have been determined for the reaction of 2,6-diphenyl- and
4-substituted (OMe and Ph) 2,6-diphenyl-pyrylium ions with
methoxide ions in methanol. In the case of 2,4,6-
-triphenylpyrylium salts, the 2*H* adduct is only formed to a
slight extent in dipolar aprotic solvents and not at all in
methanol solution. The equilibrium constants have been
evaluated for the formation of the 4*H* adducts (G. Doddi *et al.*,
J. Org. Chem., 1982, **47**, 960).

The reaction of pyrylium salts with primary amines is an
important route to the synthesis of *N*-substituted pyridinium
salts (6). The kinetics of the reaction have been studied
(*idem, ibid.*, 1983, **48**, 5268).

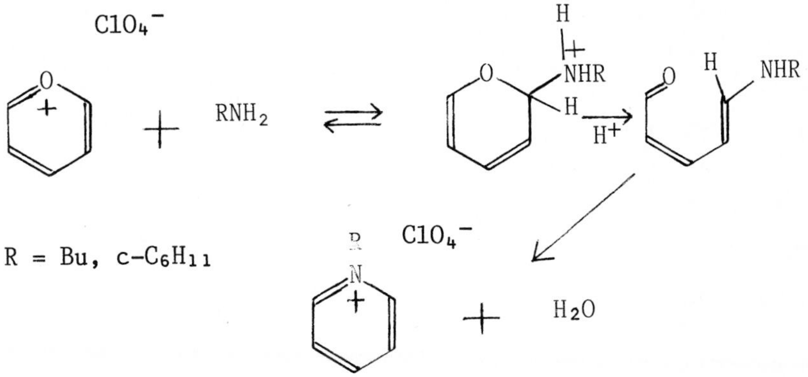

R = Bu, c-C$_6$H$_{11}$

(6)

2,3,5,6-Tetraphenylpyrylium perchlorate (7) reacts with sodium azide to yield the 2-azido-derivative, which on warming loses nitrogen to give the oxazepine (9), *via* the isolable azirine (8) (J-P. Le Roux, J-C. Cherton and P.L. Desbene, Compt. rend., 1975, 280C, 7).

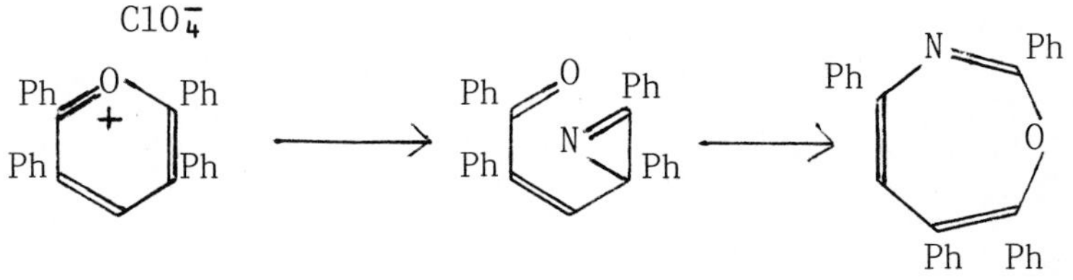

(7)   (8)   (9)

*(b) 2H-Pyran-2-ones (α- or 2-pyrones) and 4H-pyran-4-ones
(γ- or 4-pyrones)*

(i) 2H-Pyrones

α-Oxoketene dithioacetals have been used for the
synthesis of annulated and simple 2*H*-pyrones. The α-
oxoketene dithioacetal (1) with lithium methyl(phenylthio)-
copper in THF gives the enone (2), which on addition of
tert-butyl lithioacetate, followed by quenching with 2M HCl
yields the δ-keto ester (3), affording the 2*H*-pyrone (4) on
treatment with trifluoroacetic acid in trifluoroacetic
anhydride. A number of 2*H*-pyrones have been obtained by this
method (R.K. Dieter and J.R. Fishpaugh, J. org. Chem., 1983,
<u>48</u>, 4439).

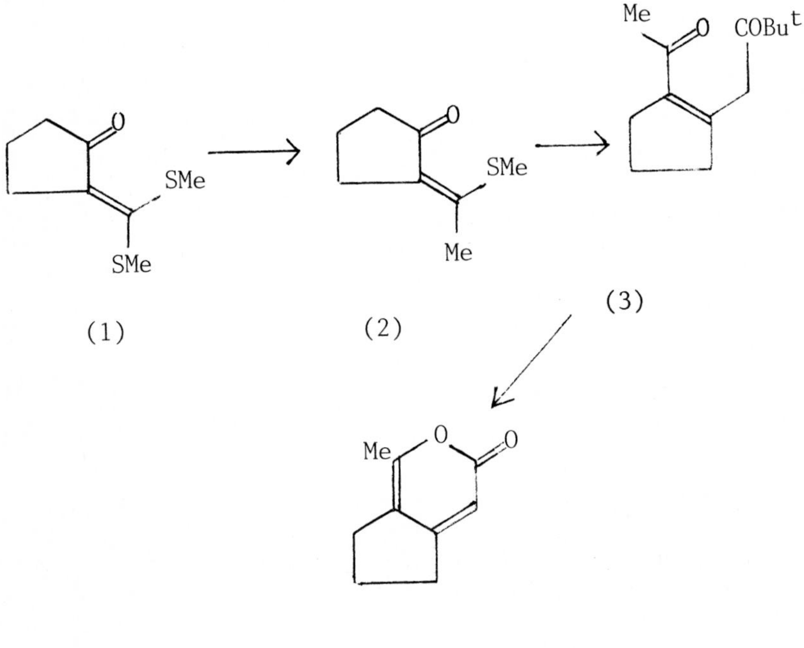

(1)　　　　　　　(2)　　　　　　　(3)

(4)

The [4+2] cycloaddition reaction between $\alpha\beta$-unsaturated carbonyl compounds (5) and phosphacumulene ylides (6) results in the formation of pyran-substituted phosphoranes (7), which on benzoic acid catalyzed Hofmann degradation yield $2H$--pyrones (8) (H.J. Bestmann and G. Schmid, Tetrahedron Letters, 1984, 25, 1441).

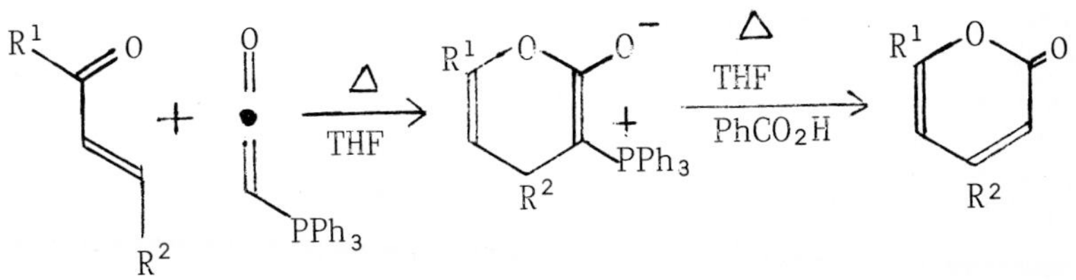

R^1 = Ph, R^2 = Ph, Bz, 2-Furyl
R^1 = CH=CHPh, R^2 = Ph

(5) (6) (7) (8)

A number of $3(R^1)$, $4(R^2)$, $5(R^3)$-trisubstituted-$2H$--pyrones [R^1 = Me, pentyl; R^2 = Me, Et; R^3 = Et, Pr; R^2R^3 = $(CH_2)_5$, o-phenylene ethene] have been obtained in high yields by the esterification of $(EtO)_2P(O)CHR^1CO_2H$ with $HOCH=CR^3COR^2$ using DMF/oxalyl chloride, followed by cyclization of the resulting product $(EtO)_2P(O)CHR^1CO_2CH=CR^3COR^2$ via the Wittig-Horner-Emmons reaction (H. Stetter and H.J. Kogelnik, Synth., 1986, 140).

It has been reported that the ruthenium-catalyzed oxygenation of 3,5-di-*tert*-butylcatechol gives a mixture of a di-*tert*-butylmuconic acid and 4,6-di-*tert*-butyl--2*H*-pyrone (M. Matsumoto and K. Kuroda, J. Amer. chem. Soc., 1982, **104**, 1433). A process initiated by the photochemical oxygenation of cyclohexylfurans (9 and 10) results in the formation of 2*H*-pyrones (11 and 12) (F.J. Jaeggi *et al.*, Heterocycles, 1982, **19**, 1839).

$$
\begin{array}{cccc}
(9) & (10) & (11) & (12)
\end{array}
$$

4-Methoxycarbonyl-6-trifluoromethyl-2*H*-pyran-2-one (16) has been synthesized starting from the 1:1-adduct (13) of the copper (I) chloride-catalyzed addition of 1,1,1-trichloro--2,2,2-trifluoroethane to dimethyl itaconate. Adduct (13) dehydrochlorinates on treatment with triethylamine in boiling toluene to give a mixture of butadienes (14 and 15), which in boiling mesitylene in the presence of a small amount of hydroquinone and copper (I) chloride, ring close to yield the 2*H*-pyrone (16) (P. Martin *et al.*, Tetrahedron Letters, 1985, **26**, 3947).

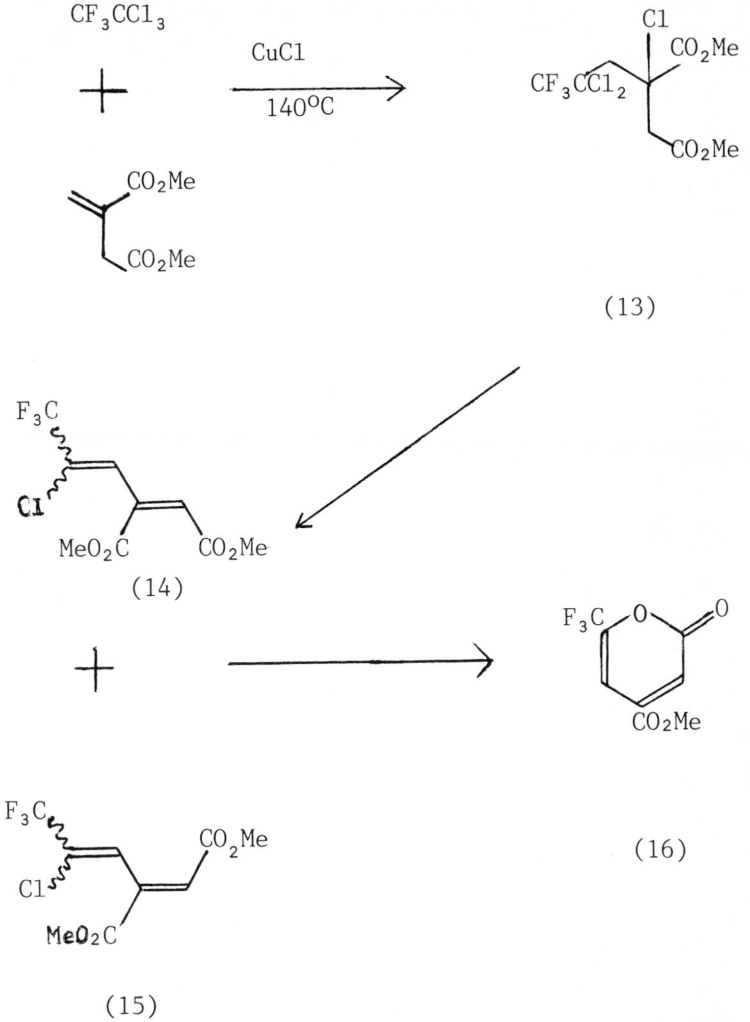

CF$_3$CCl$_3$

+

$\xrightarrow[\text{140}^\circ\text{C}]{\text{CuCl}}$

CF$_3$CCl$_2$—C(Cl)(CO$_2$Me)—CH$_2$CO$_2$Me

CH$_2$=C(CO$_2$Me)—CH$_2$CO$_2$Me

(13)

F$_3$C—CH(Cl)—CH=C(CO$_2$Me)—CH(CO$_2$Me)

(14)

+

$\longrightarrow$

F$_3$C—O—C=O pyrone, CO$_2$Me

(16)

F$_3$C—CH(Cl)—CH=C(CO$_2$Me)—CH—CO$_2$Me

(15)

A number of dienones have been oxidized using iodoxybenzene
in the presence of a catalytic amount of tetraphenylporphin-
atomanganese (III) chloride to give a mixture of epoxides and
pyrones. Thus pentadienone (17) affords a 1:1 mixture of
3,5-di-*tert*-butyl-2*H*-pyrone (18) and epoxide (19) (T. Takata,
R. Tajima and W. Ando, Chem. Letters, 1985, 665).

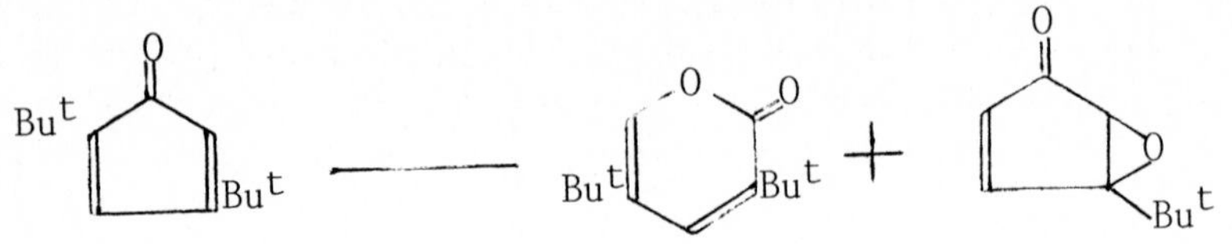

(17)   (18)   (19)

Tricarbonyliron complexes (21) of some 6-ethoxy-2H-pyrones are obtained by the addition of an alkyne to the iron carbene complex (20) and subsequent incorporation of two molecules of carbon monoxide. Rearrangement to the isomer (22) occurs on continued heating of the complex (21) (M.F. Semmelhack *et al.*, J. Amer. chem. Soc., 1984, 106, 5363).

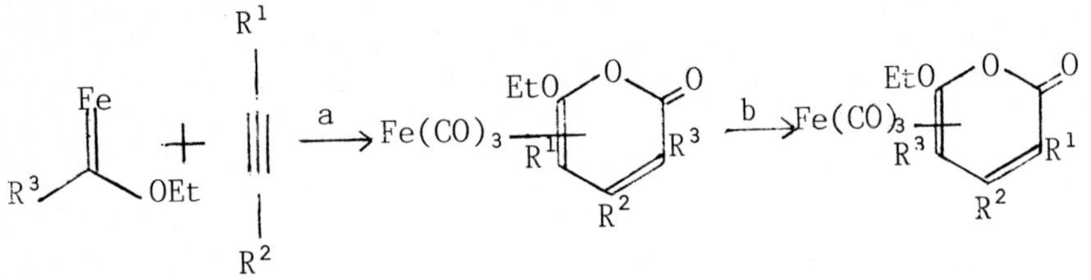

(20)   (21)   (22)

Reagents:- (a) CH$_2$Cl$_2$, CO, 70°, 55psi; (b) Δ

R^1	R^2	R^3	Yield (21) (%)
Me	Me	Ph	89
Ph	H	Ph	74
Et	H	Ph	33
Ph	Ph	Ph	95
CO$_2$Me	Me	Ph	15
Me	Me	Bu^t	60
Ph	H	Bu^t	39
Me	Me	Bu	33

The reaction between methyl propiolate and 4,6-dimethyl-2H-
-pyrone (23) affords methyl 2,4-dimethylbenzoate (24) and
methyl 3,5-dimethylbenzoate (25), in a 4:1 ratio
(R.G. Salomon, J.R. Burns and W.J. Dominic, J. org. Chem.,
1976, **41**, 2918) (see p 5).

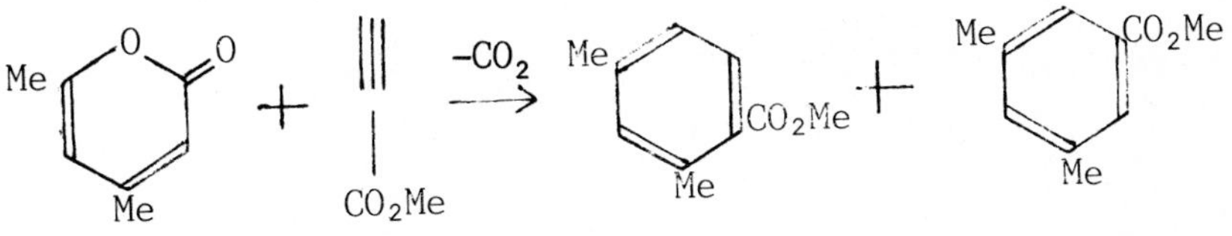

(23)   (24)   (25)

The photochemical cycloaddition of 4,6-dimethyl-2H-pyrone
to acrylonitrile   yields the isomeric [2+2] adducts (26)
and a cyclobutane derivative (27) (T. Shimo, K. Somekawa and
S. Kumamoto, Nippon Kagaku Kaiski, 1983, 394).

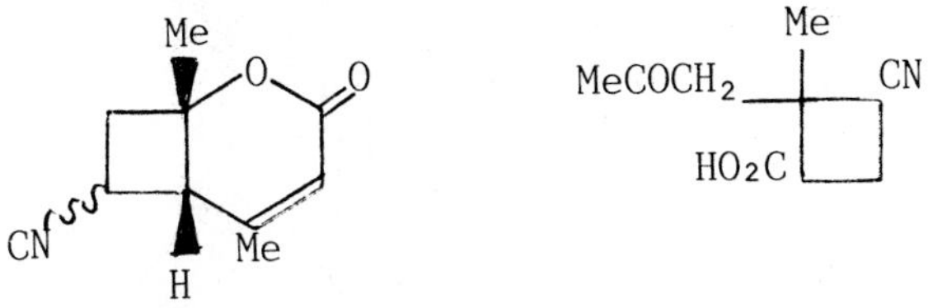

(26)   (27)

The reaction between 4,6-dimethoxy-2H-pyrone and nucleophiles (J.A. Ray, Diss. Abs. Int.B, 1984, <u>45</u>, 1474), and the cycloaddition reactions of pyrones (M.O. Agho, *ibid.*, p. 2423) have been reviewed.

(ii) 4H-Pyrones
The potassium enolate of 4-methoxybut-3-en-2-one (1) reacts with acid chlorides, anhydrides, and acylimidazoles by C--acylation to give 2-substituted 4H-pyrones (2) (A. Morgan and B. Ganem, Tetrahedron Letters, 1980, 2773).

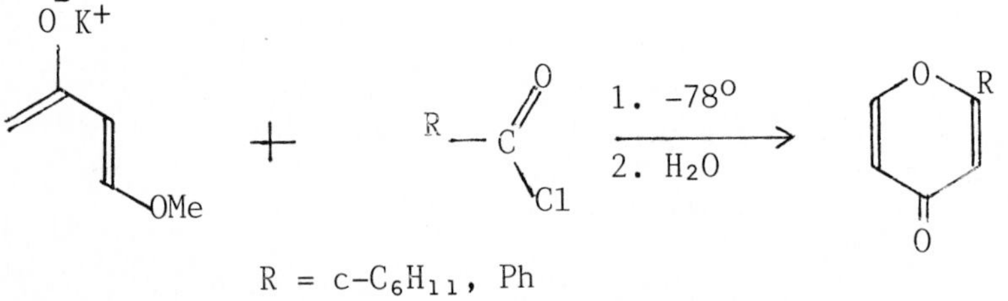

$$R = c\text{--}C_6H_{11}, \ Ph$$

(1)   (2)

Similarly, 2-mono- and 2,6-di- substituted 4H-pyrones have been synthesized from 4-methoxybut-3-en-2-one lithium enolates (M. Koreeda and H. Akagi, *ibid* ., p. 1197).

It has been shown that the photochemically reactive state of 3,5-dimethyl-4-pyrone (3) evolves sequentially into a zwitterionic intermediate (4) and 3,6-dimethyl-2-pyrone (5) (J.A. Barltrop, A.C. Day, and C.J. Samuel, Chem. Comm., 1976, 822).

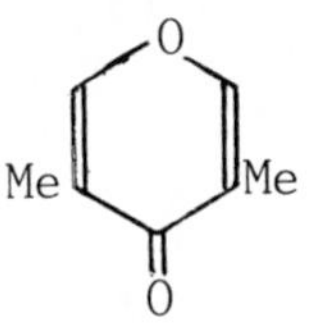

(3) 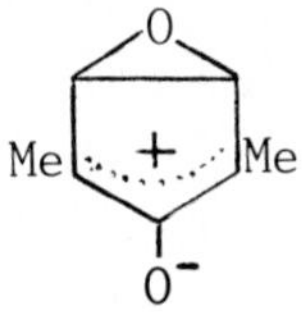   (4) 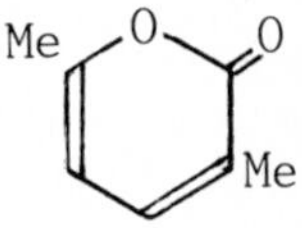   (5)

The electrochemical reduction of 4-pyrones substituted in both the 2- and the 6-position in acetonitrile-water (80:20) may lead to complex 'double dimers' (6, 7 and 8) with a degree of stereoselectivity (G. Mason, G. Le Guillanton, and J. Simonet, *ibid.*, 1982, 571).

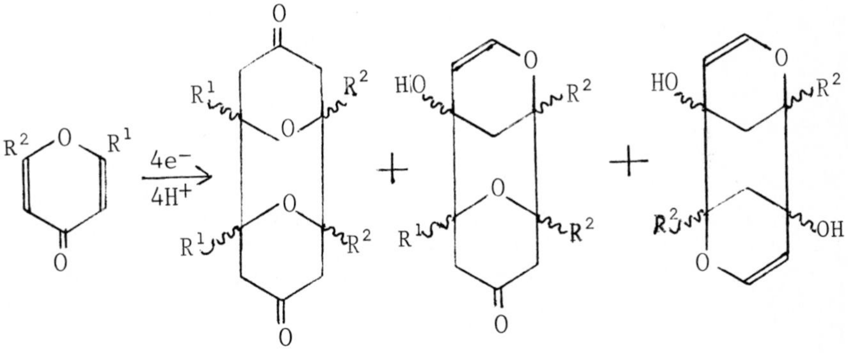

$R^1 = R^2 = Me$
$R^1 = R^2 = Ph$
$R^1 = Me, R^2 = Ph$ (6)   (7)   (8)

(iii) Hydroxy-2-pyrones (hydroxy-α-pyrones)
4-Hydroxy-6-methyl-2-pyrone is obtained in a 75% yield by
treating dehydroacetic acid with aqueous sulphuric acid with
staged addition of water, 2,6-dimethyl-4-pyrone-3-carboxylic
acid (14.5%) is also formed (T.R. Opie, Eur. Pat. Appl. EP
59,052, 1982). A series of substituted 4-hydroxy-2-pyrones
and some derivatives have been synthesized and investigated
for their inhibitory activity toward human leukocyte
elastase, procine pancreatic elastase, and chymotrypsin
(W.C. Groutas *et al*., J. heterocyclic Chem., 1985, 22, 433).

β-Oxocarboxylic acids including β-oxoglutaric half-
esters are cyclized on treatment with 1,1'-carbonyl-
diimidazole to yield 6-substituted 3-acyl-4-hydroxy-2-
-pyrones. 5-Aryl-3,5-dioxo-1-pentanoic acids afford 6-aryl-
-4-hydroxy-2-pyrones (S. Chta, A. Tsujimura, and M. Okamoto,
Chem. pharm. Bull., 1981, 29, 2762).

Acetone, mesityl oxide, and pentane-2,4-dione react with
4-hydroxy-6-methyl-2-pyrone (9) at the 3-position. Under
conditions for the Knoevenagel reaction the hydroxypyrone (9)
with acetone gives 2-acetonyl-4,4,7-trimethyl-4H,5H-
-pyrano[3,2-c]pyran-5-one (10) (44%) together with ethyl
3,5-dioxohexanoate (11) (P. de March *et al*., J. heterocyclic
Chem., 1984, 21, 1369).

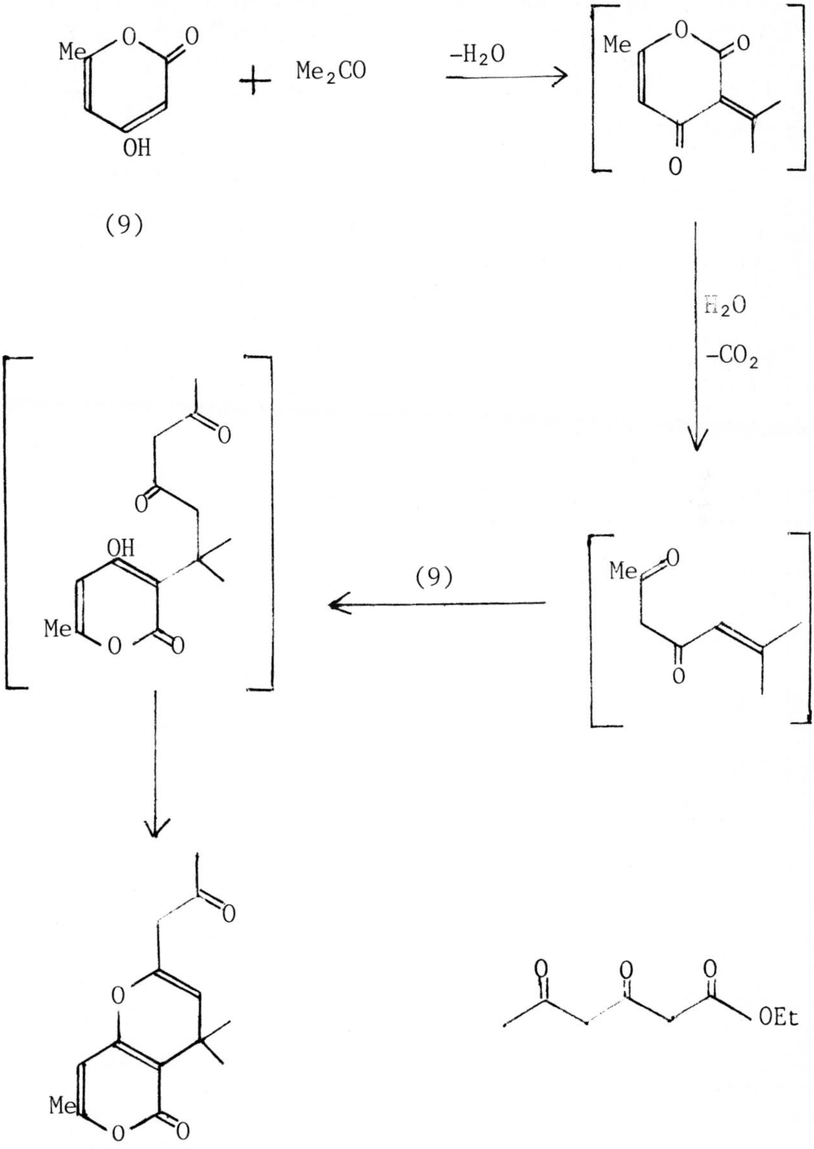

(9)

(10)   (11)

The reaction between 4-hydroxy-6-methyl-2-pyrone (12)
and aldehydes under Knoevenagel conditions yields
electrophilic intermediates, which can be trapped by more of
the original 2-pyrone or thiols to give dilactones (13) or
monolactones (14), respectively (de March *et al*., J.
heterocyclic Chem., 1982, 19, 335).

4-Methoxy-2-pyrones have been synthesized with substituents
at C-5 (R. Bacardit, M. Moreno-Manas, and R. Pleixats, *ibid*.,
p.157) and C-6 positions (J.A. Ray and T.M. Harris,
Tetrahedron Letters, 1982, 1971).

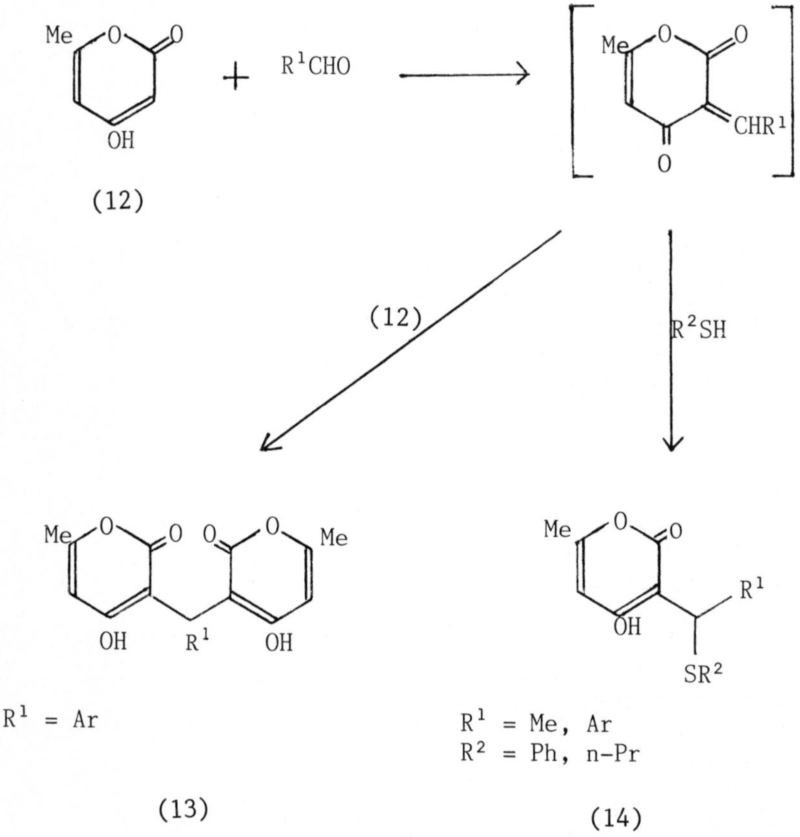

The ^{1}H-and ^{13}C-nmr spectra of 3-acetyl-6-methyl-2H-
-pyran-2,4(3H)-dione have been studied in deuteriochloroform
over a range of temperature for the elucidation of possible
tautomerism, but it only appears to exist in one form (15)
(S.-F. Tan *et al.*, J. chem. Soc., Perkin II, 1982, 513).

(15)

The ir- and uv-spectra together with the ^{1}H- and ^{13}C-nmr
spectra of some acetylpyran-2,6-diones have been discussed in
relation to their tautomeric structures and compared with
those of the acetylpyran-2,4-diones. Also reported are their
reactions with primary amines to yield Schiff bases, salts,
pyridones or open-chain products and a thermal rearrangement
of the Schiff bases (Tan, K.-P. Ang, and H. Jayachandran,
ibid., 1986, 973).

(iv) Hydroxy-4-pyrones (hydroxy-γ-pyrones)
5-Acetyl-2-hydroxy-6-methyl-4-pyrone has been synthesized
from carbon suboxide and acetylacetone in the presence of
acetylacetonate-metal complexes (L. Pandolfo and G. Paiaro,
J. mol. Catal., 1984, 27, 343).
3-Hydroxy-2-methyl-4-pyrone (maltol) (3) is obtained by the
hydrolysis of 4-halogeno-6-hydroxy-2-methyl-2H-pyran-3(6H)-
-ones (2), prepared by treating methylfurfuryl alcohol (1)
with 2 equivalents of halogen (Cl or Br). Similarly
ethylfurfuryl alcohol affords 2-ethyl-3-hydroxy-4-pyrone
(ethylmaltol) (P.D. Weeks *et al*., J. org. Chem., 1980, 45,
1109).

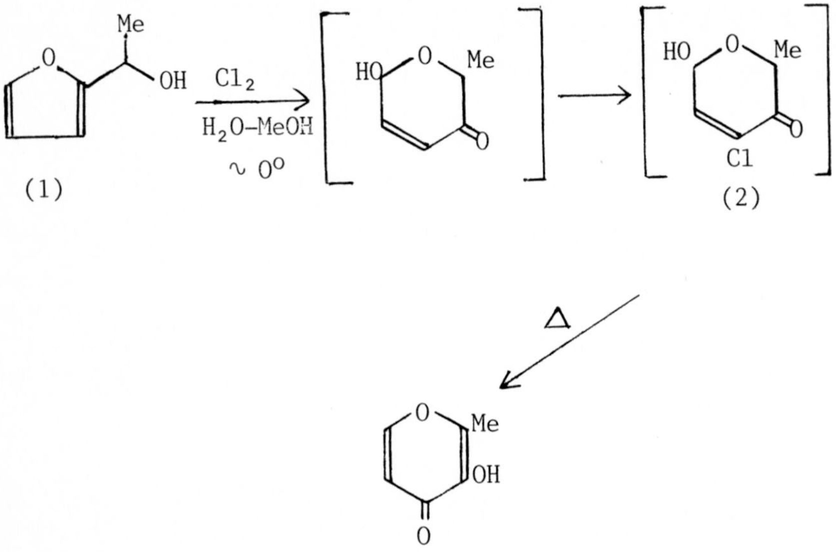

(1) (2) (3)

3-Hydroxy-2-methyl-4-pyrone has also been synthesized from D-glucose (T. Shone *et al.*, *ibid.*, 1983 **48**, 5126).

2. Hydropyrans

(a) Dihydropyrans and derivatives

(i) 2,3-Dihydro-4-pyrans
A number of insecticidal 2,3-dihydro-4-pyrans have been prepared and it was found that a hydroserricornin (1) had higher pheromone activity than serricornin in tests with *Lasioderma serricorne* (H. Levinson *et al.*, Ger. Offen. DE 3,045,909, 1982).

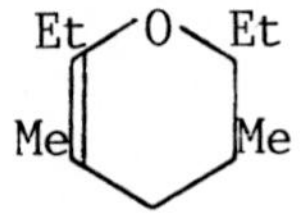

(1)

Some 2,3-dihydropyrans (2) are obtained by the aluminium chloride activated cycloaddition of α,β-unsaturated acyl cyanides to simple olefins at room temperature. The 6-cyano derivative (2; $R^1 = Pr^iO$, $R^2 = H$) is obtained by heating the appropriate starting materials at 160°C in the absence of aluminium chloride (Z.M. Ismail and H.M.R. Hoffman, Angew. Chem. intern. Edn., 1982, 21, 859).

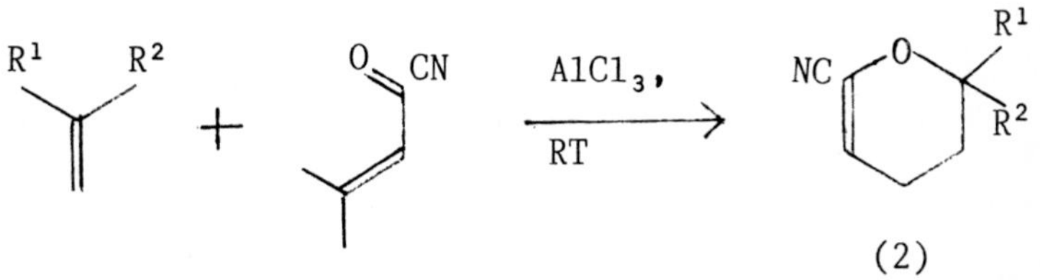

(2)

R^1	R^2
Pr^iO	H
Me_3SiCH_2	H
Ph	H
MeCH=CH	H
Me	Me

Other related 2,3-dihydropyrans have been obtained from the addition of electron-poor α,β-unsaturated carbonyl compounds to electron rich olefins (H.K. Hall *et al.*, Tetrahedron Letters, 1982, **23**, 603; L.F. Tietze *et al.*, *ibid.*, p. 1147; R.R. Schmidt and M. Maier, *ibid.*, p. 1789; D. Dvorak and Z. Arnold, *ibid.*, p. 4401).

The acid condensation of isobutylene or α-methylstyrene with an α,β-unsaturated aliphatic aldehyde affords a separable mixture of a 2,3-dihydro-4*H*-and 5,6-dihydro-2*H*--pyran (N.A. Romanov *et al.*, Zh. org. Khim., 1982, **18**, 403).

5-Alkyl-2,3-dihydro-2-ethoxy-4*H*-pyrans (3) on chlorination followed by treatment with methylmagnesium bromide yield 3--alkyl-6-ethoxy-2-methylheptan-2-ols (4) (R. Menicagli *et al.*, Tetrahedron Letters, 1982, **23**, 1937).

$$\text{(3)} \xrightarrow[\substack{2.\ \text{MeMgBr} \\ 3.\ \text{H}^+,\ \text{H}_2\text{O}}]{1.\ \text{Cl}_2,\ \text{Et}_2\text{O}} \quad \underset{R\quad\quad Me}{Me_2CCHCH_2CH_2CHOEt}$$

R = Me, Pr^i

(3)   (4)

Heating the dihydropyranyl thione (5) produces the dihydrothiopyran (6). The lack of formation of dihydrothiopyran (7) suggests that compound (6) results from a [3,3] sigmatropic shift rather than a retro-Diels-Alder reaction (K.B. Lipkowitz and B.P. Munday, *ibid.*, 1977, 3417).

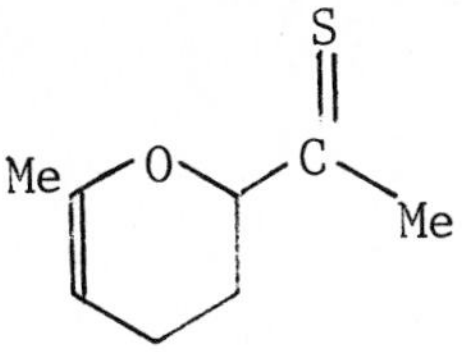

(5)

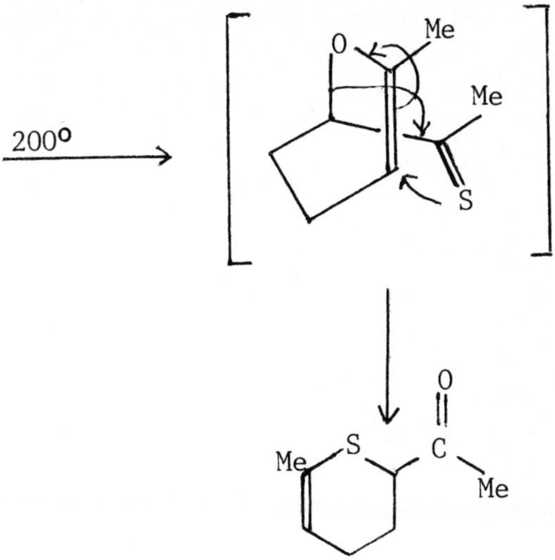

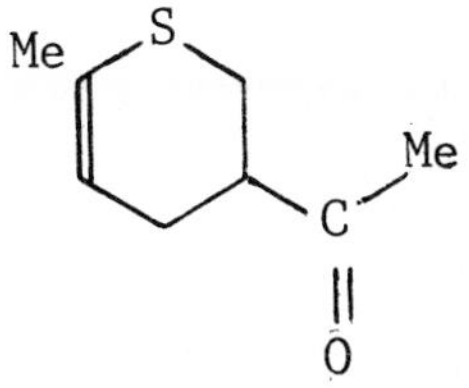

200°

(7)

(6)

The electrophilic addition reactions of alkyl- and 2-
-alkoxy-2,3-dihydro-4-pyrans have been discussed
(L.V. Wertheimer, Diss. Abs. Int. B., 1985, $\underline{45}$, 2168).

2,3 Dihydro-4-pyrones (8) are obtained by the acid-
-catalyzed [4+2]-cycloaddition between 1-alkoxy-3-
-trimethylsilyloxybuta-1,3-dienes and aldehydes in the
presence of (bornyloxy)aluminium dichloride (R.W. Aben and
H.W. Scheeren, Synthsis., 1982, 779).

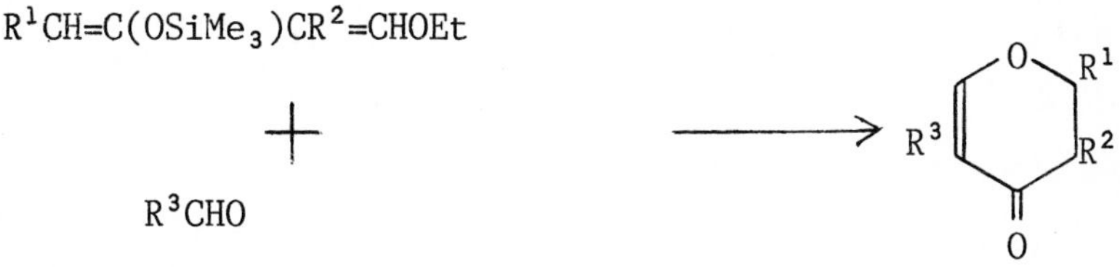

$R^1CH=C(OSiMe_3)CR^2=CHOEt$

+

R^3CHO

R^1=H, Pr^i, vinyl, 1-propenyl, Ph
R^2, R^3=H, Et, OEt

(8)

1-Methoxybut-1-en-3-yne (9) on metallation, followed by treatment with an aldehyde gives the acetylenic alcohol (10) which on acid hydrolysis affords the 2-alkyl-2,3-dihydro-4--pyrone (11) (M.T. Crimmins and D.M. Bankaitis, Tetrahedron Letters, 1983, 24, 5303).

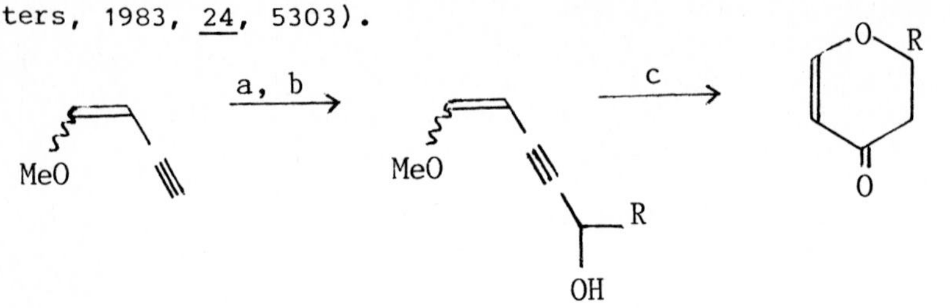

R = Me, Et, Pr, Pr^i, Bu^i

(9)   (10)   (11)

Reagents:- (a) BuLi, THF, -78°; (b) RCHO, THF, -78°
          (c) 4-TSA, THF, H$_2$O or 30% HClO$_4$

Cyclization of a 3,4,5-triaryl-5-oxopentanoic acids affords a 4,5,6-triaryl-3,4-dihydro-2-pyrone, (I.E. El-Kholy, M.M. Mishrikey and S.L. Abdoul-Ela, J. heterocyclic Chem., 1982 19, 1329).

(ii) 5,6-Dihydro-2-pyrans

3-Oxatetrahydropyrans (1,3-dioxans) (1) on heating in aliphatic alcohols in the presence of an acid catalyst afford 5,6-dihydro-2H-pyrans (2) (U.G. Ibatullin *et al.*, USSR Pat., SU 899,559, 1982).

R^1, R^2, R^3 = H, Me

(1)   (2)

The stereoselective synthesis of a number of 5,6-dihydro-
-2*H*-pyrans (5 and 6) involving [3,3]-sigmatropic rearrangement
of 6-alkenyl-4-oxatetrahydro-2-pyrones (3 and 4) has been
described. The 6-alkenyl-oxatetrahydropyrone (3 or 4) is
converted into the corresponding trimethylsilyl ketene
acetal, which after a Claisen [3,3]-sigmatropic rearrangement
yields, on subsequent hydrolysis and esterification the
dihydropyran (5 or 6) (S.D. Burke, D.M. Armistead and
F.J. Schoenen, J. org. Chem., 1984, __49__, 4320).

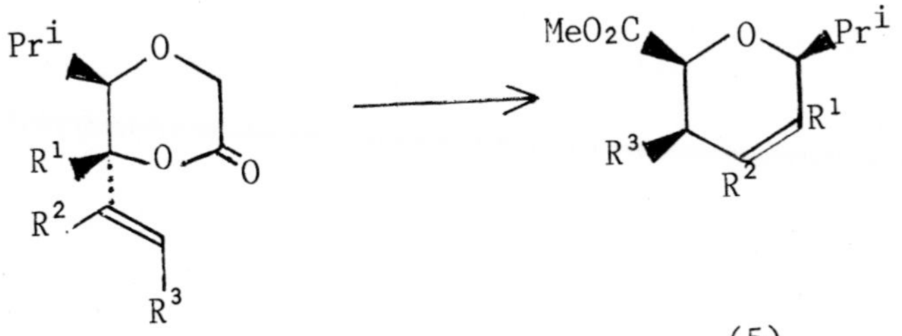

R^1, R^2 = H, Me
R^3 = H, Me, $SiMe_3$

(3)

(5)

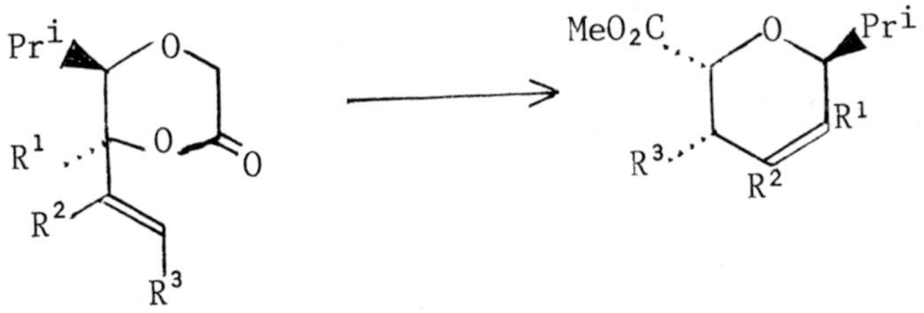

(4)

(6)

There have been several reports of the formation of 5,6-
-dihydro-2-pyrans by the Lewis acid catalyzed addition of
electron-rich dienes to carbonyl compounds (S. Danishefsky,
E.R. Larson and D. Askin, J. Amer. chem. Soc., 1982, __104__,
6457; Larson and Danishefsky, *ibid*., p. 6458). A study has

been made of some of the reactions of 5,6-dihydro-4-methyl-
-2*H*-pyran and its isomers including hydrogenation and
condensation with formaldehyde (Ibatullin *et al*., Izv. Akad.
Nauk SSSR, 1982, 2114). The Friedel-Craft reaction between
5,6-dihydro-4-methyl-2*H*-pyran and 2,3,5-trimethylhydroquinone
affords 3,4-dihydro-6-hydroxy-2,5,7,8-tetramethyl-2*H*-benzo[b]-
pyran-2-ethanol, a key intermediate in the synthesis of α-
-tocopherol (Y. Fujita *et al*., Chem. Letters, 1985, 1399).

Amination of 3,4-dibromo-4-methyltetrahydropyran gives the
3-amino derivative (7; R^1 = H,Me,Et; R^2 = cyclohexyl,
Ph, PhCH$_2$). Derivative (7; R^2 = Ph, or MeC$_6$H$_4$) undergoes
a Claisen rearrangement at 120° in nitrobenzene, containing
zinc chloride to yield 5-(2-aminoaryl)-5,6-dihydro-4-methyl-
-2*H*-pyran (8; R^3 = H,Me) (Ibatullin *et al*., Izv. Akad. Nauk
SSSR, 1985, 2753).

(7)

(8)

Dienes (9) on heating with diethyl oxomalonate in
acetonitrile containing a trace of hydroquinone yield
adducts, which on hydrolysis with potassium hydroxide in
aqueous THF yield the diacids (10), which *via* the Curtius
reaction are converted into lactones (11). Lactone (11) on
reaction with lithium tetrahydridoaluminate and then with
manganese dioxide gives the 5,6-dihydro-2-pyrone (12)
(R.A. Ruden and R. Bonjouklian, J. Amer. chem., Soc., 1975,
97, 6892). The formation of compound (11) from diene (9) is
equivalent to the [4+2] cycloaddition of carbon dioxide to
the diene.

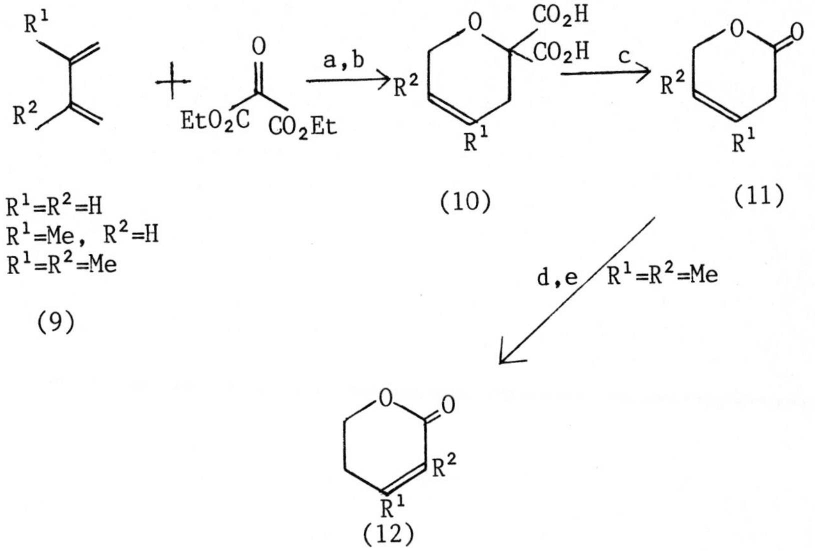
R1
R2
EtO2C
CO2Et
O
a,b
CO2H
CO2H
c
R2
R1
(10)
(11)
R1=R2=H
R1=Me, R2=H
R1=R2=Me
(9)
d,e
R1=R2=Me
(12)

Reagents:- (a) 130°; (b) OH^-; (c) Curtius; (d) $LiAlH_4$, THF, 0°; (e) MnO_2, CH_2Cl_2

The dianion (14) of methyl acetoacetate (13) reacts with the appropriate aldehyde to give the 6-substituted 5,6-dihydro-2-pyrone (15), which has been converted into derivatives (16) (W.C. Groutas *et al.*, J. heterocyclic Chem., 1985, 22, 433).

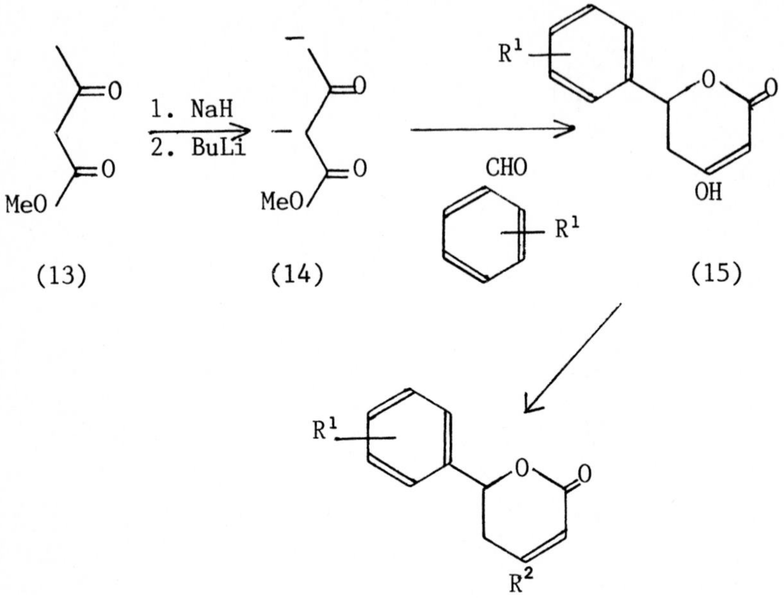

R^1=H, CF_3
R^2=OMe, 4-MeC$_6$H$_4$SO$_3$

(16)

Some 5,6-dihydro-2-pyrones have been obtained by the rearrangement of disubstituted 3-methylene-2-oxotetra-hydrofurans (G. Falsone and B. Spur, Arch. Pharm., 1982, 315, 491) and 3-substituted 5,5-dialkyl-5,6-dihydro-2-pyrones treated with diazoalkanes to yield cycloaddition products (17) (*idem, ibid*., p. 597). Some of these latter compounds showed coronary vasodilator and other cardiac depressant activity.

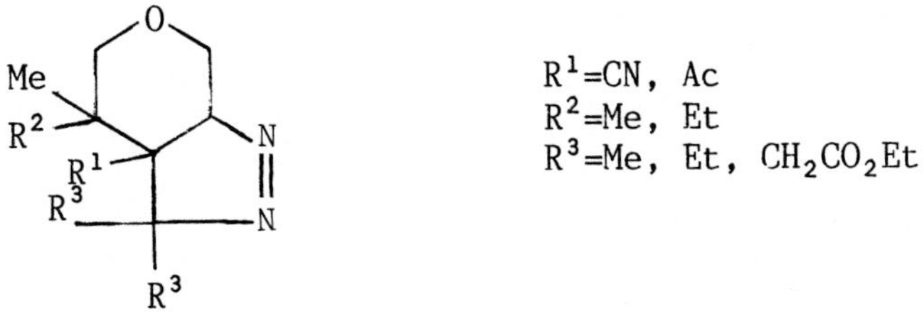

R^1=CN, Ac
R^2=Me, Et
R^3=Me, Et, CH_2CO_2Et

(17)

5,6-Dihydro-4-hydroxy-6-methyl-2-pyrone (19) is obtained by the hydrogenation of 4-hydroxy-6-methyl-2-pyrone (18) over palladium. The pyrone (18) is synthesized from Meldrum's acid (20) (J. Haeusler, Monatsh. Chem., 1982, 113, 1213).

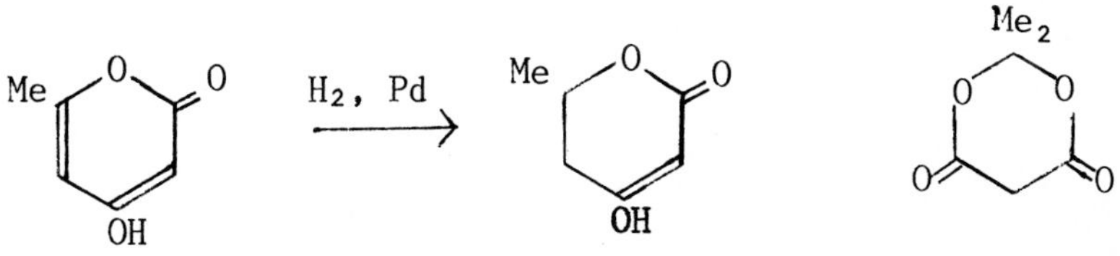

(18)   (19)   (20)

(b) Tetrahydropyrans and derivatives

Tetrahydropyran derivatives have been obtained in 23% yield by electrosynthesis (C. Degrand et al., Electrochim. Acta, 1984, 29, 625). The allyl esters (1) undergo Claisen rearrangement (via enol form of ester) to afford the tetrahydropyrans (2), which on oxidative decarboxylation in the presence of methanol give the 2-methoxytetrahydropyrans (3) (P.G.M. Wuts and C. Sutherland, Tetrahedron Letters, 1982, 23, 3987).

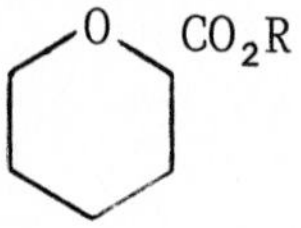

R = CH$_2$CH=CH$_2$,

(1)

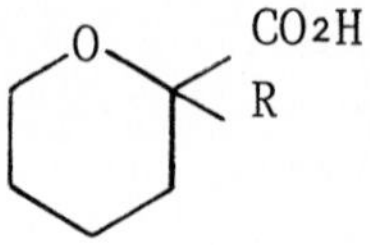

CH$_2$CMe = CH$_2$,

(2)

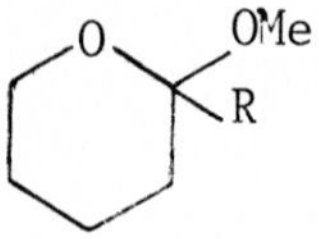

CH$_2$CH = CHMe

(3)

The diastereoisomeric 6-methyl-2-*tert*-butyltetrahydropyran-
-3-ols (4) have been synthesized (C. Anselmi, G. Catelani and
L. Monti, Gazz. 1983, 113, 167).

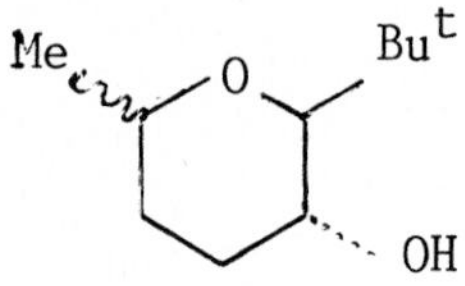

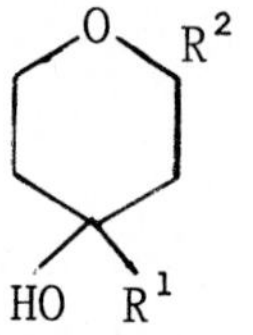

R^1 = Me, Ph
R^2 = alkyl, Ph

(4)   (5)

Acid-catalyzed isomerization of 2-R^2-4,4-dimethyl- and 2-
-R^2-4-methyl-4-phenyl-1,3-dioxanes (R^2 = substituent) yields
a mixture of products containing 2-R^2-4-methyl- and 2-R^2-4-
-phenyl-tetrahydropyran-4-ols (5), respectively (N.A. Romanov

et al., Zh. Prikl. Khim., 1982, <u>55</u>, 2778). 2,2-Dimethyl-3-
-(2-oxo-1-propyl)tetrahydropyran has been obtained from (+)-α-
-pinene and (+)-3-carene (J. Kula and J. Gora, Ann., 1984,
1860). (1'R,2S,2"E,5R,6R)-2-(1'-Bromoethyl)-2,5-dimethyl-6-
-(penta-2",4"-dienyl)tetrahydropyran (6) and (1'R,2S,5R,6R)-
-2-(1'-bromoethyl)-2,5-dimethyl-6-(pent-4"-enyl)tetrahydro-
pyran (7) have been isolated from the sponge *Haliclona sp.*
(R.J. Capon, E.L. Ghisalberti and P.R. Jefferies,
Tetrahedron, 1982, <u>38</u>, 1699).

(6)   (7)

The acid-catalyzed and enzymatic hydrolysis of the *trans*-
and the *cis* isomer of 2-methyl-3,4-epoxytetrahydropyran has
been studied (Catelani and E. Mastrorilli, J. chem. Soc.,
Perkin 1, 1983, 2717). The addition reaction between but-1-
-en-3-yldichloro-*n*-butyltin and aldehydes yields 4-chloro-
-2,6-dialkyl-3-methyltetrahydropyrans (A. Gambaro *et al.*, J.
organometal. Chem., 1984, <u>260</u>, 255). Also a number of
3(R^2), 4(R^3)-disubstituted di(R^1)tetrahydropyrans
[R^1 = Me,Et,Me$_2$CH,Et(Me)CH,Me$_3$C; R^2 = H,Me; R^3 = Cl,Br]
have been obtained from the reaction between R^1CHO and
BuR^{3}SnCH$_2$CH=CHR^2 (A. Boaretto *et al.*, *ibid.*, 1986,
<u>299</u>, 157). The stereoselective reduction of 7-chloro-7-
-phenyl-2-oxabicyclo[4.1.0]heptane with lithium tetrahydrido-
aluminate-diglyme affords endo-7-phenyl-2-oxabicyclo[4.1.0]-
heptane (8) (G.D. Hobbs, J.D. Woodyard and J.R Curtis, Org.
Prop. Proced. Int., 1981, <u>13</u>, 356).

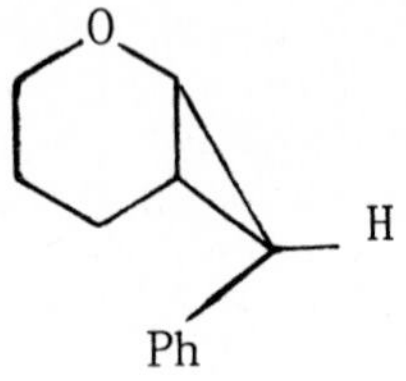

(8)

Tetrahydropyranyl ketones (11) are obtained by converting the aldehyde (9) into the nitrile (10) and treatment of the latter with the appropriate Grignard reagent (R.A. Kuroyan *et al.*, Arm. Khim. Zh., 1982, **35**, 658).

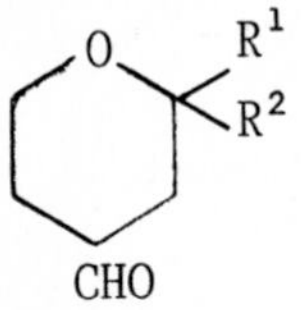

CHO

R^1 = Me, H

(9)

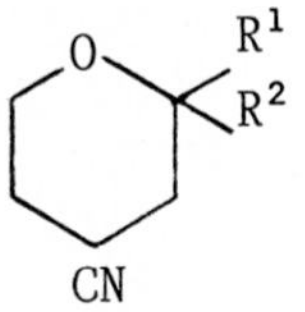

CN

R^2 = Me, Et, Pr^i

(10)

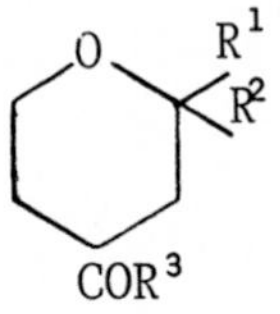

COR^3

R^3 = aryl, benzyl

(11)

$COCH_2COMe$

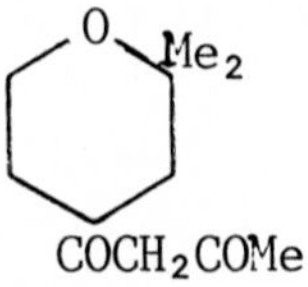

(12)

For the synthesis of 1-(2,2-dimethyltetrahydro-4-
-pyranyl)butane-1,3-dione (12) see R.S. Vartanyan,
R.S. Shaginyan and S.A. Vartanyan (*ibid.*, p. 671); for a
review of the chemistry of 4-functionally substituted
tetrahydropyrans, Vartanyan and Vartanyan (*ibid.*, 1981, **34**,
728); and for the intramolecular alkoxypalladation-
carbonylation of appropriate alkenes leading to the formation
of tetrahydropyran derivatives, M.F. Semmelhack and
C. Badurow (J. Amer. chem. Soc., 1984, **106**, 1496).

4-Aryl-6-methyl-3-phenyltetrahydro-2-pyrones (14) are
formed following the hydrolysis of the appropriate compound
(13) (S. Axiotis *et al.*, Eur. J. med. Chem.-Chim. Ther.,
1981, **16**, 431). Configurational determinations have been
made and the stereochemistry of the reactions leading to the
precursors for the above reaction has been studied (Axiotis
and J. Dreux, *ibid.*, p. 439).

$MeCH(OH)CH_2CHArCHPhCN$ $\longrightarrow$

Ar = Ph, $4-ClC_6H_4$

(13)   (14)

One of the diastereomers of 3,5-dimethyl-6-(1-methylbutyl)-
tetrahydro-2-pyrone (15), isolated from fire-ant queens has
been synthesized and named invictolide. It is responsible,
in part for "queen recognition by the workers of the species"
(R.J. Rocca *et al.*, Tetrahedron Letters, 1983, **24**, 1893).

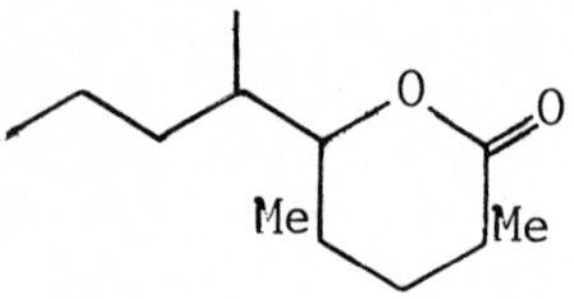

(15) 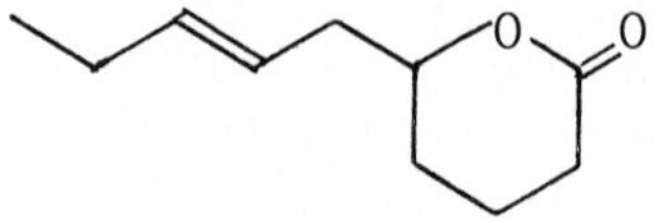   (16)

A new synthesis of jasmine lactone (16) and related δ-
-lactones by the dye-sensitized photooxygenation of 3-
-substituted cyclohexane-1,2-diones has been described
(M. Utaka *et al.*, Chem. Letters, 1983, 911).
   Treatment of six-membered cyclic anhydrides, including
glutaric and some of its alkylated derivatives with
ethoxycarbonylmethylenetriphenylphosphorane yields
tetrahydro-2-pyrone derivatives, for example, compound (17)
(A.D. Abell, I.R. Doyle and R.A. Massy-Westropp, Austral. J.
Chem., 1982, **35**, 2277).

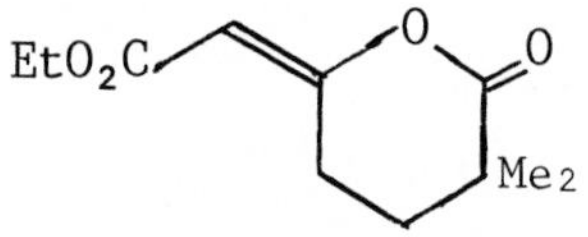

(17)

Diols of type (18) on carbonylation with formic acid in sulphuric acid or other strong acids yields 3,3-disubstituted tetrahydro-2-pyrones (19) (Y. Takahashi, H. Nagai and N. Yoneda, Chem. Abs., 1984, $\underline{101}$, 6979b.).

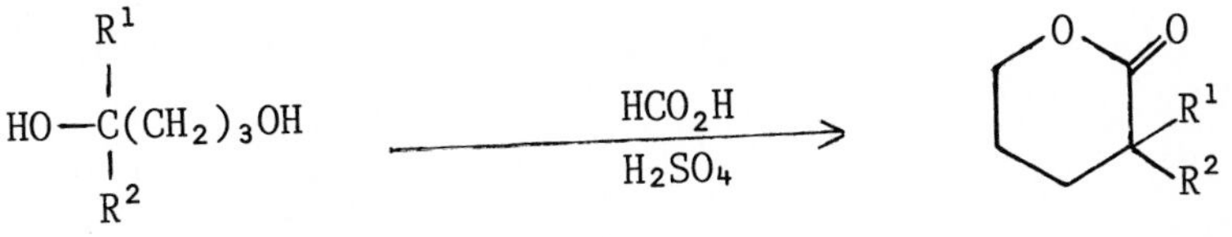

$$HO-\underset{\underset{R^2}{|}}{\overset{\overset{R^1}{|}}{C}}(CH_2)_3OH \xrightarrow[\text{H}_2\text{SO}_4]{\text{HCO}_2\text{H}}$$

R^1 = Me,
R^2 = Me, Et, Pr, Bu, pentyl, etc.

(18)

(19)

(±)-*trans*-Rosoxide (20) found in roses and geraniums has been synthesized as shown below (T. Cohen and M-T. Lin, J. Amer. chem. Soc., 1984, $\underline{106}$, 1130).

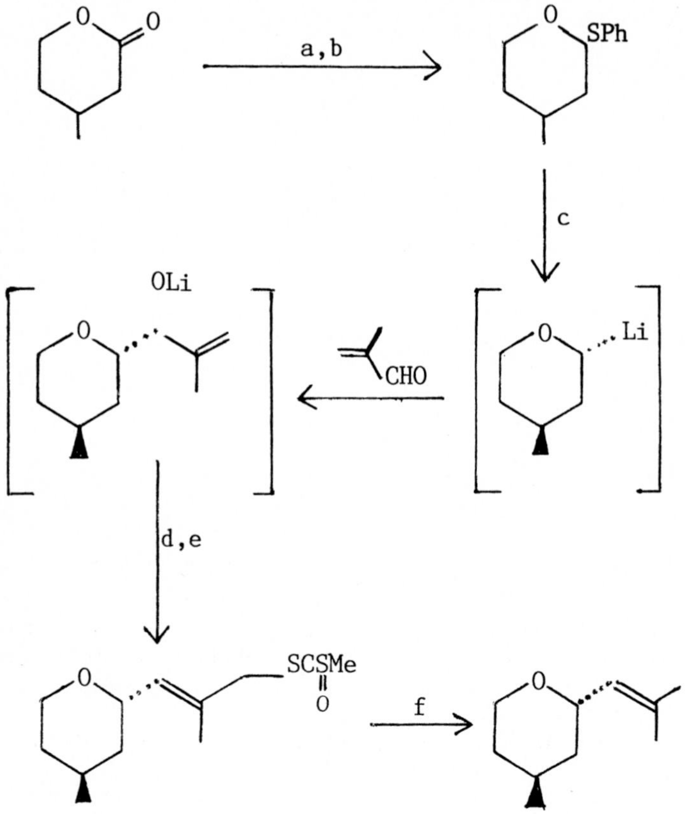

(20)

Reagents:- (a) DIBAH, PhMe, -78°, 1h; (b) PhSH, $BF_3.OEt_2$, -78°, 5 min; (c) LDMAN, -78°, 0.75h; (d) HMPA, CS_2, $-78^\circ \rightarrow RT$; (e) MeI, Δ, 1h; (f) (5 eq) n-Bu_3SnH, AlBN, PhMe, Δ, 3h

The ^{1}H- and ^{13}C-nmr spectra of 3,5-diacetyltetra-
hydropyran-2,4,6-trione and its Schiff's bases in deuteriochloroform have been studied over a range of temperatures in
order to identify any possible tautomers (S-F. Tan *et al* ., J.
chem. Soc., Perkin II, 1982, 513).

3. Benzo[b]pyrans (5,6-benzopyrans, chromenes) and their
   derivatives

In Volume IVE the trivial name chromene was used as the
parent name for all its derivatives. Since then the name
benzo[b]pyran has been used more and more in the literature
and will be adopted for naming compounds in this section, but
some trivial names still acceptable according to IUPAC rules
will be used, for example, coumarin, chromone, flavone, which
are still extensively used. Nomenclature based on 2H-1- and
4H-1-benzopyran is also used in the literature.

(a) Benzo[b]pyrans and their oxidation products

(i) Benzopyrans, alkyl- and aryl-benzo[b]pyrans and flavenes
(1) Substituted 4H-benzo[b]pyrans (2-chromenes, 4H- or β-
    -chromenes
4-Phenyl-2-phenylvinyl-4*H*-benzo[b]pyran (3) is prepared by
the reaction of the α,β-enone (1; R^1 = Ph, R^2 = PhCH=CH)
with the (2-hydroxyphenyl)mercury chloride (2) in a mixture
of methylene dichloride aqueous 3M HCl containing $PdCl_2$ and
tetra-*n*-butylammonium chloride. If R^1 = R^2 = Ph a mixture
of 2,4-dipheny-4H-benzo[b]pyran (40%) and 1,3-diphenyl-3-(2-
-hydroxyphenyl)propan-1-one (50%) is obtained. In other
examples of this reaction, dehydration to the benzopyran does
not occur and the dihydro-2-hydroxybenzopyran (chromanol) is
obtained (S. Cacchi, D. Misiti and G. Palmieri, J. org.
Chem., 1982, **47**, 2995).

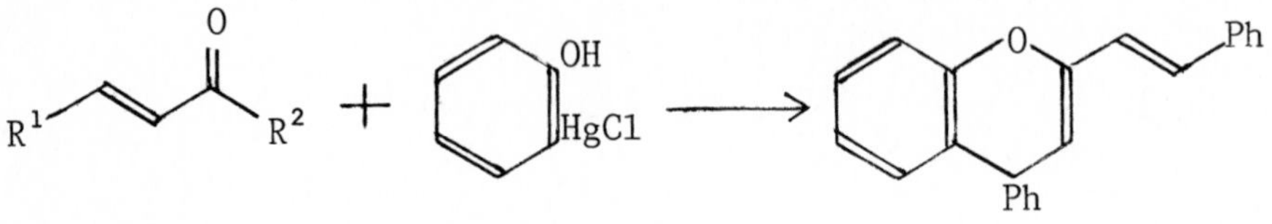

$$R^1 \diagup\!\!\!\diagdown\!\!\!\diagup\overset{O}{\diagup} R^2 \quad + \quad \underset{HgCl}{\overset{OH}{\bigcirc}} \longrightarrow$$

(1)   (2)   (3)

The cycloaddition of a phenol to 2-(4-bromophenyl)-1,1-
-dicyanoethene affords the 2-amino-4-(4-bromophenyl)-3-
-cyano-4H-benzo[b]pyran (4) (Yu.A. Sharanin and G.V. Klokol,
Zh. org. Khim., 1983, **19**, 1782).

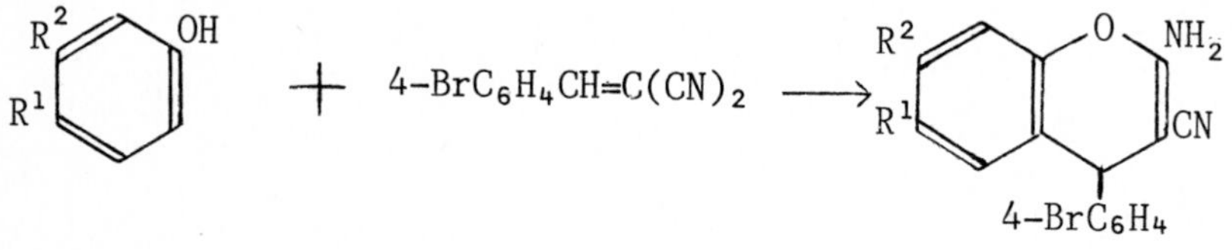

$$\underset{R^1}{\overset{R^2}{\bigcirc}}OH \quad + \quad 4\text{-}BrC_6H_4CH{=}C(CN)_2 \quad \longrightarrow$$

4-BrC₆H₄

R^1 = H, R^2 = OH, NH$_2$
R^1 = C$_6$H$_{13}$, R^2 = OH

(4)

The reaction between salicylaldehyde and ethyl cyano-
acetate yields the substituted 2-amino-4H-benzo[b]pyran (5)
(G. Renzi, P. Mascagni and E.I. Piumelli, Boll. chim.
Farm., 1981, **120**, 525).

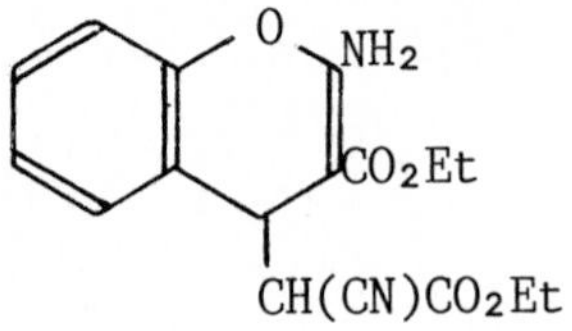

(5)

Reactions. 2-(2-Naphthyloxymethyl)-1*H*-naphtho[2,1-b]pyran
(6) in boiling *N*,*N*-diethylaniline undergoes a novel
oxidative Claisen rearrangement to give the naphthopyrano –
naphthopyran (7). Similarly 2-(4-methylphenyloxy)-1*H*-
-naphtho[2,1-b]pyran affords product (8) (S. Selvaraj *et al.*,
Tetrahedron Letters, 1983, **24**, 2509).

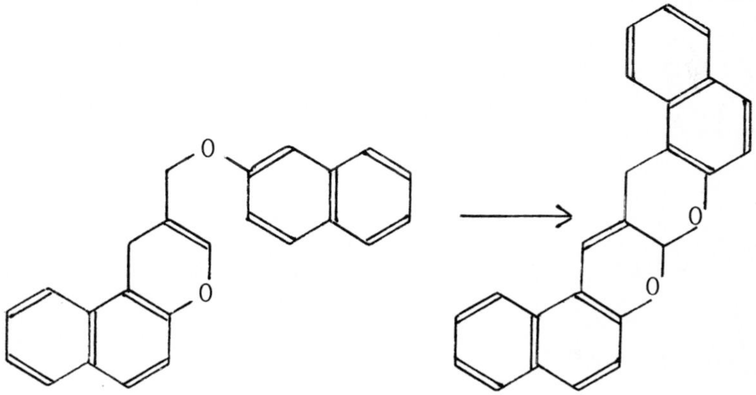

(6)   (7)

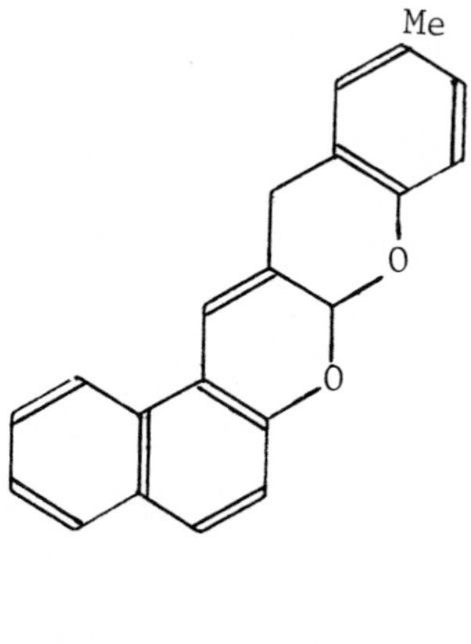

(8)

(2) Substituted 2H-benzo[b]pyrans (3-chromenes, 2H- or α-
  -chromenes)

2H-Benzo[b]pyrans (3) are synthesized by a two-step process
from phenols (1) and β-halogenopropionaldehyde acetals.
The ether-acetal (2) is formed when the potassium salt of
the phenol reacts with the chloroacetal in the presence of
a catalytic amount of tetra-n-butylammonium iodide. The
acetal (2) with a catalytic amount of 4-toluenesulphonic
acid in boiling benzene yields the 2*H*-benzo[b]pyran (3).
The cyclization is highly regioselective (P.F. Schuda and
J.L. Philips, J. heterocyclic Chem., 1984, __21__, 669).

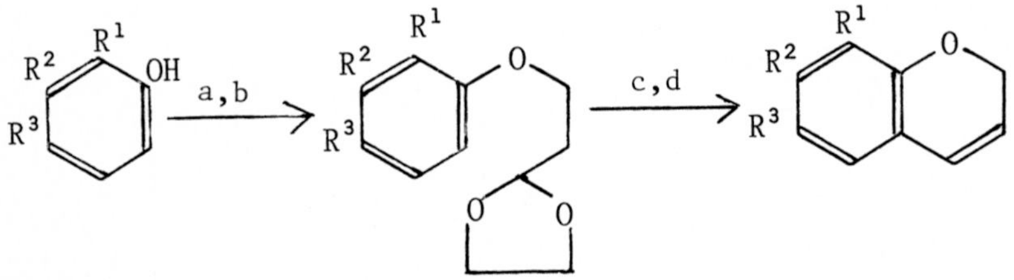

(1)   (2)   (3)

$R^1 = R^2 = R = H$ $R^1 = OMe$, $R^2 = R^3 = H$
$R^2 = OMe$, $R^1 = R^3 = H$ $R^3 = OMe$, $R^1 = R^2 = H$
          $R^2 = OCH_2Ph$, $R^1 = R^2 = H$

Reagents:- (a) KH, THF; (b) 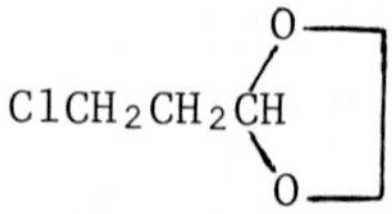 $ClCH_2CH_2CH$ (O—O) , n-Bu_4NI;

     (c) 4-$HO_3SC_6H_4Br$, C_6H_6; (d) 10% NaOH

6-Methyl-2H-benzo[b]pyran has been obtained from 4-
-methylphenylpropargyl ether (N.A. Andreev, V.I. Levashova
and L.I. Bunina-Krivorukova, Zh. org. Khim., 1985, 21,
1061) and 4-methyl-2H-benzo[b]pyran from the complex
formed by (O-butenyl)iodophenol and cobalt 'salen'
(V.F. Patel, G. Pattenden and J.J. Russell, Tetrahedron
Letters, 1986, 27, 2303).

2,2-Dimethyl-2H-benzo[b]pyran has been prepared from
salicylaldehyde and ethyl 3-methylbut-2-en-oate or ethyl α-
-bromoisobutyrate (Y. Kawase *et al*., Bull. chem. Soc. Japan,
1982, 55, 1153).

A 4- and 6-substituted 2,2-dimethyl-2H-benzo[b]pyran such
as (6) is prepared by heating either compound (4) or
compound (5) (J.J. Talley, Synth., 1983, 845).

R^1 = H, Me, OMe, Cl
R^2 = H, Me

(4)   (5)   (6)

2,2-Dimethyl-7-methoxy-4-vinyl-2H-benzo[b]pyran has been
obtained by dehydration of the alcohol resulting from the
reaction between the related chroman-4-one and
vinylmagnesium bromide (P. Anastasis, P.E. Brown and
W.Y. Marcus, J. chem. Soc., Perkin I, 1984, 2815).

6-Acetyl-2,2-dimethyl-2*H*-benzo[b]pyrans (V.K. Ahluwalia
and K. Mukherjee, Indian J. Chem., 1984, 23B, 880) and 6-
-butyryl-2,2-dimethyl-2H-benzo[b]pyrans (Ahluwalia and
R.P. Singh, Gazz., 1984, 114, 359) are obtained on
dehydrogenating the corresponding 3,4-dihydrobenzopyrans
with dichlorodicyanobenzoquinone. 2,2-Dimethyl-7-
-methoxybenzo[b]pyran-3-carboxylic acid, m.p. 183°
(decomp.), ethyl ester, syrup (Anastasis and Brown, J.
chem. Soc., Perkin I, 1983, 197). 2,2-Dimethyl-6-methoxy-
-5-methoxycarbonyl-2*H*-benzo[b]pyran is obtained by the
dehydrogenation of its 3,4-dihydro derivative, using
dichlorodicyanobenzoquinone (Ahluwalia and A.K. Tehim,
Monatsh., 1983, 114). For the preparation of some 2*H*-
-benzo[b]pyran-3-carboxamides see R.C. Gupta *et al*., (India
J. Chem., 1982, 21B, 344).

Substituted 2*H*-benzo[b]pyrans may be obtained by the
thermal cyclization of *Z*-2-(3-hydroxy-1-propenyl)phenols,
formed by the reaction between trialkylaluminium compounds
and coumarins or chromones (A. Alberola *et al*., J.
heterocyclic Chem., 1983, 20, 715). A number of
substituted 2*H*-benzo[b]pyrans, including derivatives of
2,2-dimethylbenzopyran have been reported (E.A. Faruk,
Eur. Pat. Appl. EP 93,535, 1983; J.M. Evans and
V.A. Ashwood, *ibid*., 107,423, 1984; T. Timar *et al*., *ibid*.,
118,794, 1984; Evans and F. Cassidy, *ibid*., 126,367, 1984).

3-Ethoxy-4-methoxyphenol with 3-methyl-2-butenal and
Ti(OEt)$_4$ in boiling toluene gives 2,2-dimethyl-7-ethoxy-6-
-methoxy-2*H*-benzo[b]pyran (E. Kiehlmann, J.E. Conn and
J.H. Borden, Org. Prep. Proceed. Int., 1982, 14, 337).

Spiro(2*H*-benzo[b]pyran-2,1'-cyclopentane), b.p. 116°/10mm.,
-cyclohexane, b.p. 124°/3mm; and -cycloheptane, b.p.
144°/8.5mm. have been prepared from the corresponding
cycloalkanones (P.J. Brogden and J.D. Hepworth, J. chem.
Soc., Perkin I, 1983, 827).

Salicylaldehyde (7; R = H) reacts with dimethyl penta-
-2,3-dienedioate (8) in boiling benzene in the presence of
a 5% methonolic solution of benzyltrimethylammonium
hydroxide (triton B) to yield the 2*H*-benzo[b]pyran (9;
R = H). Similarly in the presence of potassium *tert*-
-butoxide in *tert*-butanol, 2-hydroxyacetophenone (7;
R = Me) is converted into the 2H-benzopyran (9; R = Me)
(N.S. Nixon, F. Scheinmann and J.L. Suschitzky, Tetrahedron
Letters, 1983, 24, 597; J. chem. Res., 1984, 380).

$$(7) \qquad\qquad (8) \qquad\qquad\qquad (9)$$

Some 2-naphthyl-3-nitro-2H-benzo[b]pyrans have been prepared from salicylaldehydes and (nitroethenyl)naphthalenes in methylene dichloride in the presence of triethylamine (R.S. Varma, M. Kadkhodayan and G.W. Kabalka, Synth., 1986, 486).

The synthesis of some benzo[b]pyran prostaglandins has been described (P.A. Aristoff, A.W. Harrison and A.M. Huber, Tetrahedron Letters, 1984, 25, 3955). For a review on the synthesis of benzo[b]pyrans see H.J. Shue (Diss. Abs. Int. B., 1985, 45, 2924).

Reactions. The hydrogenation of 3,4-diaryl-2,2-dimethyl-7--methoxy-2H-benzo[b]pyran in the presence of W_2-Raney Ni yields the corresponding 3,4-dihydrobenzopyran as a mixture of the *cis-* and the *trans-* isomer (P.K. Arora and S. Ray, J. prakt. Chem., 1981, 323, 850).

Thallium (III) nitrate oxidation of the 2,2-dimethyl-2H--benzo[b]pyran (10) in methanol affords only the *cis--dihydrodimethoxybenzopyran (11) and in ethane-1,2-diol the fused dioxan (12) is obtained in low yield. 3-Methyl-3H--naphtho[2,1-b]pyran (13) on similar oxidation in aqueous dioxan gives a single *cis*-1,2-diol (14) unexpectedly arising from initial thalliation *syn* to the 3-methyl group (M.J. Begley *et al.*, J. chem. Soc., Perkin I, 1983, 883).

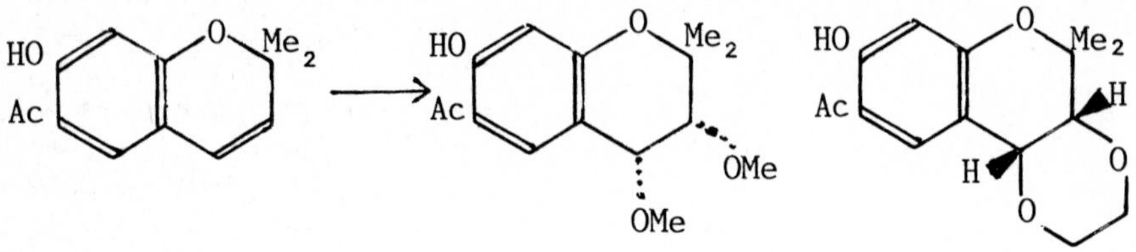

48

(10)   (11)   (12)

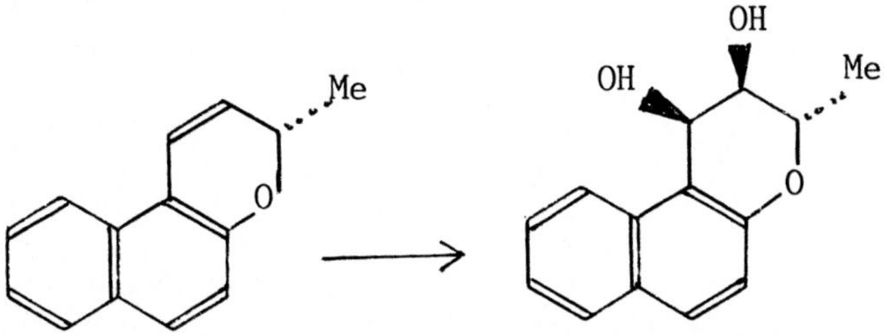

(13)   (14)

Addition of dichlorocarbene to 4-ethoxy-2-methyl-2*H*-
-benzo[b]pyran in the presence of benzyltriethylammonium
chloride yields adduct (15) (A.V. Koblik, K.F. Suzdalev and
G.N. Dorofeenko, Khim. Geterotsikl. Soedin., 1982, 163).

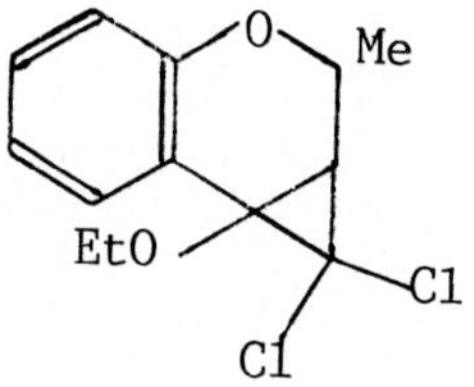

(15)

Acylation of 7-methoxy-2,2-dimethyl-2H-benzo[b]pyran affords the 6-acyl derivative, whereas 6-methoxy-2,2--dimethylbenzo[b]pyran yields the 7-acyl derivative; the corresponding 8-methoxy compound gives a mixture of the 5- and the 6-acyl derivative. The 6-methoxy compound furnishes only the 8-formyl derivative (S. Yamaguchi *et al.*, Bull. chem. Soc. Japan, 1984, 57, 442).

The thermal (B.S. Thyagarajan, K.K. Balasubramanian and K.C. Majumdar, J. heterocyclic Chem., 1973, 10, 159) and charge induced (D.K. Bates and M.C. Jones, J. org. Chem., 1978, 43, 3856) Claisen rearrangement of 4-aryloxymethyl--2H-benzo[b]pyrans and 3-aryloxymethyl-2H-benzo[b]pyrans have been investigated. The latter yield only the normal expected products (S. Selvaraj *et al.*, Tetrahedron Letters, 1983, 24, 2509).

2H-Benzo[b]pyrans (16) react with trialkylaluminium compounds by alkyl- and hydrogen transfer to C-2 and C-4 to give the 2-allylphenols (17 and 18) and the (E)-2-propenyl-phenols (19 and 20). The position attacked and the ratio of alkyl- to hydrogen-transfer depends on the alkylaluminium used and on the size of the substituent at C-2 (A. Alberola *et al.*, J. chem. Soc., Perkin I, 1983, 1209).

(16)   (17)   (18)

R^1,R^2=H or alkyl
R^3=Et or t-Bu

(19)   (20)

2,3-Dialkyl- and 2,2,3-trialkyl-2H-benzo[b]pyrans on treatment with triethylaluminium or ethylmagnesium bromide furnish substituted 2-allylphenols in good yield. The reactions of 2,3-dialkyl-2H-benzo[b]pyrans are stereoselective, giving preferentially either the E or the Z isomer, depending upon the experimental conditions. Stereoselectivity increases with the bulk of the alkyl substituent at C-3. 2,4-Dialkyl- and 2,2,4-trialkyl-2H- -benzo[b]pyrans yield only 2-allylphenols with ethylmagnesium bromide, but the mixture of the E and the Z isomer is very difficult to separate (*idem, ibid.*, 1984, 1259).

The photolysis of 3-hydroxy-4-methoxycarbonyl-2H- -benzo[b]pyran (21) in acetonitrile yields the dihydro-coumarin (25) *via* the o-quinonoid intermediate (22). However in benzene solution the valence tautomer (24) of the dihydrofuran (23) exists long enough for further light absorption to take place, resulting in loss of carbon monoxide and the formation of the 2-vinylphenol (26) as the major product (A. Padwa *et al.*, J. org. Chem., 1975, <u>40</u>, 1142; Padwa and A. Au, J. Amer. chem. Soc., 1975, <u>97</u>, 242).

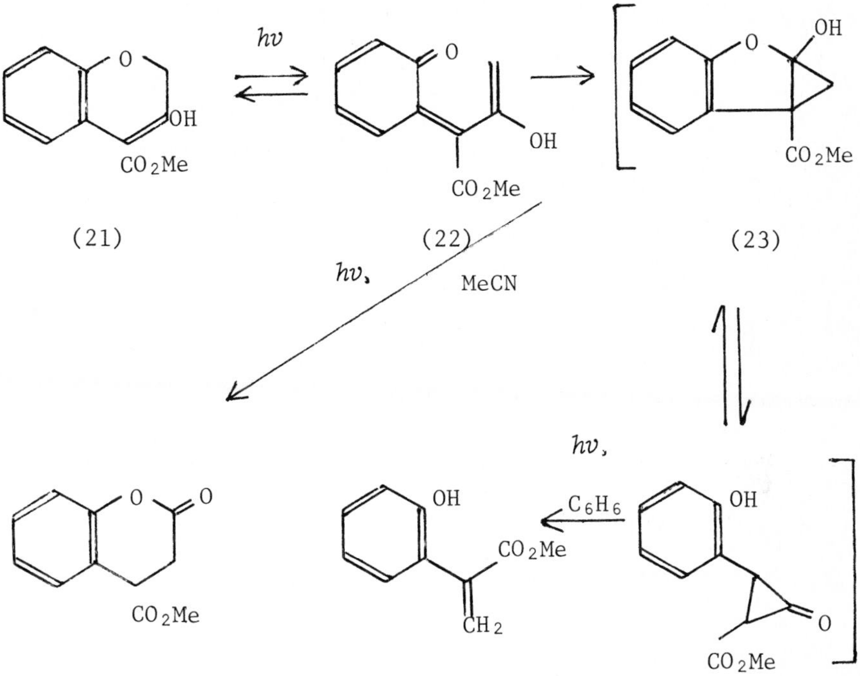

(21)   (22)   (23)

(25)   (26)   (24)

2,2-Dimethyl-3-fluoro-2*H*-benzo[b]pyrans are formed on
condensing phenols with $Me_2C=CFCH(OMe)_2$ in pyridine
(F. Camps Diez *et al.*, Span. Pat. ES 496,301, 1981). Both
4,6-dichloro-2-(methylthiomethoxy)cinnamaldehyde and 4,6-
-dichloro-2-hydroxycinnamaldehyde react with mercury (II)
chloride in hot aqueous acetonitrile to give 5,7-dichloro-
-2H-benzo[b]pyran-2-ol, m.p. 120-130° (decomp.)
(G.E. Stokker, J. heterocyclic Chem., 1984, 21, 609). For
3-chloro-6-methyl-2*H*-benzo[b]pyran see N.A. Andreev,
V.I. Levashova and L.I. Bunina-Krivorukova (Zh. org. Khim.,
1985, 21, 1061).

3-Nitro-2H-benzo[b]pyrans unsubstituted at the 2-position
are obtained when salicylaldehyde reacts with 2-nitroethanol
in the presence of dibutylamine hydrocholoride in isopentyl
acetate (D. Dauzonne and R. Royer, Synth., 1984, 348). 2-
-Alkoxy-3-nitro-2H-benzo[b]pyrans have been prepared by
condensation of substituted salicylaldehydes with β-nitro-
acetaldehyde dialkyl acetals (L. René, M. Faulques and
R. Royer, J. heterocyclic Chem., 1982, 19, 691).

(3) Flavenes (2-arylbenzo[b]pyrans)
3-Nitro-2H-flavenes (2-aryl-3-nitro-2H-benzo[b]pyrans) have
been prepared by condensation of salicylaldehyde with 3,4-
$-R^1R^2C_6H_5CH=CHNO_2$ (R^1 = H,MeO, R^2 = H,Me,MeO,OH;
R^1R^2 = OCH_2O) in the presence of triethylamine
(S.R. Deshpande, H.H. Mathur and G.K. Trivedi, Indian J.
Chem., 1983, 22B, 166). Some have also been obtained by
the cyclocondensation of the appropriate salicylaldehydes
and β-nitrostyrenes on basic alumina (R.S. Varma and
G.W. Kabalka, Heterocycles, 1985, 23, 139). 3-Nitro-2H-
-flavenes may be used in the synthesis of flavonols.

Reduction of 3-nitro-2H-flavenes with sodium tetrahydridoborate, followed by reduction of the nitro group with
hydrazine in the presence of Raney nickel affords 3-amino-
-3,4-dihydro-2H-flavenes (P.K. Arora and A.P. Bhaduri, Indian
J. Chem., 1981, 20B, 951).

3-Nitro-2H-flavene (1) undergoes a novel high yield
photoreaction in methanol to yield 3,4-dihydro-4,4-
-dimethoxy-3-hydroxyimino-2H-flavene (2) (S.T. Reid,
J.K. Thompson and C.F. Mushambi, Tetrahedron Letters, 1983,
24, 2209).

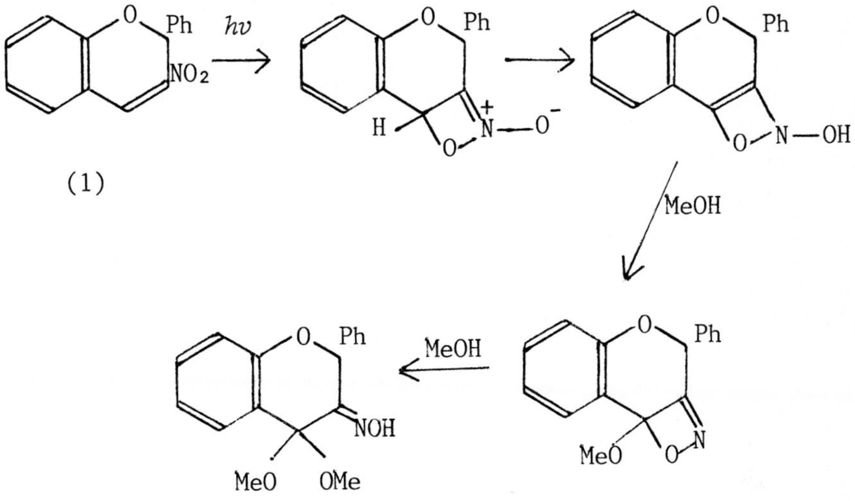

(1)

(2)

The reaction between 4H-flavenes (3) and dihalogeno-
carbenes in the presence of benzyltriethylammonium
chloride, results in addition at the pyran double bond to
give the cyclopropane derivatives (4).

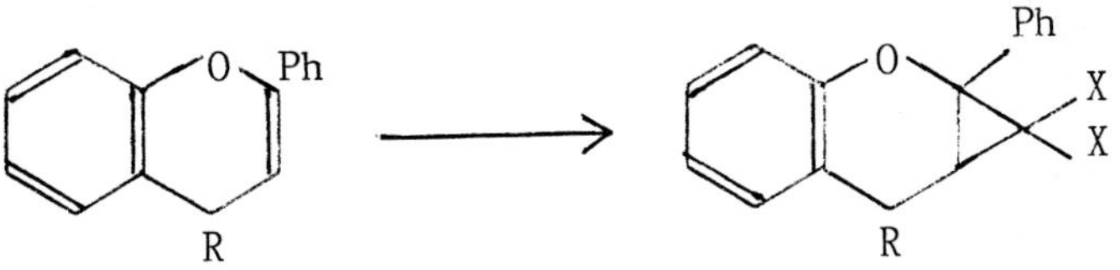

(3)   (4)

R = Ph, X = Cl
R = PhCh$_2$, X = Cl
R = 4-MeC$_6$H$_4$, X = Cl
R = Ph, X = Br
R = 2-thienyl, X = Cl

Similarly the addition of dichlorocarbene to 4-ethoxy-2H- -flavenes (5) affords the adduct (6) (A.V. Koblik, K.F. Suzdalev and G.N. Dorofeenko, Khim. Geterotsikl. Soedin, 1982, 163).

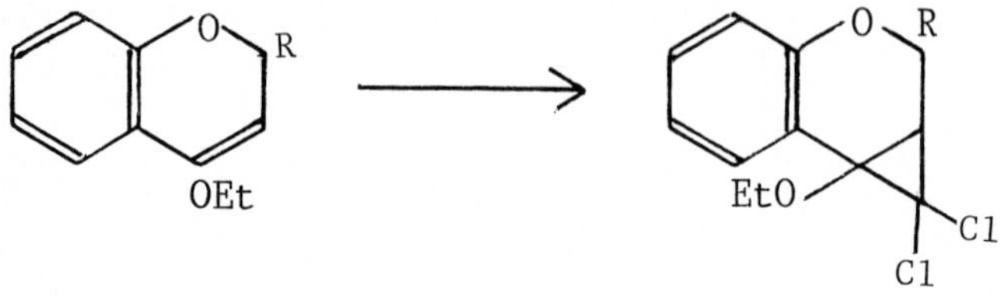

(5)   (6)

R = Ph, 2-MeC$_6$H$_4$

(ii) Naturally occurring derivatives of 2,2-dimethylbenzo-[b]pyran

Precocene II (ageratochromene) is the major component of *Eupatorium sternbergianum* (Compositae) and has been isolated by chromatography on silica gel impregnated with silver nitrate (20%). It yields three dimeric products (1, 2 and 3). Since the latter two containing a tetrahydrofuran ring are formed in the presence of SiO_2-$AgNO_3$ or SiO_2-$FeCl_3$ and not with $ZnBr_2$, it is suggested that they possibly arise by one-electron-transfer oxidation (B.M. Fraga *et al.*, J. chem. Soc., Perkin I, 1983, 2687).

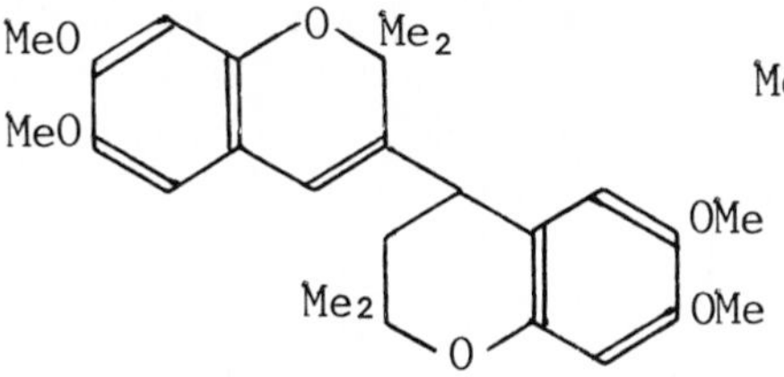

(1)

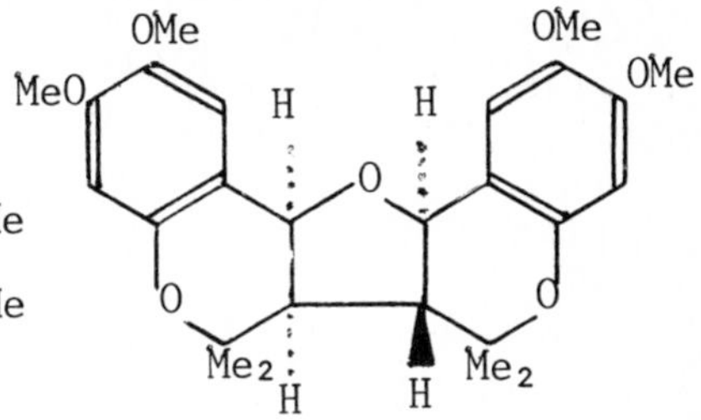

(2)

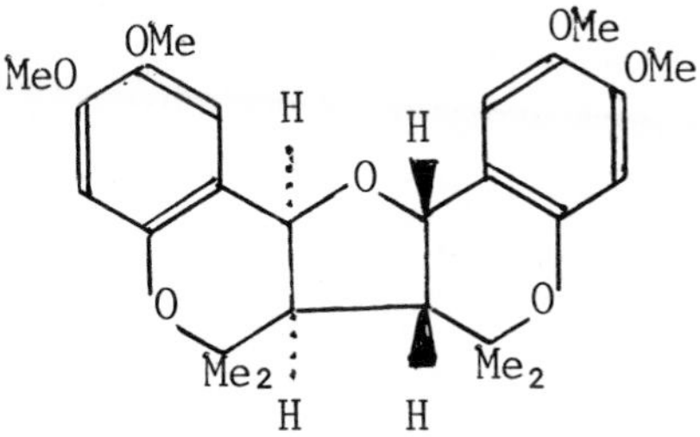

(3)

(iii) Benzopyrylium or chromylium and flavylium salts

6,7-Dihydroxy-4-carboxylflavylium (6,7-dihydroxy-4-
-carboxyl-2-phenylbenzopyrylium) chloride is formed on
passing hydrogen chloride into a mixture of pyrogallol
triacetate and benzoylpyruvic acid in acetic acid
(G.F. Tantsyura, Chem. Abs., 1982, 96, 68757q).

A number of bridged 2-arylbenzopyrylium salts have been
synthesized and characterized by spectral methods
(P. Czerney and H. Hartmann, J. prakt. Chem., 1983, 325,
161). See Czerney and Hartmann (Fr. Pat., Demande FR
2,479,223, 1981) for the preparation of amino substituted
2-(3-coumaryl)benzopyrylium salts, useful in electro-
photography. Some flavylium salts have been obtained by a
new method, which involves treating $RClC=CHCH=NMe_2ClO_4$ or
$RClC=CHCHO$ (eg. R = Ph, $4-MeC_6H_4$, $4-MeOC_6H_4$, $4-ClC_6H_4$)
with phenols in AcOH - $HClO_4$ (Czerney and Hartmann, Z.
Chem., 1982, 22, 136).

2,4-Diaryl-5-oxo-5,6,7,8-tetrahydrobenzopyrylium salts (2) are obtained by the cyclization of trione (1) in AcOH – Ac_2O with $HClO_4$ – Et_2OBF_3 (V.G. Kharchenko, L.I. Markova and K.M. Korshunova, Khim. Geterotsikl. Soedin., 1982, 708).

ClO_4^-

(1)

(iv) The anthocyanins and anthocyanidins

The well known anthocyanins and anthocyanidins, along with some modified derivatives have been isolated from new sources. Methods have been developed for the separation and identification of various anthocyanins and anthocyanidins. These include the application of thin layer, paper, and high performance liquid chromatography and the use of spectral data. A method has been described for the simultaneous separation of anthocyanins and anthocyanidins from various plant sources by thin layer chromatography using cellulose and a variety of solvent mixtures (O.M. Andersen and G.W. Francis, J. Chromatog., 1985, 318, 450).

Methods for acid hydrolysis of anthocyanins and alkaline hydrolysis of the resulting anthocyanidins have been developed. The sugars from the former are identified by paper chromatography and the phenolic acids from the latter by uv-spectroscopy. By this method malvin was identified as the anthocyanin from Berberis buxifolia, yielding malvidin and glucose with the anthocyanidin affording syringic acid (N.L.S. Diaz and N.R. Glave, Rev. Agroquim.

Tecnol. Aliment., 1981, 21, 419). Anthocyanidins have been
oxidatively cleaved into benzoic acid derivatives by acid
hydrolysis to chalcones or flavenol acetates, followed by
oxidative cleavage using ruthenium oxide (R. Mentlein,
E. Vowinkel and B. Wolf, Ann., 1984, 401). The possible
use of reversed-phase high-performance liquid chromatography
for the analysis of complex mixtures of anthocyanins,
anthocyanidins and other natural products has been
discussed (K. Vande Casteele *et al.*, Chromatogr., 1983,
259, 291).

The demethylation of methoxyflavylium salts in the
cyanidin series in pyridinium chloride under nitrogen shows
that the 3-methoxyl group is demethylated first, followed
by the 7 and 5, then the 3' and 4' groups. A number of
methyl ethers including peonidin, elodenidin (5-methyl-
cyanidin), and rosinidin (7-methylpeonidin) have been
produced by demethylation. Their spectral and chromatographic characteristics have been recorded (T. Momose and
K. Abe, Phytochem., 1982, 21, 794). Picosecond
fluorescence kinetics and polarization anisotropy
measurements have been obtained for anthocyanins isolated
from *Streptocarpus holstii* and *Anthurium andreanum* flower
petals (F. Pellegrino, P. Sekuler and R.R. Alfano,
Photobiochem. Photobiophys., 1981, 2, 15). The anthocyanins
in grapes have been determined by spectrometric methods
(S.V. Durmishidze and A.N. Sopromadze, Chem. Abs., 1984,
100, 50125y) and the effects of metabolic inhibitors on
anthocyanin accumulation in petals of *Rosa hybrida* Hort.
studied (H. Nakamae and N. Nakamura, Plant Cell Physiol.,
1983, 24, 995).

Pelargonidin derivatives. Two anthocyanins possessing the
same aglycone, pelargonidin as glycosides have been
isolated from the flowers of *Salvia splendens* [D. Albulescu
and C. Gavrila-Dinu, Farmacia (Bucharest), 1981, 29, 25].

Cyanidin derivatives. Cyanidin-3-glucoside and -3,5-
-diglucoside along with other anthocyanins have been
obtained from the leaves and stems of *Desmodium adhaesivum*
(P. Matinod *et al.*, Politecnica, 1978, 4, 74); cyanidin-3-
-(*p*-coumaroyl)-glucoside and -3-(*p*-coumaroyl)-
-rhamnoglucoside from black cherry (G. Gombkoto, Acta
Aliment., 1980, 9, 335); cyanidin-3-glucoside and -3-
-rhamnoglucoside from plum fruit (*Prunus salicina*),
variety "Sordum" (M. Tsuji, M. Harakawa and Y. Komiyama,
Nippon Shokuhin Kogyo Gakkaishi, 1981, 28, 517); cyanidin-

-3-glucoside and -3-galactoside and a mixture of two
cyanidin-3-biosides, one of which appeared to be cyanidin-
-3-rhamnosylgalactoside, from buckwheat (*Fagopyrum
sagittatum*) hypocotyls (K. Inouye, Y. Hosoyama and
T. Shimadate, Chem. Abs., 1982, **97**, 8860m); cyanidin-3-
-xyloglucoside, -3-rhamnoglucoside and -3,5-diglucoside from
black elderberries (E. Pogorzelski, Przem. Spozyw., 1983,
37, 167); and cyanidin monoglucoside from the flowers of
Veratrum nigrum var. japonicum Baker (H. Kawanobe, Akita
Daigaku Kyoikugakubu Kenkyu Kiyo, Shizen Kagaku, 1984,
131). Free cyanidin has been isolated along with cyanidin-3-
and -7-glucoside from blackberries (*Rubus ulmifolia*, Shott),
indicating that cyanidin glucosides may be relatively
unstable in this fruit (Diaz and Glave, Rev. Agroquim.
Tecnol. Aliment., 1982, **22**, 111). Cyanidin-3-rutinoside
and two novel glycosides, 6-hydroxycyanidin-3-rutinoside
and -glucoside have been obtained from the red flowers of
four cultivars of *Alstroemeria* (N. Saito *et al.*, Phytochem.,
1985, **24**, 2125). A cyanidin-3-glycoside and an acetylated
derivative have been isolated along with other products from
the flowers of *Allium aschersonianum*.

Sambicyanin (1) has been isolated from *Sambucus nigra*.
The parent aglycone is cyanidin and the accompanying
disaccharide has been named sambiose (L. Reichel and
W. Reichwald, Pharmazie, 1977, **32**, 40).

(1)

Cyanidin-3,5-diglucoside has been known for many years to be the anthocyanidin in the blue cornflower (*Centaurea cyanus*), but it has now become evident that the major anthocyanin is cyanidin-3-succinylglucoside-5-glucoside, tentatively called centaurocyanin. It appears that this anthocyanin, and not cyanin, forms the blue complex pigment protocyanin (K. Takeda and S. Tominaga, Bot. Mag., 1983, **96**, 359).

Delphinidin derivatives. Xyloglucosyldelphinidin and (*p*-
-coumarylgalloyl)glucosyldelphinidin have been identified in the seed coat pigment of *Abrus precatorius* (M.S. Karawya *et al.*, Fitoterapia, 1981, **52**, 175). Five anthocyanins having delphinidin as the aglycone have been separated, following their extraction from the bluish-purple corolla of *Petunia hybrida* cv. blue picoti (T. Takano and H. Iida, Meijo Daigaku Nogakubu Gakujutsu Hokoku, 1983, **19**, 24). Six blue acylated anthocyanins all based on delphinidin-3,3',5'-
-triglucoside have been isolated from the blue flowers of *Clitoria ternatea* (N. Saito *et al.*, Phytochem., 1985, **24**, 1583). Platyconin a diacylated anthocyanin obtained from the Chinese bellflower, *Platycodon grandiflorum*, has been shown to be 3-*O*-[6-*O*-(α-L-rhamnopyranosyl)-β-D-gluco-
pyranosyl]-7-*O*-[6-*O*-[*trans*-4-*O*-[6-*O*-[*trans*-4-*O*-
-(β-D-glucopyranosyl)-caffeoyl]-β-D-glucopyranosyl]-
caffeoyl]-β-D-glucopyranosyl]delphinidin (T. Goto *et al.*, Tetrahedron Letters, 1983, **24**, 2181).

Peonidin derivative. Peonidin-3-*O*-β-L-arabinopyranosyl-*O*-
-β-D-glucopyranoside and kaempferol have been isolated from the fresh petals of *Ipomoea fistulosa* (O.C.D. Gupta, R. Gupta and P.C. Gupta, Planta Med., 1980, **38**, 147).

Malvidin derivative. The petal extract from *Anagallis arvensis* f. *coerulea* contains malvidin-3-rhamnoside, luteolin, luteolin-7-glucoside, and quercetin-3-rhamnoside (N. Ishikura, Z. Pflanzenphysiol., 1981, **103**, 469).

Sources of more than one anthocyanin, anthocyanidin derivatives
Pelargonidin and cyanidin derivatives. The relationship between pelargonidin and cyanidin as mono- or di-glucosides with reference to the colouration of carnation flowers has been investigated (S. Maekawa and N. Nakamura, Kobe Daigaku Nogakubu Kenkyu Hokoku, 1977, 12, 161; V.K. Parshikov and I.U. Rekhviashvili, Chem. Abs., 1986, 104, 222007n). Of the four anthocyanins along with flavones isolated from the flower of *Saraca indica* two have been identified as pelargonidin-3,5-diglycoside and cyanidin-3,5-diglucoside (V. Lakshmi and J.S. Chauham, Q.J. Crude Drug Res., 1976, 14, 65). Pelargonidin- and cyanidin-3-rhamnosylglucoside have been isolated from the spathes of *Anthurium andreanum* Lind. (R.Y. Iwata, C.S. Tang and H. Kamemoto, J. Amer. Soc. hortic. Sci., 1979, 104, 464). A predominance of the latter results in pink to dark red colours, whereas predominance of the former results in coral to orange (*idem, ibid.*, 1985, 110, 383). The different fruit colours (yellow-orange and several variants of red-black) of *Cotoneaster* are due to different proportions and concentrations of pelargonidin- and cyanidin-3-galactoside and isoquercitrin (T. Kruegel and A. Krainhoefner, Chem. Abs., 1985, 103, 51155g). The anthocyanins from Bridal Pink and Samantha roses have been shown to be either pelargonidin- or cyanidin-3,5-diglucoside, along with traces of the related 3-glucoside (S. Asen, J. Amer. Soc. hortic. Sci., 1982, 107, 744).
Cyanidin and delphinidin derivatives. Cyanidin and delphinidin have been found in the leaves of *Cayratia trifolia, Cissus hastata,* and *Tetrastigma dubium* species of Malaysian Vitaceae (A. Latiff, Malays. J. Sci., 1980, 6, 95) and in leaves of some *Cichorium intybus* (E.M. Cappelletti, Plant Med. Phytother., 1984, 18, 3). Cyanidin- and delphinidin-3-glucoside and -3,5-diglucoside have been found in the seed coats of pomegranate (*Punica granatum*) (N.A. Santagati, R. Duro and F. Duro, Riv. Merceol., 1984, 23, 247). Delphinidin-3-*p*-coumaroylsophoroside-5-glucoside (cayratinin) is found in the coloured petals of *Epimedium* species, but it occurs along with cyanidin glucoside in the stems, young leaves, and autumn leaves of all the species (K. Yoshitama, Bot. Mag., 1984, 97, 429).

Cyanidin and peonidin derivatives. Cyanidin, cyanidin-3-
-galactoside, -glucoside, -araginoside, -galactoside-5-
-glucoside, and peonidin-3,5-diglucoside have been obtained
from *Rhododendron semsii* flowers (R. De Loose, Chem. Abs.,
1980, 93, 163709x). The preponderant anthocyanin in
Digitalis purpurea mutants is cyanidin-3-glucoside along
with some peonidin-3-glucoside, whereas normal plants
contain the corresponding 3,5-diglucosides (A. Braun,Bull.
Liaison-Groupe Polyphenols, 1980, 259).
Peonidin and malvidin derivatives. Peonidin-3-glucoside
and -3,5-diglucoside, and malvidin-3,5-diglucoside have
been identified as pigments in the flower of *Hedysarum
coronarium.* Varying mixtures of these derivatives are
responsible for the different flower colours in this
species (A. Chriki and J.B. Harborne, Phytochem., 1983, 22,
2322). Peonidin-3-sophoroside and malvidin-3,5-diglucoside
have been identified as pigments from the flowers of
Tibouchina grandiflora. Two other pigments have
tentatively been identified as peonidin-3-sambubioside and
malvidin-3-(*p*-coumaroyl)sambubioside (F.O. Bobbio,
P.A. Bobbio and C.H. Degaspari, Food Chem., 1985, 18,
153).
Pelargonidin, cyanidin and delphinidin derivatives.
Pelargonidin-, cyanidin-, and delphinidin-3,5-diglucoside
have been obtained from the flower of *Erythrina suberosa*
(Y.S. Verma, V.K. Saxena and S.S. Nigam, Proc. Natl. Acad.
Sci., India, 1977, 47A, 71). Thirty seven sweet pea
cultivars have been placed in five classes, according to
their pigments being derived from (1) pelargonidin,
(2) cyanidin, (3) delphinidin, (4) pelargonidin accompanied
by cyanidin, and (5) cyanidin accompanied by delphinidin
(Y. Sakata and K. Arisumi, Kagoshima Daigaku Nogakubu
Gakujutsu Hokoku, 1983, 13). Garden tulips contain
glycosides of pelargonidin, cyanidin and delphinidin
(V.K. Parshikov, Fiziol. Biokhim. Kul't. Rast., 1986, 18,
196).
Pelargonidin, cyanidin and peonidin derivatives. An
examination of ten garden varieties of morning glory
belonging to the Higo line has shown that the fifteen
anthocyanins present in the petal are mainly acylated
glucosides of the aglycones, pelargonidin, cyanidin and
peonidin. The red flowers of Higo morning glory contain
almost exclusively pelargonidin glucosides, whereas the
blue flowers contain mainly acylated peonidin glucosides
(N. Ishikura, Kumamoto J. Sci., Biol., 1981, 15, 29).

Pelargonidin, peonidin and malvidin derivatives. The blue-
-flowered cultivars of African violet (*Saintpaulia ionantha*)
contain malvidin-3-rutinoside-5-glucoside; those with red
petals, peonidin-3-rutinoside-5-glucoside; and the pink-
petalled cultivars, pelargonidin-3-rutinoside-5-glucoside
(J.A. Khokhar *et al.*, Hort. Science, 1982, 17, 810).
Cyanidin, delphinidin and peonidin derivatives. Flowers
from twelve *Anthyllis vulneraria* populations and one
Anthyllis montana population contain cyanidin-3-galactoside.
Some also contain small amounts of delphinidin and peonidin
glycosides (probably 3-galactoside), but only *A. montana*
flowers contain malvidin. Elevated levels of delphinidin
and peonidin are found in three Iberian populations
(A.A. Sterk, P. De Vlaming and A.C. Bolsman-Louwen, Acta
Bot. Neerl., 1977, 26, 349). Oranges of the Moro, Tarocco
and Sanguinello varieties besides cyanidin-3-glucoside
also contain cyanidin-3,5-diglucoside, delphinidin-3-
-glucoside, delphinidin-3,5-diglucoside, peonidin-3,5-
-diglucoside, cyanidin-3-acetylglucoside, cyanidin-3-
-ferulylglucoside, cyanidin-3-coumarylferulylglucoside,
cyanidin-3-sinapylglucoside and peonidin-3-coumaryl-
glucoside [E. Maccarone, A. Maccarone and P. Rapisarda,
Ann. Chim. (Rome), 1985, 75, 79].
Cyanidin, delphinidin and malvidin. Cyanidin, delphinidin
and malvidin have been obtained from blueberries, *Vaccinium
myrtillus* (G. Bounous *et al.*, Chem. Abs., 1983, 98, 50439u).
Delphinidin, malvidin and petunidin derivatives.
Delphinidin-3-glucoside, delphinidin-3,5-diglucoside,
malvidin-3-glucoside, malvidin-3,5-diglucoside, petunidin-
-3-glucoside and petunidin-3,5-diglucoside are contained in
the petals of *Alcea rosea* L. (S. Kohlmunzer, G. Konska and
E. Wiatr, Herba Hung., 1983, 22, 13).
Mixtures of four, five and six anthocyanidins.
Pelargonidin, cyanidin, delphinidin and petunidin along
with other products have been isolated from *Mentha
rofundifolia* (L.) Hudson (E. Marin Pares, Circ. Farm.,
1983, 41, 133). Pelargonidin, cyanidin, delphinidin,
peonidin and malvidin 3-glucosides, -galactosides and
-arabinosides are the major floral pigments in *Erica*
(Ericaceae) and a comparison has been made with the
anthocyanins found in the related family Epacridaceae
(R.K. Crowden and S.J. Jarman, Phytochem., 1976, 15, 1796).
Bilberry (*Vaccinium arctostaphylos*) juice contains
anthocyanins of the aglycones pelargonidin, cyanidin,

peonidin, malvidin and petunidin (N.A. Ugulave,
A.N. Nizharadze and I.F. Gribovskaya, Izv. Akad. Nauk Gruz.
SSR, 1978, 4, 81). Fruit of *Berberis vulgaris*, *B.*
sphaerocarpa, *B. integerrima*, *B. sieboldii* and *B. coreana*
contain pelargonidin, cyanidin, delphinidin, peonidin and
petunidin aglycones bound with glucose and rutinose. All
the species also contain cyanidin-, peonidin-, and
petunidin-3-glucose (V.V. Vereskovskii and D.K. Shapiro,
Khim. Prir. Soedin., 1985, 569). Anthocyanins identified
along with other products from geranium florets are
pelargonidin-, cyanidin-, delphinidin-, peonidin-,
malvidin-, and petunidin-3,5-diglucoside (Asen and
R. Griesbach, J. Amer. Soc. hortic. Sci., 1983, 108, 845).

Carajurin a 3-deoxyanthocyanidin appears to be unique to
the genus Arrabidaea (R. Scogin, Biochem. Syst. Ecol., 1980,
8, 273).

The absolute configuration of awobanin, shisonin and
violanin, which occur in *Commelina communis L.*, *Perilla*
ocimoides, and *Viola tricolour*, respectively, have been
determined from the nmr spectral data obtained from the
corresponding chlorides (T. Goto, S. Takase and T. Kondo,
Tetrahedron Letters, 1978, 2413).

Besides cyanidin-3-glucoside and -3,5-diglucoside two
complex anthocyanins (orchicyanins or chicyanins) have been
isolated from the flowers of European orchids, *Orchis*,
Dactylorhiza and *Gymnadenia* (W. Uphoff, Experientia, 1979,
35, 1013).

The relationship of flower colour to anthocyanin content
and other parameters has been studied for *Azalea indica*
[R. De Loose, Sci. Hortic. (Amsterdam), 1978, 9, 285),
artichoke (*Cynara scolymus L.*) [S. Aubert and C. Foury,
Studi Carciofo, (Congr. Int.), 3rd 1979 (Pub. 1981), 57],
five species and thirty four hybrids of *Camellia* (Y. Sakata
and K. Arisumi, Mem. Fac. Agric., Kagoshima Univ., 1985,
21, 149), and thirteen species and one hundred and thirty
nine hybrids of rhododendron (K. Arisumi, Y. Sakata and
I. Miyajima, *ibid.*, p. 133).

Anthocyanins and in some cases other products have been
identified in six species of *Amelanchier* genus
(D.K. Shapiro, T.I. Narizhnaya and L.V. Anikhimovskaya,
Vestsi Akad. Navuk BSSR, Ser. Biyal. Navuk, 1980, 57);
Euphorbia hirta, *E. nerifolia* and *E. thymifolia* (R.K. Baslas
and R. Agarwal, Indian J. Chem., 1980, 19B, 717);
Amelanchier vulgaris fruit (E.G. Martynov *et al.*, Khim.

Prir. Soedin., 1980, 725); leaves of agrimony (*Agrimonia eupatoria L.*) [P. Bohinc and O. Humar, Farm. Vestn. (Ljubljana), 1982, 33, 243]; and in the Polygonaceae (K. Yoshitama, M. Hisada and N. Ishikura, Bot. Mag., 1984, 97, 31).

A study has been made showing the widespread occurrence in nature of anthocyanins as zwitterions (J.R. Harborne and M. Boardley, Z. Naturforsch., Biosci., 1985, 40C, 305) and of the kinetics of the hydrolysis of strawberry anthocyanins by a thermostable anthocynanin-β-glycosidase from *Aspergillus niger* (H. Blom and M.S. Thomassen, Food Chem., 1985, 17, 157).

Evidence has been provided for the biosynthetic sequence of the predominant pigment, pelargonidin-3-p-coumaroyl-glucoside-5-glucoside, in developing petals of the red--flowered genotype (llhhpp) of *Impatiens balsamina* (D. Strack and R.I. Mansell, Z. Pflanzenphysiol., 1979, 91, 63). It has been found that in the biosynthesis of anthocyanins in *Petunia hybrida*, anthocyanidin-3-glucosides and quercetin-7-glucoside are synthesized if dihydro-quercetin, quercetin-7-glucoside or dihydroquercetin-4'--glucoside are used as precusors (A.W. Schram *et al.*, Planta, 1981, 153, 459). The activities of some enzymes concerned in the biosynthesis and accumulation of anthocyanins during the development of buds and flowers of *Matthiola incana* have been studied (B. Dangelmayr *et al.*, Z. Naturforsch., Biosci., 1983, 38C, 551). It has been shown that the biosynthetic pathways for the plant pigments, anthocyanins and betacyanins are incompatible (H.N. Srivastav and M.M. Laloraya, Plant Biochem. J., 1978, 5, 136). Investigations have been carried out into the possible participation of anthocyanins in the photoreactions of photosynthesis (V. Sharma and D. Banerjee, *ibid.*, 1981, 8, 76).

The chemistry of anthocyanins, anthocyanidins and related flavylium salts (G.A. Iacobucci and J.G. Sweeny, Tetrahedron, 1983, 39, 3005), the chromatographic separation of anthocyanins, glycosylfavones, flavonol glycosides, isoflavonoids and related compounds [J.B. Harborne, Adv. med. Plant Res., Plenary Lect. Int. Congr., 32nd, 1984 (Pub. 1985), 135], and the anthocyanins of the Orchidaceae (M.H. Fisch, Orchid Biol., 1977, 1, 117) have been reviewed.

(v) Coumarins (2H-benzo[b]pyran-2-one, 5,6-benzo-2-pyrone, 5,6-benzo-α-pyrone) and its derivatives

Preparations. Coumarin has been prepared in 20% yield by the photooxygenation of 2-allylphenol (S. Pandita and S.S. Chibber, Curr. Sci., 1982, 51, 411) and by vapour phase catalytic dehydrogenation of dihydrocoumarin at 400-500° (V.G. Cherkaev *et al.*, U.S.S.R. Pat., SU 1,030,363, 1983).

Some phenols which give no, or very little, coumarin under Peckmann conditions afford coumarins in overall yields varying from 50 to 75% by the following route. The phenol, for example, guaiacol (1) and triethyl orthoacrylate (2) are heated together in boiling toluene containing a catalytic amount of pivalic acid to give, *via* a Claisen rearrangement and ring closure 2,2-diethoxy-3,4--dihydro-2H-benzo[b]pyran (3). Hydrolysis of derivative (3) with 10% hydrocholoric acid at room temperature affords the ester (4), which with boiling ethanolic potassium hydroxide yields the dihydrocinnamic acid (5). Boiling with acetic anhydride converts acid (5) into the dihydrocoumarin (6), which is dehydrogenated on heating with 10% palladium/carbon in boiling diphenyl ether to furnish 8--methoxycoumarin (7) (70%). Also prepared by this route are 8-methyl- (76%) and 6-chloro-8-methyl-coumarin (50%) and 3H-naphtho[2,1-b]pyran-3-one (5,6-benzocoumarin) (J.A. Panetta and H. Rapoport, J. org. Chem., 1982, 47, 946).

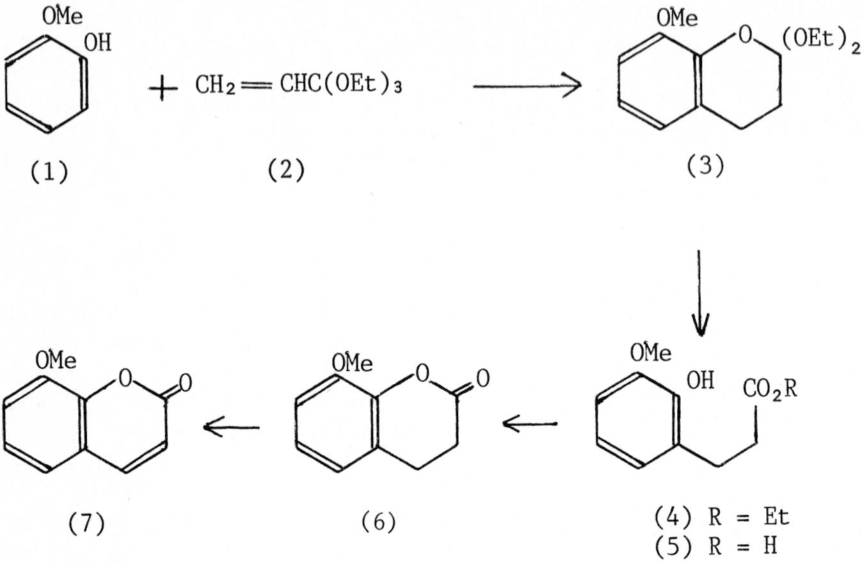

$$
\begin{array}{ccc}
(1) & (2) & (3)
\end{array}
$$

$$
\begin{array}{ccc}
(7) & (6) & \begin{array}{l}(4)\ R = Et \\ (5)\ R = H\end{array}
\end{array}
$$

Cyclization of trimethylsilylketene with 2-acylphenols affords coumarins (8) (R.T. Taylor and R.A. Cassell, Synth., 1982, 672).

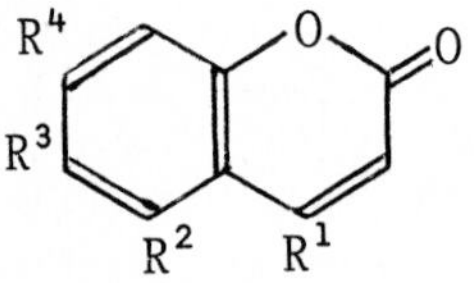

R^1 = H, Me, Ph
R^2 = H, NO_2, Cl
R^3 = H
R^2R^3 = CH = CHCH = CH
R^4 = H, OH

(8)

The use of trimethylsilylketene in the synthesis of coumarin derivatives has been reviewed (Cassell, Diss. Abs. Int. B, 1984, 44, 2424).

Substituted coumarins (10) are obtained from a mixture of salicylaldehyde or 2-hydroxyacetophenone and an appropriately substituted acetic acid (9) in the presence of triethylamine and phenyl dichlorophosphate (J. Gallastegui, J.M. Lago and C. Palomo, J. chem. Res., 1984, 170).

R^1 = H, Me

R^2 = Ph, CO_2Et, PhO, $PhCH_2$, $3,4-(MeO)_2C_6H_3$

(9)   (10)

Reagents:- (a) $PhOPCl_2$, NEt_3, CH_2Cl_2, or $(ClCH_2)_2$, Δ

An improved synthesis of 3-methycoumarins involves the reaction of a substituted salicylaldehyde with the Wittig reagent $Ph_3P=CMeCO_2Et$ (R.S. Mali, S.N. Yeola and B.K. Kulharni, Indian J. Chem., 1983, 22B, 352). It has been found that polyethylene glycol is a superior solvent for Claisen rearrangement of α-(aryloxy)methylacrylic acids to 3-methylcoumarins (K. Sunitha, K.K. Balasubramanian and K. Rajagopalan, J. org. Chem., 1985, 50, 1530). 3-(1-Oxo--2-alkenyl)coumarins are obtained by the reaction of salicylaldehyde with $R^1R^2C=CHCOCH_2CO_2Me$ [R^1 = Ph, $4-NO_2C_6H_4$, Me, R^2 = H; R^1 = R^2 = Me; R^1R^2 = $(CH_2)_5$] (J.A.M. Van den Goorbergh, M. Van der Steeg and A. Van der Gen, Synth., 1984, 859). 3-Acetyl-4--methylcoumarin has been obtained by rearrangement of

benzopyran (11) in boiling 50% aqueous acetic acid. This
involves hydrolysis, decarboxylation and rearrangement
(N.S. Nixon, F. Scheinmann and J.L. Suschitzky, Tetrahedron
Letters, 1983, <u>24</u>, 597; J. chem. Res., 1984, 380).

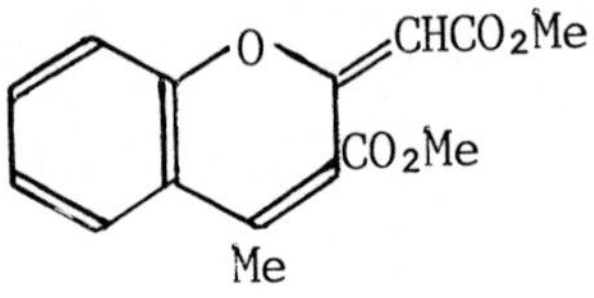

(11)

The Pechmann reaction between resorcinol and methyl
acetoacetate affords 7-hydroxy-4-methylcoumarin (72%) and a
dimer (12) (M. Kamezawa, K. Kohara and H. Tachibana, Nippon
Kagaku Kaishi, 1982, 1956).

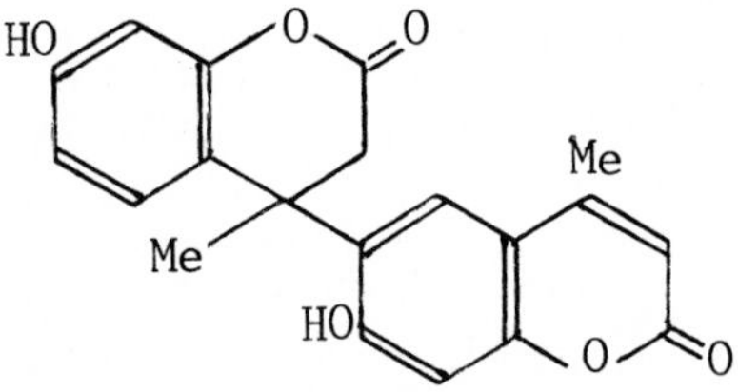

(12)

2-Hydroxyacetophenone reacts with phenylacetyl chloride in acetone in the presence of potassium carbonate to give exclusively 4-methyl-3-phenylcoumarin. Similarly salicylaldehydes afford 3-phenylcoumarins (S. Neelakantan, P.V. Raman and A. Tinabaye, Indian J. Chem., 1982, 21B, 256). 4-Ethyl-, 4-benzyl-, -3-phenylcoumarin and 3,4- -diphenylcoumarin are obtained from the appropriate 2- -hydroxy-acylbenzene (P.P. Rao and G. Srimannarayana, Synth., 1981, 887). Salicylaldehydes on treatment with N,N-diethyl(methyl or phenyl)acetamide and phosphoryl chloride afford 3-methyl- and 3-phenyl-coumarins, respectively (C.P. Phadke *et al*., Synth. Comm., 1984, 14, 407). Preparations of 3-phenyl- and 6-methyl-3-phenyl- -coumarin (M. Natarajan and V.T. Ramakrishnan, Indian J. Chem., 1984, 23B, 720) and of 4,6,7,- and 4,6,8- -trimethylcoumarins (R.I. Al-Bayati, M.T. Ayoub and R. Al-Hamdany, J. Iraqi chem. Soc., 1985, 10, 141) have been reported. 8-Substituted 5,6-dimethoxy-4-phenylcoumarins are prepared by condensation of the appropriate 2- -substituted (eg. ethyl, propyl, n-pentyl) 3,5-dimethoxy- phenols with ethyl benzoylacetate in ethanol in the presence of dry hydrogen chloride (V.K. Ahluwalia, I. Mukherjee and N. Rani, Indian J. Chem., 1981, 20B, 918).

Heating methyl 4'-O-methyl- and 4'-O-benzyl-coumarates yields the corresponding linear prenylated umbelliferones (13) directly *via* sequential *para*-Claisen rearrangement and relactonisation. The total syntheses of suberosin (13; $R^1 = R^2 = R^3 = Me$), a constituent of *Pastinaca* species (Umbelliferae), demethylsuberosin (13; $R^1 = H$, $R^2 = R^3 = Me$), and ostruthin (13; $R^1 = H$, $R^2 = Me$, $R^3 = Me_2C:CHCH_2CH_2$) (isolated from *Pastinaca ostruthium*) are achieved by this method (N. Cairns, L.M. Harwood and D.P. Astles, Chem. Comm., 1986, 1264).

(13)

4-Styrylcoumarins are conveniently prepared by the condensation of the appropriate 2-hydroxychalcones with the Wittig reagent $Ph_3P=CHCO_2Et$ (U.K. Joshi, R.M. Kelkar and M.V. Paradkar, *ibid.*, 1983, **22B**, 1151). The following have been reported: the synthesis and anthelmintic activity of 4-methyl-7-(substituted phenylaminocarbonyl- or disubstituted aminocarbonyl-methoxy)coumarins (R.R. Mohan, C. Agarwal and V.S. Misra, Indian Drugs, 1984, **21**, 389); the synthesis of 7-(2-hydroxyethyloxy)-4-methylcoumarin (J. Zawadzka *et al.*, Acta Pol. Pharm., 1984, **41**, 641); coumarin-3-carboxyanilides (A.K. Mittal and O.P. Singhal, J. Indian chem. Soc., 1982, **59**, 373); 3-ethoxycoumarins (E.R. Bissell, Synth., 1982, 846); 3-aryloxycoumarins (M. Lacova, J. Chovancova and V. Konecny, Chem. Pap., 1986, **40**, 121); fluorocoumarins (C.A. Gandolfi *et al.*, Eur. Pat. Appl. EP 185,319, 1986); some arylaminocoumarins (M.D. Bhavsar and D.J. Bhaisa, Chem. Abs., 1987, **106**, 49967s); 7-diethylamino-4-methylcoumarin (K. Pastalka, Czech. Pat. CS 231,025, 1986); and 4-substituted-3--(aminomethyl)coumarins (V.L. Savel'ev *et al.*, Khim. Geterotsikl. Soedin., 1982, 896). The cyclization of $EtO_2CC(NH_2)=C(CN)CO_2Et$ with salicylaldehyde yields a coumarin derivative (14) (S. Abdou *et al.*, Monatsh., 1982, **113**, 985).

(14)

Biochromophoric compounds containing a nucleotide base and coumarin linked with a polymethylene chain have been synthesized (G. Wenska and S. Paszyc, Canad. J. Chem., 1984, **62**, 2006).

Reactions. Coumarin, its isomers and derivatives are
considered as aromatic or potentially aromatic chemical
species and have been analyzed as such by use of standard
and parameter-varying graph-theoretical procedures (P. Ilié
et al., J. heterocyclic Chem., 1982, 19, 625).

Hydrogenation of 3,4-diarylcoumarins and 3,4-diaryl-2,2-
-dimethyl-2H-benzo[b]pyrans in the presence of W_2-Raney
nickel affords the dihydro derivatives, generally as a
mixture of the *cis-* and the *trans*-isomer (P.K. Arora and
S. Ray, J. prakt. Chem., 1981, 323, 850). Hydrogenation of
4,7-dimethylcoumarin in an alkaline medium in the presence
of Raney nickel under pressure and at elevated temperature
produces a mixture of octahydro-4,7-dimethylcoumarins. The
stereochemistry of the major product (±)-*trans-*
-4aβ(*H*),8aα(*H*)-octahydro-4α,7β-dimethylcoumarin (15) has
been established by transforming it into *p*-menthane
derivatives (Y.S. Sanghvi and A.S. Rao, J. heterocyclic
Chem., 1982, 19, 1377).

(15)

Treatment of coumarins with triethyl- or tri-isobutyl-
-aluminium causes ring opening to yield mixtures containing
cinnamyl alcohols (A. Alberola, F. Alonso Cermeno and
A. Gonzaler Ortega, An. Quim., 1982, 78C, 9). The reaction
between a 3,4-diarylcoumarin and methylmagnesium iodide,
followed by treatment of the product with acid furnishes a
3,4-diaryl-2,2-dimethyl-2*H*-benzo[b]pyran (I. Sharma *et al.*,
Indian J. Chem., 1984, 23B, 567). On treatment with
lithium di-isopropylamide coumarins tend to behave simply
as esters and give amides, but 4-methoxycoumarin is readily
lithiated at position 3 (A.M.B.S.R.C.S. Costa *et al.*, J.
chem. Soc., Perkin I, 1985, 799).

Cycloaddition of coumarin with 1,1-disubstituted alkenes (16) gives a mixture of lactones (17 and 18). However under more vigorous conditions a small amount of regioisomer (19) is produced along with lactones (17 and 18) (M.E. Jung *et al.*, Tetrahedron Letters, 1984, 25, 3659).

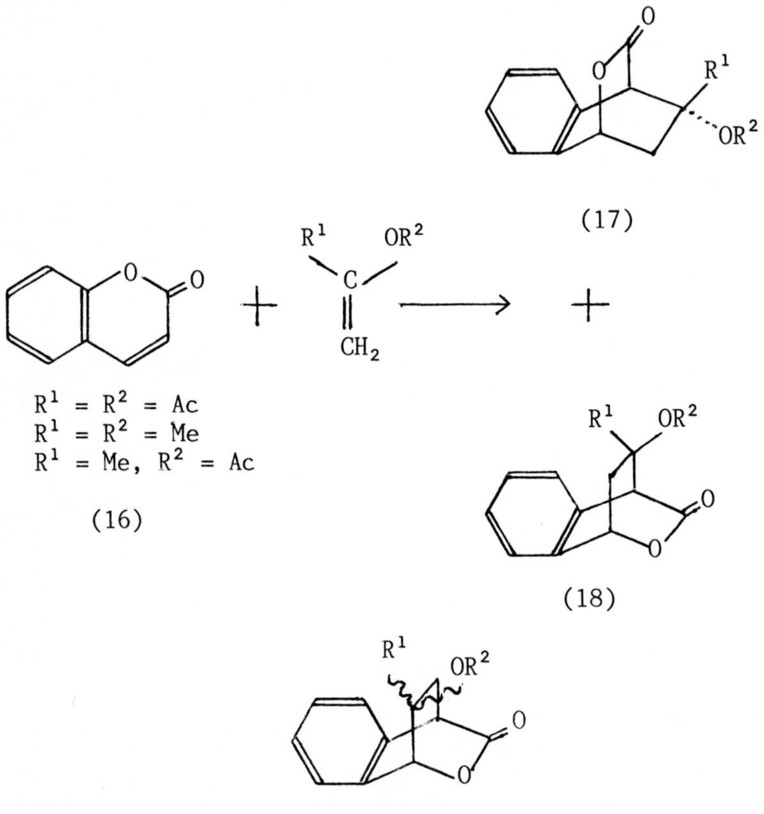

(17)

$R^1 = R^2 = Ac$
$R^1 = R^2 = Me$
$R^1 = Me, R^2 = Ac$

(16)

(18)

(19)

The reaction of diethyl malonate with coumarin in the presence of activated barium hydroxide, as catalyst, furnishes benzo[b]pyran (20) by an unusual 1,2-addition--elimination process at the carbonyl bond, instead of the expected Michael addition product (J.V. Sinisterra and J.M. Marinas, Monatsh., 1986, 117, 111).

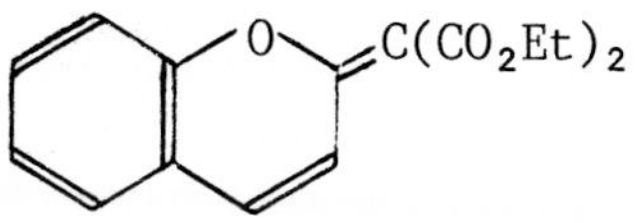

$$\text{(20)}$$

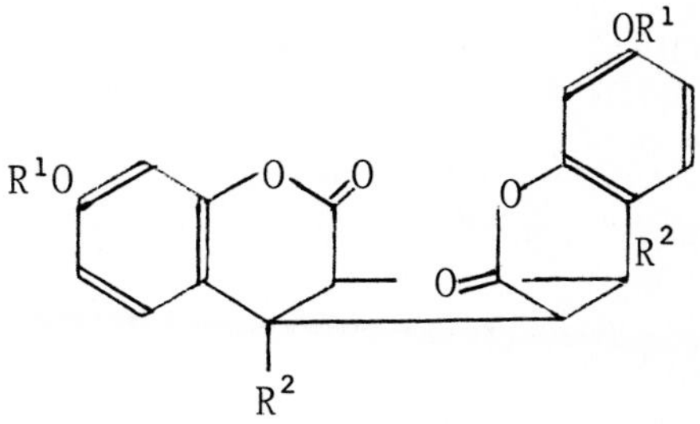

$$R^1 = \text{alkyl}, \; R^2 = H, \; Me$$

$$\text{(21)}$$

Lewis acid induced photochemical [2+2] cycloaddition reactions of coumarins have been discussed (J.D. Oxman, Diss. Abs. Int. B, 1984, _45_, 564). 7-Alkoxy- and 7-alkoxy--4-methyl-coumarins photodimerize in organic solvents to yield the _syn_ head-tail dimer (21) (K. Muthuramu, N. Ramnath and V. Ramamurthy, J. org. Chem., 1983, _48_, 1872).

The fluorescence behaviour of inclusion complexes of 7--substituted coumarin derivatives with β-cyclodextrin have been studied (A. Takadate _et al_., Yakugaku Zasshi, 1983, _103_, 193; S. Scypinski and J.M. Drake, J. phys. Chem., 1985, _89_, 2432).

The (±)-*anti* head-to-head coumarin dimer has been resolved, and the absolute configurations of the optically active forms determined on the basis of CD and [1]H-nmr spectral data (K. Saigo *et al.*, Tetrahedron Letters, 1983, 24, 5381). The thermal behaviour of head-to-head coumarin dimers has been studied (N. Yonezawa *et al.*, Bull. chem. Soc. Japan, 1984, 57, 1608).

The solid-state photochemical behaviour of a number of substituted coumarins has been investigated and some of these underwent photodimerization, which is remarkable in view of the inertness of coumarin itself in the solid state (K. Gnanaguru *et al.*, J. org. Chem., 1985, 50, 2337).

The synthesis of some 3-aroylcoumarins has been described (Yu.S. Andreichikov *et al.*, U.S.S.R. Pat., SU 707,197, 1982).

2,6-Dichlorocinnamic acid (22) and its esters on uv irradiation undergo $E \rightarrow Z$ photoisomerism, followed by a facile photocyclization with elimination of HCl or RCl to yield 5-chlorocoumarin (23) (R. Arad-Yellin, B.S. Green and K.A. Muszkat, *ibid.*, 1983, 48, 2578).

R = H, alkyl

(22)   (23)

The reaction between salicylaldehydes and methyl nitroacetate to furnish 3-nitrocoumarins is catalyzed by dialkylammonium chlorides (D. Dauzonne and R. Royer, Synth., 1983, 836). It has been found that 4-chloro-3- -nitrocoumarins react with nucleophiles to afford a number of novel substituted coumarins (K. Tabakovic *et al.*, Ann., 1983, 1901). In 4-chloro-3,6-dinitrocoumarin the chlorine atom readily and quantitatively reacts with amines and amino acids, making it a useful derivatizing agent in thin layer chromatography (M. Kules, M. Trkovnik and A. Juric, Acta Pharm. Jugosl., 1983, <u>34</u>, 81). A number of 3,8- -dinitro- and 6,8-dichloro-3-nitro-coumarins have been prepared (A. Cascaval, Synth., 1985, 428).

4-Bromomethylcoumarins are obtained by the cyclo- condensation of phenols with ethyl γ-bromoacetoacetate (M.V. Kulkarni, B.G. Pujar and V.D. Patil. Arch. Pharm. 1983, <u>316</u>, 15). 5,7-Diacetoxy-4-bromomethylcoumarin, m.p. 167-9°, on heating with sodium acetate in acetic anhydride is converted into 4-acetoxymethyl-5,7-diacetoxycoumarin, m.p. 167-7°. A study of these compounds has established a basis for the synthesis of 5-hydroxycoumarins containing a 4-(α-acetoxyalkyl)group and hence mammein (24) and surangin B (25) (C.W. Bird, H.I. Butler and A. Hawi, Chem and Ind., 1984, 145). The photo Fries reaction of 7-acetoxy-and 7- -furoyloxy-coumarins yields 6-acylcoumarins (S.K. Kulshrestha, P. Dureja and S.K. Mukerjee, Indian J. Chem., 1984, <u>23B</u>,1064).

(24) R^1 = isopentyl, R^2 = s–Bu, i–Bu

(25) R^2 = geranyl, R^2 = s–Bu

For the preparation of 3-(trichlorovinyl)coumarin see N. Yonezawa *et al.*, (J. org. Chem., 1985, <u>50</u>, 3026).

4-(Arylamino)coumarins may be alkylated at the 3-position by using benzyltrimethylammonium chloride as phase-transfer catalyst (N. Ajdini and O. Leci, Chem. Abs., 1985, 103, 53920b). For an easy route to 3-cyano-7-N,N-dialkylamino--2-[(β,β-dicyanovinyl)imino]coumarins see P. Czerney and H. Hartmann (Z. Chem., 1982, 22, 406); for the synthesis of 3-cyano-7-diethylamino-4-perfluoroalkylcoumarins, A.Ya. Il'chenko *et al*., (Zh. org. Khim., 1981, 17, 2630); and for the synthesis and properties of 4-substituted 3--aminomethylcoumarins, V.L. Savel'ev *et al*., (Khim. Geterotsikl. Soedin., 1982, 896).

Several derivatives of 3-ketocoumarin, including, 3--aroylcoumarins and 3,3'-carbonylbiscoumarins, have been prepared and shown to have many of the photophysical properties required for efficient triple sensitizers (D.P. Specht, P.A. Martic and S. Farid, Tetrahedron, 1982, 38, 1203).

Coumarin-3-carboxylic with amines in 1:1-1:1.5 molar ratio furnishes the corresponding ammonium salt, but with a 1:4 molar ratio salicylaldehyde and $RNHCOCH_2CO_2RNH_3^+$ (R = CH_2Ph, Bu) are obtained. With the former conditions the ethyl ester gives the corresponding amide and with the latter ring opening occurs (A.A. Avetisyan *et al*., Arm. Khim. Zh., 1981, 34, 876). N-Monosubstituted coumarin-3--carboxamides undergo cycloaddition reactions with 2,3--dimethylbuta-1,3-diene and 1,2-bis(methylene)cyclohexane to afford polycyclic compounds (H. Gotthardt and N. Hoffman, Ann., 1985, 901). The ring opening of coumarin-3--carboxamides with hydrazine has been reported (A.M. Islam *et al*., Egypt. J. Chem., 1983, 26, 233).

Coumarin-3-carboxylic acid hydrazides are obtained by the condensation of 2-hydroxyphenylhydrazones with carbon suboxide (L. Bonsignore *et al*., J. heterocyclic Chem., 1985, 22, 463). For the preparation of several coumarin-3--carboxylic acids see E.A. Shirokova, G.M. Segal and I.V. Torgov (U.S.S.R. Pat., SU 1,145,020, 1985). A number of ethyl 7-substituted coumarin-3-carboxylates, for example, ethyl 7-ethylamino-6-methylcoumarin-3-carboxylate, have been prepared by the condensation of phenols with $EtOCH=C(CO_2Et)_2$ in the presence of a Lewis acid (E.R. Bissell, U.S. Pat. Appl. U.S. 340,958, 1984).

Ethyl coumarin-3-carboxylates on reduction with sodium tetrahydridoborate in methanol or ethanol afford 2'--hydroxybenzylmalonic esters (B.S. Kirkiacharian, J.D. Brion and D. Billet, Compt. rend., 1982, 294, 181) and ethyl

coumarin-3-carboxylate, coumarin-3-carboxylic acid, or 3-
-cyano-2-imino-coumarin on treatment with hydrazine in
boiling ethanol yield salicylaldehyde azine
(F.S.G. Soliman, I.M. Labouta and W. Stadlbauer, Arch.
pharm. Chemi, Sci. Ed., 1985, 13, 49). A number of
coumarinimides have been synthesized and some show
antifungal activity (M.R. Manrao *et al.*, Indian J. Chem.,
1984, 23B, 1130).

The chlorosulphonylation of 4,5,7-trimethylcoumarin
yields 4,5,7-trimethylcoumarin-8-sulphonyl chloride, which
has been converted into a number of derivatives (K.B. Gupta
and R.C. Srivastava, J. Indian. chem. Soc., 1982, 59,
1193). A number of 4-(sulphonamidomethyl)coumarins have
been prepared and found to exhibit better antibacterial
activity against *Staphylococcus aureus* and *Escherichia coli*
than sulphanilamide (S.S. Hanmantgad, M.V. Kulkarni and
V.D. Patil, Indian J. Chem., 1985, 24B, 459).

Diethyl coumarin-3-phosphonates (27) (Table 1) are formed
when salicylaldehydes condense with triethyl phosphono-
acetate (26), using titanium tetrachloride and pyridine as
catalyst (R.K. Singh and M.D. Rogers, J. heterocyclic
Chem., 1985, 22, 1713).

(26)   (27)

Table 1
Diethyl coumarin-3-phosphonates

Substituent R	M.p^{o}C	Yield, %
H	65-66 B.p. 203-205/0.1mm.	67
6-Br	110-111	67
6-Cl	70-75	61
6,8-DiCl	118-122	83
7-OMe	86-88	23
5,6-Benzo (Diethyl 3-oxo-3H-naphtho- [2,1-b]pyran-2-phosphonate)	138-140	47

(vi) Hydroxycoumarins

3-Phenoxycoumarins are prepared when substituted salicylaldehydes react with *N,N*-diethylphenoxyacetamide in the presence of phosphoryl chloride (P.S. Salunke *et al.*, Synth., 1985, 111). Allyl ethers of 3-hydroxycoumarin (1) undergo a [3,3] sigmatropic rearrangement on pyrolysis or in boiling methylaniline or diphenyl oxide to give 4--allyl-3-hydroxycoumarins (2) (A.K. Mitra *et al.*, Indian J. Chem., 1982, 21B, 834).

R = CH_2═$CHCH_2$, CH_2═$CMeCH_2$,
2-cyclohexenyl

(1)   (2)

Autoclaving 2-hydroxyacetophenone with carbon monoxide in
the presence of 1,8-diazabicyclo[5.4.0]undec-7-ene and
sulphur or selenium, results in cyclocondensation and the
formation of 4-hydroxycoumarin (100% using Se). Some 6,7-
-disubstituted 4-hydroxycoumarins have been similarly
prepared (Jpn. Kokai Tokkyo Koho JP 57,145,870 [82, 145,
870], 1982). A number of 4-hydroxycoumarins with substituents on the benzene ring have been prepared by heating
malonic acid and the appropriate phenol with phosphoryl
chloride and zinc chloride (A.K. Shah *et al.*, Curr. Sci.,
1984, 53, 1241). 4-Hydroxycoumarin-[phenyl-u-^{14}C] has been
synthesized (A.K. Nadian J. labelled Compd. Radiopharm.,
1984, 21, 307).
   (+)-4-Hydroxy-3-[3-oxo-2-*p*-menthylmethyl]coumarin
(S.R. Paraskar and P.H. Ladwa, Indian J. Chem., 1983, 22B,
829), coumarin-4-yloxyalkanoic acid esters (D.R. Shridhar
et al., *ibid.*, p. 835), 3-(1-tetralinyl)-4-hydroxy-
coumarins (I.D. Entwistle and P. Boehm, Eur. Pat. Appl. EP
98,629, 1984; 177,080, 1986), analogues of 3-(1-phenyl-3-
-oxobutyl)-4-hydroxycoumarin (Warfarin) (A.O. Obaseki,
J.E. Steffen and W.R. Porter, J. heterocyclic Chem., 1985,
22, 529), and some 4-hydroxy-3-(3-oxoalkyl)coumarins
(P. Bravo, C. Ticozzi and G. Cavicchio, Synth., 1985, 894),
have been synthesized.

Studies on the tautomerism of 4-hydroxycoumarin using ir-
-spectroscopy have provided no evidence for the existence of
the 2-hydroxychromone tautomer, except in the case of
anhydrous 4-hydroxycoumarin in the solid state (Obaseki,
Porter and W.F. Trager, J. heterocyclic Chem., 1982, 19,
385). A related study has been made of 3-substituted-2-[13]C-
-4-hydroxycoumarins and 3-substituted-4-alkoxycoumarins
(Porter and Trager, *ibid*., p. 475).

4-Hydroxycoumarins on anodic oxidation in methanol
containing lithium perchlorate or sodium cyanide afford the
corresponding methoxy or cyano derivative. In basic media
biscoumarols are produced (A. Nishiyama *et al*., Chem.
pharm. Bull., 1983, 31, 2853).

3-Bromo-4-hydroxycoumarins on treatment with thiols (RSH;
R = Et, Pr, Pr[i], Bu, Bu[i], Ph) yield alkyl or aryl 4-
-hydroxycoumarinyl sulphides (G.M. Vishnyakova,
T.V. Smirnova and O.N. Fedorova, Izv. Vyssh. Uchebn.
Zaved., Khim. Khim. Tekhnol., 1983, 26, 554). 3-Acetyl-4-
-hydroxycoumarins with antibacterial, anticoagulant and
insecticidal properties are conveniently synthesized by
acetylating 4-hydroxycoumarins using acetic anhydride/
pyridine (V.K. Ahluwalia, D. Kumar and M.C. Gupta, Indian
J. Chem., 1978, 16B, 527). 4-Hydroxycoumarin is converted
by dimethyl sulphoxide and acetic anhydride into the
acetate, but at 120° this is further transformed into the
ylide 3-dimethylsulphoniochroman-2,4-dionate (3). At
160° the reaction yields dicoumarol (4) and other products
(K-Z. Khan *et al*., J. chem. Soc., Perkin I, 1983, 841).

(3)

(4)

Condensation of 4-hydroxy-7-methoxy-8-methylcoumarin (5;
R = Me) with cinnamyl alcohol in the presence of orthophosphoric acid at 50-55° gives a mixture of pyrano[3,2-c]-
benzo[b]pyran-5-one (6) and 3-cinnamyl-4-hydroxy-7-methoxy-
-8-methylcoumarin (7). At 75-80° only the former product
is obtained. Similar results are obtained with coumarin
(5; R = H, MeO) (Ahluwalia, K.K. Arora and I. Mukherjee,
Indian J. Chem., 1985, 24B, 298).

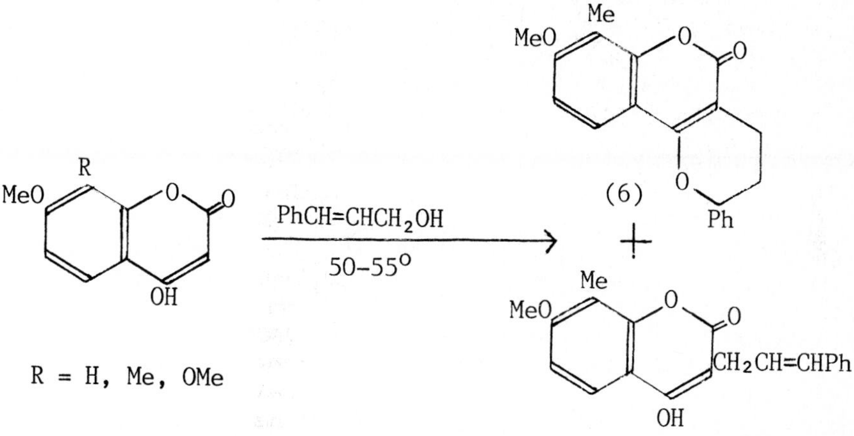

R = H, Me, OMe

(5)

PhCH=CHCH$_2$OH

50-55°

(6)

(7)

The reaction between the 4-hydroxycoumarins, shown below,
and butadiene or but-1-en-3-ol in the presence of
phosphoric acid gives the 3-(but-2-ene-1-yl)-4-hydroxy-
coumarins (8) (Ahluwalia, Mukherjee and R.P. Singh,
Heterocycles, 1984, 22, 229).

$$R^1 = R^2 = H$$
$$R^1 = R^2 = MeO$$
$$R^1 = MeO, \quad R^2 = H$$

(8)

The addition of 4-hydroxycoumarin to α,β-unsaturated carbonyl derivatives and Mannich base methiodides, derived from cyclic ketones, has been investigated (V. Khanna *et al.*, Indian J. Chem., 1986, 25B, 102).

Ring opening occurs on treating 3-allyl- or 3-phenyl--coumarins with dimethyl sulphoxide/acetic anhydride to give 2-substituted 3-(2-hydroxyphenyl)-1-methylthiopropan-3-ones (Khan *et al.*, Indian J. Chem., 1985, 24B, 42). 4-Hydroxy-coumarin on reaction with 3,3-dimethylallyl bromide gives a mixture of 4-(3,3-dimethylallyloxy)coumarin and 3,4--dihydro-3,3-bis(3,3-dimethylallyl)-2H-benzo[b]pyran-2,4--dione (A. Gonzalez *et al.*, An. Quim., 1981, 77C, 63).

3-Arylcarbamoyl-4-hydroxycoumarins are active against Gram-positive bacteria and 3-acylamino-4-hydroxycoumarins have moderate antibacterial and antifungal effects (T. Patonay *et al.*, Pharmazie, 1984, 39, 86). For the preparation of a number of 3-carbamoyl-4-hydroxycoumarins see E. Franz, E. Klauke and I. Hammann (Ger. Offen. Pat. DE 3,012,642, 1981).

The [1]H-nmr, ir and uv spectral data of a series of 3--arylazo-4-hydroxycoumarins indicate that they exist predominantly in the keto hydrazone form (9) both in solid state and in solution and not in the hydroxyazo form (10) (A.S. Shawali, N.M.S. Harb and K.O. Badahdah, J. heterocyclic Chem., 1985, 22, 1397).

(9)

(10)

6-Allyl-7-hydroxycoumarin, a key intermediate for the preparation of linear coumarins is obtained from 7-alloxy-coumarin *via* a regioselective boron halide catalyzed *ortho*--Claisen rearrangement of methyl 4'-allyloxy-2'-methoxy-cinnamate (N. Cairns *et al.*, Chem. Comm., 1986, 182).

7-Methoxycoumarin-4-carboxaldehyde tosylhydrazone on decomposition with sodium hydroxide yields 4-diazomethyl-7--methoxycoumarin a new fluorescent labelling reagent for alcohols and carboxylic acids (A. Takadate *et al.*, Chem. pharm. Bull., 1982, 30, 4120). Carbon-14-, deuterium-, and tritium- labelled 7-ethoxycoumarins have been synthesized (J.S. Walsh, H.E. Mertel and G.T. Miwa, J. labelled Compd. Radiopharm., 1983, 20, 731).

7-Hydroxy-4-phenylcoumarin (11; R^1 = H, R^2 = Ph) on heating with butyl bromide and potassium carbonate in acetone affords 7-butoxy-4-phenylcoumarin (12; R^1 = H, R^2 = Ph, R^3 = Bu) (A.K. Roa, M.S. Raju and K.M. Raju, Acta Cienc. Indica, [Ser.] Chem., 1983, 9, 200). A study has been made of the photodimerization of 7-methoxy- and 7--chlorocoumarin (K. Gnanaguru *et al.*, J. Photochem., 1984, 27, 355).

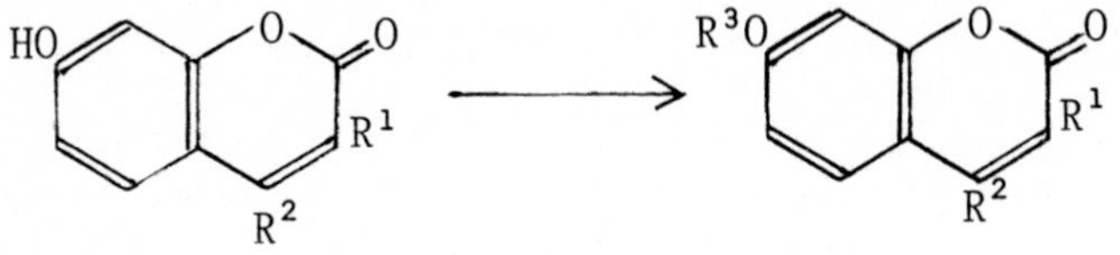

(11)   (12)

Heating resorcinol and ethyl α-fluoroacetate with
sulphuric acid gives 3-fluoro-7-hydroxy-4-methylcoumarin
(13) (Daikin Kogyo Co., Ltd., Japan. Pat. 60 32,783 [85,
32,783], 1985).

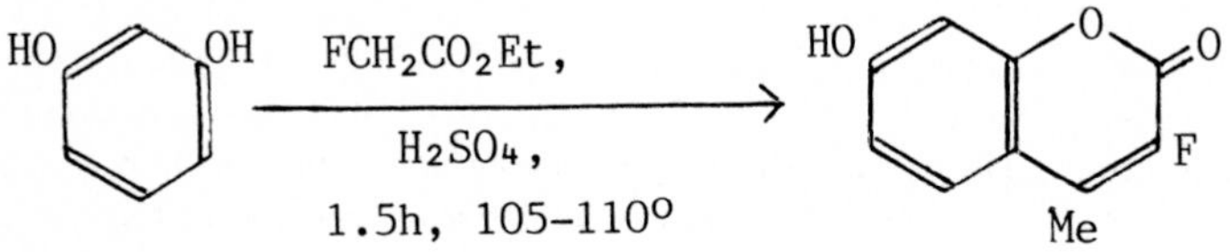

(13)

3-Bromo-4-bromomethyl-7-methoxycoumarin and 3-bromo-4-
-methyl-7-methoxycoumarin with boiling diethylaniline give
the debrominated derivative and an anilinomethyl derivative
(V.G.S. Box and C.G. Clement, Heterocycles, 1983, 20,
1049).
   The synthesis of 3-substituted-aminomethyl-7-ethoxy-
coumarins (G. Sailaja, K.M. Raju and M.S. Raju, Indian J.
Chem., 1985, 24B, 206); some 7-alkoxy-3,4-substituted
coumarins, screened for bactericidal and fungicidal
activity (A.K. Rao, Raju and Raju, Acta Cienc. Indica
[Ser.] Chem., 1982, 8, 224); and N-benzylidene derivatives

of 7-hydroxy-4-methyl(or phenyl)coumarin as potential
fungicides and bactericides (Ahluwalia *et al*., Chem. Abs.,
1984, 100, 191697y) have been reported; and the reactions
between ethyl 7-hydroxycoumarin-3-carboxylate and
hydrazine, amines and Grignard reagents have been described
(M.M. Mohamed, M.A. Hassan and M.El-Borai, J. chem. Soc.
Pak., 1983, 5, 263).

Mannich bases of the type (14, for example, R = PhEtN)
have been derived from 7-hydroxy-4-phenylcoumarin (Rao,
Raju and Raju, J. Indian chem. Soc., 1981, 58, 1021).

CH₂R

HO

O

O

Ph

(14)

Synchronous luminescence spectrometry has been employed
to study the fluorescence behaviour of 7-hydroxycoumarins
(C.G. Colombano and O.E. Troccoli, Anal. Chem., 1985, 57,
1907). For the ^{13}C-nmr spectrum of 7-methoxycoumarin see
B-Z. Ahn (Chem. Abs., 1980, 92, 93555x). The [1]H-nmr
chemical shifts of coumarin and all its monomethoxy and six
possible dimethoxy (substituted on benzene ring) derivatives
have been recorded (P. Joseph-Nathan, M. Dominguez and
D.A. Ortega, J. heterocyclic Chem., 1984, 21, 1141).

A new natural coumarin angustifolin [3-(1,1-dimethylprop-
-2-en-1-yl)-7-hydroxycoumarin] (15) has been isolated from
the aerial parts of *Ruta angustifolia* (J.B. Del Castillo
and F. Rodriguez Luis, Phytochem, 1984, 23, 2095).

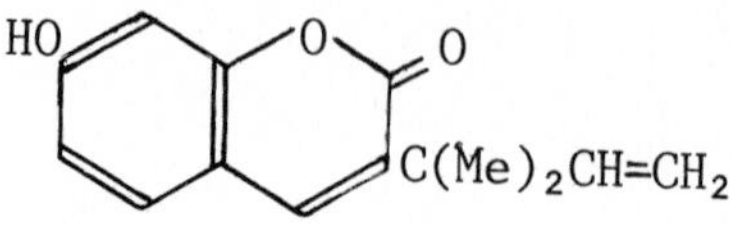

(15)

(vii) Dihydroxycoumarins

Topological resonance energies of coumarin, 3- and 4-
-hydroxy-, 3,4- and 4,6-dihydroxy-, and 4,5,7-trihydroxy-
-coumarins have been reported (P. Ilic, A. Juric and
N. Trinajstic, Croat. Chem. Acta, 1981, <u>53</u>, 587; Ilic *et
al*., J. heterocyclic Chem., 1982, <u>19</u>, 625). The hydrolysis
of iodonium ylide (1) affords 3,4-dihydroxycoumarin
(E. Pongratz and T. Kappe, Monatsh., 1984, <u>115</u>, 231). 3,7-
-Dihydroxycoumarin, m.p. 249°, 3-hydroxy-7-methoxycoumarin,
m.p. 225°, and 6,7-dihydroxycoumarin (esculetin) have been
isolated from the powdered root of *Euphorbia terracina* and
the 3-hydroxy-7-methoxy derivative from the aerial parts of
E. paralias (Z.F. Mahmoud and N.A. Abdel Salam, Pharmazie,
1979, <u>34</u>, 446).

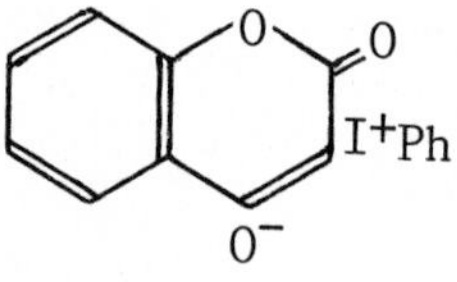

(1) 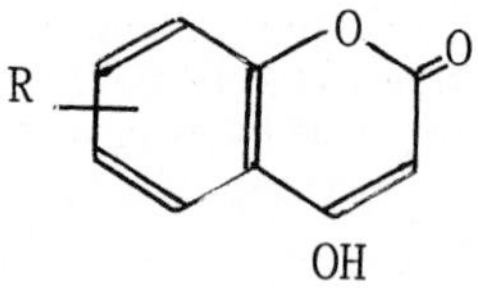   (2)

Methylation of the 4-hydroxy group in di- and tri-hydroxy-
coumarins [2; R = 5-OH, 7-OH, 5,7-(OH)$_2$] occurs on boiling
them with methanolic hydrogen chloride (V.K. Ahluwalia and
D. Kumar, Indian J. Chem., 1977, 15B, 945). Phenolic
metabolites of warfarin have been synthesized from 4,6- and
4,7-dihydroxycoumarin (E. Bush and W.F. Trager, J. pharm.
Sci., 1983, 72, 830) and methyl and ethyl 2,5-dioxo-8-
-hydroxy-2H,5H-pyrans[3,2-c]-benzopyran-3-carboxylate
from 4,7-dihydroxycoumarin. Both esters exhibit strong
fluorescence in the visible (O.S. Wolfbeis, Monatsh., 1976,
107, 783). For 7-hydroxy-4-methoxy-3-phenylcoumarin see
Ahluwalia and S. Mehta, (Indian J. Chem., 1977, 15B, 1097);
7-aryloxyalkyloxy-4-hydroxy-3-nitrocoumarins, D.R. Buckle
et al., (J. med. Chem., 1979, 22, 158); and for 7-*N*-benzyl-
piperazino derivatives of 4-hydroxy-3-nitrocoumarin, (*idem*,
ibid., 1984, 27, 1452). 4,7-Bis, 4- and 7-diethylamino-
alkoxycoumarins are able to form complexes with DNA
(C. Antonello, S.M. Magno and O. Gia, Chem. Abs., 1979, 90,
80722h).

Partial methylation of 5,6-dihydroxycoumarin with
dimethyl sulphate affords 6-hydroxy-5-methoxycoumarin,
whereas the isomeric 5-hydroxy-6-methoxycoumarin is
obtained by partial benzylation, followed by
methylation and reductive debenzylation. The latter
compound is different from that isolated from the bark of
Fraxinus floribunda and previously reported as 5-hydroxy-6-
-methoxycoumarin (Ahluwalia, M. Khanna and N. Rani, Indian
J. Chem., 1981, 20B, 343).

5,7-Dihydroxycoumarin has been separated from phenolic
acids by the trichloroacetate silica gel plate
chromographic method (W. Zhang and Z. Sun, Chem. Abs.,
1983, 99, 218706a); 5,7-diacetoxycoumarin has been
selectively C-8 isoprenylated (P. Rodighiero *et al*., Gazz.,
1980, 110, 167); and 5,7-diallyloxycoumarin and 5,7-
-diallyloxy-4-methylcoumarin have been converted into
difuranocoumarins (K. Singh, S.K. Banerjee and C.K. Atal,
Indian. J. Chem., 1981, 20B, 108). Coumurrayin, *5,7-
-dimethoxy-8-prenylcoumarin* has been synthesized
(R.D.H. Murray and Z.D. Jorge, Tetrahedron, 1984, 40,
3129).

A study has been made of the ^{13}C-nmr spectra of 6,7-
-dihydroxycoumarin (esculetin) and 7,8-dihydroxycoumarin
(daphnetin) (B-Z. Ahn, Chem. Abs., 1980, 92, 93555x) and
the ^{13}C-H long-range couplings and the H-bonding and
stereochemical effects on them (C-J. Chang, H.G. Floss and

W. Steck, J. org. Chem., 1977, <u>42</u>, 1337). The structure of bungeidol (3) a monoterpenoid coumarin isolated from *Haplophyllum bungei* has been confirmed by its ^{13}C-nmr spectrum and those of its transformation products, for example, 7-hydroxy-6-methoxycoumarin (scopoletin), 6,7--dimethoxycoumarin (scoparone) and 6-hydroxy-7-methoxy-coumarin (isoscopoletin) (A.Z. Abyshev and V.P. Zmeikov, Khim. Prir. Soedin., 1982, 294).

$$Me_2C(OH)CH_2CH_2CH(OH)CMe=CHCH_2O$$

(3)

7-Hydroxy-6-methoxycoumarin (38%) and 6-hydroxy-7-methoxycoumarin (30%) are obtained from 6,7-dimethoxycoumarin and aluminium chloride in boiling benzene. Treatment of 6,7-dihydroxycoumarin with acetic anhydride in pyridine affords 6-acetoxy-7-hydroxycoumarin, which on methylation using dimethyl sulphate followed by hydrolysis with hydrochloric acid yields 6-hydroxy-7--methoxycoumarin (X. Liang *et al.*, Chem. Abs., 1983, <u>99</u>, 122158x). 6,7-Dihydroxycoumarin has been isolated along with other products from *Artemisia siversiana* (I.I. Chemesova, L.M. Belenovskaya and L.P. Markova, Khim. Prir. Soedin., 1982, 521), *Campanula rotundifolia* and *C. persicifolia* (L.S. Teslov, L.N. Koretskaya and G. Tsareva, *ibid.*, 1983, 387), and along with 7-hydroxy-6--methoxycoumarin from *Canthium didymum* (S. Dan, S.S. Dan and P. Mukhopadhayay, J. Indian chem. Soc., 1982, <u>59</u>, 419). The air-dried fronds of *Microsorium fortunei* contain 3--carboxy-6,7-dihydroxycoumarin (T. Murakami *et al.*, Yakugaku Zasshi, 1985, <u>105</u>, 655). Obliquetol, *6,7--dihydroxy-8-prenylcoumarin* along with seven other derivatives of 6,7-dihydroxycoumarin and nieshoutol (see

p. 94) have been isolated from the heartwood of sneeze
wood, *Ptaeroxylon obliquum* (R.D.H. Murray and Z.D. Jorge,
Tetrahedron, 1983, **39**, 3163). Coumarin, 7-hydroxy-6-
-methoxy-, 6-hydroxy-7-methoxy-, and 6,7-dihydroxy-coumarin
have been found in brandy (E.Ya. Martynenko, N.V. Lobko and
N.F. Komissarenko, Chem. Abs., 1982, **96**, 197837q).
Titanium (IV) and niobium (V) complexes with 6,7- and 7,8-
-dihydroxycoumarins and their 4-methyl and 4-phenyl
derivatives have been prepared and characterized
[H.B. Singh, S. Srivastava and S. Maheshwari, J. Inst.
Chem. (India), 1982, **54**, 276] and binary and ternary
complexes of copper (II) with some dihydroxycoumarins have
been described (Singh, R.K. Negi and Srivastava, Proc.
Indian Acad. Sci., 1981, **90**, 141).

7,8-Dihydroxycoumarin has been isolated along with other
products from *Euphorbia dracunculoides* seeds (H.M. Chawla
et al., Sci. Cult., 1982, **48**, 203) and from the aerial
parts of *Stellera chamaejasme* (L.D. Madonova *et al.*, Khim.
Prir. Soedin., 1985, 709) and 7-hydroxy-8-methoxycoumarin
with other products from the roots of *Daphne tangutica*
(L. Zhuang *et al.*, Planta Med., 1982, **45**, 172). 7-Hydroxy-
-8-prenyloxy- (4) and 8-hydroxy-7-prenyloxy-coumarin (5)
have been isolated from the leaves of *Melampodium
divaricatum* and synthesized (J. Borges-del-Castillo *et al.*,
Phytochem., 1984, **23**, 859).

(4)   (5)

Celerin (6), m.p. 154-156°, isolated from *Apium graveolens* has been synthesized and shown to be 5-(1,1--dimethylallyl)-8-hydroxy-7-methoxycoumarin (R.D.H. Murray and Z.D. Jorge, Tetrahedron Letters, 1983, 24, 5897).

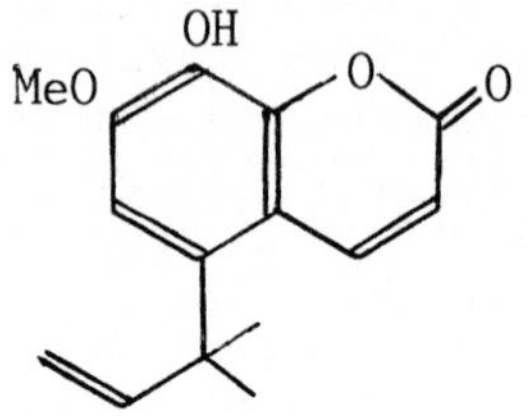

(6)

The gas chromatographic separation of the monomethyl ethers of 5,7-, 6,7-, and 7,8-dihydroxycoumarins has been described (S. Kato and K. Yamamoto, J. Chromatogr., 1985, 333, 175) and some methoxy substituted 4-methylcoumarins [6-OMe, 7-OMe, 5,7-(OME)$_2$, 6,7-(OMe)$_2$] have been prepared (R.I. Al-Bayati, M.T. Ayoub and R. Al-Hamdany, J. Iraqi chem. Soc., 1985, 10, 141).

(viii) Trihydroxycoumarins

Trihydroxycoumarins and other phenolic compounds
including 3,5,7-, 4,6,7-, 5,7,8-, and 6,7,8-trihydroxy-
coumarins have been separated (K. Vande Casteele, H. Geiger
and C.F. Van Sumere, J. Chromatogr., 1983, 258, 111;
Vande Casteele *et al.*, *ibid.*, 259, 291; H.J. Thompson and
S.A. Brown, *ibid.*, 1984, 314, 323).

3,6,7-Trihydroxycoumarin has been found in *Dasycladus
vermicularis* and other members of the siphonalean green
algal family Dasycladaceae (D. Menzel, R. Kazlauskas and
J. Reichelt, Chem. Abs., 1983, 98, 140544d). 4,8-Dimethoxy-
-7-hydroxy-5-methylcoumarin has been obtained from
Platymiscium praecox and 4,7-dihydroxy-8-methoxy-5-methyl-
coumarin has been synthesized (V.K. Ahluwalia and D. Kumar,
Indian J. Chem., 1977, 15B, 18). 5,6,7- and 6,7,8-
-Trimethoxy-, 6,7-dimethoxy-, and 5,7-dimethoxy-6-hydroxy-
-coumarin (fraxinol) have been isolated along with
alkaloids and other compounds from *Xanthoxylium
integrifoliolum* (Rutaceae) root wood (H. Ishii *et al.*,
Yakugaku Zasshi, 1982, 102, 182). 5,6-Dimethoxy-7-hydroxy-
coumarin (umckalin) has been obtained from the root bark of
Allamanda blanchetti (Apocynaceae) and its ^{13}C-nmr spectrum
and that of its methyl ether reported. This is the first
isolation of umckalin from a source other than the genus
Pelargonium (J. Bhattacharyya and M. de S. De Morais, J.
nat. Prod., 1986, 49, 370). 6,7,8-Trimethoxycoumarin along
with angustifolin and scoparone, and the alkaloid,
graveolin has been obtained from *Ruta angustifolia*
(J.B. Del Castillo and F. Rodriguez Luis, Phytochem., 1984,
23, 2095).

The 4-hydroxy group in both 4,5,7- and 4,7,8-trihydroxy-
coumarin is methylated on boiling them with methanolic
hydrogen chloride. Similar results are also obtained with
4,5- and 4,7-dihydroxycoumarin (Ahluwalia and Kumar, Indian
J. Chem., 1977, 15B, 945). Trihydroxycoumarins and some
other coumarins, containing a 4-hydroxy group, when
subjected to dye-sensitized photooxygenation by singlet
oxygen undergo cleavage of the heterocyclic ring
(S.S. Chibber and R.P. Sharma, *ibid.*, 1979, 18B, 538).

(ix) Furocoumarins

The hydrolysis, acetylation, methylation, benzoylation, and/or Schiff base formation of bergaptol, isophellopterin and some hydroxycoumarins are conveniently carried out on thin-layer chromatoplates [B.A.H. El-Jawil and F.K.A. El-Beih, J. Chin. chem. Soc. (Jaipei), 1983, 30, 127].

The irradiation of the fused cyclobutanol (1) in benzene in the presence of HgO-I_2 results in a coumarin-chromone transformation and formation of furocoumarin (2) and furochromone (3) (H. Suginome, C.F. Liu and A. Furusaki, Chem. Letters, 1985, 27).

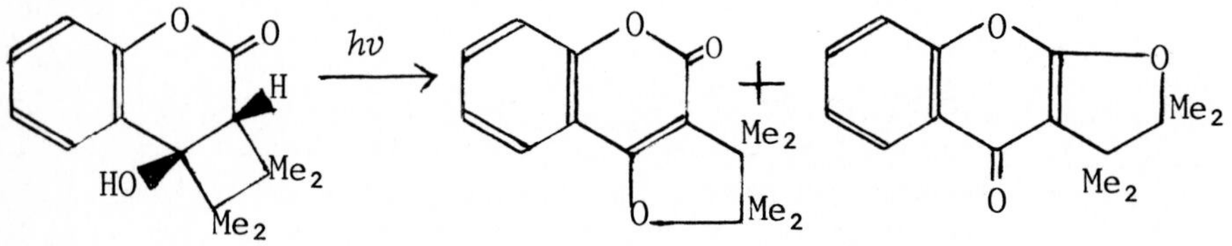

(1)   (2)   (3)

The photochemistry and photobiology of psoralenes, *2H-*
-furo[3,2-g]benzo[b]pyran-2-ones (E. Ben-Hur and P.S. Song,
Adv. radiat. Biol., 1984, **11**, 131) and the photochemical
and photophysical properties of furocoumarins [F. Dall-Acqua,
S. Caffieri and G. Rodighiero, NATO ASI Ser., Ser. A, 1985
(Primary Photo-Processes Biol. Med.), 241] have been
reviewed. The triplet and singlet excited state of furocoumarin have been discussed (R.V. Bensasson, *ibid.*, p. 241).
   The ^{13}C-nmr spectra of a number of mono-, di- and
tri-methyl derivatives of allopsoralen (4), isopseudopsoralen
(5), pseudoisopsoralen (6) and pseudopsoralen (7) have been
reported. The assignment of each signal has been achieved
by using carbon-proton coupling constants, relaxation
efficiency considerations and shift effects caused by the
introduction of methyl groups at various positions of the
furocoumarin nucleus (A. Guiotto *et al.*, J. heterocyclic
Chem., 1985, **22**, 649).

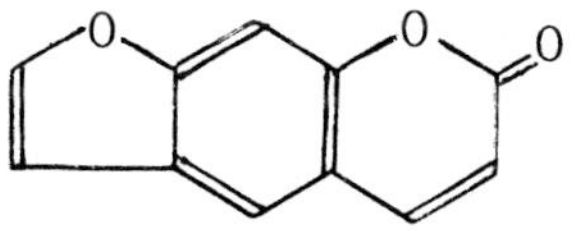

(4) 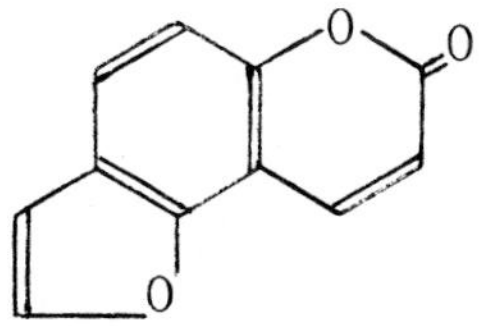   (5)

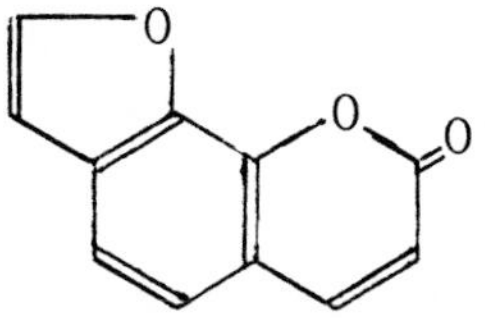

(6) 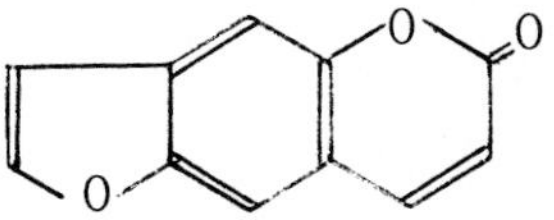   (7)

Hortiolone, *5-hydroxy-7-isopropenyl-9-prenyl-2H-furo[3,2-g]-benzo[b]pyran-2-one* (8) has been isolated from *Hortia arborea* and its structure determined (F. Delle Monache *et al.*, Gazz., 1976, **106**, 681; 1977, **107**, 399). It has been synthesized from 7-hydroxy-5-prenyloxycoumarin (R.D.H. Murray, Z.D. Jorge and K.W.M. Lawrie, Tetrahedron, 1983, **39**, 3159).

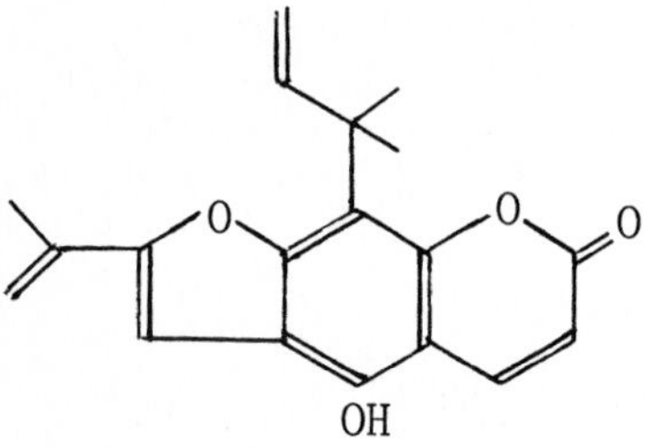

(8) 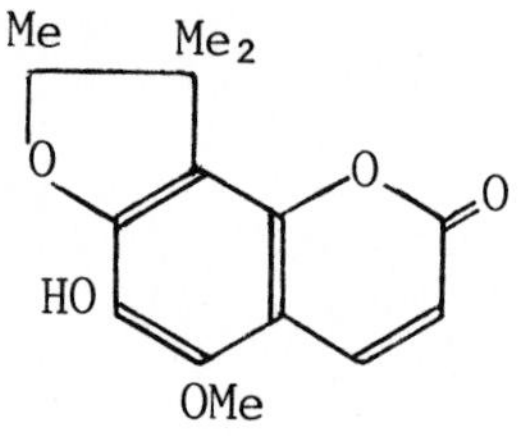   (9)

Nieshoutol isolated from *Ptaeroxylan obliquum* has been shown, from chemical and spectral evidence, to possess structure (9) (Murray and M.M. Ballantyne, Tetrahedron, 1970, **26**, 4473).

Some benzofurano derivatives, 4-methyl-6-nitro-8,9,10,11--tetrahydro-2H-benzofuro[2,3-h]benzo[b]pyran-2-one, 4--ethyl-11-nitro-6,7,8,9-tetrahydro-2H-benzofuro[3,2-g]-benzo[b]pyran-2-one, and 6,7,dihydro-4,13-dimethylbisbenzo-[b]pyrano[6,7-d:7',6'-d']-2H,11H-benzo[1,2-b:3,4-b']-difuran-2,11-dione have been synthesized as possible antifertility agents (S. Sinha *et al.*, J. heterocyclic Chem., 1985, **22**, 235).

(x) Pyranocoumarins

Clausarin (1), m.p. 123–124°, isolated from *Clausens pentaphylla* roots, co-occurs with dentatin (2), m.p. 93–94°, whose structure has been revised and confirmed by synthesis. Clausarin has also been found in *C. excavata* where it co-occurs with nordenatin (3), m.p. 183–184°. It is unique in possessing two 1,1-dimethylallyl groups and is the only known coumarin in which all the carbons of the nucleus, except C-4, are substituted. Nordentatin (3), dentatin (2) and clausarin (1) have been synthesized (R.D.H. Murray and Z.D. Jorge, Tetrahedron, 1984, 40, 3133).

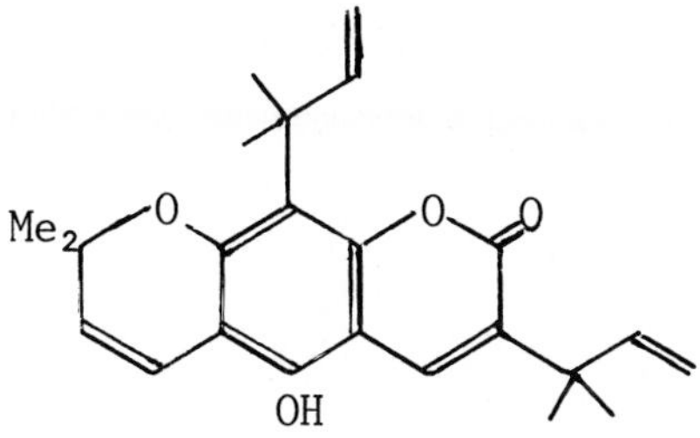

(1)

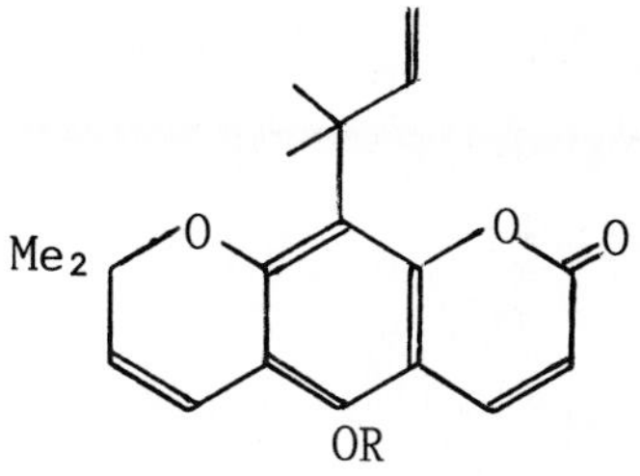

(2) R = Me
(3) R = H

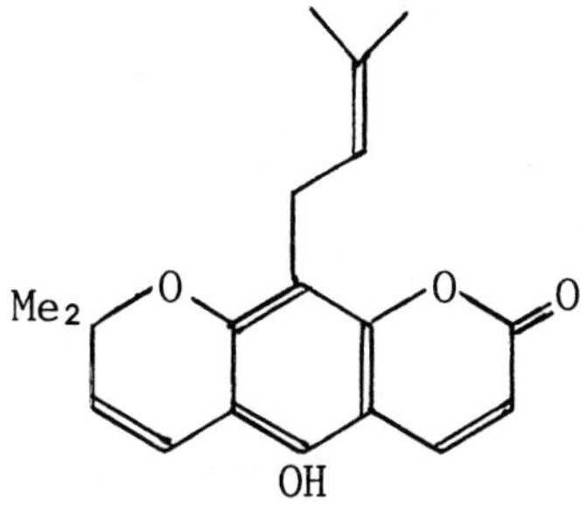

(4)

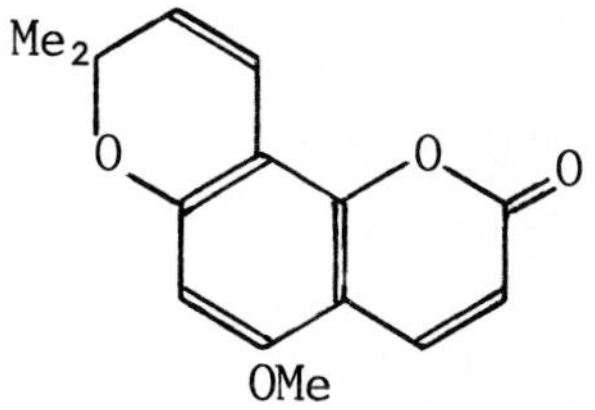

(5)

Trachyphyllin (4) has been isolated from *Eriostemon trachyphyllus* and synthesized from 5,7,-diacetoxycoumarin (*idem*, Tetrahedron Letters, 1983, 24, 3403). The structure of 5-methoxyseselin (5) and that of trachyphyllin (4) have been confirmed by syntheses from 7-acetoxy-5-prenyloxy-coumarin (6) and a convenient route to xanthoxyletin (7) has been established (*idem*, Tetrahedron, 1984, 40, 3129).

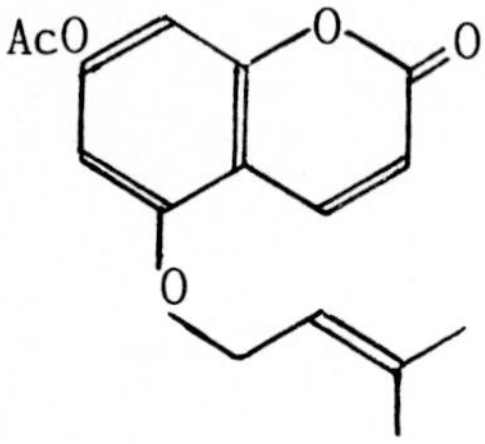

(6) 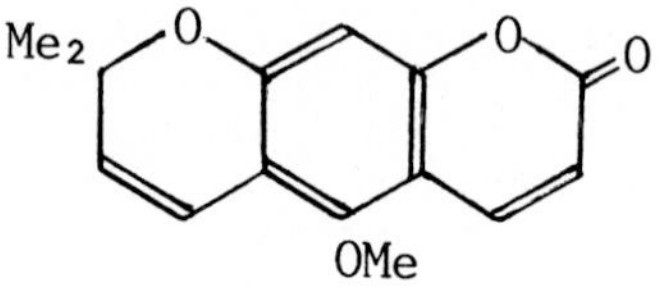   (7)

The Claisen condensation of ethyl acetate with 6- and with 8-acetyl-7-hydroxy-4-methylcoumarin, followed by cyclization affords pyranocoumarins (8) and (9) respectively (M. El-Kady *et al*., Rev. Roum. Chim., 1983, 28, 915).

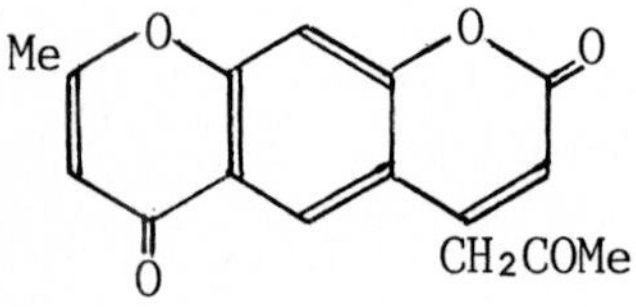

(8) 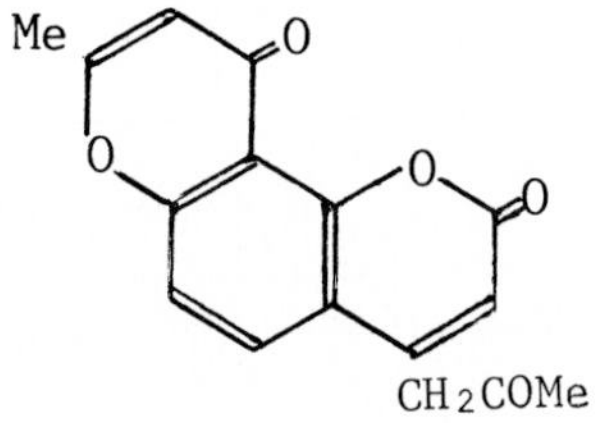   (9)

10-Acetoxy-9,10-dihydro-8,8-dimethyl-2H,9H-benzo[1,2-b:-3,4-b']dipyran-2,9-dione (10) rearranges on alkaline hydrolysis to yield the furocoumarin (11) (V.S. Kamat *et al.*, Indian J. Chem., 1985, **24B**, 547).

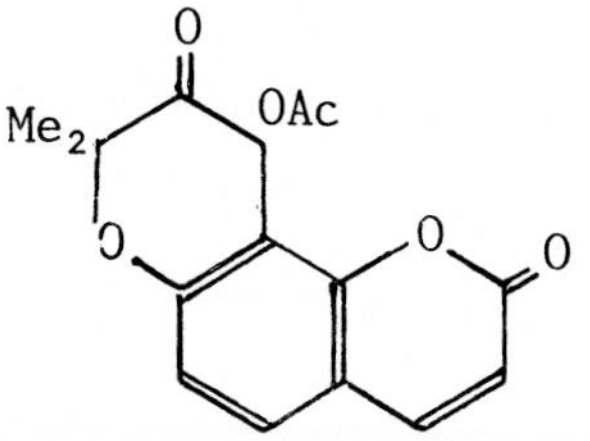

(10) 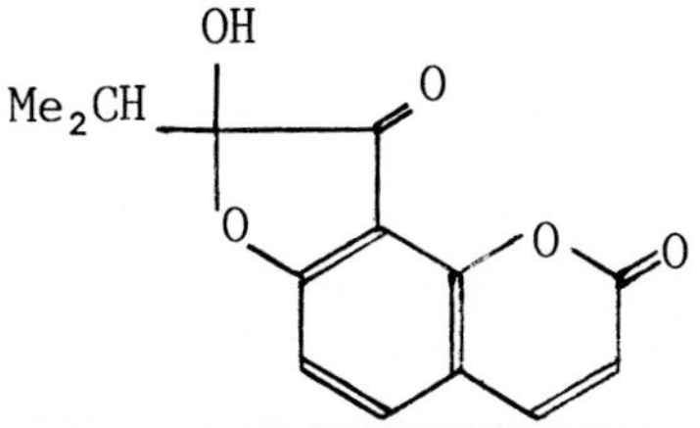   (11)

(xi) Chromones (4H-benzo[b]pyran-4-one, 5,6-benzo-4-pyrone,
*5,6-benzo-γ-pyrone)**

There has been an increase in the interest in the chemistry of chromones, due largely to their physiological properties.

*   G.P. Ellis, Chemistry of Heterocyclic Compounds, 1977, **31**, (Chromenes, Chromanones and Chromones), (Wiley, New York, N.Y.).

Methods of synthesis

2-Methylchromone (4) may be obtained by the hydrolysis and decarboxylation of chromone (3), prepared by the reaction of dimethyl penta-2,3-dienedioate (1) with methyl salicylate in *t.*-butanol in the presence of potassium *t.*-butoxide. The intermediate (2) is not isolated but can be prepared by carrying out the reaction in benzene in the presence of Triton B and methanol. Cyclization of (2) to chromone (3) is effected by lithium diisopropylamide in tetrahydrofuran.

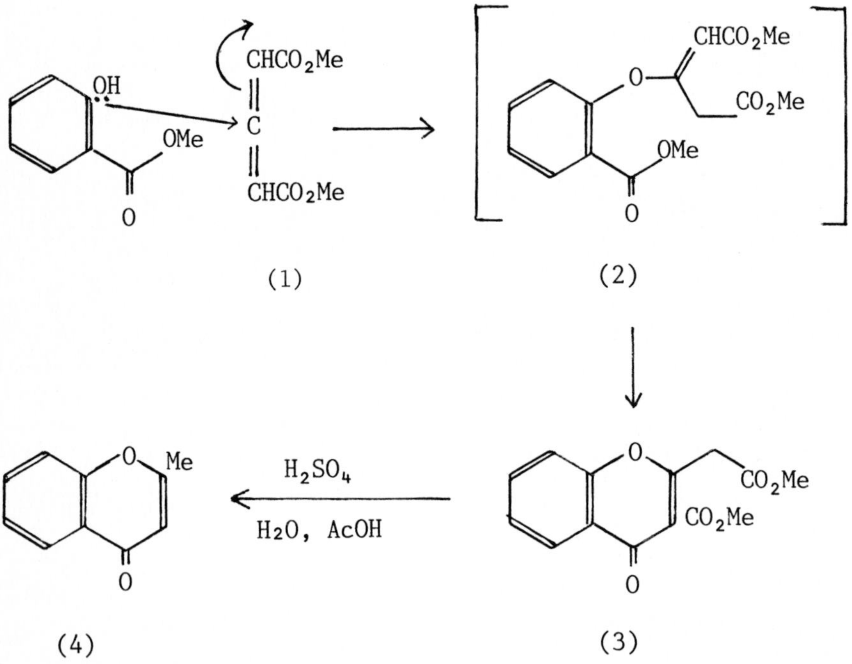

(1)   (2)

(4)   (3)

2-Methychromone (4) is obtained with chromone-2-acetic acid (7) when phenol reacts with dimethyl penta-2,3- -dienedioate (1) in benzene with Triton B as catalyst, and the resulting enol ether (5) hydrolysed to the dibasic acid (6) which is cyclized in the presence of polyphosphoric acid (N.S. Nixon and F. Scheinmann, Tetrahedron Letters, 1983, 24, 597).

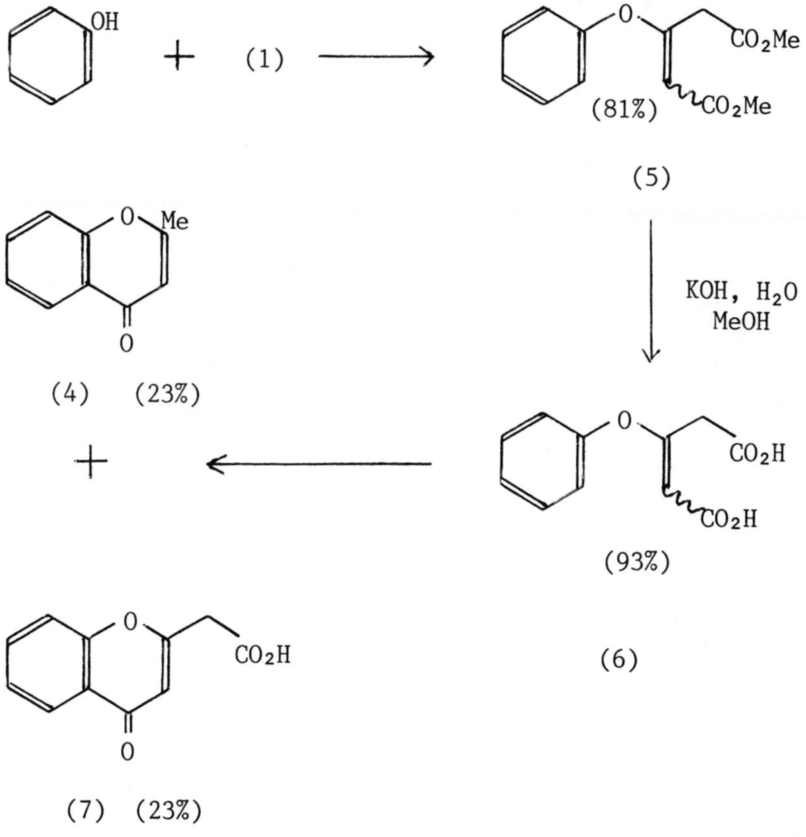

2-(2-Methoxy-1-naphthyl)chromone (9) is prepared by cyclizing ketone (8). It has been converted into a 6,8- -dibromo derivative and then the latter hydrolysed to 3,5- -dibromo-2-hydroxybenzoic acid (Y.B. Vibhute and M.H. Jagdale, J. Indian chem. Soc., 1981, 58, 1110).

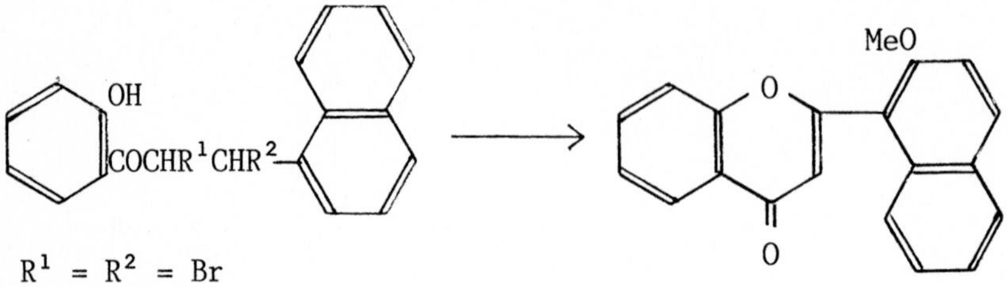

$R^1 = R^2 = Br$
$R^1 = R^2 = bond$

(8)   (9)

A simple synthesis of chromones involves the
cyclocondensation of phosphorane (10) with an acid chloride
or acid anhydride. Thus phosphorane (10) with benzoyl
chloride in boiling toluene containing pyridine affords 2-
-phenylchromone (11) (A. Hercouet, M. Le Corre and
Y. Le Floch'h, Synth., 1982, 597).

(10)   (11)

Acylsalicylic acids (12) are converted into chromones (13)
when they react with $Ph_3\overset{+}{P}-\overset{-}{C}=C=NPh$ (H.J. Bestmann and
G. Schade, Chem. Letters, 1983, 997).

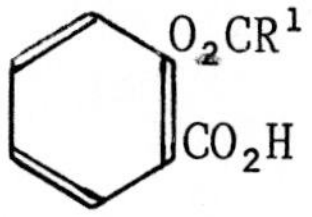

R^1 = Me, Ph, 4-MeOC$_6$H$_4$

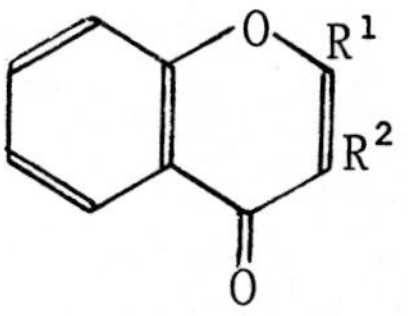

R^1 = Me, R^2 = CONHPh
or
R^1 = Me, Ph, 4-MeOC$_6$H$_4$
R^2 = H

(12)   (13)

3-Substituted chromones (15; R^3 = H) are prepared by the treatment of 2-hydroxy-ω-substituted acetophenones (14; R^1 = Me, MeO, HO, Cl; R^2 = MeO, Me, Ph, HO) with triethyl orthoformate and perchloric acid. This method provides a convenient route to 3-hydroxychromone, which is otherwise tedious to prepare.

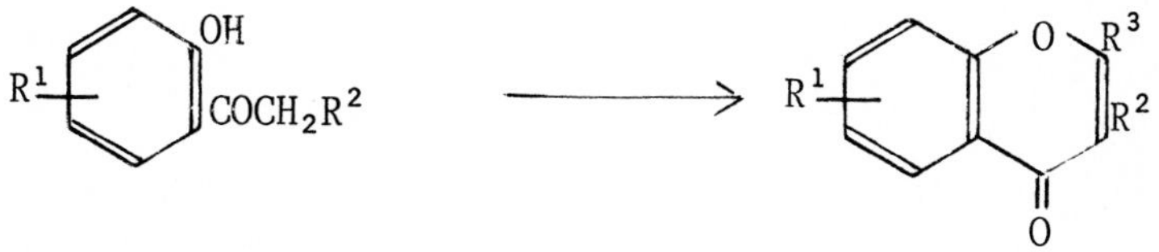

(14)   (15)

Chromone may be obtained in good yield from 2-hydroxy-ω-
-bromoacetophenone (14; R^1 = H, R^2 = Br), the bromine
being lost during cyclization. Similarly 6-bromochromone
(15; R^1 = 6-Br, R^2 = R^3 = H) is formed from ω,4-dibromo-2-
-hydroxyacetophenone (14; R^1 = 4-Br, R^2 = Br). Using
triethyl orthoacetate instead of the orthoformate affords
2-methychromones (15; R^3 = Me).

2-Hydroxy-ω-nitroacetophenone (14; R^1 = H, R^2 = NO_2)
is not cyclized to 3-nitrochromone (15; R^1 = R^3 = H,
R^2 = NO_2) under the above conditions, but cyclises on
treatment with acetic formic anhydride and sodium formate.
These conditions are used to convert some hydroxyacetophenones
(14; R^1 = H, NO_2, OMe, R^2 = MeCO, NO_2, COPh, SOMe, SO_2MeO)
into chromones. The choice between the two methods for the
preparation of chromones is determined by the electronic
character of the substituent(s). The 3-nitro group may be
reduced to a 3-amino group, diazotized and converted into a
3-hydroxy group [G.J.P. Becket, G.P. Ellis and
M.I.U. Trindade, J. chem. Res. (S), 1978, 47]. Acetic
formic anhydride has been used as a cyclizing reagent in
the presence of sodium formate or triethylamine, in the
synthesis of 3-heterylchromones from ω-substituted 2,4-
-dihydroxyacetophenones (V.G. Pivovarenko, V.P. Khilya and
F.S. Babichev, Chem. Abs., 1985, 103, 141792s).

Benzo[h]chromone (4H-naphtho[1,2-b]pyran-4-one) (19) is
obtained by the reaction of the boron difluoride complex
(16), from 2-acetyl-1-hydroxynaphthalene with
dimethylformamide dimethylacetal, to give the intermediate
(17). The latter on alkaline hydrolysis yields the 2-
-hydroxychroman-4-one derivative (18), which with sulphuric
acid yields (19) (G.A. Reynolds and J.A. Van Allan, J.
heterocyclic Chem., 1969, 6, 375).

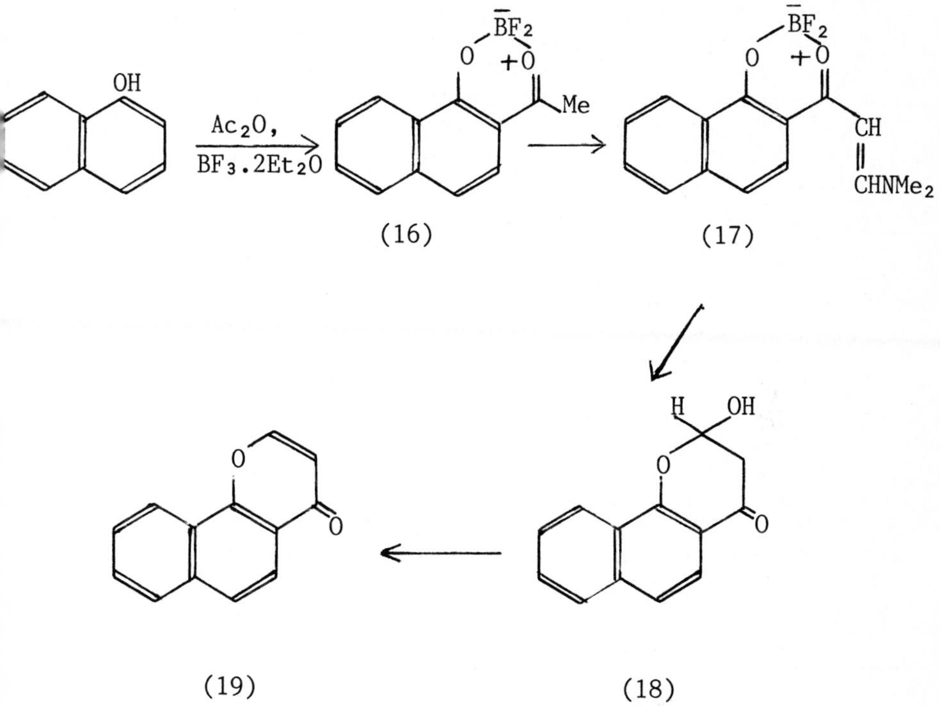

(16)　　　　　　　　　　(17)

(19)　　　　　　　　　　(18)

　　The above procedure does not give satisfactory yields of chromone (24) or benzo[f]chromone (1H-naphtho[2,1-b]pyran--1-one) (25). The best results are obtained by treating the boron difluorides (20 and 21) with N,N-dimethylthio-formamide in acetic acid to give intermediates (22 and 23). The complexes (17, 22 and 23) with 35% perchloric acid in acetic acid or alcohol, are directly converted into the chromones (19, 24 and 25). This general method can be adapted for the preparation of 2-substituted 4-pyrones (Reynolds, Van Allan and A.K. Seidel, *ibid.*, 1979, **16**, 369).

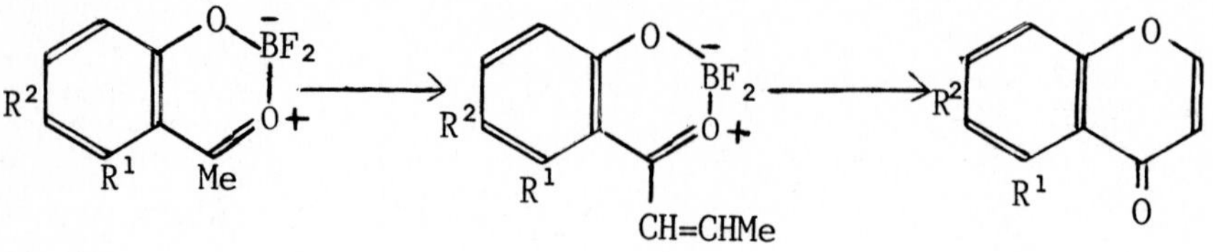

(20) $R^1 = R^2 = H$ (22) $R^1 = R^2 = H(85\%)$ (24) $R^1 = R^2 = H$
(21) $R^1_,R^2 = benzo$ (23) $R^1_,R^2 = benzo$ (93%) (25) $R^1_,R^2 = benzo$

Some chromones (27) (Table 2) have been obtained by the cyclization of the appropriate 2-hydroxyphenyl arylalkyl or alkyl ketone (26) using boron trifluoride etherate and methanesulphonyl chloride in dimethylformamide (R.J. Bass, Chem. Comm., 1976, 78).

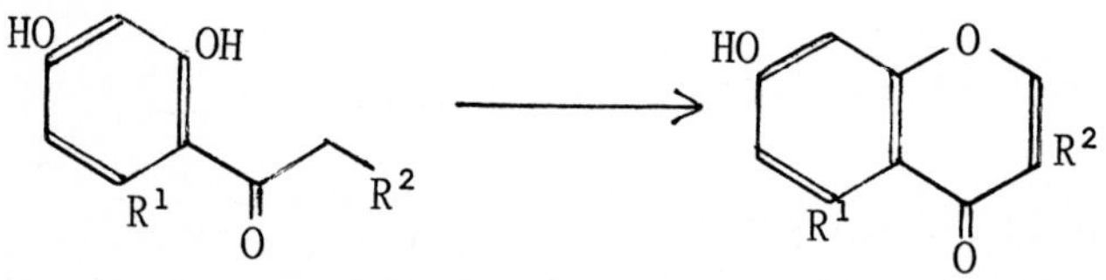

(26)   (27)

Table 2   Chromones (27)

Substituents		M.p. (°C)	Yield (%)
R^1	R^2		
OH	PhCH2	177 - 178	89
OH	MeO	229 - 231	66
H	c-C$_5$H$_9$	202 - 203	60

6-Bromo-2-methylbenzo[h]chromone (6-bromo-2-methyl-4*H*-
-naphtho[1,2-b]pyran-4-one) is prepared by the cyclization
with sulphuric acid of 2-acetoacetyl-4-bromo-1-naphthol,
obtained by the Claisen condensation of 2-acetyl-4-bromo-
-1-naphthol with ethyl acetate. This method has been
adapted for the preparation of 2-methylbenzo[h]chromone, 2-
-methyl-, 2,7- and 2,8-dimethyl-chromone (S.M. Sami,
A.A. Sayed and S.S. Ibrahim, Egypt. J. Chem., 1980, 23,
337).
    Both 2- and 2,3-substituted chromones are synthesized by
the internal Claisen condensation of 1-acyl-2-acyloxyarenes,
followed by cyclization of the resulting product, for
example, 2-acyloxyacetophenone (28) affords 2-substituted
chromone (29) (I. Hirao, M. Yamaguchi and M. Hamada, Synth.,
1984, 1076).

R = Me, Et, Pr^i, n-nonyl, Ph(CH$_2$)$_3$, Ph

    (28)   (29)

Reagents:- (a) NaH, Me$_2$SO, R.T.; (b) AcOH, HCl, 100°

Chromanones (30) on treatment with sodium methoxide and
ethyl formate yield 3-hydroxymethylene derivatives (31) of
the 2-substituted chroman-4-ones, which rearrange under
acid——base conditions to give 3-alkenylchromones (32)
(Table 3). The behaviour of spiro[chroman-2,1'-cyclopentan]-
-4-one differs from that of the other examples in that it
yields 3-cyclopentenylchromone directly (C.D. Gabbutt and
J.D. Hepworth, Tetrahedron Letters, 1985, 26, 1879).

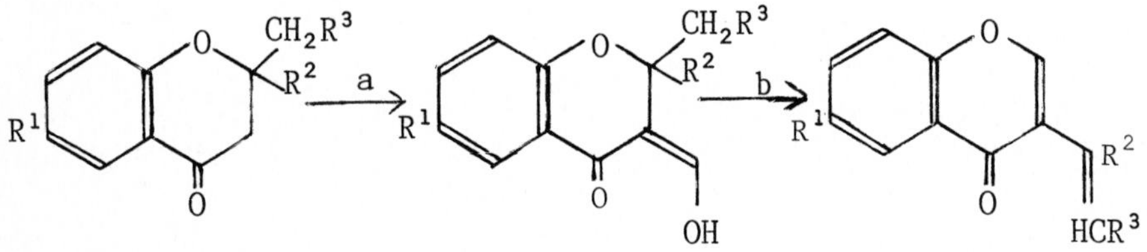

     (30)   (31)   (32)

Reagents:- (a) HCO_2Et, NaOMe;   (b) C_5H_5N, 4-TsOH, boil

107

Table 3   3-Alkenylchromones (32)

Substituents			M.p. (°C) [B.p. (°C)mmHg]*
R^1	R^2	R^3	
H	── $(CH_2)_3$ ──		115
H	── $(CH_2)_4$ ──		94 - 95
H	Me	H	57 - 59
H	Me	Me	89 - 91
Me	H	H	[125/0.1]
H	H	Me	63
H	H	Et	[150 - 155/0.1]

*   B.p. refers to Kugelrohr oven temperature at which distillation occurs.

Irradiation of alkenoylbenzoquinones (33; R^1 = Ph, 4-MeC$_6$H$_4$, 4-MeOC$_6$H$_4$, 4-ClC$_6$H$_4$) in an alcohol (R^2OH; R^2 = Me, Et, Pr^i, Bu^t) under anaerobic conditions yields chromone isomers (34 and 35) and also benzofuranones (36) and benzoates (37). The product ratio varies with the substituents R^1 and R^2 (K. Maruyama *et al*., Chem. Letters, 1985, 595).

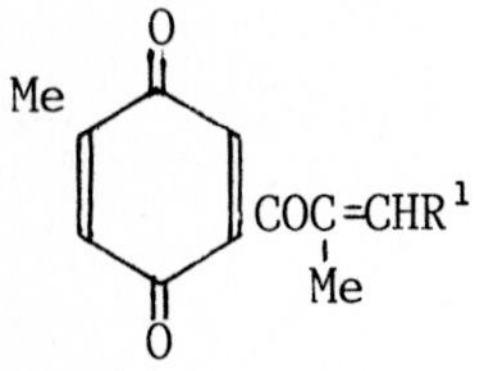

(33) 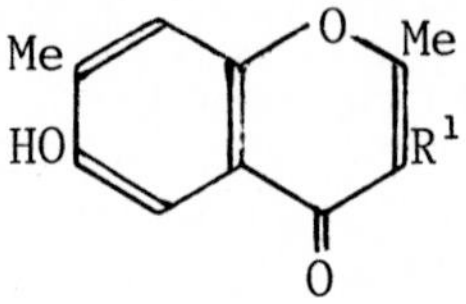   (34) 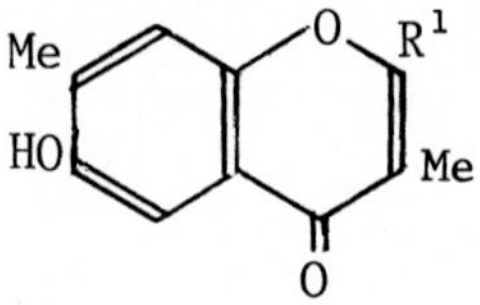   (35)

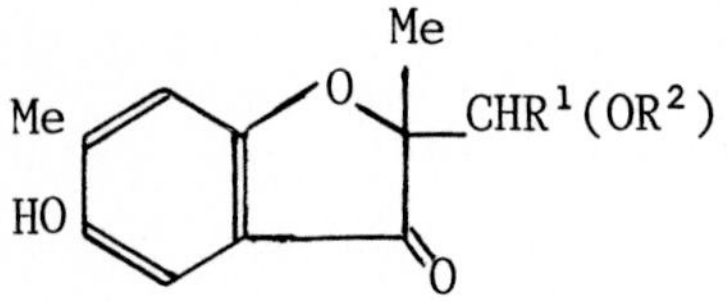

(36) 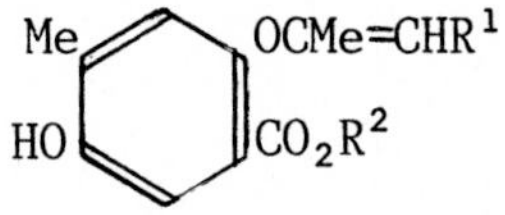   (37)

2-Aryloxymethylchromones (40) are obtained by the acid-
-catalyzed cyclization of the diketones (39), formed by the
Baker——Venkataraman transformation of the aryloxymethyl 2-
-acetylbenzoates (38) initially prepared from 2-hydroxy-
acetophenones and aryloxyacetic acids (S. Giri, Nizamuddin
and A.K. Mishra, Indian J. Chem., 1981, 20B, 1013).

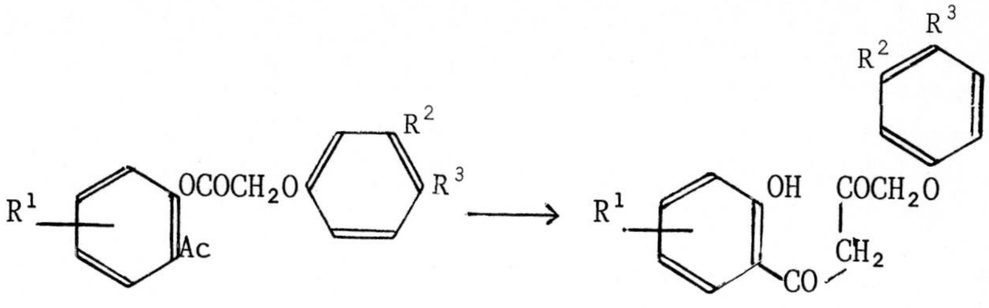

R^1 = H, 3-, 4-, 5- Me
R^2 = H, Me
R^3 = F, Cl, Me

(39)

(38)

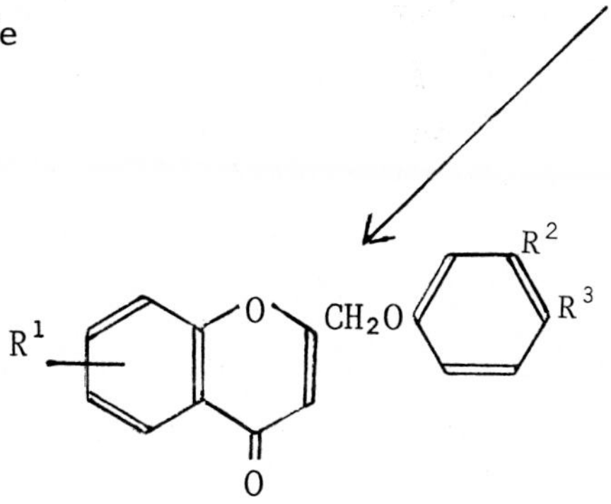

(40)

Reactions

Chromic acid oxidation of the products formed by treating chromones (1) with borane give chroman-4-ones (2) (B.S. Kirkiacharian, Compt. rend., 1980, <u>291C</u>, 73).

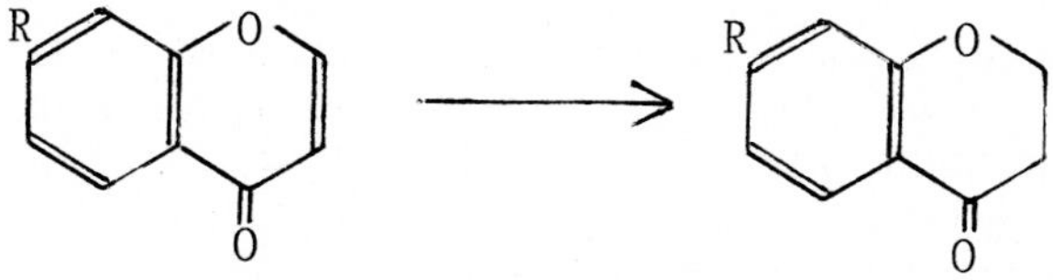

R = H, OMe

(1)   (2)

The hypervalent iodine oxidation of chromone using iodobenzene diacetate——potassium hydroxide in methanol affords the α-hydroxy-β-methoxydimethylacetal (3), which on treatment with concentrated hydrochloric acid in acetone yields 3-hydroxychromone (4) (R.M. Moriarty and O. Prakash, J. heterocyclic Chem., 1985, 22, 583).

(3)   (4)

The proton-transfer spectroscopy of 3-hydroxychromones (D. McMorrow and M. Kasha, J. Amer. chem. Soc., 1983, 105, 5133), and the 2-hydroxychromone/4-hydroxycoumarin tautomerism (W.R. Porter and W.F. Trager, J. heterocyclic Chem., 1982, 19, 475) have been discussed.

Chromones (5) react with aluminium trialkyls to give, for example, 1,3-diethylated, 1,3-di*iso*butylated, or 3--*iso*butylated-1-reduced substituted Z-2-(3-hydroxy-1--propenyl)phenols (6, 7 and 8), which in boiling mesitylene containing silica gel are cyclized to 2*H*-benzo[b]pyrans (9 and 10). Cyclization in other acidic solutions does not always lead to high yields (A. Alberola *et al*., An. Quim., 1982, 78C, 15; J. heterocyclic Chem., 1983, 20, 715).

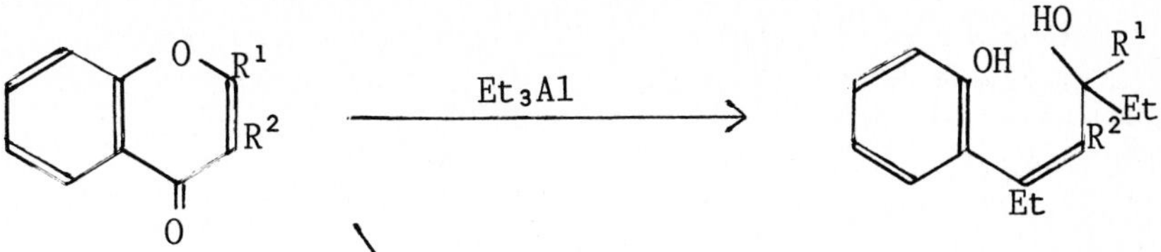

(6)

Et$_3$Al

R^1	R^2	R^3
H	H	Et
H	H	Bu^i
Ph	H	Et
H	Ph	Et
PhCH$_2$	H	Bu^i

(5)

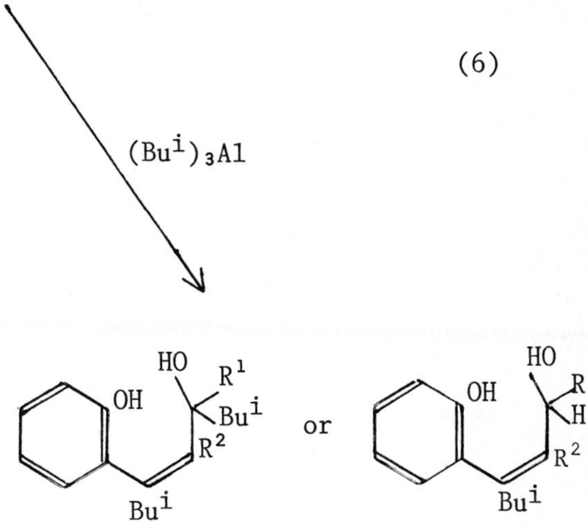

(Bu^i)$_3$Al

(7)   or   (8)

(6) and (7) 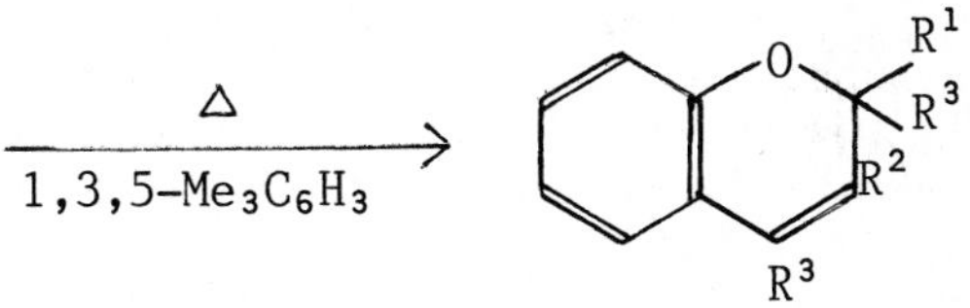
$\xrightarrow[\text{1,3,5-Me}_3\text{C}_6\text{H}_3]{\Delta}$
(9)

(8) 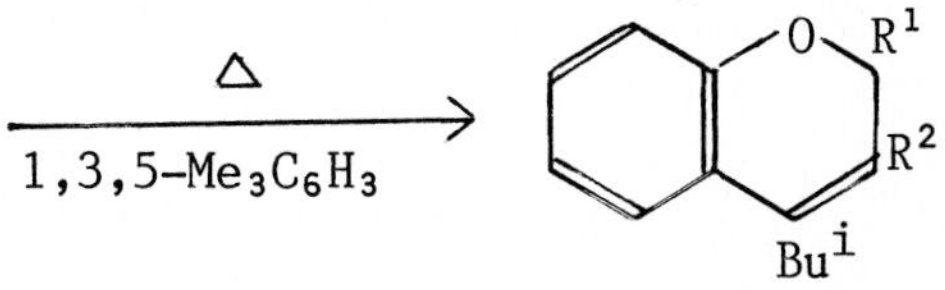
$\xrightarrow[\text{1,3,5-Me}_3\text{C}_6\text{H}_3]{\Delta}$
(10)

Treatment of an aqueous solution containing chromone, sodium dodecylbenzenesulphonate, silver nitrate, and an alkanoic acid, with ammonium peroxydisulphate gives a 2-
-alkylchromone and a 2-alkylchroman-4-one. It is suggested that the reaction involves the relatively stable 2-
-substituted chroman-4-one-3-yl radical and that the sulphonate-detergent has a micellar catalytic activity on the free radical alkylation process (J.F.J. Engbersen, A. Koudijs and H.C. van der Plas, J. heterocyclic Chem., 1982, 19, 1281).

Chromones activated by electron-withdrawing groups at C-3 are in general alkylated at C-2 by diazoalkanes. Thus 6-
-methylchromone-3-carbonitrile (11) is converted by diazoethane into 2-ethyl-6-methylchromone-3-carbonitrile (12).

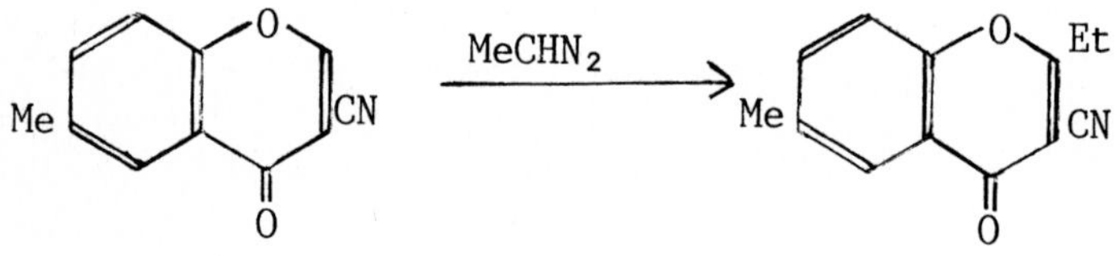

(11)   (12)

2-Diazopropane affords cyclopropane by-products, and a 3-
-formyl group usually suffers homologation to the
appropriate ketone, as when 2-diazopropane converts 6-
-methylchromone-3-carboxaldehyde (13) into 2-isopropyl-6-
-methyl-3-(methylpropanoyl)chromone (14). In contrast to
coumarin chemistry, there is no ring expansion into a 1-
-benzoxepin.

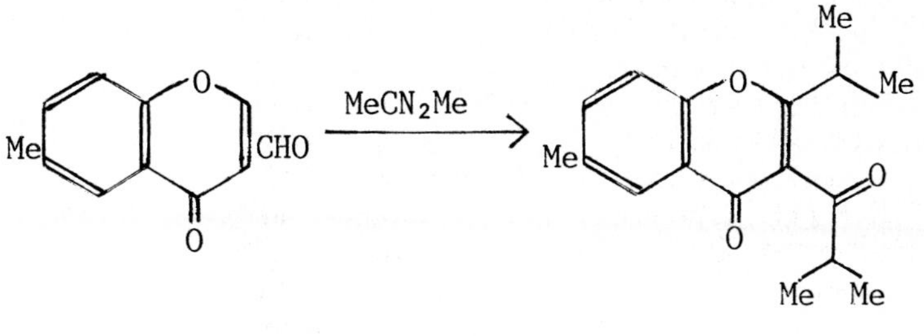

(13)   (14)

Chromones activated at C-2 show diverse reactions.
Although nitriles are usually considered to be unreactive
towards diazoalkanes, 6-methylchromone-2-carbonitrile (15)
reacts slowly to give a triazole; subsequent protropy and
further alkylation yields what is provisionally considered
to be the 2-(1,2,3-triazol-4-yl)chromone (16).

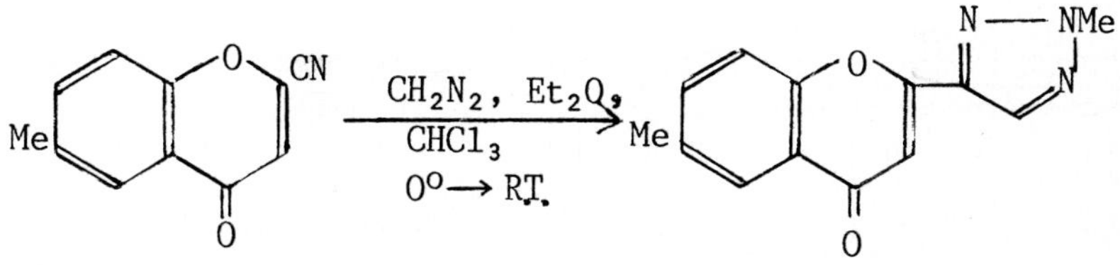

(15)   (16)

Chromone-2-carboxaldehyde with diazomethane affords a
mixture of 2-acetylchromone and 2-oxiranylchromone, but
with diazoethane and 2-diazopropane it undergoes ring
expansion to give a benzoxepin derivative, for example,
(17). Ethyl chromone-2-carboxylate reacts slowly but
furnishes a benzoxepin derivative besides a 3-alkylated
chromone (F.M. Dean and R.S. Johnson, J. chem. Soc., Perkin
I, 1981, 224).

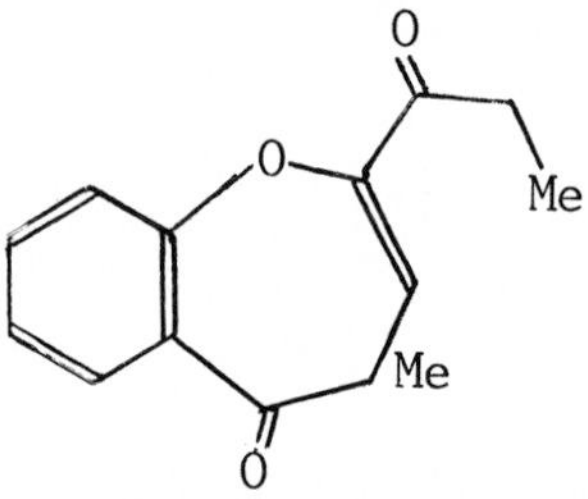

(17)

3-Bromoacetylchromone (18) on reaction with sodium
hydroxide yields benzoxepin (19), also obtained by
treatment of compound (20) with alkali. Compound (20) is
prepared by the reaction of the bromo derivative (18) with
ammonium hydroxide. Benzoxepin (19) on treatment with acid
affords benzoxepin-3,5-dione and 2-acetylphenoxyacetic
acid. The salt from 3-chloromethyl-8-methoxychromone and
pyridine on treatment with base furnishes 3-hydroxymethylene-
-8-methoxychroman-4-one (S. Klutchko and M. Van Strandtmann,
Synth., 1977, 61; U.S. Pt. 3,991,082, 1976; 4,001,224,
1977).

(18)   (19)   (20)

Chromones are epoxidized by alkaline hydrogen peroxide
and the 3-substituted chromone epoxides are, by a
considerable amount, the most stable. Acid-catalyzed ring
opening of the epoxides occurs regioselectively affording
3-hydroxychromanones, also base-catalyzed ring opening
occurs regioselectively, but at the 3-position. It is not
possible, using alkaline hydrogen peroxide, to convert 2-
-methylchromone (21) completely into its epoxide (22)
because the epoxide partially rearranges to 2-methyl-
chromonol (25), presumably *via* a 2,3-diol (23). The
epoxidation, therefore, is interrupted before chromonol
formation is observed and consequently reactions of the
epoxide are carried out in the presence of 2-
-methylchromone. Epoxide (22) does not give a fluorohydrin
on treatment with boron trifluoroetherate in benzene; it
rearranges to 2-methylchromonol (25). With sodium ethoxide
it yields 3-ethoxy-2-methylchromone (26), thus confirming
that base-catalyzed ring opening occurs at the 3-position.
Under acid conditions epoxide (22) undergoes rearrangement
to give chromonol (25); the intermediate cation (24)
eliminating a proton from the 3-position rather than from
the 2-methyl group (J.A. Donnelly, J.R. Keegan and
K. Quigley, Tetrahedron, 1980, 36, 1671).

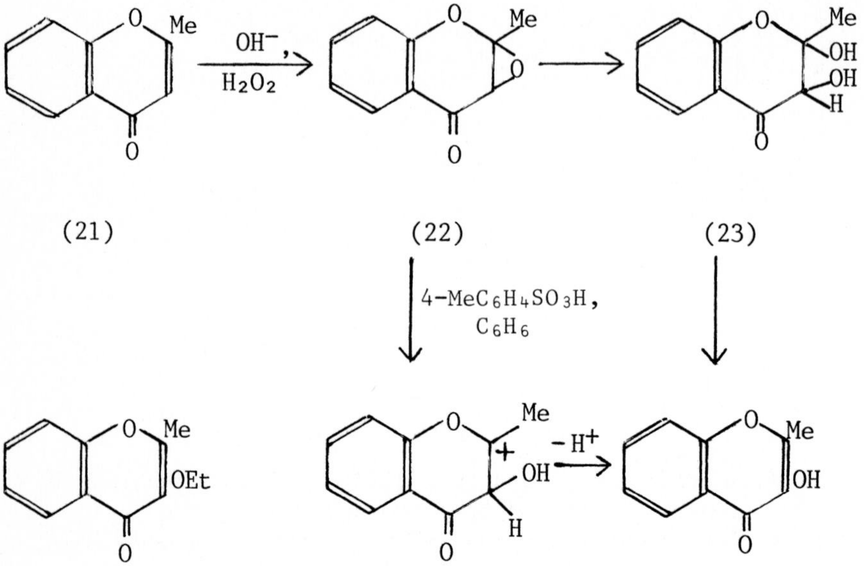

The reaction of chromone and 3-substituted chromones (27) with hydroxylamine in aqueous solution yield the isoxazoles (29) *via* the intermediate (28). Treatment of the isoxazoles (29) with sodium hydroxide solution (between 1 and 4 molar) yields either the nitrile (30) or the 2H-benzo-[b]pyran (31), depending on the strength of the sodium hydroxide solution and thus the pH of the reaction medium. Hydrolysis of derivative (31) affords the 4-hydroxycoumarin (32), thus providing a route from chromones to 4-hydroxy-coumarins (V. Szabó and J. Borda, Acta Chim. Acad. Sci. Hung., 1980, 103, 271; 1977, 95, 333).

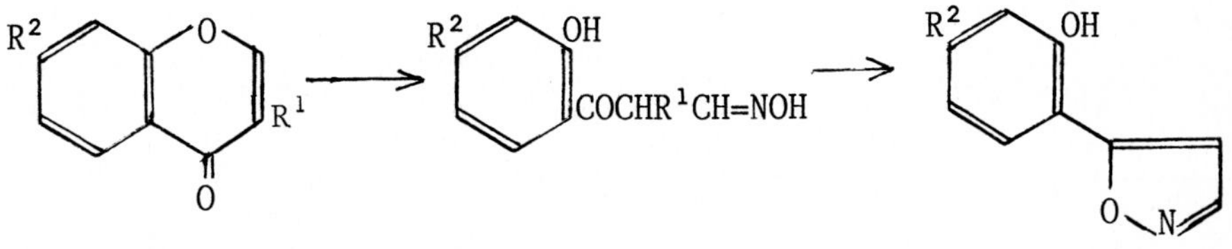

R^1 = H, Me, Ph, 4-MeOC$_6$H$_4$, 4-NO$_2$C$_6$H$_4$
R^2 = H, Me, MeO

(27)   (28)   (29)

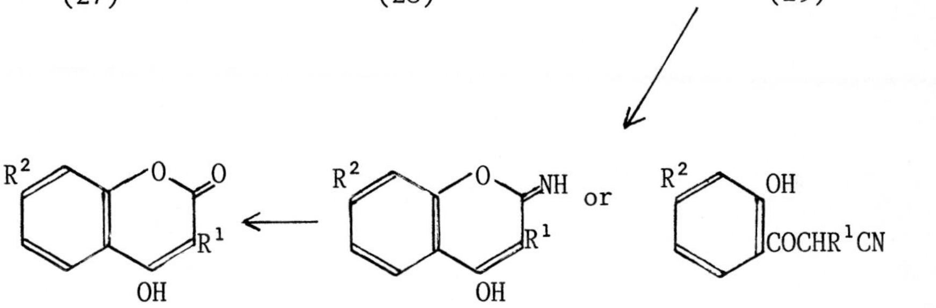

(32)   (31)   (30)

Compound (30; R^1 = R^2 = H) obtained from chromone cyclizes in acid media to give 4-hydroxycoumarin (32; R^1 = R^2 = H) (Szabó, Borda and E. Theisz, Magy. Kem. Foly., 1978, 84, 134).

The nucleophilic reaction of a 6-substituted chromone (33) with hydroxylamine hydrochloride in anhydrous methanol leads to a 2-methoxychroman-4-one (34), as an intermediate in the formation of a chromone oxime (36) and a 2-methoxychroman-4--one oxime (37). ^{1}H-Nmr spectroscopic examination of the reaction mixture shows the presence of compounds (34, 35, 36, 37 and 38). The isoxazole (39) is derived from the dihydromethoxyisoxazole (38) (Szabó *et al.*, Tetrahedron Letters, 1982, 23, 5347; R. Beugelmans and C. Morin *ibid.*, 1976, 2145).

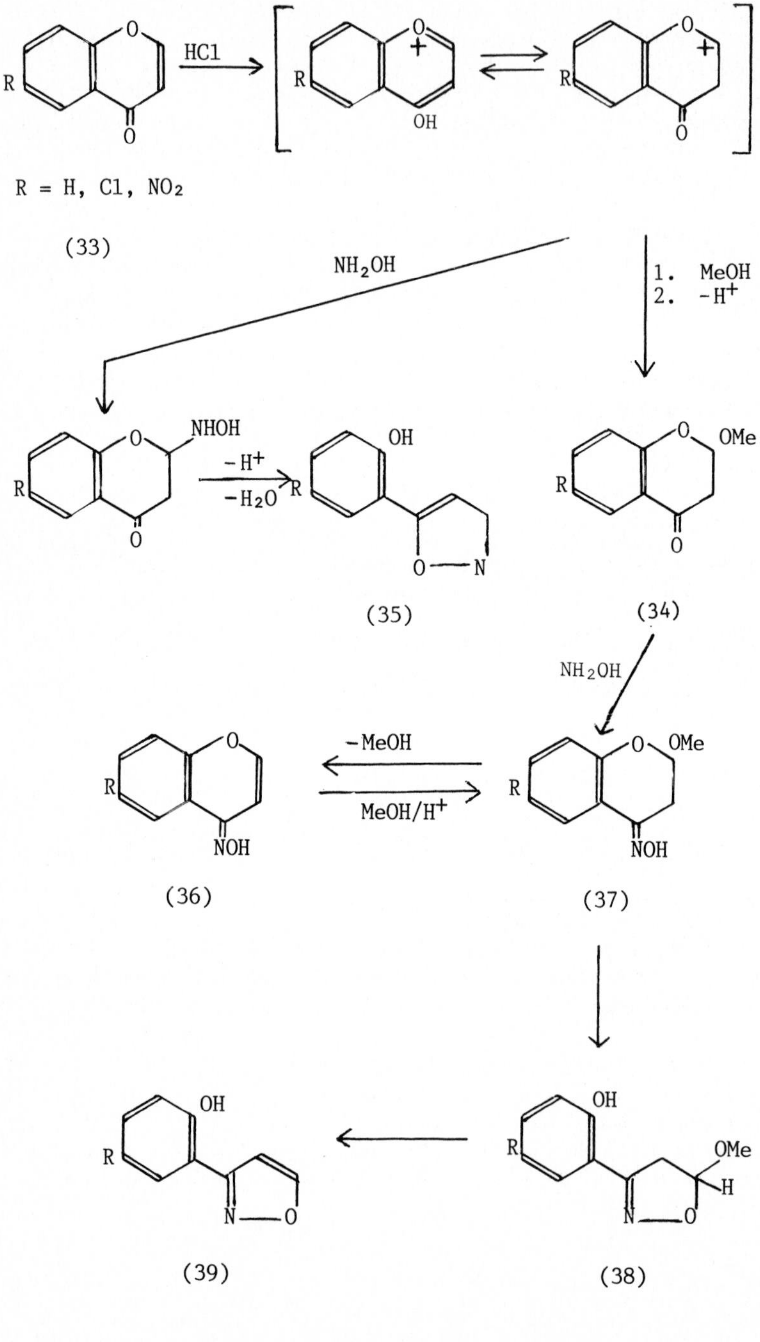

HCl
R
O
OH
R = H, Cl, NO2
(33)
NH2OH
1. MeOH
2. -H+
NHOH
-H+
-H2O
OMe
(35)
(34)
NH2OH
-MeOH
MeOH/H+
NOH
(36)
(37)
(38)
(39)
H
N

When treated with hydroxylamine under standard conditions chromone does not yield the oxime (36; R = H), but rather two isoxazoles (35 and 39; R = H) (Beugelmans and Morin, J. org. Chem., 1977, 42, 1356). The isoxazoles (39; R = H, Me, NHOH) undergo a colour reaction with iron (III) chloride, whereas isoxazoles (35; R = H, Et) do not (W. Basinski and Z. Jerzmanowska, Pol. J. Chem., 1979, 53, 229). The reactions of cyano-, 3-N,N-dimethylhydrazomethyl-, and oximomethyl-chromones with nucleophiles containing nitrogen have been studied (C.K. Ghosh, N. Tewari and C. Bandyopadhyay, Indian J. Chem., 1983, 22B, 1200).

The isoxazole-carboxylic acid (40) in acetic anhydride/ sodium acetate or the oxime (41) in acetic anhydride/ pyridine gives acetyl derivative (42), which rearranges to 2-cyano-1-(2-hydroxyphenyl)butan-1,3-dione (43) and undergoes ring closure to afford 3-cyano-2-methylchromone (44) (Jermanowska and Basinski, Rocz. Chem., 1977, 51, 2283).

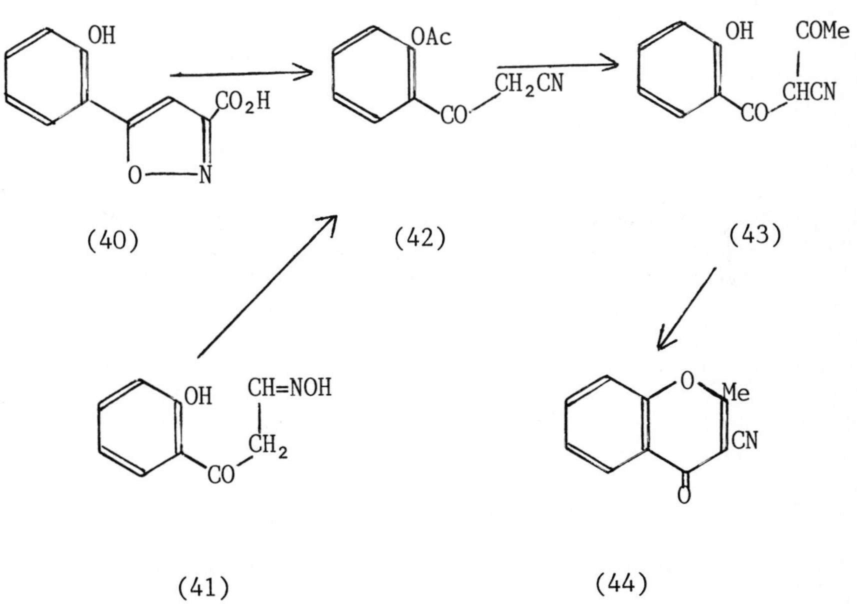

(40)   (42)   (43)

(41)   (44)

3-Aminochromones (45) are obtained in high yields by treating 3-bromochroman-4-ones with sodium azide in aqueous dimethylformamide at 51°C. When the reaction is carried out at lower temperatures or in acetone, ethanol or aceto-nitrile lower yields are obtained, probably due to the competitive formation of the chromone by dehydrobromination. The amines (45) have been converted into their hydrochlorides and acetyl derivatives (Szabó and L. Nemeth, Magy. Kem. Foly., 1978, <u>84</u>, 164).

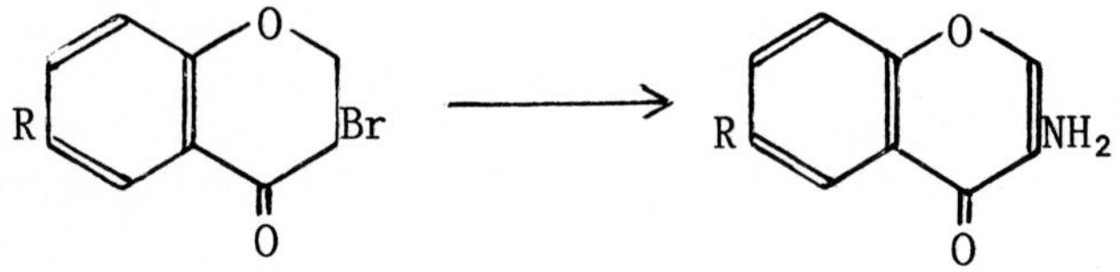

R = H, Me, MeO, Cl

(45)

The addition of butyl-, benzyl- or cyclohexyl-amine to 3--bromochromone (46) in the presence of potassium carbonate in acetonitrile yields the appropriate ring contraction product (47), rather than the respective N-substituted 3--aminochromone. The substituted chromones (48) are formed from (46) with the secondary amines, pyrrolidine and piperidine. These results clearly imply that some of the 3-aminochromones reported by Winter and Hamilton might, in fact, be ring contraction products.

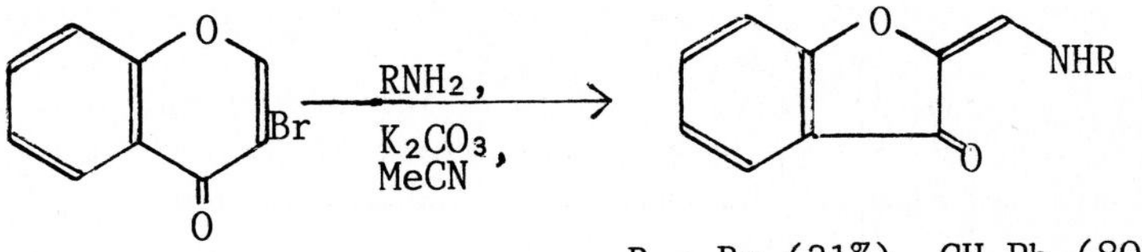

$$R = Bu\ (31\%),\ CH_2Ph\ (80\%)$$
$$c-C_6H_{11}\ (90\%)$$

(46)   (47)

(46)

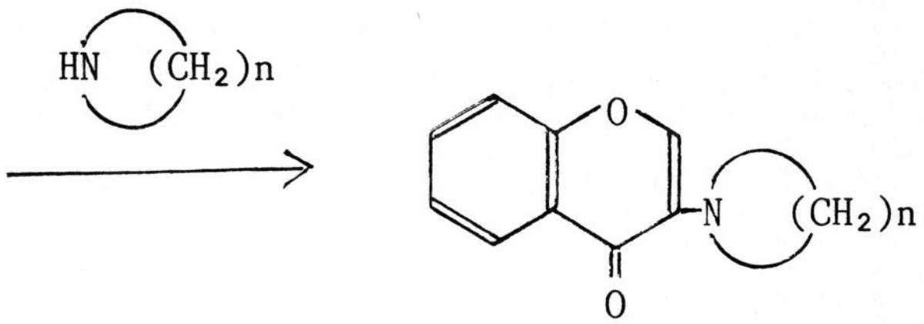

$$n = 4\ (43\%)$$
$$n = 5\ (70\%)$$

(48)

3-Bromochromone (46) with two equivalents of hydroxylamine in methanol furnishes the benzofuran (49). This reaction occurs in the absence of carbonate and in protic solvent (R.B. Gammill, S.A. Nash and S.A. Mizsak, Tetrahedron Letters, 1983, 24, 3435).

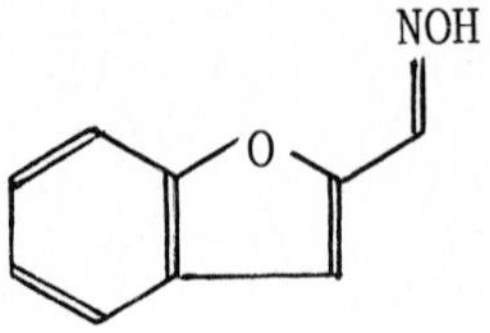

NOH

(49)

The reaction of chromone with the carbanion from acetophenone and that from propiophenone, followed by acidification yields the diones (50) and (51) respectively, which react with ammonium acetate to yield the phenylpyridyl derivatives (52 and 53), respectively (F. Eiden and C. Herdeis, Arch. Pharm., 1977, <u>310</u>, 744).

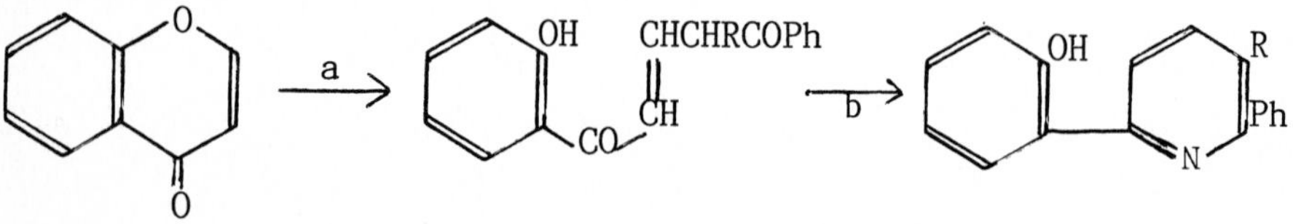

(50) R = H	(52) R = H
(51) R = Me	(53) R = Me

Reagents:- (a) PhCOMe or PhCOEt, Bu^{t}OK; (b) AcONH$_4$

The cleavage of the pyrone ring in chromone, flavone,
isoflavone, 3-methylchromone and 3-phenoxychromone in
aqueous alkali solution has been studied (M. Zsuga *et al.*,
Acta Chim. Acad. Sci. Hung., 1979, 101, 73).

The chromone-3-mercury (II) chlorides (55) have been
obtained by the solvo-mercuration of the 2-methoxybenzoyl-
acetylenes (54) with mercury (II) acetate in acetic acid
(R.C. Larock and L.W. Harrison, J. Amer. chem. Soc., 1984,
106, 4218).

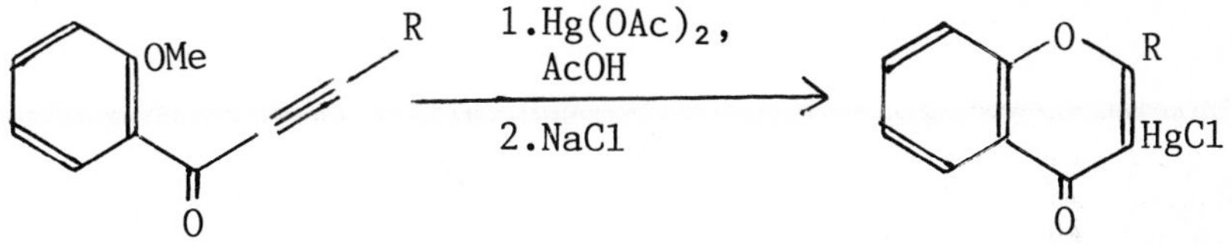

R = n-Pr, Ph

(54)   (55)

3-Chloromercurio-2-propylchromone, m.p. 136-138°;   3-
-chloromercurio-2-phenylchromone (3-chloromercurioflavone),
m.p. 254-255°.

3-(Phenylthio)chromones (57) are prepared by treating 2-
-hydroxy-ω-(phenylthio)acetophenones (56) with methyl
formate and dimethylformamide. 3-(Phenylthio)chromone (57;
$R^1 = R^2 = H$) is obtained by the reaction of acetophenone
(56; $R^1 = R^2 = H$) with methyl formate and sodium *t*-butoxide
(V. Szabó and A. Kiss, Acta Chim. Hung., 1983, 113, 193).

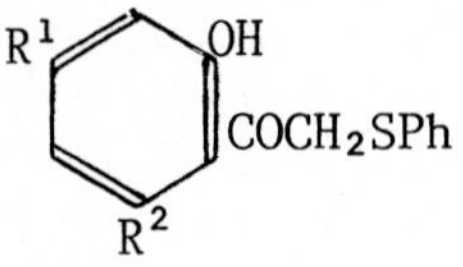

R^1, R^2 = H, OH, OMe

(56)   (57)

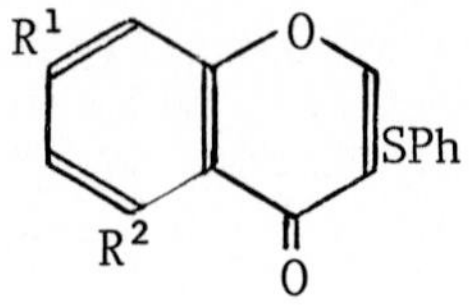

Cyclization of the enamino ketone (58) with thionyl chloride gives chromone-3-sulphinic (59) acid, which can be oxidized to chromone-3-sulphonic acid (60) (W. Loewe and G. Berthold, Arch. Pharm., 1982, <u>315</u>, 892).

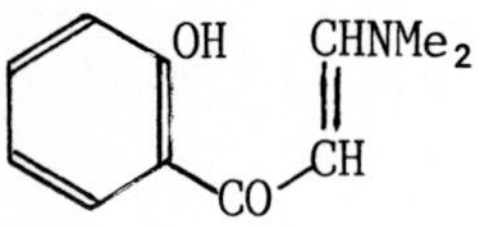

(58) 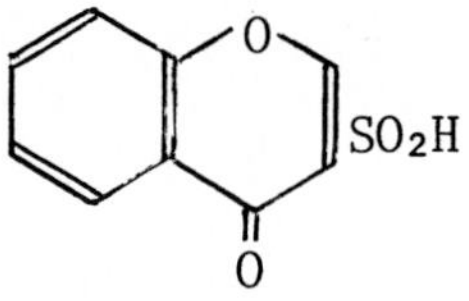   (59) 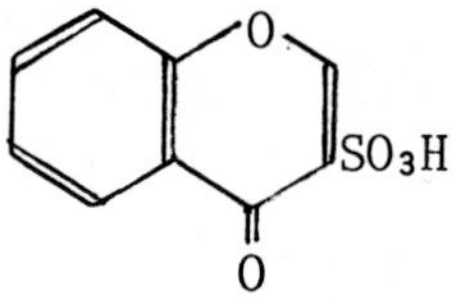   (60)

A number of 2-aryl-3-arylsulphonyl-6-methylchromones have been synthesized (K.P. Jadhav and D.B. Ingle, Indian J. Chem., 1983, <u>22B</u>, 150).

Treatment of the appropriate chromone-3-sulphinic acid with hydrogen bromide/acetic acid, or reduction of the

corresponding thiosulphate ester with dimethyl sulphoxide affords the disulphide (61), which reacts with ethan-2-ol-
-1-thiol to give the chromone-3-thiol (62) (Loewe and A. Kennemann, Arch. Pharm., 1985, 318, 239).

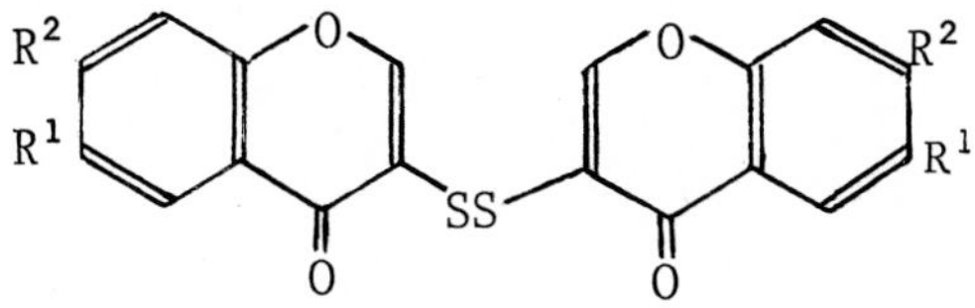

R¹, R² = H, H; Me, H; H, MeO

(61)   (62)

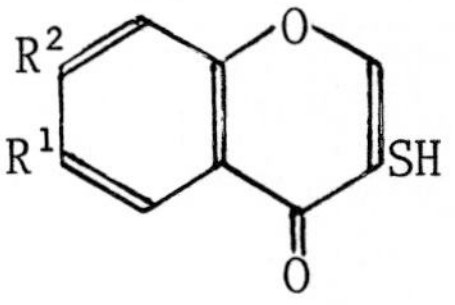

Chromones is lithiated at either the 2- or the 3- position depending upon the substitution pattern and whether the substituents are activating (A.M.S.R.C.S. Costa *et al*., J. chem. Soc., Perkin I, 1985, 799).

Chromone undergoes a Michael reaction with 2,5-bis-(trimethylsiloxy)-3,4-dimethylfuran to furnish the dihydrochromone derivative (63), m.p. 180-182°
(P. Brownbridge and T-H. Chan, Tetrahedron Letters, 1980, 21, 3431).

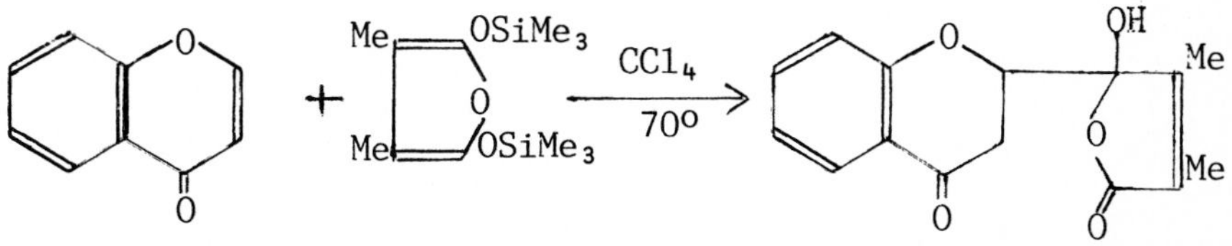

(63)

Irradiation of chromones (64) in methanolic hydrogen chloride induces the homolytic addition of methanol to the pyrone ring double bond to yield the 2-hydroxymethylchroman--4-ones (65). The analogous reaction of chromones (66) yields methylated chromones (67), presumably *via* acid--catalyzed dehydration of the hydroxymethylchromanones (I. Yokoe, Y. Shirataki and M. Komatsu, Chem. pharm. Bull., 1978, **26**, 2277).

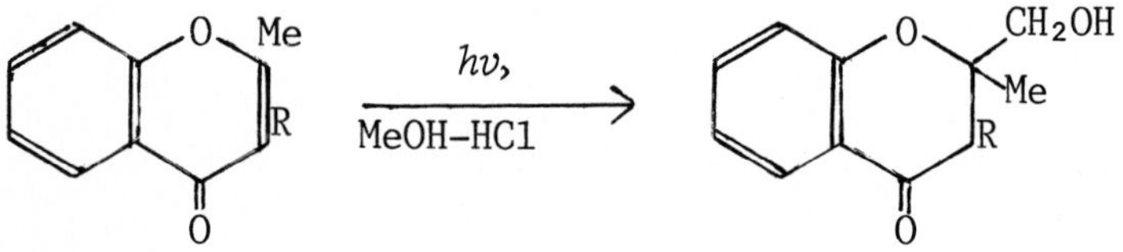

R = H, Me

(64)   (65)

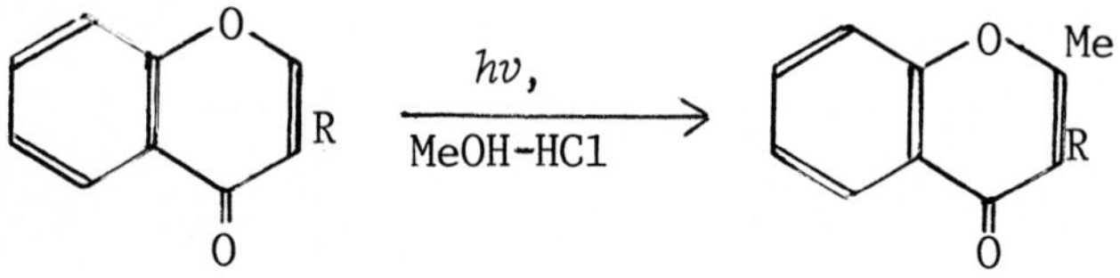

R = H, Me

(66)   (67)

2-Cyanochromone on uv-irradiation in the presence of an alkene, gives rise to a [3+2] cycloadduct together with formation of the normal [2+2] cycloadduct. For example, with isobutene cycloadducts (68; X = O) and (69) are obtained. The ketone (68; X = O) is formed from the hydrolytically sensitive imine (68; X = NH) during work-up. The product ratio is temperature dependent and the product structures follow from their spectral properties and by their conversion into compounds (70 and 71), which have been prepared in an unambiguous manner. A mechanism involving a vinyl nitrene intermediate has been proposed (I. Saito, K. Shimozono and T. Matsuura, Tetrahedron Letters, 1982, $\underline{23}$, 5439).

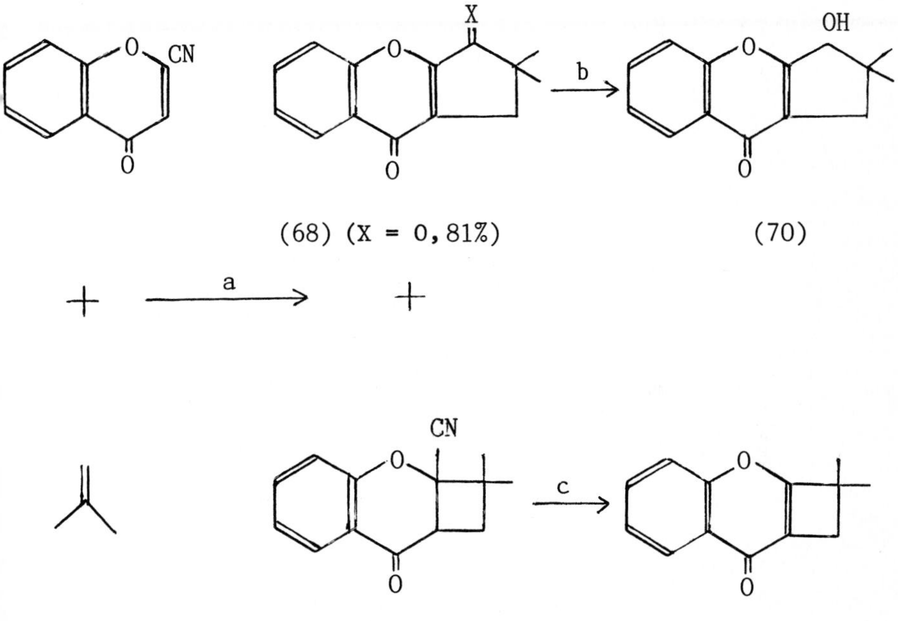

(68) (X = O, 81%)   (70)

(69) (3%)   (71)

Reagents:- (a) $h\nu$, MeOH, N_2; (b) (X = O), $NaBH_4$, MeOH;
            (c) NaH, C_6H_6

The reaction between chromone and diphenylnitrilimine
(72) gives the cycloadduct (73) (A.S. Shawali, B.A. Eltawil
and H.A. Albar, *ibid.*, 1984, <u>25</u>, 4139).

(72)   (73)

[13]C-Nmr chemical shift assignments of some chromones and
isoflavones substituted by OH, AcO and MeO groups (H.C. Jha,
F. Zilliken and E. Breitmaier, Canad. J. Chem., 1980, <u>58</u>,
1211), and assignments of ^{13}C-nmr chemical shifts and some
C-H coupling constants for chromone, 2-, 3-, and 7-
-methylchromones, flavone, and isoflavone (T.N. Huckerby and
G. Sunman, J. mol. Struct., 1979, <u>56</u>, 87) have been
reported. Hydrogen transfer in the retro-Diels-Alder mass
spectra fragmentation of chromones has been determined by
deuterium labelling studies (S. Eguchi, Org. mass. Spec.,
1979, <u>14</u>, 345). The energy intensity and localization of
transitions in the electronic spectra of chromone and some
flavones have been calculated (A.I. Rybachenko *et al.*, Khim.
Prir. Soedin., 1981, 307). Calculated π-electron densities
and orbital energy values have been used quite successfully
to correlate the [1]H-nmr chemical shifts and polargraphic
half-wave potentials (V.K. Ahuja, Indian J. Chem., 1978, <u>16A</u>,
631). A study has been made of the laser excitation
fluorescence of the ground- and excited-state proton transfer
in 3-hydroxychromone and 3-hydroxyflavone (M. Itoh and
Y. Fujiwara, J. phys. Chem., 1983, <u>87</u>, 4558).

Ab initio calculations of ground-state wavefunctions of chromones and thiochromones have been carried out and correlated with the observed reactivity of these molecules. Quantum mechanical electrostatic potentials correctly predict nucleophilic attack to occur preferentially at C-2 in chromones. Frontier-orbital considerations correctly predict the site selectivity of the photochemical cycloaddition to the pyrone double bond of chromone (J.P. Huke *et al.*, J. chem. Soc., Perkin II, 1984, 2119). The electronic structures of a number of molecules related to chromone have been studied by recording their He[1] photoelectron spectra and interpreting these data with the aid of *ab initio* calculations (Huke and I.H. Hillier, *ibid.*, 1985, 1191). The protonation of chromones in concentrated sulphuric acid has been studied by uv-spectroscopy and the Bunnett-Olsen (1966) equation is found to be the most suitable for the estimation of reactivities and for comparisons with quantum chemical calculations (M. Zsuga, T. Nagy and V. Szabó, Magy. Kem. Foly., 1980, **86**, 108).

Carboxaldehydes, ketones and carboxylic acids

Chromones possessing a carboxaldehyde group at the 2- or
the 3- position are useful intermediates in the synthesis
of a wide range of heterocycles (C.K. Ghosh, J. heterocyclic
Chem., 1983, 20, 1437), for instance, the transformation of
chromone-3-carboxaldehydes into pyridines and pyrroles
(P.D. Clarke *et al.*, J. chem. Soc., Perkin I, 1985, 1747).

Chromone-3-carboxaldehydes (1) undergo [4+2]-cycloaddition
with ethyl vinyl ether to afford the benzopyrans (2), which
are hydrolyzed to give the chromonylacroleins (3)
(C.K. Ghosh, N. Tewari and A. Bhattacharya, Synth., 1984,
614).

R = H, Me, Cl, Br   (2)

(1)

(3)

Chromone-3-carboxaldehydes (4) react with aromatic amines
(5) to yield 3-aryliminomethylchromones (6). The latter
interact with thioglycollic acid to give first 3-arylamino-
methylene-2-(carboxmethylthio)chroman-4-ones (7) (1,4-
-adducts) and then 3-(3-aryl-4-oxothiazolidin-2-yl)chromones
(8) on prolonged heating [A.O. Fitton *et al.*, J. chem. Res.,
(S), 1984, 248].

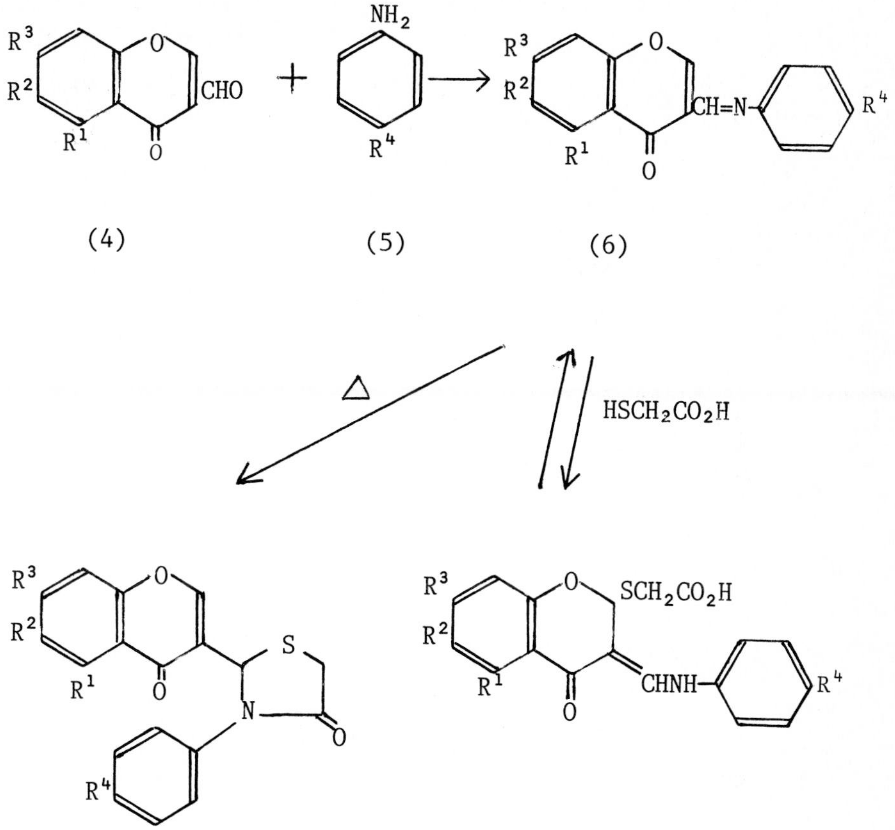

(4)   (5)   (6)

(8)   (7)

Treatment of chromone-3-carboxaldehyde with hydroxylamine hydrochloride affords oxime (9), isoxazoles (10 and 11), pyrazolone (12) and benzopyranopyrazoledione (13), depending on the reaction conditions (W. Basinski and Z. Jerzmanowska, Pol. J. Chem., 1983, 57, 471).

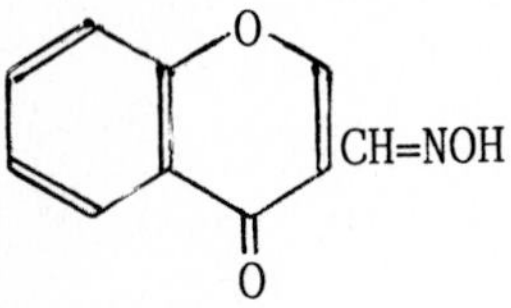

(9) 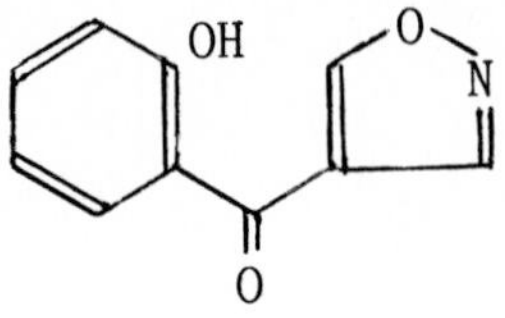   (10) 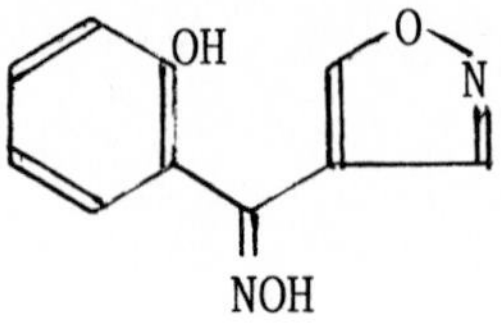   (11)

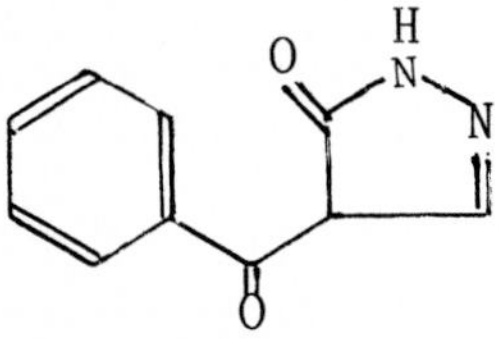

(12) 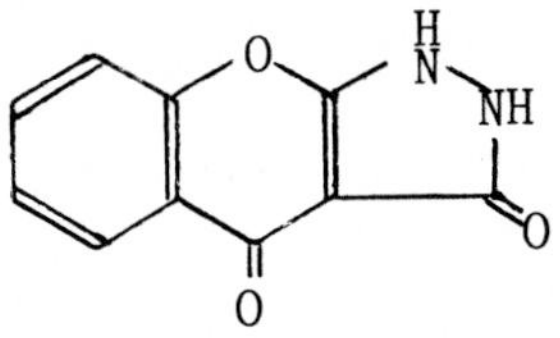   (13)

The reactions between some 6-substituted chromone-3-
-carboxaldehydes and secondary and tertiary alkyl and aryl
amines have been studied (Ghosh, C. Bandyopadhyay and
Tewari, J. org. Chem., 1984, 49, 2812). The dipole moments
and ir-spectra of a number of chromone-3-carboxaldehydes
with substituents at the 6-,7- and 8- positions
(V.K. Polyakov, R.G. Shevtsova and S.V. Tsukerman, Zh.
obshch. Khim., 1979, 49, 1560); polargraphic studies of
chromone-3-carboxaldehyde (L.V. Kononenko *et al.*, *ibid.*,
p.2693); and investigations into the proton-acceptor-
-capabilities of substituted chromone-3-carboxaldehydes
(Polyakov *et al.*, *ibid.*, 1978, 48, 2273) have been
reported.

1-(2-Hydroxyphenyl)butane-1,3-dione (14) with
dimethylformamide dimethylacetal in benzene yields the
enamino diketone (15), but a xanthone derivative in methanol
medium. The diketone (15) on treatment with acid affords

3-acetylchromone (16) (Ghosh and Bhattacharya, Indian J.
Chem., 1984, 23B, 668).

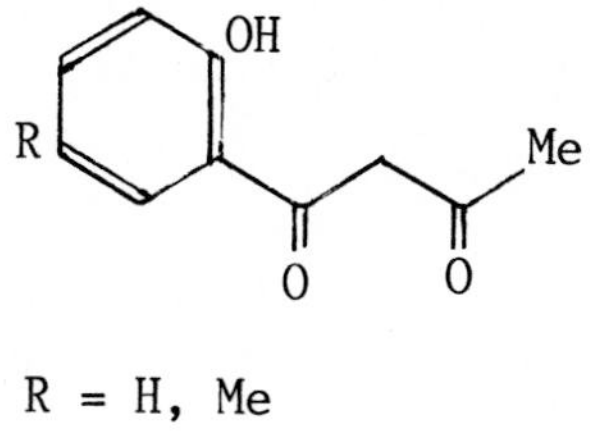

R = H, Me

(14)

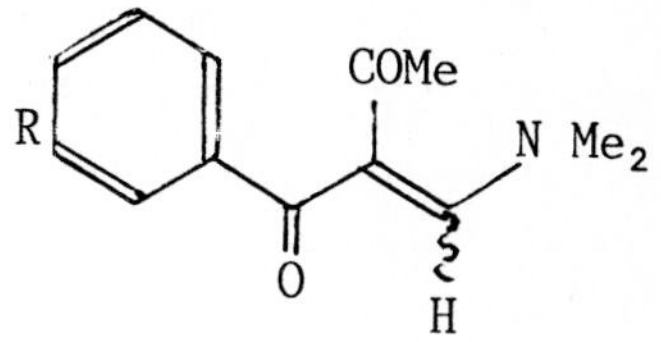

(15)

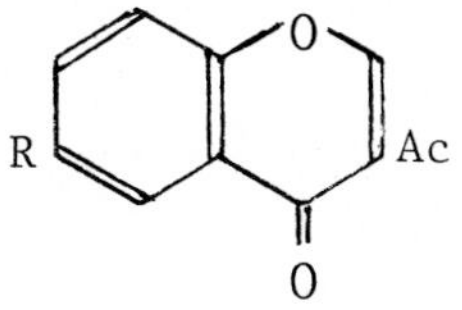

(16)

Treatment of 3-acylchromones with amidines furnishes
acylpyrimidines. The reaction between 3-acetylchromone and
acetamidine yields the pyrimidine (17; R = H) and the
benzopyranopyrimidine (18), whereas 3-acetyl-2-methylchromone
yields only the pyrimidine (17; R = Me) (W. Loewe, Arch.
Pharm., 1977, 310, 559; Ann., 1977, 1050).

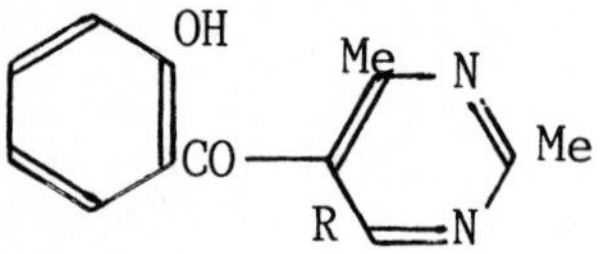

(17) 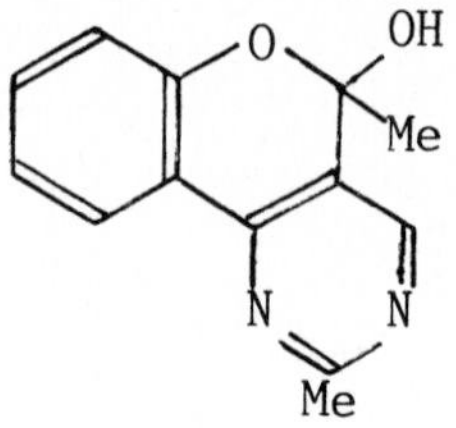   (18)

2-Diazoacetylchromones (20) are prepared from the
appropriate acid chloride (19; R^3 = Cl) and diazomethane.
The reaction of diazoacetylchromone (20; R^1 = R^2 = H) with
hydrogen chloride or hydrogen bromide affords the acid halides
(19; R^1 = R^2 = H, R^3 = Cl or Br), whereas hydrogen iodide
yields chromone-2-carboxaldehyde (19; R^1 = R^2 = R^3 = H).
Methyl and ethyl chromone-2-carboxylate have been obtained
from 2-diazoacetylchromone (M. Payard *et al.*, Bull. Soc.
Chim. Fr., 1977, 505).

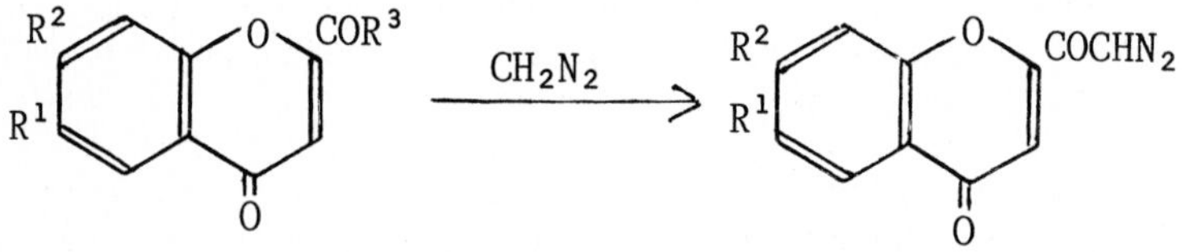

R^1 = H, F, Cl, Br, NO_2, Me, R^2 = H
R^1 = H, R^2 = OMe

(19)   (20)

3-Methoxymethyl-6-methyl-2-pivaloylmethylchromone (21; R = H) when treated with lithium diisopropylamide in tetrahydrofuran at $-70°$, followed by pivaloyl chloride gives 3-methoxymethyl-6-methyl-8-pivaloyl-2-pivaloylmethyl-chromone (21; R = t-BuCO), m.p. 145-148° (A.M.S.R.C.S. Costa *et al*., Chem. Comm., 1983, 1098).

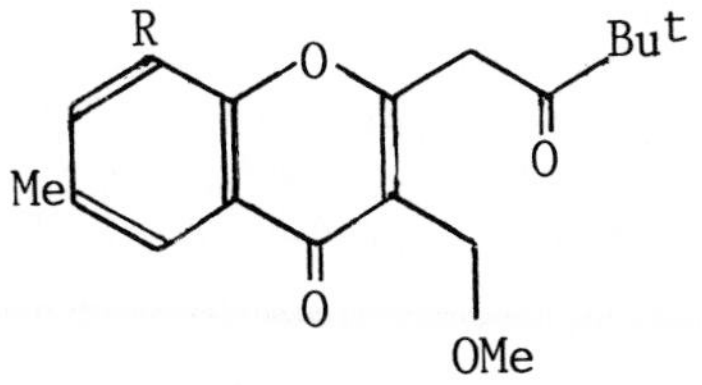

(21)

Chromones (22) activated by carbonyl substituents at C-3 are transformed into the corresponding 2-methylchroman-4-ones (23 and 24) on treatment with lithium dimethylcuprate. The β-ketoester (23; R^1 = OMe, R^2 = Me) on heating at 155° (4h) with sodium chloride in wet dimethylsulphoxide is smoothly decarbomethoxylated to give 2,2-dimethylchroman-4--one (25) (T.W. Wallace, Tetrahedron Letters, 1984, 25, 4299).

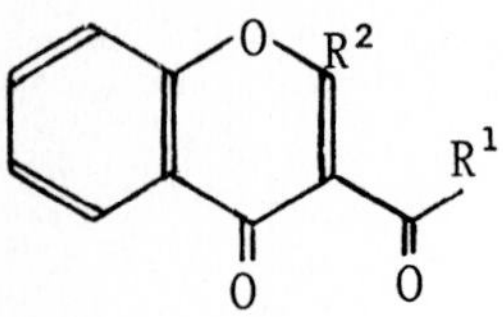

R^1	R^2
Ph	H
OMe	H
Me	H
H	H
Me	Me
OMe	Me

(22)

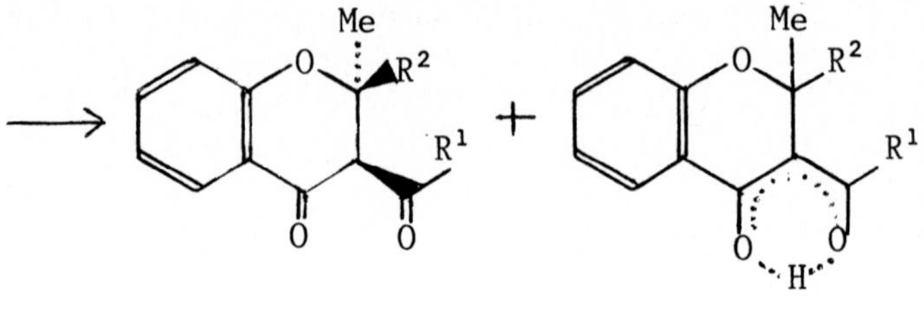

(23)

(24)

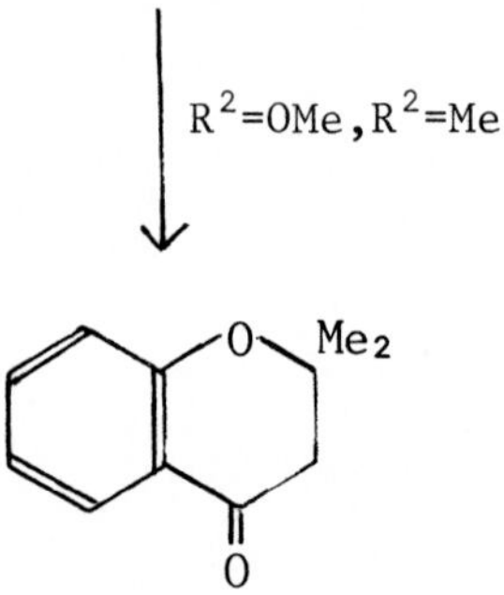

$R^2 = OMe, R^2 = Me$

(25)

3-Acetylchromone (26) undergoes acyl-acyl rearrangement in the presence of a basic catalyst to the isomeric 2--methylchromone-3-carboxaldehyde (30), which because of its active methyl group remains in equilibrium with the tautomer (31) at least in basic medium. The tautomer (31) undergoes a facile Diels-Alder reaction with the pyrano--dienophile (26) to give adduct (32), which undergoes base catalyzed elimination and deacylative elimination to yield xanthone (33). The proposed mechanism is justified by the fact that aldehyde (30) on treatment with any of the chromones (26 - 29) yields xanthone (33) in quantitative yield (Ghosh, Bhattacharya and C. Bandyopadhyay, Chem. Comm., 1984, 1319).

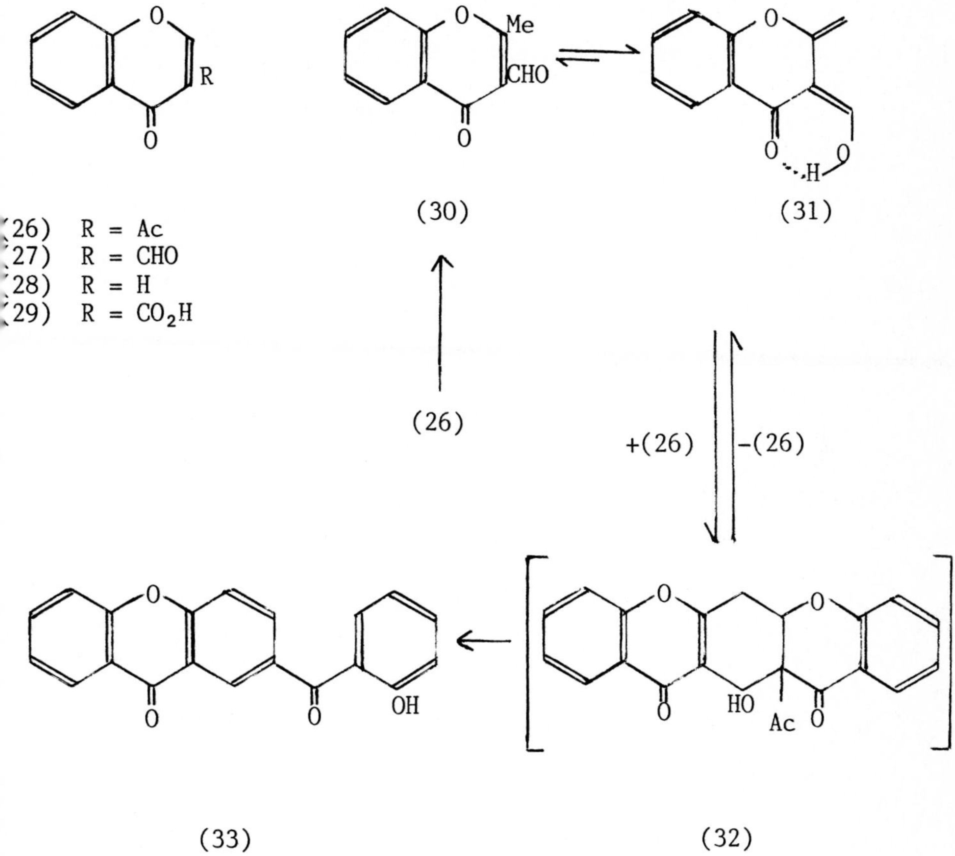

(26)   R = Ac
(27)   R = CHO
(28)   R = H
(29)   R = CO_2H

(30)

(31)

(26)

+(26)   −(26)

(33)

(32)

Chromone-3-carboxaldehyde in the presence of ethyl
acetoacetate and ethanolic piperidine undergoes a novel
transformation to the 2-hydroxybenzophenone (34)
(W.D. Jones and W.L. Albrecht, J. org. Chem., 1976, 41,
706).

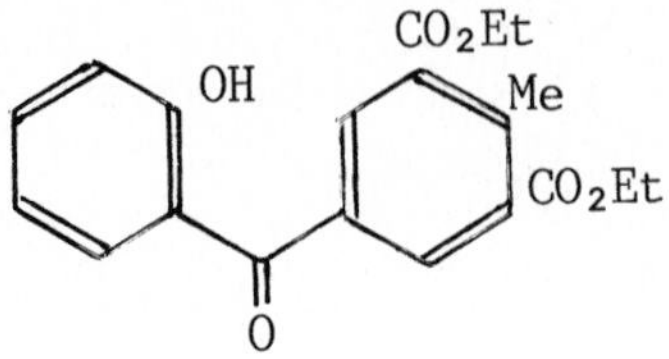

(34)

5-(2-Hydroxy-3-thiopropoxy)chromone-2-carboxylic acids
and their esters [36; R^1 = (un)substituted alkyl,
cycloalkyl; R^2 = H, alkyl, alkali metal] have been
prepared. Thus, compound (35) on thiomethylation and
cyclization with diethyl oxalate gives the ethyl chromone-
-2-carboxylate (36; R^1 = Me, R^2 = Et), which on
hydrolysis yields acid (36; R^1 = Me, R^2 = H)
(A. Pedrazzoli and S. Boveri, U.S. Pt. 4,282,247, 1981).

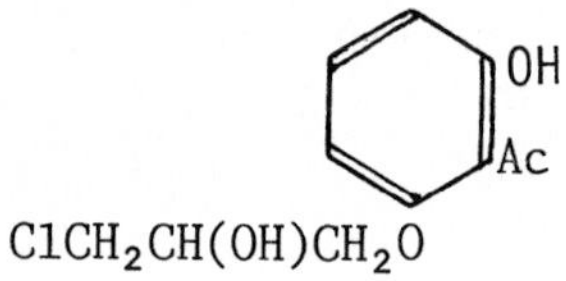

(35) 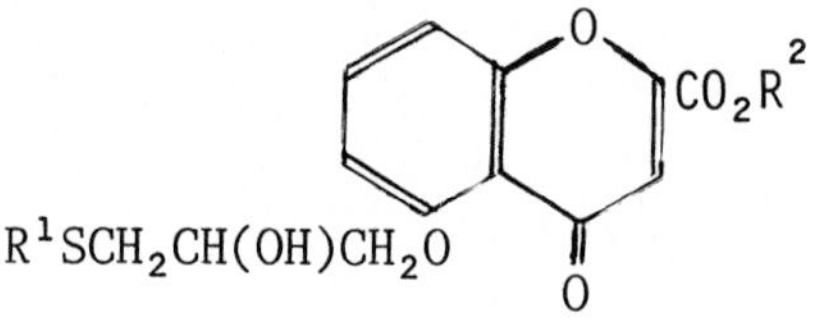   (36)

Solvent effects on reactivity trends for the hydrolysis
of chromone-2-carboxylate in methanol-water mixtures have
been studied and analyzed into initial state and transition
state components [J. Burgess and E-E. Abu-Gharib, J. chem.
Res. (S), 1984, 8]. Ethyl 5-hydroxy-8-propylchromone-2-

-carboxylate (37) reacts with chloroacetone under basic
conditions (dimethylformamide in presence of either
potassium carbonate or sodium hydride) to give
(1α,1aα,7aα)- and (1β,1aα,7aα)-ethyl 1-acetyl-1,1a,7,7a-
-tetrahydro-6-hydroxy-3-propylbenzo[b]cyclopropa[e]pyran-
-1a-carboxylate (38 and 39), products of cyclopropanation
of the chromone 2,3-double bond, as well as ethyl 1,7-
-diacetyl-7,7a-dihydro-5-propyl-6aH-cyclopropa[b]furo-
[4,3,2-de]benzo[b]pyran-6a-carboxylate (40). Similarly
tetrahydronaphtho[2,3-b]pyran (41) gives analogous products
and with dimethylsulphoxonium methylide affords the
isomeric naphthofuranones (42 and 43) (I.D. Dicker,
J. Shipman and J.L. Suschitzky, J. chem. Soc., Perkin I,
1984, 487).

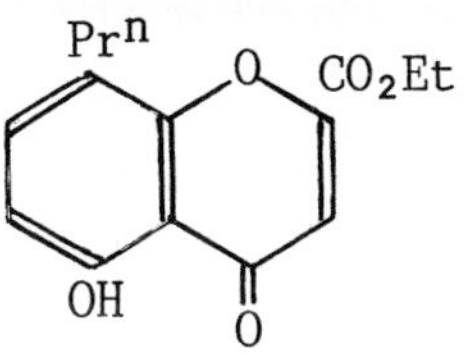

(37) 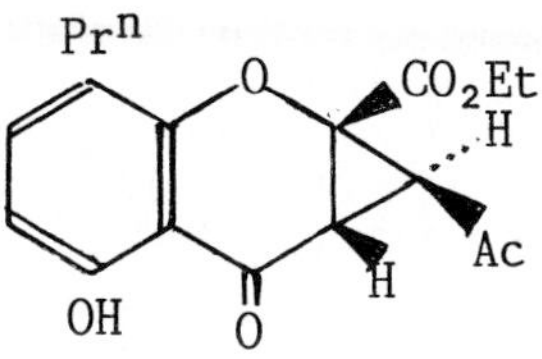   (38) 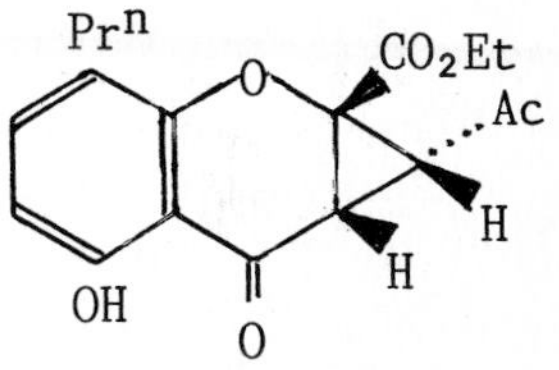   (39)

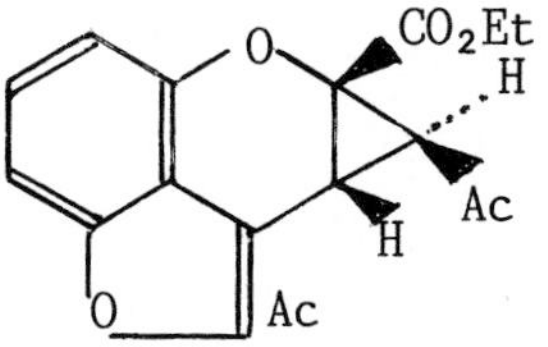

(40) 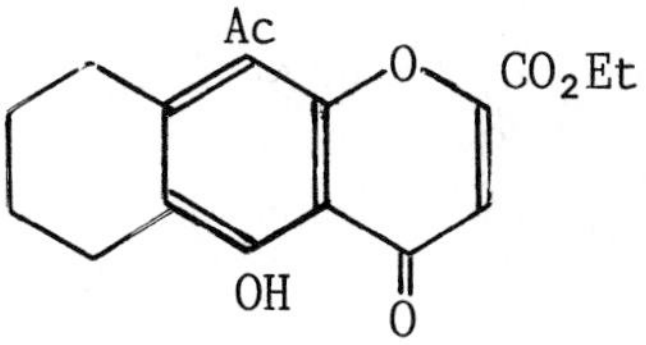   (41)

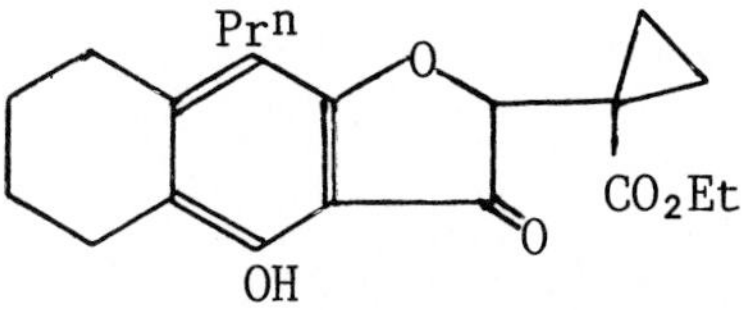

(42) 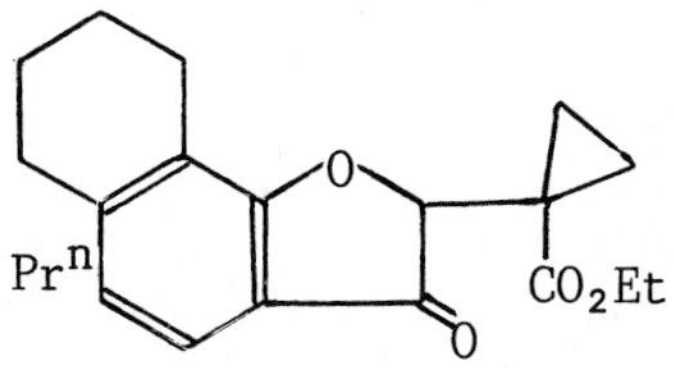   (43)

5-(2-hydroxypropoxy)-8-propylchromone-2-carboxylic acid
and some derivatives have been prepared (Fisons PLC Switz.
Pat. 637,650, 1983).

Chromone-2-acetic acids are synthesized by the Michael
addition of phenols to allene-1,3-dicarboxylic esters,
followed by cyclization of the adducts. Phenols (44), (45)
and (46) with dimethyl allene-1,3-dicarboxylate (47) in the
presence of Triton B in boiling benzene yield adducts (48),
(49) and (50) respectively, which are cyclized in
polyphosphoric acid to their respective chromones (51),
(52) and (53). The enol ethers (48), (49) and (50) have E-
-stereochemistry as indicated by their ^{1}H-nmr spectral
data and on hydrolysis yield acids (54), (55) and (56).
Acids (54) and (55) on treatment with polyphosphoric acid
are converted into chromone-2-acetic acids (57) and (58)
(N.S. Nixon, F. Scheinmann and J.L. Suschitzky, J. chem.
Res., 1984, 380).

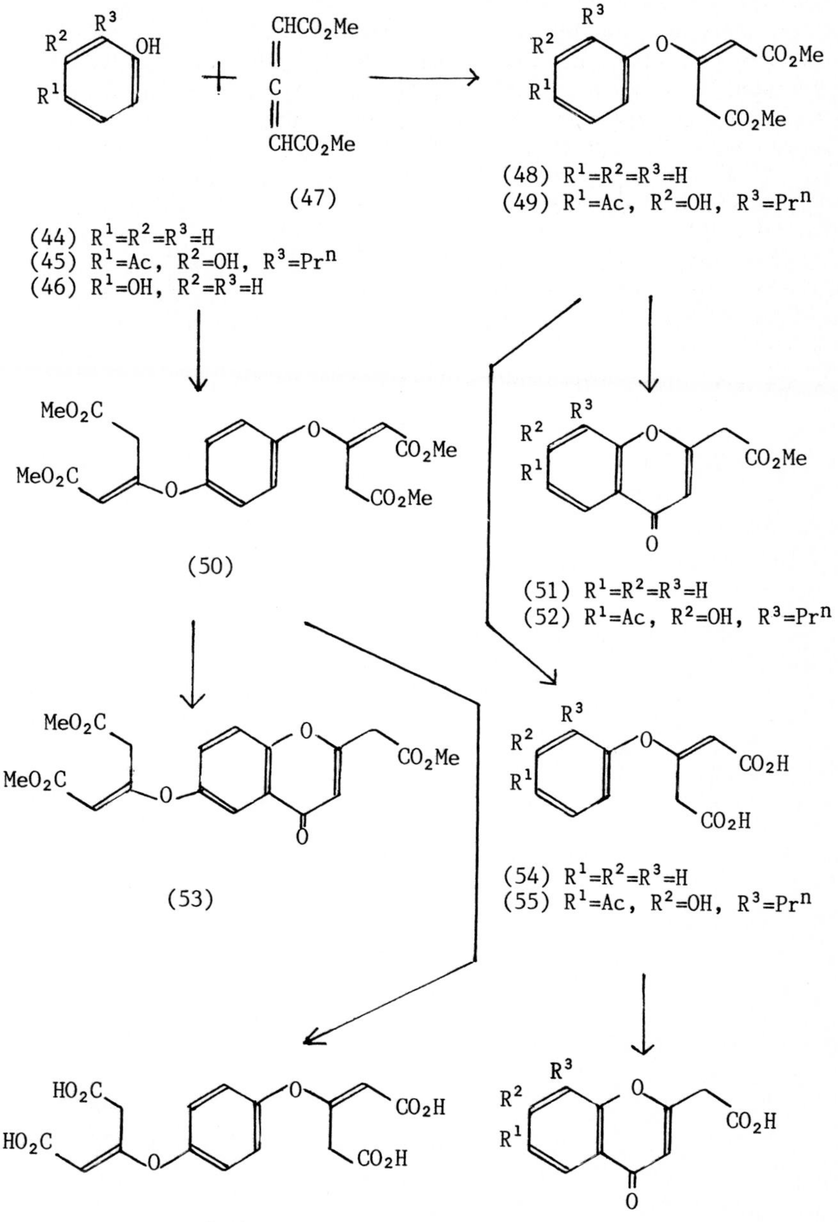

(47)

(44) $R^1=R^2=R^3=H$
(45) $R^1=Ac$, $R^2=OH$, $R^3=Pr^n$
(46) $R^1=OH$, $R^2=R^3=H$

(48) $R^1=R^2=R^3=H$
(49) $R^1=Ac$, $R^2=OH$, $R^3=Pr^n$

(50)

(51) $R^1=R^2=R^3=H$
(52) $R^1=Ac$, $R^2=OH$, $R^3=Pr^n$

(53)

(54) $R^1=R^2=R^3=H$
(55) $R^1=Ac$, $R^2=OH$, $R^3=Pr^n$

(56)

(57) $R^1=R^2=R^3=H$
(58) $R^1=Ac$, $R^2=OH$, $R^3=Pr^n$

Three routes to 7-methoxychromone-3-carboxylic acid and 6,7-methylenedioxychromone-3-carboxylic acid (chromone-3- -carboxylic acid 6,7-ketene acetal) have been described. The latter is most effectively obtained *via* a Kostanecki- -Robinson reaction and the former by *N*-bromosuccinimide oxidation of the corresponding aldehyde (T. Hoegberg *et al.*, Acta Chem. Scand., 1984, <u>B38</u>, 359). A number of chromone-3-carboxylic acids [59; R = H, 7-Me, 7-MeO, 7-Cl, 6,7-(AcO)$_2$, 6-7-(EtOCO$_2$)$_2$, 6,7-(Cl$_3$CCH$_2$OCO$_2$)$_2$] have been prepared by treating the corresponding 3-carboxaldehyde with *N*-halogenosuccinimide. Thus, chromone-3-carboxylic acid (59; R = H) is obtained in 93% yield when the appropriate aldehyde reacts with *N*-bromosuccinimide in carbon tetrachloride with exposure to a 150W lamp, followed by removal of the solvent and treating the residue with water (Eisai Co. Ltd., Jap. Pat. 82 59, 883, 1982).

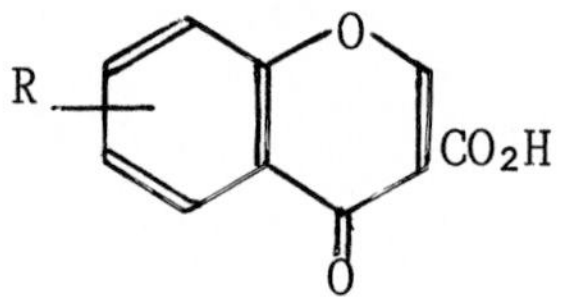

(59)

Natural Products

Two minor lipid components of the brown seaweed *Zonaria tournefortii* have been characterized as (all Z)-5',7'-
-dihydroxy-2'-nonadeca-4,7,10,13,16-pentenylchromone (1)
and 5',7'-dihydroxy-2'-pentadecylchromone (2) (C. Tringali
and M. Piattelli, Tetrahedron Letters, 1982, 23, 1509).

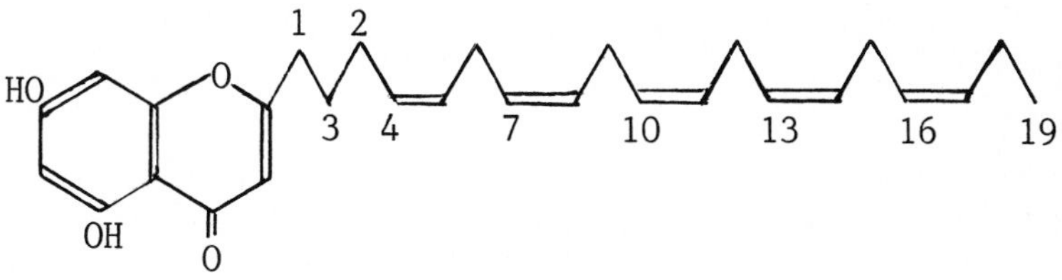

(1)

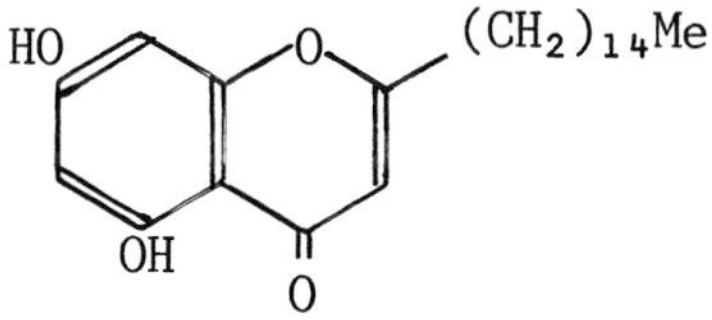

(2)

(xii) Furochromones
Khellin (1), the lipid-altering and anti-atherosclerotic
furochromone has been synthesized by two different routes.
The key in both approaches is the cyclo-addition of the
chromium carbene complex (2) to an alkoxyalkyne to yield
the benzofuran derivative, carrying the functional groups
necessary for the 4-pyrone ring. The reaction between the
complex (2) and the alkoxyalkynes (3) in the presence of
acetic anhydride and triethylamine furnishes the benzofurans
(4), which are converted into khellinone (5) and
khellinquinone (6) by four and five steps, respectively. The
routes from khellinone (5) and khellinquinone (6) to khellin
(1) are known (Y. Yamashita, J. Amer. chem. Soc., 1985, 107,
5823).

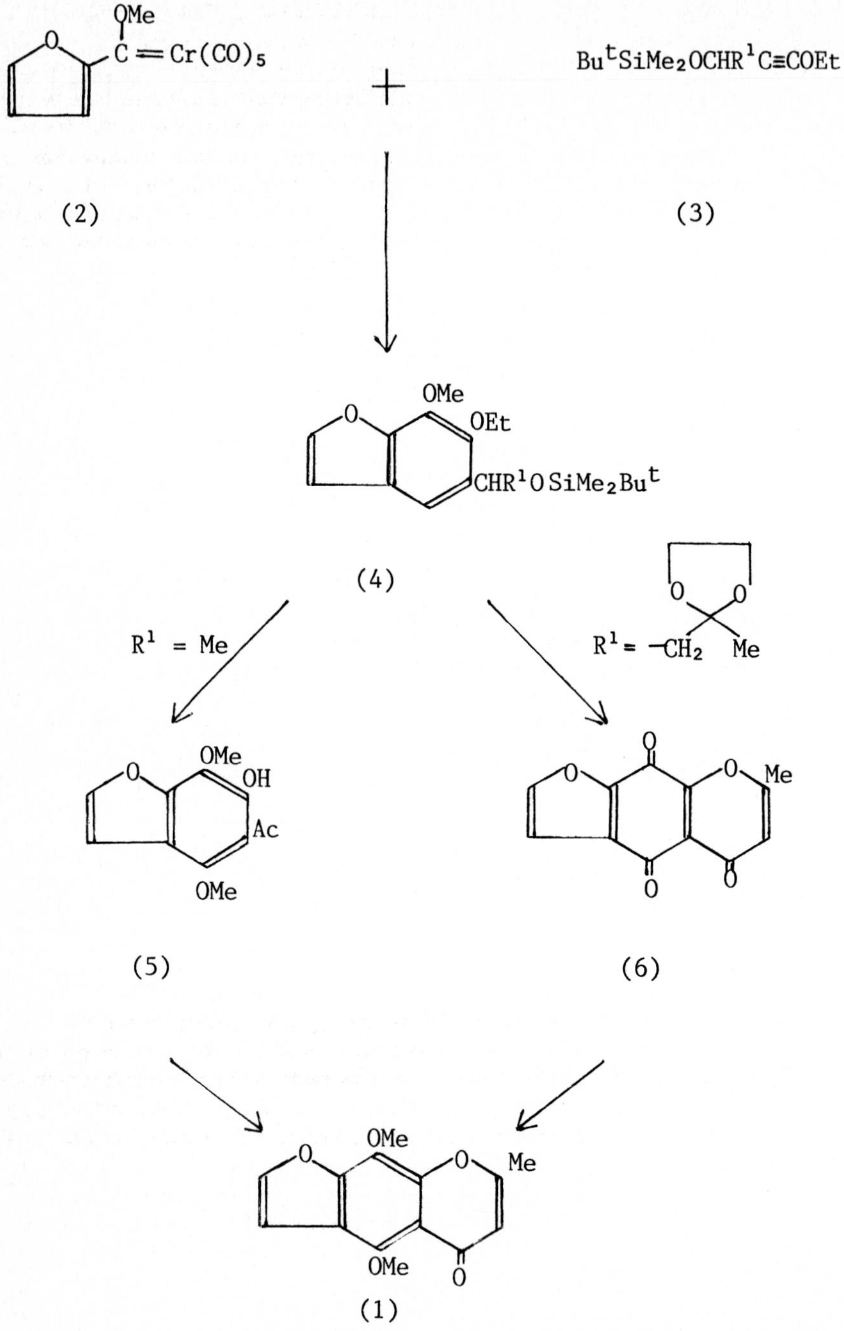
OMe
C = Cr(CO)5
(2)
Bu^tSiMe2OCHR^1C≡COEt
(3)
OMe
OEt
CHR^1OSiMe2Bu^t
(4)
R^1 = Me
R^1 = –CH2 Me
OMe
OH
Ac
OMe
(5)
Me
(6)
OMe
Me
OMe
(1)

A commercial method has been discussed for the isolation of khellin from the fruit of *Ammi visnaga* (T. Singh, K.L. Handa and P.R. Rao, Res. Ind., 1977, 22, 11).

The benzofuran derivative, 6-acetoxy-4,7-dimethoxybenzo-furan-5-carbonyl chloride, derived from khellin (1), on condensation with appropriate enamines or lithium enolates, followed by treatment with acid yields the khellin analogues (8). Better yields are obtained with lithium enolates than with enamines. (T. Watanabe *et al.*, J. chem. Soc., Perkin I, 1978, 726).

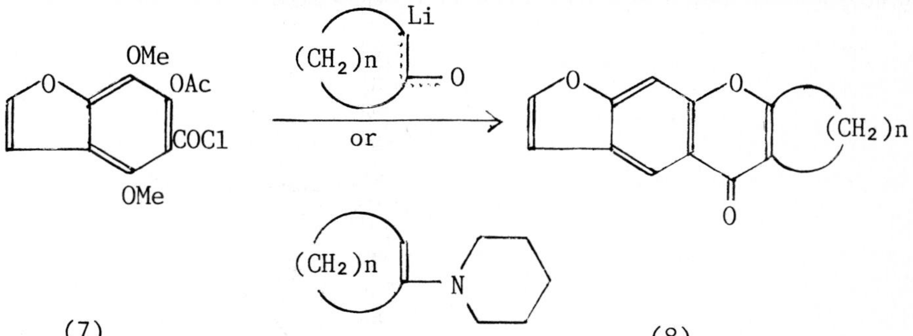

Khellin (1) is converted into ammiol (11) following cleavage with potassium hydroxide to the benzofuryl ketone (9), recyclization with ethyl 2-(methylthio)acetate to give the methylthio derivative (10), and then subjection to another four preparative stages (R.B. Gammill, J. org. Chem., 1984, 49, 5035).

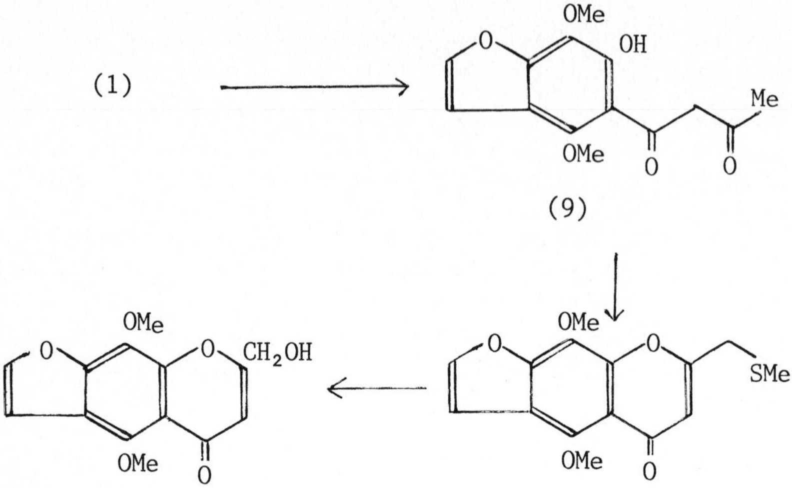

(1) $\longrightarrow$

(9)

(11)   (10)

The chromic acid oxidation of visnagin (12) is known to
give the hydroxyaldehyde (13) in good yield. Unfortunately
the analogous oxidation of khellin is not successful.

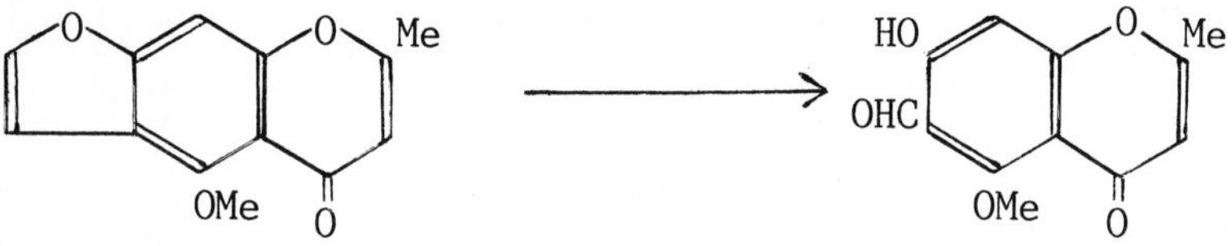

(12)   (13)

Catalytic osmylation of khellin (1) in THF at 50° in the
presence of sodium periodate (2.2 equiv.) furnishes the
hydroxyaldehyde (14), which on treatment with chloroacetone
in boiling THF, in the presence of potassium carbonate and
18-crown-6 yields 2-acetylfurochromone (15), m.p. 184-185°
(53%). The slow addition of *o*-sulphonic acid hydroxylamine
to the hydroxyaldehyde (14) in a two phase system (water/
methylene dichloride) containing sodium hydrogen carbonate
(2 equiv.) affords the isoxazole (16) (79%).

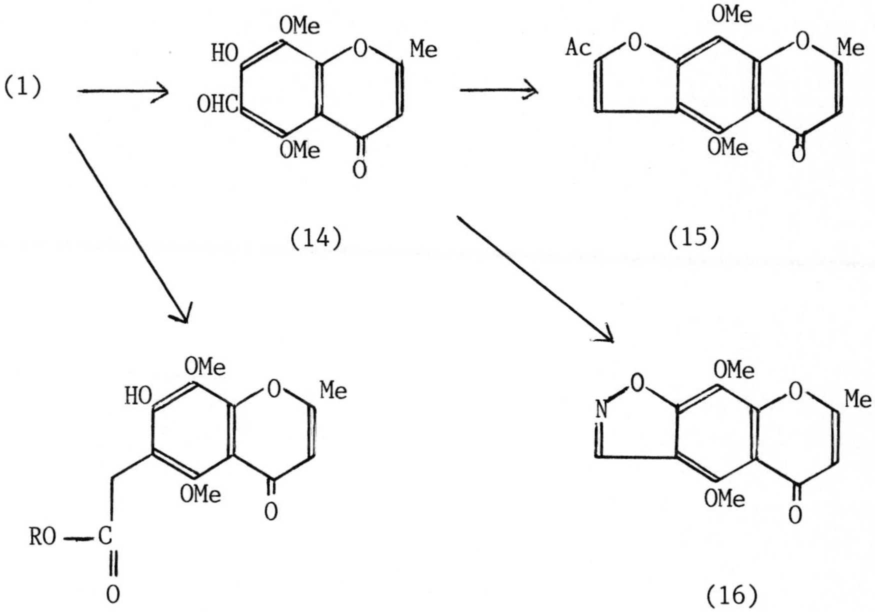

(1) ⟶

(14)

(15)

(17) R = Me
(18) R = H

(16)

Catalytic oxidation of khellin (1) with palladium chloride
(CuCl/O₂/30psi) in methanol yields the hydroxy ester (17)
(73%), another versatile analogue synthon (Gammill and
S.A. Nash, Tetrahedron Letters, 1984, 25, 2953). Hydrolysis

of the ester (17) affords acid (18), which on treatment with acetic anhydride gives the lactone (19). Hydroxyaldehyde (14) undergoes smooth Dakin oxidation with sodium hydroxide/ hydrogen peroxide to afford diol (20), which on alkylation with methylene diiodide gives the methylenedioxy analogue. Oxidation of khellin (1) with thallium (III) nitrate in methanol results in the addition of two mols of methanol across the 2,3-double bond in the furan ring to give the dihydrotetramethoxyfurochromone (21). Hydroxyaldehyde (14) is also obtained by the oxidation of khellin with mercuric (II) nitrate in aqueous THF followed by treatment with sodium periodate (*idem*, J. org. Chem., 1986, **51**, 3116).

(19)   (20)

(21)

Khellin (1) reacts with hydroxylamine under the usual conditions to yield isoxazoles (22) and (23), but no oxime. The oxime is obtained when the reaction is performed with hydroxylamine hydrochloride in anhydrous methanol. ^{13}C-Nmr spectral data of compounds related to khellin have been reported (R. Beugelmans and C. Morin, *ibid.*, 1977, **42**, 1356).

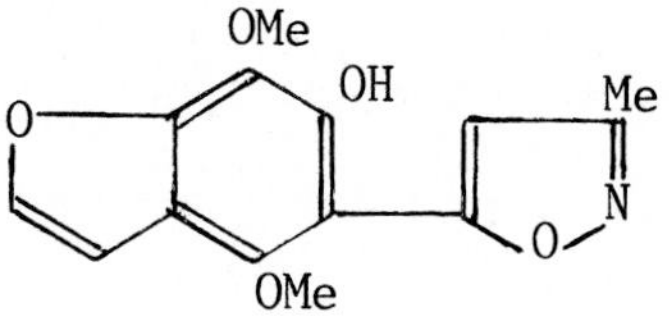

(22)

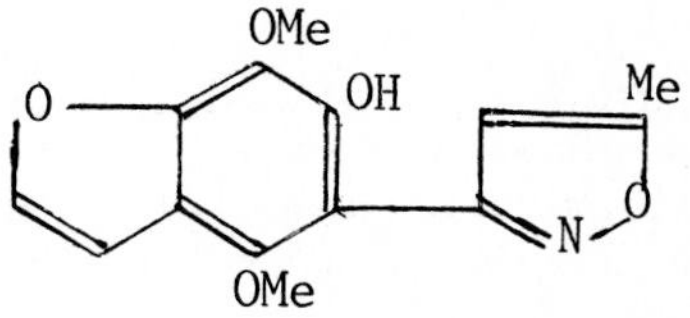

(23)

The reaction of 6-bromo-4,9-dimethoxyfurochromone (24) with amines at 100°C gives an extremely complex mixture. If the reaction is carried out in acetonitrile, at a lower temperature in the presence of potassium carbonate to neutralize the hydrogen bromide formed during the reaction, then the addition of pyrrolidine to bromo derivative (24) furnishes the novel ring contraction product (25). No 6--aminofurochromone is detected. Analogues of (25) are obtained by the addition of morpholine, benzylamine, and cyclohexylamine to (24) (Gammill, Nash and S.A. Mizsak, Tetrahedron Letters, 1983, 24, 3435).

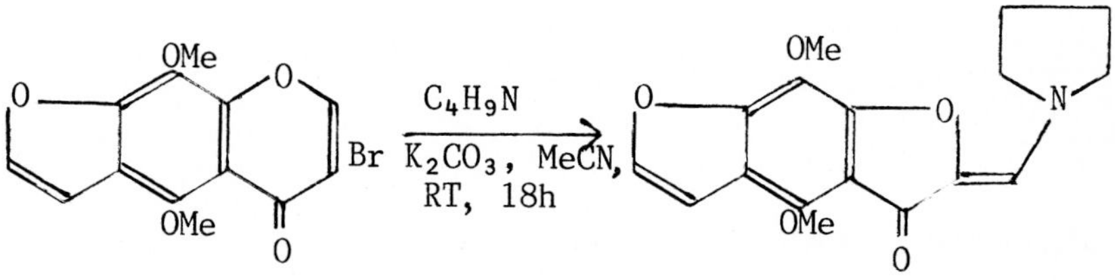

(24)

(25)

Crystal structure determination of khellin shows that it is monoclinic and the electron density distribution shows the molecule is essentially aromatic, *ie.* the rings are coplanar (S.M. Salem *et al.*, Egypt. J. Phys., 1984, 15, 121).

Boiling 4,9-dimethoxy-7-(methylthio)methyl-5H-furo[3,2-g]-benzopyran-5-one {5,9-dimethoxy-2-(methylthio)methylfuro-[3,2-g]chromone} with hydrogen bromide in chloroform yields the 4-hydroxyfurochromone (26; R^1 = Me, R^2 = H). 4-Hydroxyfurochromones [26; R^1 = alkyl, (un)substituted Ph; R^2 = H, alkanoyloxy; n = 0,1,2] have also been prepared and are useful as intermediates for compounds possessing anti-atherosclerotic properties (Gammill, U.S. Pt. 4,542,228, 1985).

(26)

A number of isotopically labelled linear furochromones (27; R = ^{2}H, ^{3}H, X = ^{12}C; R = H, X = ^{13}C, ^{14}C) (R.S.P. Hsi, Gammill and E.G. Daniels, J. labelled Compd., 1985, 22, 1273) and some biologically active visnagin-9-sulphonyl amino acids and dipeptide derivatives (A.M. El-Naggar *et al.* Pol. J. Chem., 1981, 55, 793) have been prepared.

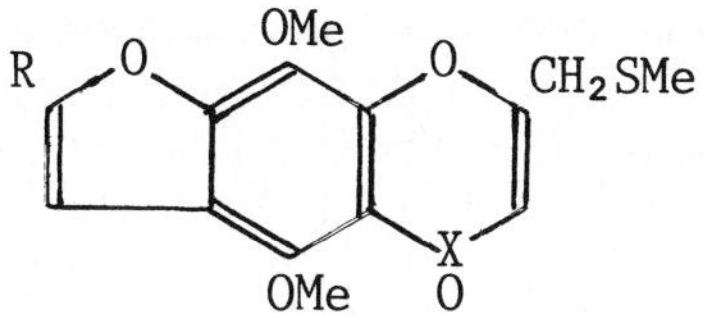

(27)

5-Methyl- and 5-ethyl-furo[2,3-b]chromone (28) along with other compounds have been isolated from the aerial parts of *Bothriodine laxa* and their structure elucidated by chemical and spectroscopic methods (F. Bohlmann and C. Zdero, Phytochem., 1977, 16, 1261).

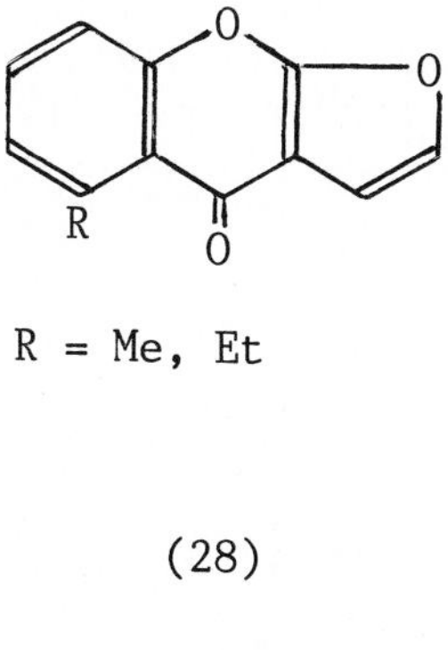

R = Me, Et

(28)

(2*S*)-2-(1-Hydroxy-1-methylethyl)-4-methoxy-7-β-D-glucosyl-oxymethyl-2,3-dihydro-5H-furo[3,2-g]benzo[b]pyran-5-one {6,7-dihydro-2-β-D-glucosyloxymethyl-(7*S*)-7-(1-hydroxy-1--methylethyl)-5-methoxyfuro[3,2-g]chromone} along with two new pyranochromones has been isolated from the dried roots of *Angelica japonica* (K. Baba *et al.*, Chem. pharm. Bull., 1981, 29, 2565).

6,7-Dihydro-2-hydroxymethyl-7-(2-hydroxy-2-propyl)-5-
-methoxyfuro[3,2-g]chromone (29) has been obtained from
Peucedanum austriaca (Jacq.) (Umbelliferae) and identified
mainly from spectral data and by chemical transformations
(M. Stefanovic *et al.*, Glas. Hem. Drus. Beograd, 1984, **49**, 5).

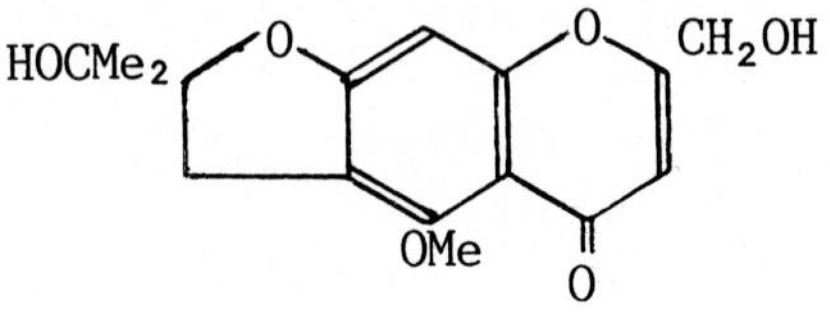

(29)

2-Methyl-7H-furo[2,3-h]benzo[b]pyran-7-one (8-methylfuro-
[2,3-h]-chromone) (32) is prepared by dehydration of 7-
-allyloxy-2,3-dihydro-2-hydroxychromone (30), followed by
Claisen rearrangement to 8-allyl-7-hydroxychromone (31),
which on treatment with osmium tetroxide/potassium
periodate and subsequent cyclization with polyphosphoric
acid affords furochromone (32) (R.J. Patolia and
K.N. Trivedi, J. Indian chem. Soc., 1981, **58**, 62).

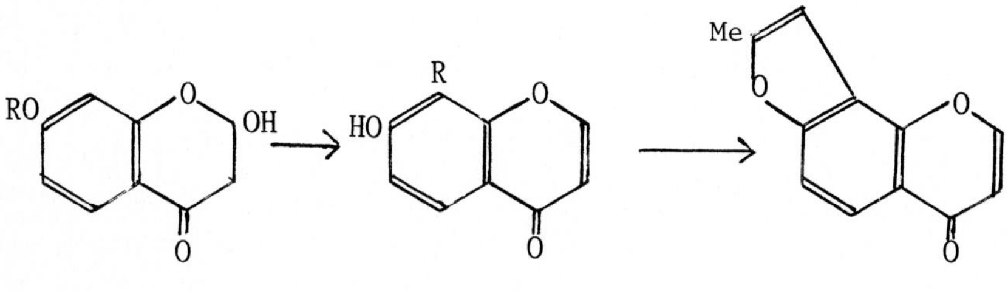

R = CH₂=CHCH₂

(30)   (31)   (32)

^{13}C-Nmr spectral data for some furochromones,
furocoumarins, and dihydrofurocoumarins have been reported
(M.H.A. Elgamal *et al.*, Phytochem., 1979, <u>18</u>, 139) and
their use in structural elucidation discussed (*idem,*
IUPAC Int. Symp. Chem. Nat. Prod., 11th, 1978, <u>2</u>, 271).

(xiii) Pyranochromones
   The 2,3-dihydro-6H-pyrano[3,2-g]benzo[b]pyran-6-one (7,8-
-dihydropyrano[3,2-g]chromone) (1), along with other
compounds has been isolated from *Peucedanum austriaca*
(Jacq) (Umbelliferae) (M. Stefanovic *et al.*, Glas. Hem.
Drus. Beograd, 1984, <u>49</u>, 5).

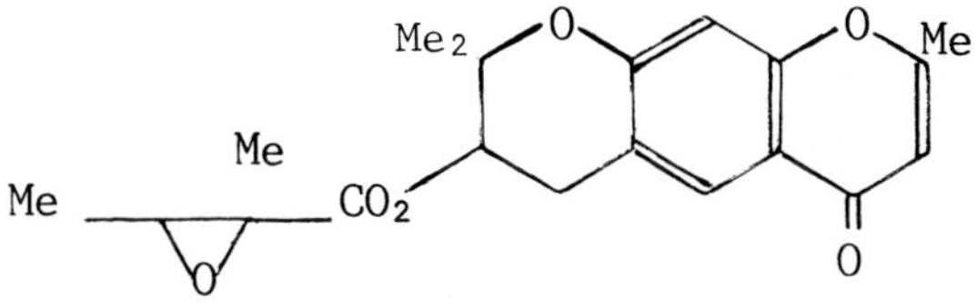

(1)

   (3S)-2,2-Dimethyl-3,5-dihydroxy-8-hydroxymethyl-3,4-
-dihydro-2H,6H-benzo[1,2-b:5,4-b']dipyran-6-one and (3S)-
-2,2-dimethyl-3-β-D-glucosyloxy-5-hydroxy-8-methyl-3,4-
-dihydro-2H,6H-benzo[1,2-b:5,4-b']dipyran-6-one (*sec-O-*
-glucosylhamaudol) have been obtained from the dried roots
of *Angelica japonica* (K. Baba *et al.*, Chem. pharm. Bull.,
1981, <u>29</u>, 2565).

(xiv) Flavones (2-phenylchromones, 2-phenyl-4H-benzo[b]pyran
  -4-ones)*

Preparations. 5-Hydroxychromone (1) undergoes an Elbs
persulphate oxidation to give 5,8-dihydroxychromone
(primetin) (2), which on lead tetraacetate oxidation
affords 5,8-quinoflavone (3). Addition of hydrogen
chloride to the quinone (3) yields 6-chloro-5,8-dihydroxy-
flavone (4), also obtained by the Elbs persulphate
oxidation of 6-chloro-5-hydroxyflavone (5), synthesized
from 3-chloro-6-hydroxy-2-methyoxyacetophenone. The ^{13}C-
-nmr spectrum of 6-chloro-5,8-dihydroxyflavone (4) has been
reported (J.H. Looker, J.R. Edman and C.A. Kingsbury, J.
org. Chem., 1984, **49**, 645).

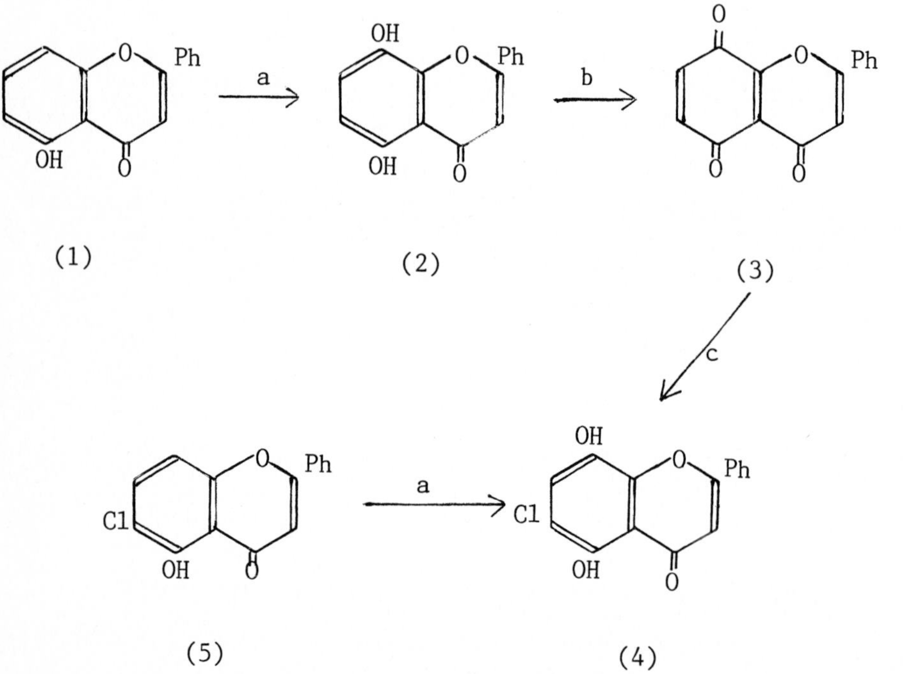

(1)   (2)   (3)

(5)   (4)

Reagents:- (a) $K_2S_2O_8$, $(Et)_4N^+OH^-$, C_5H_5N; (b) $Pb(OAc)_4$, AcOH;
          (c) HCl, AcOH

*   G.P. Ellis, Heterocyclic Chem., 1980, **1**, 329.

Cyclization of 2-hydroxydibenzoylmethanes in boiling
benzene containing toluene-4-sulphonic acid with azeotropic
distillation of the water eliminated affords flavones.
This method gives 5-methoxyflavones in quantitative yield
(P.K. Jain, J.K. Makrandi and S.K. Grover, Curr. Sci.,
1981, 50, 857). The synthesis of 5- and/or 7-hydroxyflavones
using a modified phase-transfer catalyzed Baker/Venkataraman
transformation, followed by cyclization of the resulting
propanediones with acid or base has been described
(S. Saxena, Makrandi and Grover, Synth., 1985, 697).
Magnesium chelates have been utilized in the synthesis of
3-nitro- and 3-methoxycarbonyl-flavones. This method
avoids preformed flavones, as well as, the troublesome
dehydrogenation of 3-substituted flavanones. For example,
2-hydroxy-ω-nitroacetophenones (6; R = NO$_2$) are converted
into their magnesium enolates and condensed with aroyl
chlorides to give 3-nitroflavones (7; R = NO$_2$, Ar = Ph)
(M. Cushman and A. Abbaspour, J. org. Chem., 1984, 49, 1280).

R = NO$_2$, CO$_2$Me

(6)

(7)

Reactions. Flavone on hypervalent iodine oxidation, followed by acid/catalyzed hydrolysis of the resulting product yields 3-hydroxyflavone (flavonol) (see p. 110). Similarly 2-phenyl-4*H*-naphtho[1,2-b]pyran-4-one gives 3--hydroxy-2-phenyl-4*H*-naphtho[1,2-b]pyran-4-one. Flavones can be conveniently reduced to flavanones by sodium hydrogen telluride (P. Shanmugan and N. Shobana, Proc. int. Conf. org. Chem. Selenium, Tellurium, 4th, 1983, 273).

The reaction of phenols with iron (III) chloride usually leads to the formation of biaryls, but 7,4'-dimethoxy-5--hydroxyflavone (1) undergoes selective nuclear chlorination at positions 6 and 8 by iron (III) chloride in strong acid to furnish, products (2) and (3). These on methylation yield 6-chloro-4',5,7-trimethoxyflavone (6), m.p. 242°; 8-chloro-4',5,7-trimethoxyflavone (5), m.p. 234°; and 6,8-dichloro-4',5,7-trimethoxyflavone (4), m.p. 260° (H.I. Siddiqui, N-U. Khan and W.A. Shaida, Chem. and Ind., 1982, 908).

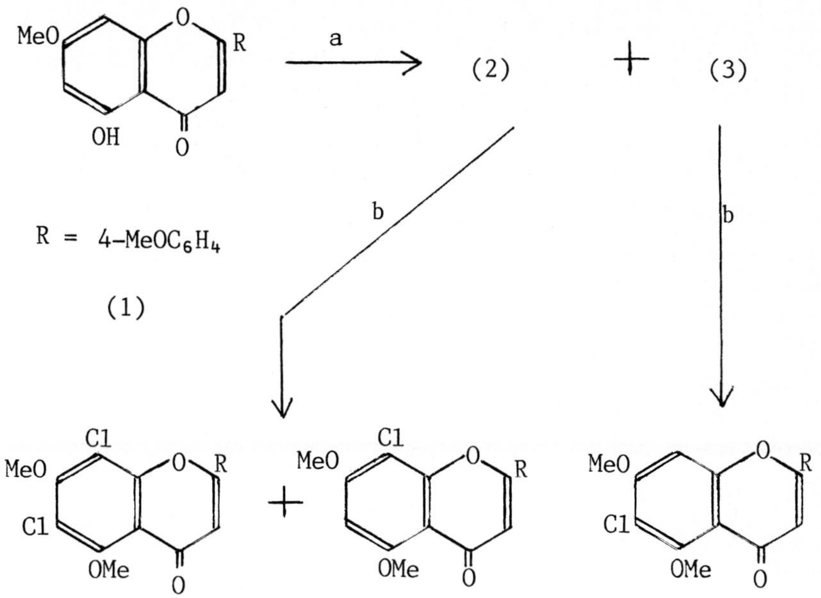

R = 4-MeOC$_6$H$_4$

(1)

(4)   (5)   (6)

Reagents:- (a) FeCl$_3$, HClO$_4$, AcOH, boil 4.5h; (b) methylation

5-Acetoxy-6-chloro-4',7-dimethoxyflavone, m.p. 250-252°.
3-Chloromercurioflavone, m.p. 254-255°, also recorded are
its ir- and nmr-spectral data (R.C. Larock and J.W. Harrison,
J. Amer. chem. Soc., 1984, 106, 4218). Lithiation of
flavones by lithium di-isopropylamide in tetrahydrofuran at
-78° occurs at the 3-position and the products are stable
at that temperature. Appropriate reagents replace the
lithium by carboxy, ethoxycarbonyl, mercapto, methylthio,
trimethylsilyl, hydroxy, and other groups
(A.M.B.S.R.C.S. Costa *et al.*, J. chem. Soc., Perkin I, 1985,
799).

3-Methoxyflavone (7) on heating with hydroxylamine hydrochloride and sodium hydroxide in ethanol gives the isoxazole derivative (8) (P. Maib and W. Basinski, Pol. J. Chem., 1981, 51, 1527).

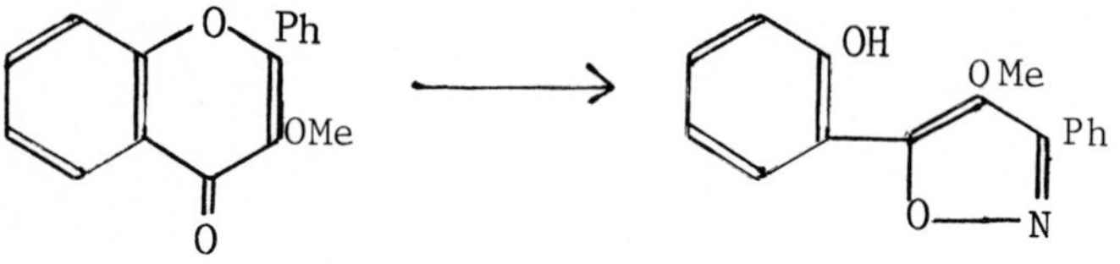

(7)   (8)

7-Hydroxy-3-phenylflavone reacts with epichlorohydrin and sodium hydroxide to give a glycidyl ether which with isopropylamine yields 7-(2-hydroxy-3-isopropylaminopropoxy)-
-3-phenyl-flavone (9; R^1 = Pr^i, R^2 = Ph). Derivatives (9; R^1 = Pr^i, Pr; R^2 = H, Ph) have also been prepared (E.S.C. Wu, Eur. Pat. Appl. EP 81,621, 1983).

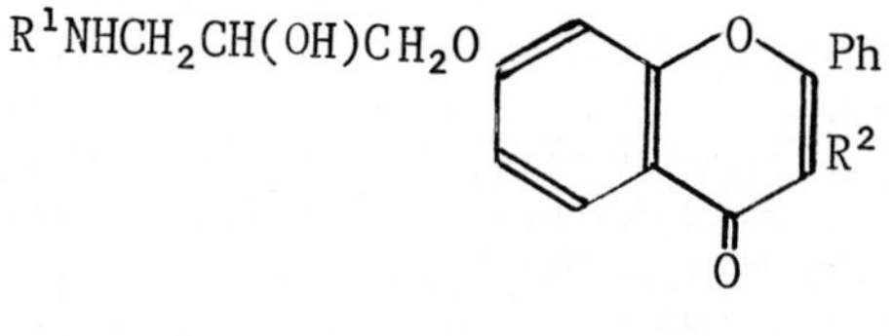

(9)

A number of 3-methylflavone-8-carboxylic acid esters (10;
e.g., R = *N*-methylpiperidyl, tropyl, quinuclidinyl) have been
prepared by esterification of 3-methylflavone-8-carbonyl
chloride (10; OR = Cl). These esters are reported to
possess muscle relaxant, anaesthetic, anti-inflammatory,
and antispastic activities (D. Nardi *et al*., *ibid*., 72,620,
1983).

$$CO_2R$$

(10)

The cleavage of the pyran ring of flavone in aqueous
alkali solution (M. Zsuga *et al*., Acta Chim. Acad. Sci.
Hung., 1979, 101, 73); the ^{13}C-nmr spectra of flavones
(J.N. Huckerby and G. Sunman, J. mol. Struct., 1979, 56,
87), and the $\pi \longrightarrow \pi^*$ electron transitions in the absorption
spectra of flavone and some of its hydroxy-substituted
derivatives (A.I. Rybackenko *et al*., Khim. Prir. Soedin.,
1981, 307) have been discussed.
    5,6,7,8-Tetrahydroflavone (11) is obtained either by
heating together equimolar amounts of cyclohexanone and ethyl
benzoylacetate or from ethyl 2-oxocyclohexanecarboxylate
(12) and acetophenone (B.S. Kirkiacharian, Compt. rend.,
1981, 293, 149).

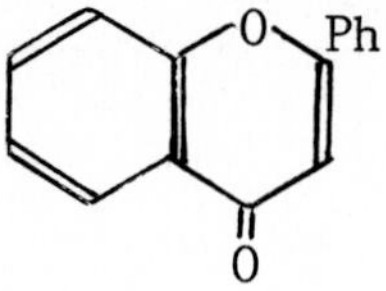

(11) 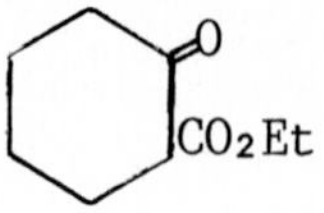   (12)

(xv) Flavonols (3-hydroxyflavones, 3-hydroxy-2-phenyl-
-4H-benzo[b]pyran-4-ones)

Preparation. Photolysis of a 2-aryl-3-nitro-2H-benzo[b]-
pyran (1) in methanol, followed by acid hydrolysis of the
product yields the corresponding 8-substituted flavonol (2)
(T.S. Rao, A.K. Singh, and G.K. Trivedi, Heterocycles,
1984, **22**, 1377).

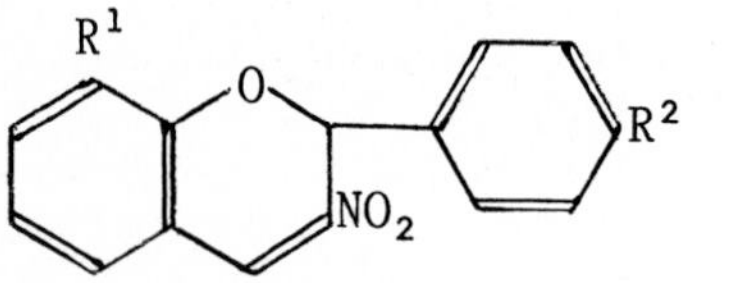

R^1 = H, Me, Cl, OMe
R^2 = H, OMe

(1)   (2)

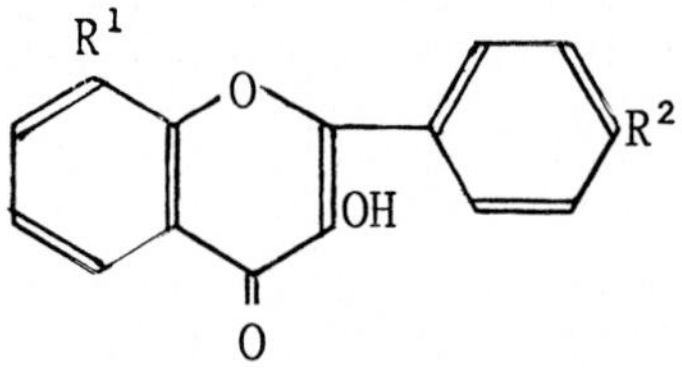

Both 6-and-7-methoxy-2-aryl-3-nitro-2H-benzo[b]pyran (3) on treatment with alkaline hydrogen peroxide afford the related flavonol (4). The 6-methoxyflavonols were previously unknown (Rao *et al.*, *ibid.*, p.1943).

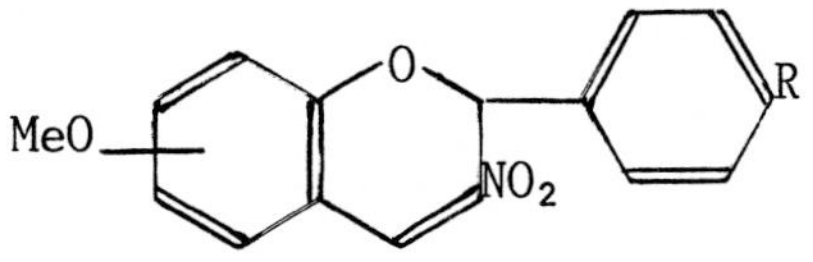

R = H, Me, OMe, Cl

(3)   (4)

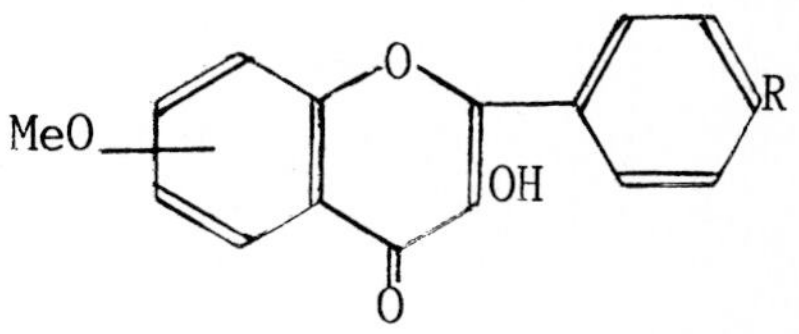

6-Methoxyflavonols are formed when the appropriate 2-aryl--6-methoxy-3-nitro-2H-benzo[b]pyran reacts with chromous chloride solution in tetrahydrafuran under nitrogen (Rao, H.H, Mathur, and Trivedi, Tetrahedron Letters, 1984, 25, 5561).

The cyclocondensation of ω-methoxy-2,4,6-trihydroxy-acetophenone (5) with 2-benzyloxy-3-chlorobenzoic acid anhydride (6) affords the 3-O-methylflavonol derivative (7), which following successive C-8 hydroxylation, methylation, and hydrogenolysis to remove the benzyl group, furnishes 3'-chloro-5,2'-dihydroxy-3,7,8-trimethoxyflavone (chlorofavonin) (8) (A.L. Tokes and R. Bognar, Acta Chim. Acad. Sci. Hung., 1981, 107, 365).

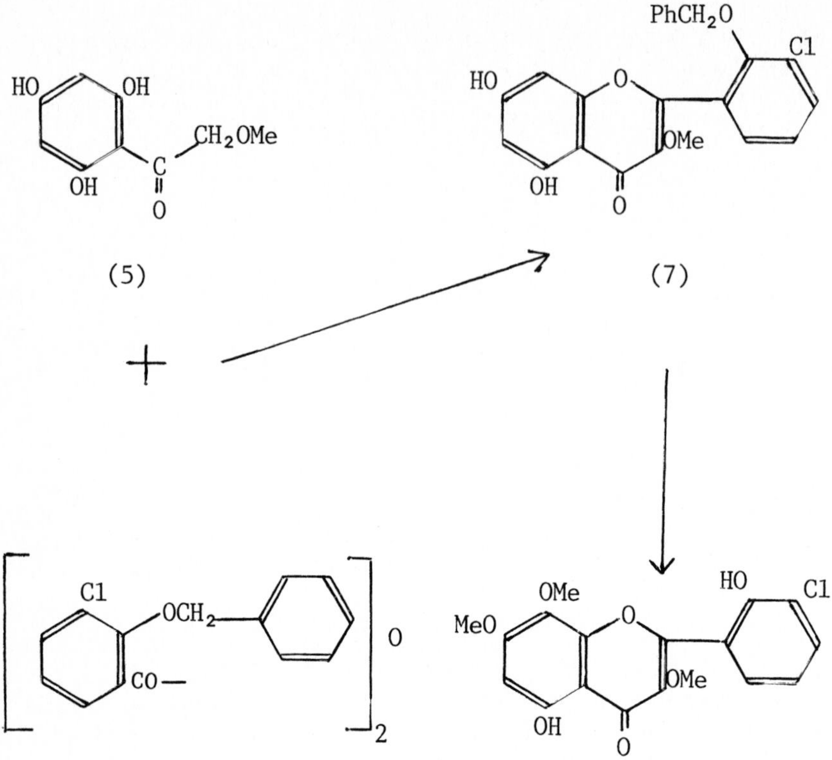

(5)

+

(7)

(6)

(8)

Reactions. Flavonols (1) are oxidized by periodic acid in methanol to the 2-methoxyflavan-3,4-diones (2), isolated as the hemiketals (3), which undergo a straightforward Wittig reaction with the stabilized phosphorane, ethyl triphenylphosphoranylideneacetate (4), in boiling ethanol to give (*E*)-3-(carbethoxymethylene)-2-methoxyflavanones (5). The methoxyflavanones (5) are readily reduced by zinc and acetic acid to give the 3-(carbethoxymethyl)flavones (6) and conjugate addition of bromide and cyanide to the enone system converts them into the substituted (carbethoxymethyl)flavones (7), for example, boiling with hydrogen bromide in ethanol gives the bromo derivative (7; X = Br). The cyano compound (7; X = CN) is readily methylated to derivative (8) or decarboxylated to the 3--cyanomethylflavone (9) (M.A. Smith *et al.*, J. org. Chem., 1982, **47**, 1702).

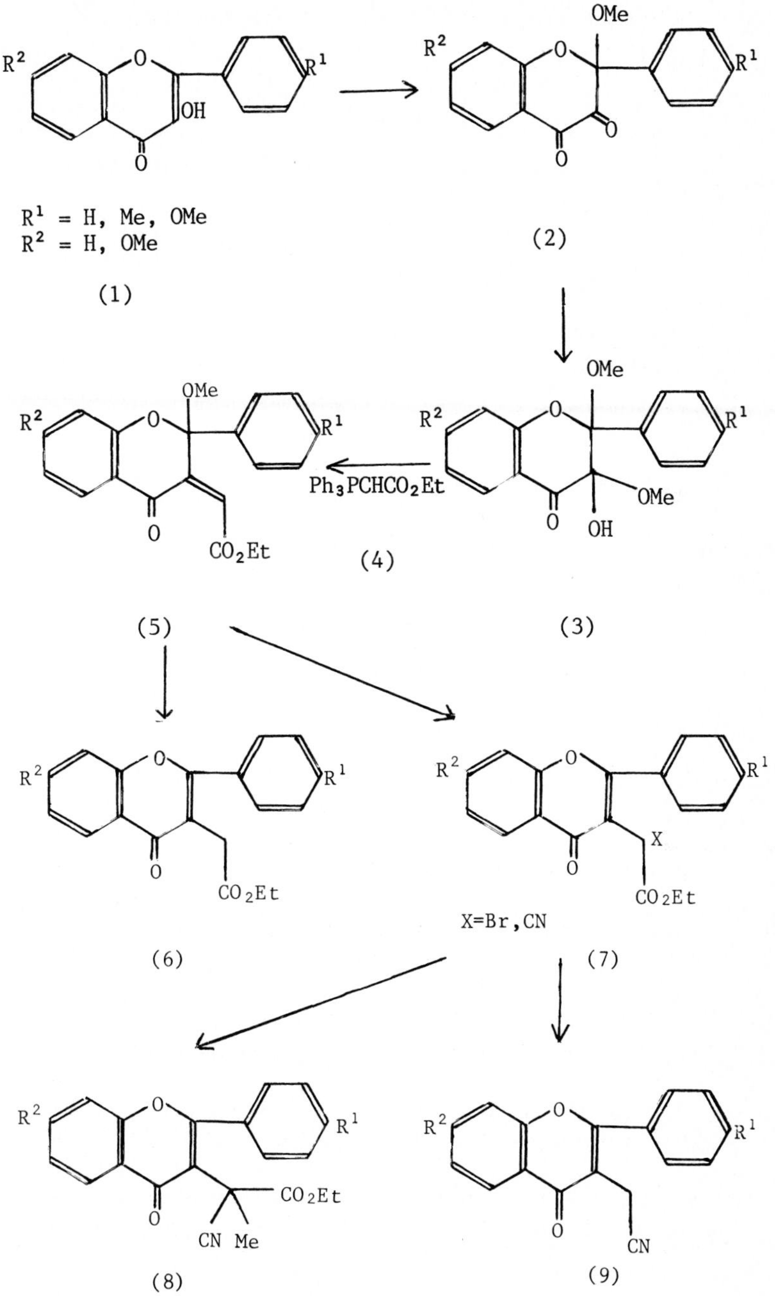

R^1 = H, Me, OMe
R^2 = H, OMe

(1)

(2)

Ph₃PCHCO₂Et

(4)

(5)

(3)

(6)

X=Br,CN

(7)

(8)

(9)

The excited-state dynamics (A.J.G. Standjord *et al*., J. phys. Chem., 1983, **87**, 1125), and the two step laser excitation fluorescence ground- and excited-state proton transfer in flavonol (M. Itot and Y. Fujiwara, *ibid*., p.4558); and the proton-transfer spectroscopy of flavonols (D. McMorrow and M. Kasha, J. Amer. chem. Soc., 1983, **105**, 5133; J. phys. Chem., 1984, **88**, 2235) have been studied, and the ^{13}C-nmr spectral data of a number of flavonols (10; R^1 = H, OH, OMe, R^2 = H; R^1 = R^2 = OMe) (A. Pelter *et al*., J. chem. Soc., Perkin I, 1981, 3182) have been reported. The application of thin layer chromatography, uv-spectroscopy, and acidic treatment have been used in the differentiation of 5,6- and 5,8-dihydroxy-flavones, 3-methoxyflavones, and flavonols trisubstituted on the benzene ring of the benzopyranone (F.A.T. Barberán, F. Ferreres, and F. Tomas, Tetrahedron, 1985, **41**, 5733).

(10)

(xvi) Flavone and flavonol pigments

The flowers, leaves, stems, roots, seeds, and other parts of a large number of plants and trees have been extracted by various solvents to furnish, besides known flavones and flavonols, a number of new ones. Since there have been a large number of such investigations recorded, only a selection of them are now reported.

The syntheses of some flavones and flavonols have been accomplished in order to confirm the structure of new flavonoids, and to provide model compounds to assist in the elucidation of the structure of others. Several studies have been made relating to the application of physico--chemical methods to the identification and quantitative estimation of flavones and flavonols in samples obtained from natural sources. These include the use of various

types of chromatography and spectroscopy, examples of which
are, the fluorescence spectra of various 2'-, 6-, and 8-
-hydroxy-, methoxy-, and glucosyloxy-flavones absorbed on
cellulose alone and with added shift reagents, and the
correlation between fluorescence and structure (H. Geiger
and H. Homberg, Z. Naturforsch., anorg. Chem., org. Chem.,
1983, 38B, 253); uv-spectral studies on chelation of
aluminium chloride with C-methylated derivatives of 2'-
-hydroxyacetophenone and related flavonoids (K.E. Malterud,
Chem. Scr., 1982, 19, 23); structural elucidation of
polymethoxyflavones from shift reagent [1]H-nmr spectroscopic
measurements (P. Joseph-Nathan, D. Abramo-Bruno, and
M.A. Torres, Phytochem., 1981, 20, 313); ^{13}C-nmr spectral
data of apigenin, quercetin, quercitrin, rutin and some
dihydroflavones and dihydroflavonols (K.R. Markham and
B. Ternai, Tetrahedron, 1976, 32, 2607), and of some
flavonoids (P.K. Agrawal and R.P. Rastogi, Heterocycles,
1981, 16, 2181); the mass spectra of a number of
trimethylsilylated flavone and flavanone mono- and di-
-glycosides (H. Schels, H.D. Zinsmeister, and K. Pfleger,
Phytochem., 1978, 17, 523); use of mass spectrometry in the
structural determination of C-glycosylflavones by study of
their permethyl ethers (M.L. Bouillant *et al.*, *ibid.*, p.527);
separation of acetates of flavones, flavonols, and flavanone
aglycones and glycosides on silica gel using different liquid
systems (R. Galensa and K. Herrmann, J. Chromatogr., 1980,
189, 217); and separation of flavonoids by reversed-phase
high-performance liquid chromatography (K. Vande Casteele,
Geiger, and C.F. Van Sumere, *ibid.*, 1982, 240, 81).
Chromatographic methods have been designed for the
determination of nobiletin, *5,6,7,8,3',4'-hexamethoxyflavone*,
tangeretin, *3,5,6,7,4'-pentamethoxyflavone*, tetramethyl-
scutellarein, *5,6,7,4'-tetramethoxyflavone*, sinensetin,
5,6,7,3',4'-pentamethoxyflavone, and heptamethoxyflavone in
orange juice (R. Rouseff and S-V. Ting, Chem. Abs., 1979, 91,
191470q).

Hydroxyflavones with no 3-hydroxyl group
   The following flavones have been isolated and their
structures determined. Conyzorigun, *3',4'-methylenedioxy-
-5,6,7,8,5'-pentamethoxy-flavone* (2), isolated from *Ageratum
conyzoides* plants was originally reported to contain a
novel stable ketene acetal system (1), by virtue of the
oxygen linkage between the chromone and phenyl moieties.

Evidence has now been presented showing that it is a flavone of known structure, namely eupalestin (2) (A.V. Vyas and N.B. Mulchandani, J. chem. Soc., Perkin I, 1984, 2945), previously isolated along with 5'-methoxy-nobiletin, *5,6,7,8,3',4',5'-heptamethoxyflavone*, from the aerial parts of *Eupatorium coelestinum* (Ngo Le-Van and Thi Van Cuong Pham, Phytochem., 1979, 18, 1859).

$$(1) \qquad\qquad (2)$$

6,7-Dimethoxy-5-hydroxyflavone and 5,7-dihydroxy-6--methoxyflavone have been isolated, along with, quercetin, quercetin 3,3'-dimethyl ether, quercetin 3-methyl ether 7--O-glucoside, and 8-methoxyapigenin, from *Carthamus glaucus alexandrinus* (S. Khafagy *et al.*, Acta Pharm. Jugosl., 1979, 29, 161). The geranylated [3; R = Me$_2$C=CH(CH$_2$)$_2$CMe=CHCH$_2$], $C_{26}H_{28}O_6$, m.p. 185-186°, and prenylated (3; R = Me$_2$C=CHCH$_2$) $C_{21}H_{20}O_6$, m.p. 230-231°, derivatives of chrysoeriol, *3'--methoxy-5,7,4'-trihydroxyflavone* (3; R = H), occur in Thailand cannabis (L. Crombie, W.M.L. Crombie, and S.V. Jamieson, Tetrahedron Letters, 1980, 21, 3607). Bausplendin *5,6,3',4'-dimethylenedioxy-7-methoxyflavone*, has been obtained from the wood of *Bauhinia splendens*, an Amazonian creeper. The isomeric 7,8,3',4'-dimethylene-dioxy-5-methoxyflavone has been synthesised (D.O. Laux, G.M. Stefani, and O.R. Gottlieb, Phytochem., 1985, 24,

(3)

(4)

1081), as have 5-hydroxy-7-(3-methyl-2,3-epoxybutoxy)-
flavone, 3,8-dimethoxy-5-hydroxy-7-(3-methyl-2,3-
-epoxybutoxy)flavone, and 4'-hydroxy-5-methoxy-7-(3-methyl-
-2,3-epoxybutoxy)flavone, along with other products, from
the aerial parts of *Achyrocline flaccida* (C. Norbedo,
G. Ferraro, and J.D. Coussio, *ibid.*, 1984, **23**, 2698).   A
di-C-pentosylflavone, from *Molluge pentaphylla*, has been
identified as 6-C-β-D-xylopyranosyl-8-C-α-L-arabino-
pyranosylapigenin, in spite of the 6-C-arabinosyl structure
suggested by the mass spectrum of its permethyl derivative
(J. Chopin, *ibid.*, 1982, **21**, 2367).   In order to verify the
structure of kanzakiflavone-1, *5,8-dihydroxy-6,7-methylene-
dioxy-4'-methoxyflavone*, and kanzakiflavone-2, *5,4'-
-dihydroxy-6,7-methylenedioxyflavone*, from *Iris
unguicularis*, their positional isomers (4; $R^1=R^2=R^5=H$,
$R^3=R^4=OH$, $R^3R^4=OCH_2O$; $R^1=R^5=H$, $R^2=R^3=OH$, $R^4=H$, OH;
$R^1=R^5=Me$, $R^2R^3=OCH_2O$, $R^4=OMe$; $R^1=R^5=Me$, $R^2=OMe$, $R^3R^4=OCH_2O$)
have been synthesized (M. Iinuma, T. Tanaka, and
S. Matsuura, Chem. pharm. Bull., 1984, **32**, 1006).   7,4'-
-Dimethoxy-5,7,3'-trihydroxyflavone has been prepared
(V.K. Sharma, S.K. Garg, and S.R. Gupta, Indian J. Chem.,
1981, **20B**, 991) and is not identical to nuchensein isolated
from *Teucrium nuchense* and previously given than structure
(O.V. Slyun'kova *et al.*, Khim. Prir. Soedin, 1978, 268).

Integrin (5), cyclointegrin (6), and oxyisocyclointegrin
(7) have been isolated from the heartwood of *Artocarpus
integer* (A.D. Pendse *et al.*, Indian J. Chem., 1976, **14B**,
69).

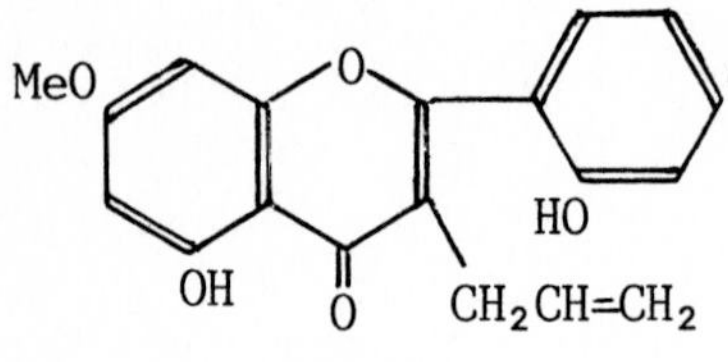

(5)

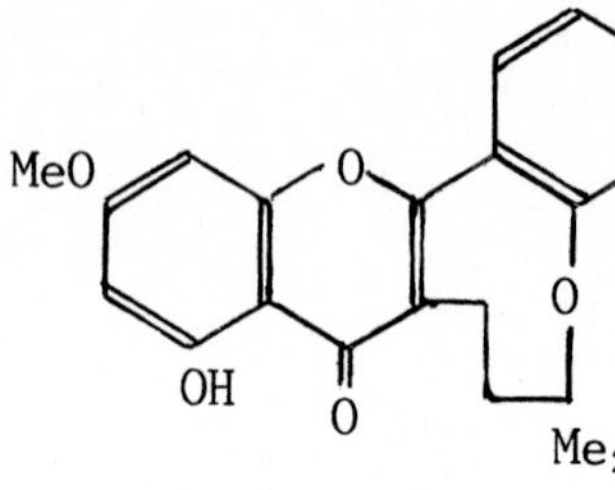

(6)

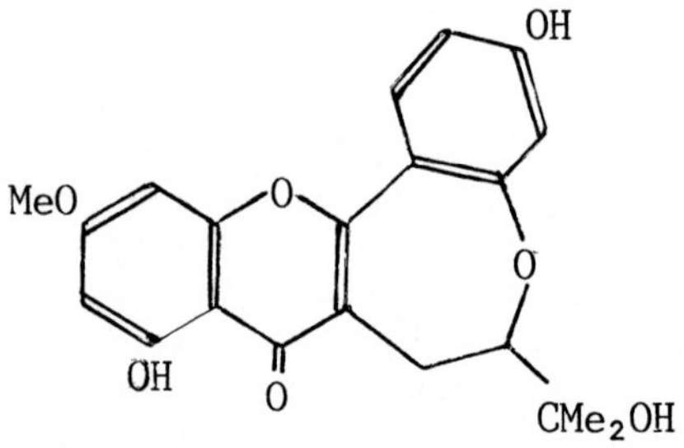

(7)

5,3'-Dihydroxy-7,8,4',5'-tetramethoxyflavone (10) has been synthesized and found to be different from lychnophora flavone-B obtained from *Lychnophora affinis* and given this structure by P.W. LeQuesene *et al*. 2-(3-Benzyloxy-4,5--dimethoxybenzoyloxy)-3,4,6-trimethoxyacetophenone (8) on Baker-Venkataraman rearrangement followed by dehydrocyclization of the resultant β-diketone affords the pentamethoxyflavone (9), which on debenzylation followed by selective demethylation gives (10) (D.K. Bhardwaj *et al*., *ibid*., 1980, **19B**, 309).

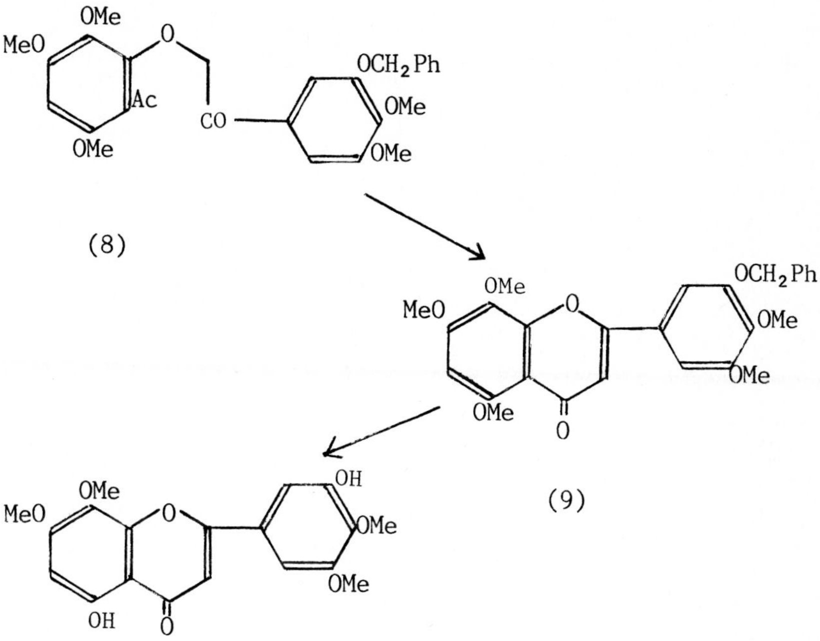

(8)

(9)

(10)

Two *O*-glucosides of C-glucoside flavones, isoorientin
2",4'-di-O-β-D-glucoside and isovitexin *2",4'-di-O-β-D-*
-glucoside have been isolated from the leaves of *Gentiana*
asdepiadea. These are the first examples of di-*O*-
-glucosides of C-glucoside flavones occurring in nature
(M. Goetz and A. Jacot-Guillarmod, Helv., 1977, 60, 1322).
It has been reported that in most *Achillea* species
formation of C-glycosylflavones based either on apigenin or
luteolin rather than of 7-*O*-glycosides occurs, although
only the latter represents a more advanced chemical
character [K. Valant-Vetschera, Stud. org. Chem.
(Amsterdam), 1981 (Pub. 1982)]. Schaftoside (11), m.p.
226°, $[\alpha]_D^{22}$ + 63.7° (Py), and isoschaftoside, have

been isolated as the major di-C-glycosylflavones of the leaves and stems of *Artemisia frigida* (Y.L. Liu *et al.*, Rev. Latinoam. Quim., 1982, **13**, 56). Schaftoside is also a constituent of *Silene schafta* (M.J. Chopin *et al.*, Phytochem., 1974, **13**, 2583).

(11)

Two highly oxygenated flavones, agehoustin C, *5,6,7,8,-2',4',5'-heptamethoxy-3'-hydroxyflavone*, and agehoustin D, *5,3'-dihydroxy-6,7,8,2',4',5'-hexamethoxyflavone*, have been isolated from the aerial parts of *Ageratum houstonianum* (L. Quijano *et al.*, *ibid.*, 1985, **24**, 1085).

The following are some examples of known flavone pigments, which have been isolated from additional sources. 5,7,8-Trimethoxyflavone, 7-hydroxy-3,5,8-trimethoxyflavone, and 3,5,7,8-tetramethoxyflavone have been obtained from *Achyrocline satureoides* var *albicans*, cirsimaritin from *Baccharis eleagnoides*, genkwanin, *apigenin 7-methyl ether*, from *Symphyopappus polystachyus* (A.A.L. Mesquita *et al.*, *ibid.*, 1986, **25**, 1255); apigenin 7,4'-dimethyl ether, genkwanin, and cirsimaritin from the aerial parts of *Salvia sapinae* (R. Pereda-Miranda, G. Delgado, and A. Romo de Vivar, *ibid.*, p.1931); apigenin from *Thuja orientalis* (M. Khabir, F. Khatoon, and W.A. Ansari, Curr. Sci., 1985, **54**, 1180); acacetin, apigenin, apigenin 7-O-glucoside, chrysoeriol, esculetin, luteolin, luteolin 7-O-glucoside, 6-methoxy-luteolin, and scutellarein from *Asphodeline globifera* and

Asphodeline damascena (A. Ulubelen and E. Tuzlaci,
Phytochem., 1985, 24, 2923); apigenin, apigenin 7-
-*O*-glucoside, luteolin, luteolin 7-*O*-glucoside, 6-
-hydroxyluteolin, 6-hydroxyluteolin 7-*O*-glucoside,
scutellarein 7-*O*-glucoside, 4'-methoxy-5,6,7-trihydroxy-
flavone 7-*O*-glycoside, and 7-methoxy-5,6,4'-trihydroxy-
flavone from *Rhinanthus angustifolius* (L. Toth, M. Bulyaki,
and G. Bujtas, Pharmazie, 1986, 41, 72); acacetin, an *O*-
-methylacacetin, and apigenin from *Calea divaricata*
(A.G. Ober, F.R. Fronczek, and N.H. Fischer, J. nat. Prod.,
1985, 48, 302); apigenin, chrysoeriol, *3'-methoxy-5,7,4'-
-trihydroxyflavone*, also from *Eridictyon glutinosum* Benth,
cirsiliol, *6,7-dimethoxy-5,3',4'-trihydroxyflavone*,
cirsimaritin, luteolin, xanthomicrol, *5,4'-dihydroxy-6,7,8-
-trimethoxyflavone*, and 5,4'-dihydroxy-6,7,8,3'-tetra-
methoxyflavone from fifteen species of Spanish *Sideritis*
(F.A.T. Barberan, J.M. Nunez, and F. Tomas, Phytochem.,
1985, 24, 1285); apigenin, luteolin, luteolin 7-*O*-
-glucoside, and vitexin from *Launaea asplenifolia*
(D.R. Gupta, R. Bhushan, and B. Ahmed, Khim. Prir. Soedin.,
1985, 408); 6-methoxygenkwania, salvigenin, and tambuletin
from the leaves of *Achillea depressa* collected in central
Bulgaria (E. Tsankova and I. Ognyanov, Planta Med., 1985,
180); lanarin, *acacetin 7-β-rutinoside*, from the leaves of
Cirsium albescens, *C. japonicum* var *australe* and *C. kawakam*
f. *variegatum*, the leaves of *C. hosogawa* also afford luteolin
and luteolin 7-*O*-glucoside besides linarin (T.S. Wu,
C.S. Kuoh, and S.I. Jen, T'ai-wan Yao Hsueh Tsa Chih, 1981,
32, 88); 6-methoxy-4'-methylapigenin along with flavonols
from the leaves of *Centaurea collina* L. (K. Kamanzi,
J. Raynaud, and B. Voirin, Plant. med. Phytother., 1982, 16,
30); the di-C-glycosylflavones, lucenin-2, carlinoside,
isoviolanthin, schaftoside, isoschaftoside, and vicenin-1,
and the mono-C-glycosylflavones, orientin, isoorientin,
vitexin, swertiajaponin, isoswertiajaponin, and isoswertisin,
besides luteolin, and luteolin 7-*O*-glucoside from the leaves
of *Passiflora sexflora* (S. McCormick and T.J. Mabry, J. nat.
Prod., 1982, 45, 782); apigenin, luteolin, and luteolin 7-*O*-
-glucoside from *Achillea pannonica* during flowering
(N.A. Kaloshina and A.V. Mazulin, Khim. Prir. Soedin., 1983,
105); cirsilineol, chrysoeriol and tricin from the leaves
plus stems of flowering *Artemisia rutifolia* (I.I. Chemesova,
L.M. Belenovskaya, and L.P. Markova, *ibid.*, 1984, 249);
luteolin and tricin, and in addition a variety of methyl
ethers from the leaves and inflorescences during a survey of

170 Australian species of Cyperaceae belonging to 35 genera, besides luteolin 5-methyl ether, luteolin 7-methyl ether, diosmetin, and acacetin were detected for the first time in the Cyperacaea (J.B. Harborne, C.A. Williams, and K.L. Wilson, Phytochem., 1985, 24, 751); 8-hydroxydiosmetin 8-glucoside and the potassium salt of 3'-sulpho-8-hydroxy-diosmetin 8-glucoside from the leaves of *Althaea officinalis* (J. Gudej, Acta Pol. Pharm., 1981, 38, 385); and the C-glycosylflavones, schaftoside, isoschaftoside, neoschaftoside, neoisoschaftoside, isovitexin, isoorientin, and sweritisin 2"-*O*-α-L-rhamnoside from the leaves of *Gemmingia chinensis* (S. Shirane *et al*., Agric. biol. Chem., 1982, 46, 2595).

Flavone and several methoxyflavones of unusual oxygenation pattern have been isolated from *Pimela simplex* and *Pimela decora*. The former yielded 2'-methoxyflavone and 7-methoxy-flavone, and the latter, 3'-methoxyflavone, 6-methoxyflavone, 6,3'-dimethoxyflavone, and 6,2',3'-trimethoxyflavone. The structures have been confirmed by synthesis from the appropriate chalcones and the ^{13}C-nmr spectral data of the flavones and chalcones have been reported (P.W. Freeman *et al*., Austral. J. Chem., 1981, 34, 1779).

Hydroxyflavones with a 3-hydroxyl group, hydroxyflavonols

Milimorin (1), $C_{16}H_{12}O_7$ (M^+, 316), m.p. 235-7°, has been isolated from the aerial parts of *Euphorbia millii* along with kaempferol. The tetramethyl ether of milimorin and the pentamethyl ether of morin are identical. Thus milimorin is 4'-*O*-methylmorin and its mass spectrum shows a peak at m/z 299 due to ion (2) (R.L. Khosa, N. Bhatia, and M. Sahai, Chem. and Ind., 1984, 881).

(1)   (2)

Gomphrenol, *5,4'-dihydroxy-6,7-methylenedioxyflavonol*, obtained from the leaves of *Gomphrena globosa* (Amaranthaceae) (M.L. Bouillant *et al.*, Phytochem., 1978, **17**, 2138). Sexangularetin, *8-methoxy-5,7,4'-trihydroxyflavonol*, obtained from the flowers and young fruit of *Epilobium angustifolium* L. along with kaempferol, quercetin, and myricetin (J. Reynaud *et al.*, Plant. med. Phyother., 1982, **16**, 120). 5,7-Dihydroxy-3,6,8-trimethoxyflavone has been isolated from the flowers of *Gnaphalium elegans* (R.D. Torrenegra *et al.*, Phytochem., 1980, **19**, 2795); two novel flavonols, 5,7,4'-trihydroxy-3,6,8-
-trimethoxyflavone and its methyl ether, 5,4'-dihydroxy-
-3,6,7,8-tetramethoxyflavone from *Baccharis sarothroides* (E. Wollenweber *et al.*, Z. Naturforsch., Biosci., 1986, **41C**, 87); a rarely methylated new flavonol, 3,5-di-*O*-
-methylkaempferol along with 3,7-di-*O*-methylkaempferol from *Linaria dalmatica* (R. Kapoor, A.K. Rishi, and C.K. Atal, Fitoterapia, 1985, **56**, 296); quercetagetin 6,3',4'-trimethyl ether, a new derivative of quercetagetin, and its 3-potassium sulphate salt along with three other known methoxylated flavonols from *Decachaeta haenkeana* (M. Miski, D.A. Gage, and T.J. Mabry, Phytochem., 1985, **24**, 3078); a novel flavonol, 5,4',5'-trihydroxy-3,7,3'-trimethoxyflavone and 5,4'-dihydroxy-3,7,3',5'-tetramethoxyflavone from the leaves of *Solanum pubescens* (G.N.K. Kumari, L.J.M. Rao, and

N.S.P. Rao, *ibid.*, 1984, <u>23</u>, 2701); two new flavonol
glucosides, herbacetin 8-methyl ether 3-*O*-glucoside-7-*O*-
-rutinoside and herbacetin 7-*O*-(6"-quinylglucoside), along
with known flavonoids from *Ephedra alata* (M.A.M. Nawwar,
H.I. El-Sissi, and H.H. Barakat, *ibid.*, p.2937); for the
first time in nature, 5,7-dihydroxy-6,3'4'-trimethoxyflavonol
from *Arnica chamissonis chamissonis* (I. Merfort, Planta Med.,
1985, 136); quercetin 3-rhamnoside 2"-gallate (3) from
Polygonum filiforme (T. Isobe *et al.*, Bull. chem. Soc. Japan,
1981, <u>54</u>, 3239); 3-*O*-methylquercetagetin, 7-*O*-glucoside and

(3)

3,3'-di-*O*-methylquercetin 7-*O*-glucoside and fifteen known
glycosides from *Desmanthodium fruticosum, D. hondurense,*
and *D. perfoliatum* (B.A. Bohm and T.F. Stuessy, Phytochem.,
1981, <u>20</u>, 1573); and quercetagetin 3,5,7-trimethyl ether
and 3,5,7,3'-tetramethyl ether from *Pulicaria arabica*
leaves and flowers (S.I. El-Negoumy, R.M.A. Mansour, and
N.A.M. Saleh, *ibid.*, 1982, <u>21</u>, 953). Comparison of all
known flavonol triosides with those from four Rhamnus

species, using ^{13}C-nmr spectroscopy, shows that all
flavonol triosides so far isolated contain the sugar moiety
rhamnose. Thus catharticin is identical to alaternin,
xanthorhamnin A to xanthorhamnin B, and xanthorhamnin C is
rhamnazin-3-O-β-rhamninoside (I. Riess-Maurer and H. Wagner,
Tetrahedron, 1982, __38__, 1269).

A number of known flavonols and some of their derivatives
have been obtained from new sources, including kaempferol,
its 3-arabinoside and 3-galactoside, quercetin and its 3-
-arabinoside from the leaves of *Aganosma caryophyllata* along
with other products (K.S. Ramana *et al.*, Indian J. pharm.
Sci., 1985, __47__, 165); quercetin, quercetin 3,7-dimethyl
ether, quercitrin, rutin, kaempferol, ayanin, 5,7-dihydroxy-
-3,3',4',5'-tetramethoxyflavone and 5,3',5'-trihydroxy-
-3,7,4'-trimethoxyflavone from the aerial parts of
Gutierrezia alamanii var. *megalocephala* (A. Lenherr,
N. Fang, and Mabry, J. nat. Prod., 1986, __49__, 185); sixteen
6-methoxylated and non-6-methoxylated flavonol derivatives
including, 6-methoxykaempferol, patuletin, eupatolitin,
quercetin, and their 3-O-glucosides and 3-O-galactosides,
penduletin, ombuin, and eupalitin and its 3-O-galactoside
from *Eupatorium areolare* var. *leiocarpum* (S.G. Yu *et al.*,
ibid., p.181); quercetin following the hydrolysis of
flavonoids from *Arnica* flowers [Z.P. Kostennikova,
G.A. Panova, and R.M. Dolotenkova, Farmatsiya (Moscow),
1985, __34__, 51]; the rarely occurring flavonoid aglycones,
laciniatin and dillenetin, *5,7-dihydroxy-3',4'-dimethoxy-
flavonol* [originally isolated from *Dillenia indica* L., m.p.
290-292° (L. Jurd, J. org. Chem., 1962, __27__, 1294;
M. Krishnamoorthy *et al.*, Indian J. Chem., 1965, __3__, 270;
G. Pavanasasivam and M.U.S. Sultanbawa, J. chem. Soc.,
Perkin I, 1975, 612)] and other known flavonols from *Arnica
montana, A. chamissonis foliosa* variety incana, and
A. chamissonis chamissonis (Merfort, *loc. cit.*);
kaempferol, quercetin, and quercetagentin derivations from
eight species of *Clibadium*, including O-methylated
quercetagetin derivatives in several taxa, 6-methoxykaempferol,
tentatively identified in one collection, kaempferol and
quercetin, as 3-O-glucosides, galactosides, rhamnosides,
rutinosides, and diglucosides, but all glycosides did not
occur in each taxon, and quercetagetin derivations
occurring as 7-O-glucosides (Bohm and Stuessy, Phytochem.,
1985, __24__, 2134); kaempferol and quercetin along with
anthocyanins from the petals of flowers of the cultivars
and hybrids of cultivated carnations (V.K. Parshikov, Fiziol.

Biokhim. Kul't. Rast., 1984, 16, 483); isoquercitrin,
myricetin, and quercetin along with other compounds from
the leaves of *Salix pantosericea* and *S. pentandroides*
(V.A. Kompantsev, Khim. Prir. Soedin., 1984, 654);
hispidulin, axillarin, and 3,4'-dimethoxy-5,7,3'-trihydroxy-
flavone from *Tanacetum sibiricum* (T.A. Stepanov *et al.*,
Khim. Prir. Soedin., 1981, 519); quercetagetin 6,7,3',4'-
-tetramethyl ether and quercetagetin 6,7,4'-trimethyl ether
from *Artemisia annua* L., a medicinal plant with antipyretic
activity (H-M. Liu, G-L. Li, and H-Z. Wu, Yao Hsueh Hsueh Pao,
1981, 16, 65); kaempferol 3-glucoside, kaempferol 3-β-D-(6"-
-*O*-*p*-hydroxycinnamoyl)-glucose, and quercetin 3-glucoside
from the leaves of *Althaea officinalis* (J. Gudej, *loc. cit.*);
quercetagetin 3,7-dimethyl ether, quercetin 3-glucoside, and
quercetin 3-glucuronide from leaves and flowers of *Pulicaria
arabica* (El-Negoumy, Mansour, and Saleh, *loc. cit.*);
kaempferol and quercimeritrin from flowers and roots of
Cephalaria kotschyi and *C. nachiczevanica* (A.M. Aliev and
I.S. Movsumov, Azerb. Med. Zh., 1981, 58, 36); quercetin 3,7-
-*O*-diglucoside, quercetin 3-*O*-glucoside-7-*O*-rhamnoside,
quercetin 3-*O*-glucoside-7-*O*-rutinoside, rutin, and quercetin
7-*O*-rhamnoside from the leaves of *Orchis papilionacea* L.
(F. Pagani, Boll. chim. Farm., 1982, 121, 174); quercetin,
kaempferol, and 6-methoxyquercetin from the leaves of
Centaurea collina L. (Kamanzi, Raynaud, and Voirin, *loc.
cit.*); quercetin and avicularin from the leaves of *Polygonum
sachalinense* (S.S. Kang, Soul Taehakkyo Saengyak Yonguso
Opjukjip, 1981, 20, 55); quercetin 5-*O*-glycoside,
isorhamnetin 5-*O*-glycoside, and quercetin 3,3'-dimethyl ether
from the roots of *Rhaponticum carthamoides* (E. Varga,
K. Szendrei, and J. Reisch, Fitoterapia, 1982, 53, 9);
isorhamnetin, kaempferol, rutin, and trifolin from *Achillea
pannonica* during flowering (Kaloshina and Mazulin, *loc. cit.*);
chrysosplenetin, *5,4'-dihydroxy-3,6,7,3'-tetramethoxyflavone*
from leaves plus stems of flowering *Artemisia rutifolia*
(Chemesova, Belenovskaya, and Markova *loc. cit.*); and
quercetin, kaempferol 3-*O*-rhamnoside, kaempferol 3-*O*-
-rhamnoglucoside, and rutin along with other products from
leaves of *Jasminum azoricum* (S.A. Ross and M.A. Abdel-Hafiz,
J. nat. Prod., 1984, 47, 736).

A study has been made of the flavonoids in flowers of
White Masterpiece, Bridal Pink, and Samantha roses (*Rosa*
spp). They were either kaempferol or quercetin 3-
-glycosides (glucoside, xyloside, arabinoside, rhamnoside,
glucuronide, rutinoside, a rhamnosylglucoside, an acylated
form of the rhamnosylglucoside, and galloylglucoside), and
kaempferol 4'-glucoside obtained almost exclusively from
old Scottish roses, *R. foetida* and *R. spinossisima* was only
present in White Masterpiece (S Asen, J. Amer. Soc. hortic.
Sci., 1982, 107, 744). Methylation of myricetin with
dimethyl sulphate affords 3,7,3',4'-tetramethylmyricetin
and 3,7,4'-trimethylmyricetin, which on conversion into its
triethyl ether and subsequent alkali hydrolysis yields
2,4'-dimethoxy-2'-ethoxy-6'-hydroxyacetophenone (4) and
3,5-diethyl-4-methylgallic acid. Similar methylation of
quercetin gives 7,4'-dimethyl-, 3,7,4'-trimethyl-, and
3,7,3',4'-tetramethyl-quercetin, and ethylation of the
trimethyl derivative followed by hydrolysis yields the
acetophenone (4) and *O*-ethylisovanillic acid (L. Chand,
R. Maurya, and A.B. Ray, J. Indian chem. Soc., 1982, 59,
1001).

(4)

R = H   54%
R = OH   64%

(5)   (6)

The Birch reduction of pentamethylquercetin (5; R = OMe)
yields the dihydrochalcone (6; R = H), whereas similar
reduction of tetramethylquercetin (5; R = OH) affords α-
-hydroxydihydrochalcone (6; R = OH) (J.G. Sweeny,
T. Radford, and G.A. Iacobucci, Heterocycles, 1977, 7, 57).

(xvii) Biflavonyls

The following biflavonyls have been reported and their structures determined, in most cases, by the application of chemical and spectroscopic methods. 4' Or 4''', 7-di-*O*--methylcupressuflavone, 7,7'',4'''-tri-*O*-methylagathisflavone, and the biflavones (1) are obtained from the leaves of *Araucaria excelsa* along with 7,7''-di-*O*-methylagathisflavone, 7,4',7'',4'''-tetra-*O*-methylcupressuflavone, biflavonyl (1; $R^1=R^2=R^3=Me$), and other known biflavonyls (N. Ilyas *et al.*, Phytochem., 1978, 17, 987).

$$R^1 = R^2 = H, Me, R^3 = H$$
$$R^1 = Me, R^2 = R^3 = H$$

(1)

Strychnobiflavone (2) is isolated from the leaves of *Strychnos pseudoquina* (Loganiaceae) together with isorhamnetin and the alkaloids, diaboline and 11-methoxy-diaboline (M. Nicoletti *et al.*, J. nat. Prod., 1984, 47, 953).

Taiwaniaflavone (3) and its methyl ethers (4) are obtained from the leaves of *Taiwania cryptomerioides* Hayata (Taxodiaceae) and their structures determined from spectral data and by the preparation of the taiwaniaflavone ethers (5). Isolated from the same source are the known biflavonyls, amentoflavone, 7,7''-di-*O*-methylamentoflavone, sequoiaflavone, hinokiflavone, and isocryptomerin (M. Kamil *et al.*, J. chem. Soc., Perkin 1, 1981, 553).

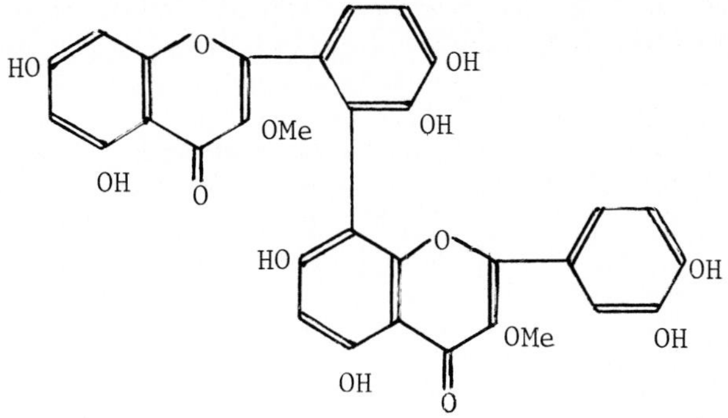

(2)

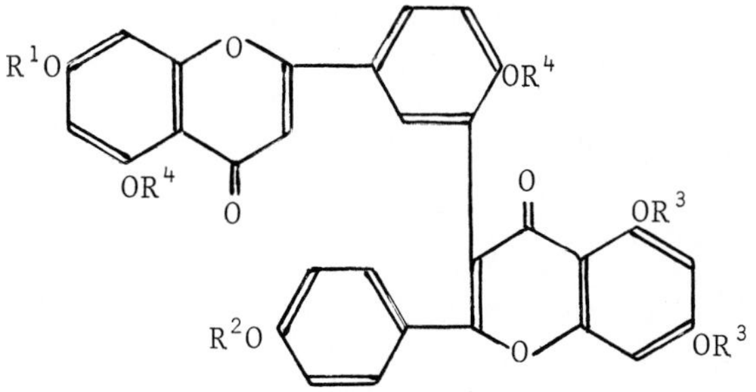

(3) $R^1=R^2=R^3=R^4=H$
(4) R^1=Me, R^2=H, Me, $R^3=R^4$=H
(5) $R^1=R^2=R^4$=Me, R^3=Me, Et

A biflavonyl (6) isolated from the aerial parts of
Hypericum aucheri is a 3,8"-bisapigenin, *5,7,4',5",7",4'''-
-hexahydroxy-3,8"-biflavonyl* and is the first biflavonyl
discovered in the genus *Hypericum*, suggesting a connection
between the Hypericoideae and other subfamilies of the
family Guttiferae, which are known to contain biflavonyls
[G. Kitanov, Farmatsiya (Sofia), 1985, **35**, 13].

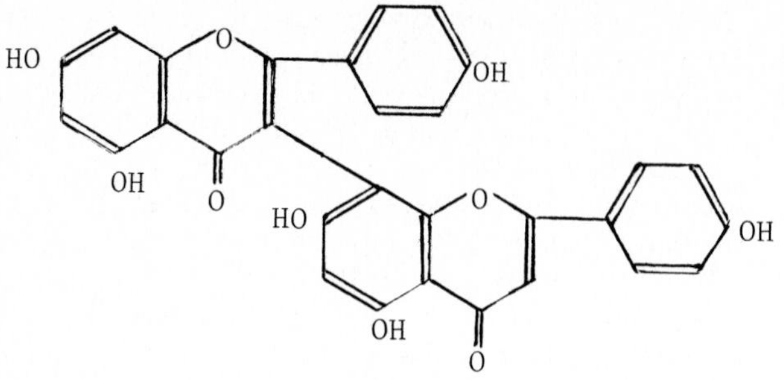

(6)

Podocarpusflavone A (7) and bilobetin (8) together with amentoflavone and isoginkgetin are found in the leaves of *Podocarpus elongatus* (J.S. Prasad and H.G. Krishnamurty, Indian J. Chem., 1976, 14B, 727). 5'-Methoxybilobetin has been isolated from *Ginkgo biloba* (M. Joly, M. Haag-Burrurier and R. Anton, Phytochem., 1980, 19, 1999).

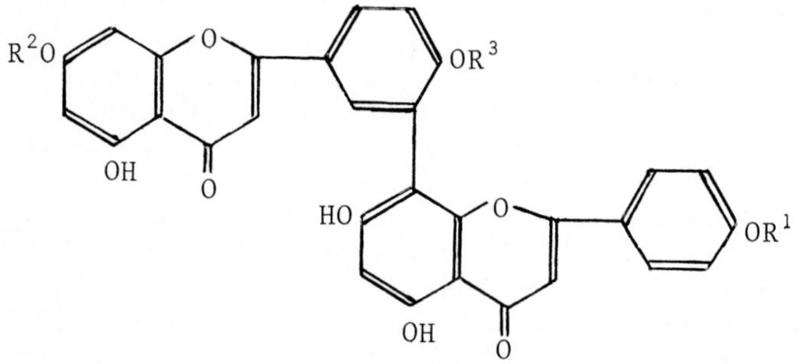

(7) R^1=Me, R^2=R^3=H
(8) R^1=R^2=H, R^3=Me

A number of known biflavonyls including the following, have
been obtained from new sources; amentoflavone, cupressuflavone,
hinokiflavone, isocryptomerin, and two unidentified biflavonyls
from the dry leaves of *Juniperus indica* (Syn. *J. pseudosabina*)
(F. Khatoon *et al.*, J. Indian chem. Soc., 1985, <u>62</u>, 410);
amentoflavone, sciadopitysin, and two partially identified
biflavonyls, mono- and di-*O*-methylamentoflavones from the
leaves of Himalayan yew, *Taxus wallichiana* (N. Parveen,
H.M. Taufeeq and N.U.D. Khan, J. nat. Prod., 1985, <u>48</u>, 994);
amentoflavone, cupressuflavone, and agathisflavone, as their
hexa-*O*-methyl ethers from the leaves of *Picea morinda* Linn
(A. Azam, M.A. Qasim, and M.S.Y. Khan, J. Indian chem. Soc.,
1985, <u>62</u>, 788); also along with hinokiflavone and isocryptomerin from the leaves of *Juniperus virginiana* (S.K. Roy *et al.*,
ibid., 1984, <u>61</u>, 172); amentoflavone, sequoiaflavone, and
cupressuflavone from the leaves of *Cupressus gracilis* and *C.
macrocarpa* (Qasim *et al.*, *ibid.*, 1985, <u>62</u>, 170); amentoflavone
and its mono-, di-, tri-, and tetra-methyl ethers from the
leaves of *Encephalartos woodii*, *E. transvenosus*, and *E.
villosus* Lam. (F. Mohammad *et al.*, Indian J. Chem., 1983,
<u>22B</u>, 184); also its mono- and di-methyl ethers from the
fresh leaves of *Ochrocarpus longifolius*, which is the first
report of the occurrence of biflavonyls in the genus
Ochrocarpus (Roy *et al.*, *ibid.*, p.609); biflavonyls based on
amentoflavone, cupressuflavone, and hinokiflavone, form the
major ones from all eleven genera of the Callitroideae, the
uneven distribution allows the distinction of five groups
within the genera (P.A. Gadek and C.J. Quinn, Phytochem.,
1983, <u>22</u>, 969); podocarpusflavone A, sciadopitysin, hinokiflavone, cryptomerin A, isocryptomerin, and cryptomerin B
from the leaves of *Taxodium mucronatum* (K. Ishratullah *et al.*,
ibid., 1978, <u>17</u>, 335); amentoflavone, cupressuflavone, 4'-*O*-
-methylcupressuflavone, and isocryptomerin from the leaves of
Cupressus govaniana (Cupressaceae) (Taufeeq, F. Mohd, and
M. Ilyas, Indian J. Chem., 1979, <u>17B</u>, 535); amentoflavone,
cupressuflavone, mono-*O*-methylamentoflavone, mono-*O*-methyl-
hinokiflavone and robustaflavone from the leaves of *Cupressus
australis (Callitris rhomoboidea)*, the latter being found for
the first time in the genus *Cupressus* of the family
Cupressaceae (N.A. Khan, Kamil, and Ilyas, *ibid.*, p.536);
cupressuflavone along with other flavonoids from the leaves
of *Thuja orientalis* (M. Khabir, Khatoon, and W.H. Ansari,
Curr. Sci., 1985, <u>54</u>, 1180); and volkensiflavone and
morelloflavone with other compounds from the heartwood of
Garcinia indica (P.J. Cotterill, F. Scheinmann, and
G.S. Puranik, Phytochem., 1977, <u>16</u>, 148).

Amentoflavone, bilobetin, isoginkgetin, ginkgetin, and sciadopitysin have been determined in *Ginkgo biloba* L. by high performance liquid chromatography with spectrophoto-metric detection and details of the best two separation systems, out of several different stationary phases used, have been reported (F. Briancon-Scheid, A. Lobstein-Guth, and R. Anton, J. Chromatogr., 1982, <u>245</u>, 261; Planta Med., 1983, <u>49</u>, 204). Amentoflavone and cupressuflavone have been converted into their benzyl ethers and separated by chromatography on silica gel and their ir-, nmr-, and mass--spectral data have been given (Taufeeq *et al.*, Proc. natl. Acad. Sci., India, 1982, <u>52A</u>, 43). The fluorescence excitation and emission spectra of cupressuflavone hexamethyl ether, hinokiflavone pentamethyl ether, and amentoflavone hexamethyl ether in a few typical solvents (A. Sharma, M.K. Machwe, and V.V.S. Murti, Curr. Sci., 1983, <u>52</u>, 858) have been discussed.

The interflavonoid linkage in biflavonyls can be determined by ^{13}C-nmr spectroscopy, on the basis of the chemical shifts and their multiplicities in the off--resonance spectra, provided that the A ring is involved, for example, in cupressuflavone and agathisflavone (V.M. Chari *et al.*, Phytochem., 1977, <u>16</u>, 1273). The ^{13}C-nmr signals for most of the carbon atoms of morelloflavone as well as methylated derivatives have been assigned by comparison with the spectra of the corresponding monomers (H. Duddeck, G. Snatzke, and S.S. Yemul, *ibid.*, 1978, <u>17</u>, 1369).

A new biflavonyl ether isolated from the leaves of *Ochna pumilla* by acetone extraction has been assigned structure (9) on the basis of its [1]H-nmr spectrum (Kamil *et al.*, Indian J. Chem., 1983, <u>22B</u>, 608).

(9)

(xviii) Furoflavones

Pongaglabol, *5-hydroxyfuro[2,3-h]flavone* (1), a new hydroxyfuroflavone has been obtained from *Pongamia glabra* (S.K. Talapatra, A.K. Mallik, and B. Talapatra, Phytochem., 1980, <u>19</u>, 1199).

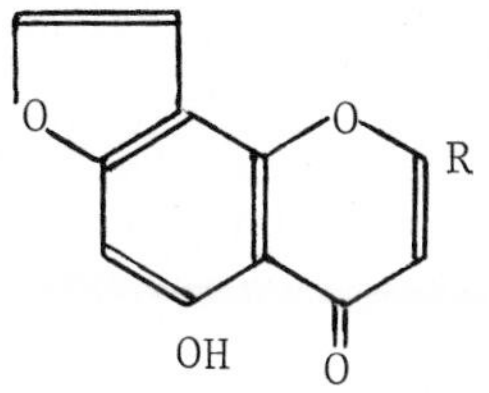

(1) R=Ph
(4) R=3-HOC$_6$H$_4$

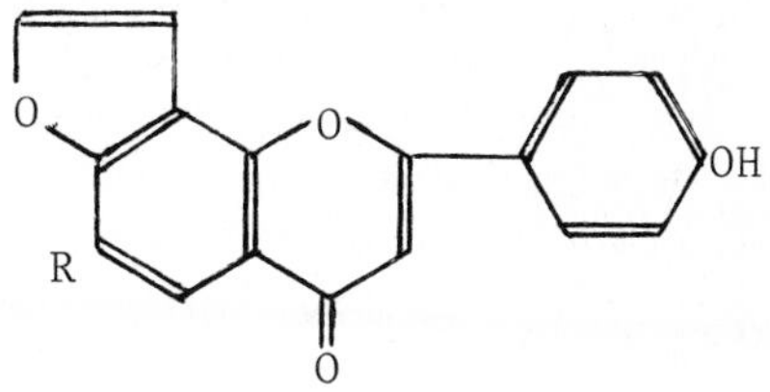

(2) R=H
(3) R=OMe

Other new furoflavones obtained from *Pongamia glabra* include two new hydroxy derivatives, isopongaglabol (2) and its methoxy derivative (3), whose structures have been confirmed by synthesis (*idem, ibid.*, 1982, <u>21</u>, 761); pongol, *3'-hydroxyfuro[2,3-h]flavone* (4), structure confirmed by synthesis, and also similarly prepared 2'-hydroxyfuro[2,3-h]-flavone (D. Roy and R.N. Khana, Indian J. Chem., 1979, <u>18B</u>, 525); and glabra-II (5) along with the benzofuran derivative, glabra-I (6) (G.P. Garg, N.N. Sharma, and Khanna, *ibid.*, 1978, <u>16B</u>, 658). Glabra-I on treatment with hydrochloric acid is converted into pongaglabrone and the hydrolysis of glabra-II affords myristicic acid and 5--acetyl-4-hydroxycoumarone.

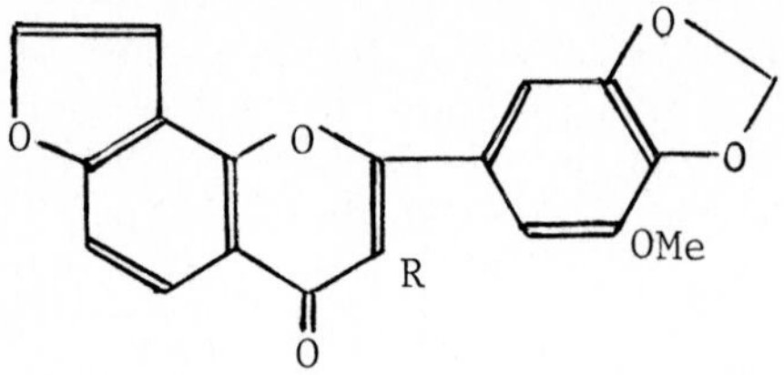

(5) R=H
(7) R=OMe

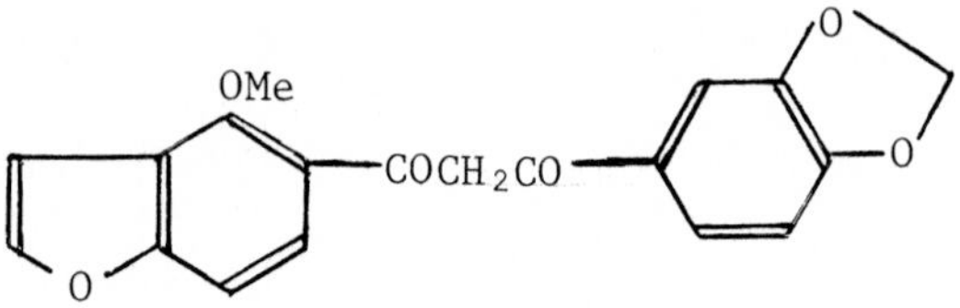

(6)

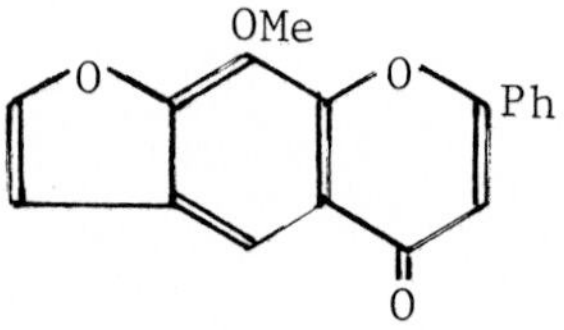

(8)

6-Methoxyisopongaglabol (3) is synthesized from 4-
-allyloxy-2-hydroxy-5-methoxyacetophenone, which on
esterification with 4-benzyloxybenzoyl chloride, followed
by Baker migration and cyclization affords 7-allyloxy-4'-
-benzyloxy-6-methoxyflavone. The latter compound on Claisen
migration furnishes 8-allyl-4'-benzyloxy-7-hydroxy-6-methoxy-
flavone, which on oxidation with osmium tetroxide/potassium
periodate, followed by cyclodehydration using polyphosphoric
acid gives 6-methoxyisopongaglabol (3) (U. Arora, *ibid*.,
1984, 23B, 373).

Also isolated from *Pongamia glabra* are 3'-methoxyponapin
(7) and 9-methoxyfuro[3,2-g]flavone (8). The latter has
been synthesized by condensing Karajanin ketone with
myristic anhydride in the presence of sodium myristicate
using the Allan-Robinson method (S.B. Malik, P. Sharma, and
T.R. Seshadri, *ibid*., 1977, 15B, 536).

Karanjin, pongaglabrone, pongapin, lanceolatin B, and two
new furoflavones, 4'-methoxyfuro[2,3-h]flavone (9), $C_{18}H_{12}O_4$,
m.p. 218-219°, and 3,4'-dimethoxyfuro[2,3-h]flavone (10),
$C_{19}H_{14}O_5$, m.p. 170-171°, have been isolated *Derris mollis*
roots. Alkaline hydrolysis of (9) affords anisic acid and
karanjic acid together with a diketophenol, $C_{18}H_{14}O_5$,
m.p. 148-149°, and that of (10) yields anisic and
karanjic acids together with karanjin ketone. Mass
spectral fragmentation data have been given for
furoflavones (4 and 5), karanjin, pongaglabrone, pongapin
and lanceolatin B (D.A. Lyra *et al.*, Gazz., 1979, <u>109</u>, 93).

(9) R=H
(10) R=OMe

Karanjin and lanceolatin-B have been isolated from the
seeds of *Tephrosia purpurea* (B. Sinha, A.A. Natu, and
D.D. Nanavati, Phytochem., 1982, <u>21</u>, 1468) and the latter
along with other compounds from the roots of *Tephrosia
falsiformis* (A. Ghanim and I. Jayaraman, Indian, J. Chem.,
1979, <u>17B</u>, 648).

For the selective demethylation of 5-methoxyfuroflavones
see Talapatra, Mallik, and Talapatra (J. Indian chem. Soc.,
1985, <u>62</u>, 408) and for the synthesis of some 9-substituted
5-methoxyfuro[3,2-g]flavones see O.H. Hishmat *et al.*, [Egypt.
J. Chem., 1984 (Pub. 1985), <u>27</u>, 831]. The ^{13}C-nmr spectra
of the angular furoflavones, lanceolatin-B, karanjin,
kanjone, and 5-methoxyfuro[2,3-h]flavone and the linear
furoflavones, 5-methoxyfuro[3,2-g]flavone have been analyzed

and chemical shifts assigned (Talapatra, Mallik, and Talapatra, J. Indian chem. Soc., 1982, <u>59</u>, 534). Also reported are the ^{13}C-nmr spectral data of the furo[2,3-h]-flavones (11) (S.C. Gupta *et al*., Org. mag. Reson., 1982, <u>20</u>, 222).

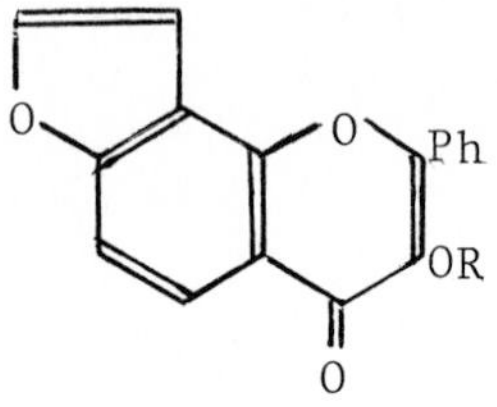

R=H,Me,Ac,CH$_2$=CHCH$_2$

(11)

(xix) Pyranoflavones

Isopongaflavone, *8,8-dimethyl-5-methoxy-2-phenylpyrano-[2,3-h]flavone* (1), is obtained from immature seeds of *Pongamia glabra* along with the known furoflavones karanjin, pongapin, lanceolatin B, and kanjone. Its structure has been determined from spectral data and confirmed by synthesis from chrysin. Chrysin undergoes prenylation to furnish 8-C-prenylchrysin, which on cyclization affords isopongaflavone (1). 8,8-Dimethyl-5-methoxy-2-phenylpyrano-[3,2-g]flavone has been synthesized (2) (D. Roy, N.N. Sharma, and R.N. Khanna, Indian J. Chem., 1977, <u>15B</u>, 1138).

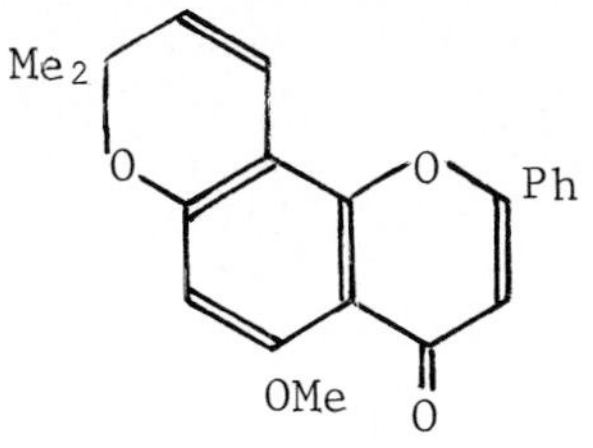

(1) 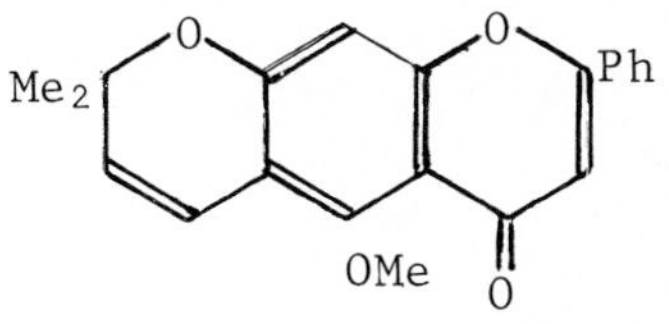   (2)

Morusinol (3), an isoprenoid flavone has been isolated from *Morus* root barks and its structure elucidated by comparison of its ^{13}C-nmr spectrum with that of other isoprene-substituted flavonoids (C. Konno, Y. Oshima, and H. Hikino, Planta Med., 1977, 32, 118). Erythrisenegalone (4) occurs in the stem bark of *Erythrina senegalensis* (Z.T. Fomum, J.F. Ayafor, and J. Wandji, Phytochem., 1985, 24, 3075).

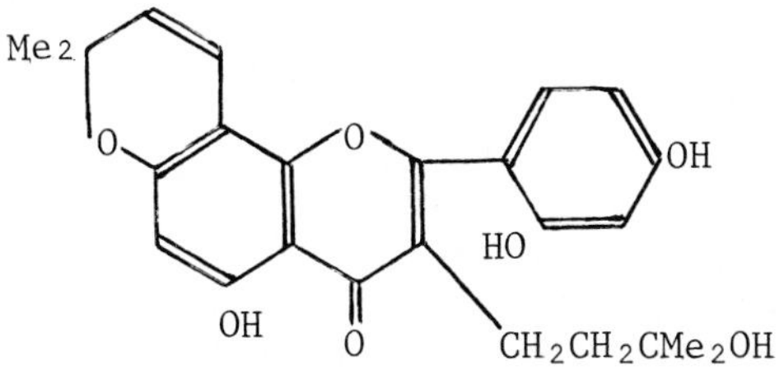

(3)

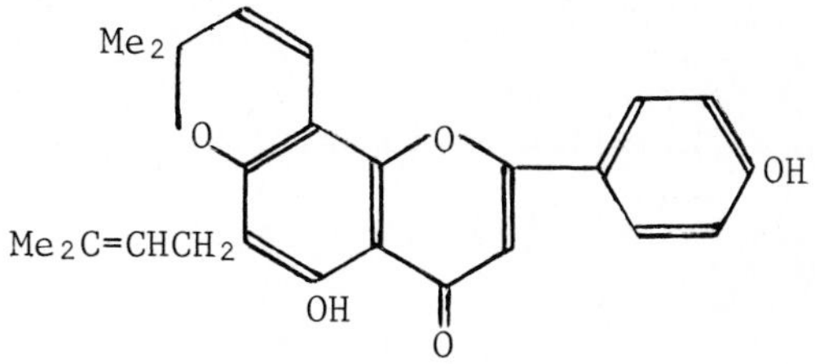

(4)

Two isoprenylated pyranoflavones, cudraflavone A (5) and cudraflavone B (6) have been isolated from the root bark of *Cudrania tricuspidata* (Moraceae) and photooxidative cyclization occurs with the latter flavone (6) leading to the formation of a hydroperoxide with a dihydrooxepin ring (T. Fujimoto *et al.*, Planta Med., 1984, 50, 161).

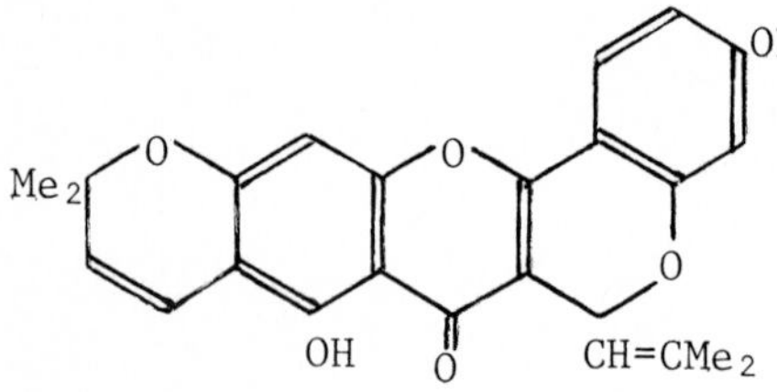

(5)

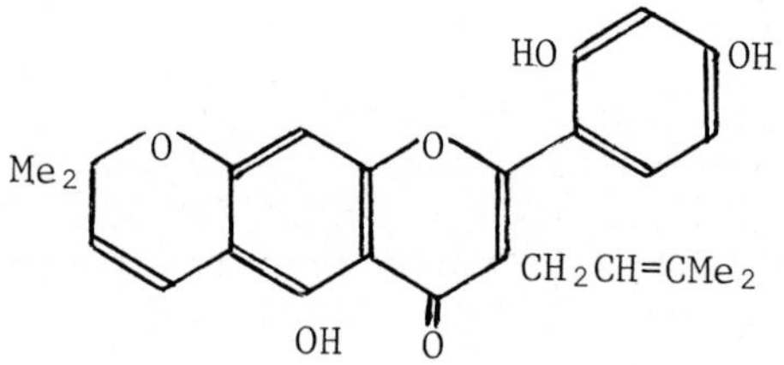

(6)

Carpchromene (7) has been synthesized from apigenin, which on treatment with 2-methylbut-3-en-2-ol yields 6-*C*- and 7-*O*-prenyl derivatives. The former on oxidative cyclization using dichlorodicyanoquinone yields carpachromene (7) (Roy, Sharma, and Khanna, Indian J. Chem., 1978, 16B, 463).

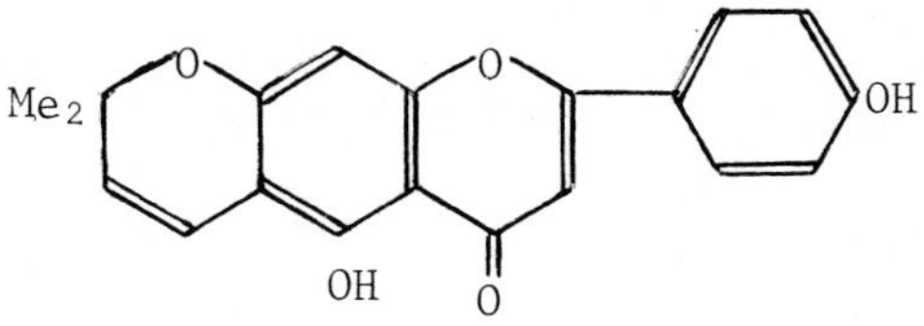

(7)

(xx) *Isoflavones, 3-phenylchromones, 4-oxo-3-phenyl-4H--benzo[b]pyrans*

Preparations. A novel synthesis of isoflavones from chalcones has been reported with reference to the biosynthesis of isoflavones. The synthesis is based on the oxidation of chalcones with thallium (III) acetate in methanol *via* a 1,2-aryl migration to give 1,2-diaryl-3,3--dimethoxypropan-1-ones, which on acid hydrolysis afford the isoflavones. Although this reaction provides a laboratory analogy for the biosynthesis of isoflavones, from chalcones, it cannot be used for a direct synthesis of isoflavones. Thus the oxidation of 4,4'-dimethoxy-2'--hydroxychalcone, for example, does not give an isoflavone, but the corresponding flavone and a coumaranone. However, the method has been adapted for the synthesis of isoflavones by the use of suitable protecting groups and applied to the synthesis of the naturally occurring isoflavones, 7,4'--dimethoxyisoflavone (1), milldurone (2), and lettadurone (3) (W.D. Ollis *et al.*, J. chem. Soc., C, 1970, 125).

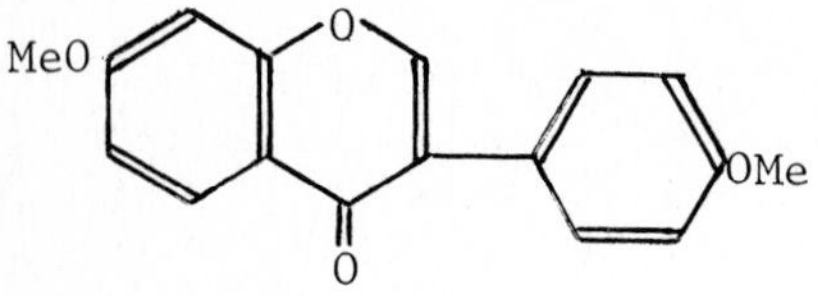

(1)

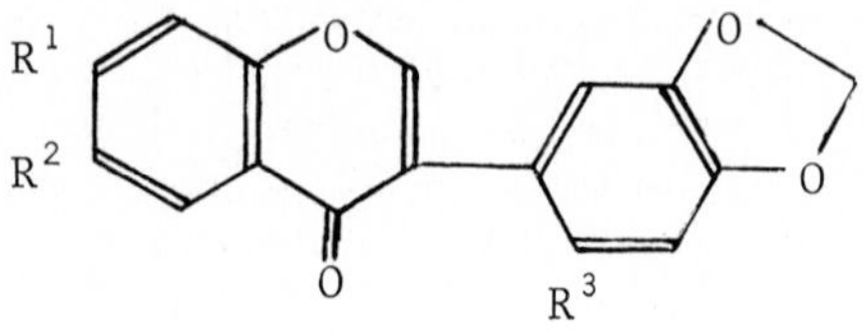

(2) $R^1=R^2=R^3=OMe$
(3) $R^1=R^2=OMe, R^3=H$

The isoflavones (5) are obtained in 65-80% yield on cyclizing chalcones (4) in the presence of thallium (III) nitrate (J.O. Oluwadiya, Niger. J. Pharm., 1979, 10, 84).

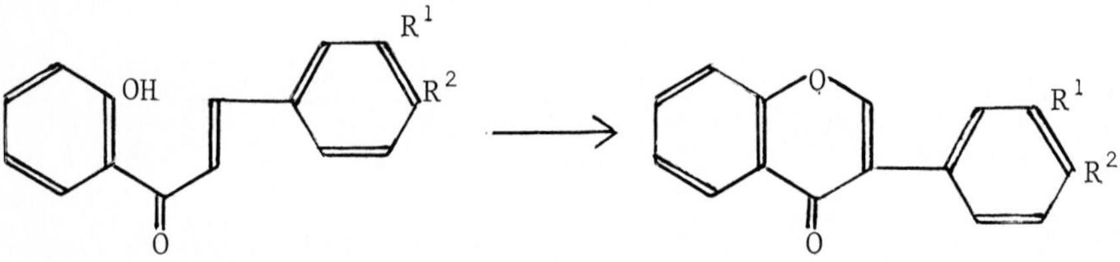

$R^1=R^2=H,OMe$
$R^1=H,R^2=OMe$

(4)   (5)

Despite the discovery that thallium (III) nitrate is more efficient for the oxidative rearrangement involved and the advantage of using more accessible chalcones, it is found that when their solubility in methanol is low, or when they have a substituent in the position *para* (5') to the free hydroxyl group, only a low yield is obtained. Thus 2'- -hydroxychalcone after treatment with thallium (III) nitrate in methanol affords the acetal (6), which on treatment with 10% hydrochloric acid gives isoflavone (7) (*ca*. 25%). The reaction mixture also yields two other compounds, one formulated as 2-dimethoxymethyl-2-phenylcoumaran-3-one(8) and the other suggested as methyl 2-(2'-hydroxy-3'-nitrophenyl)- -3-phenyl-3-methoxypropanoate (9) (R.S. Varma, Chem. and Ind., 1982, 56).

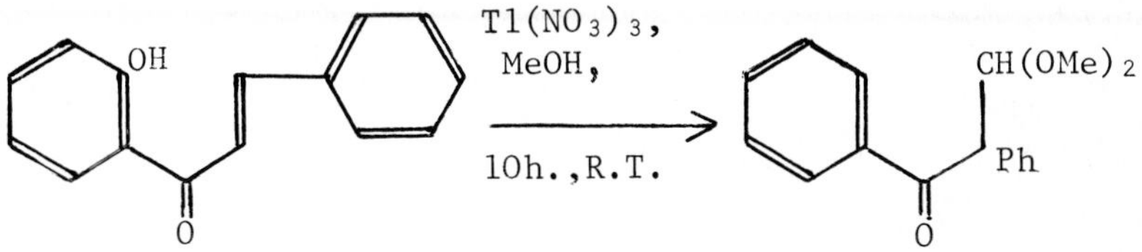

(6)

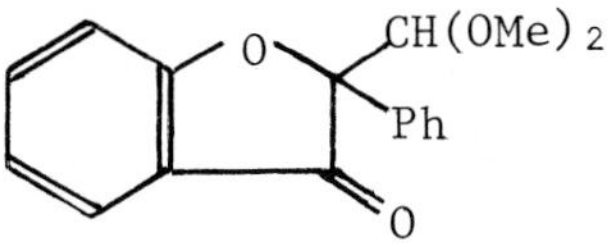

10% HCl

(8)

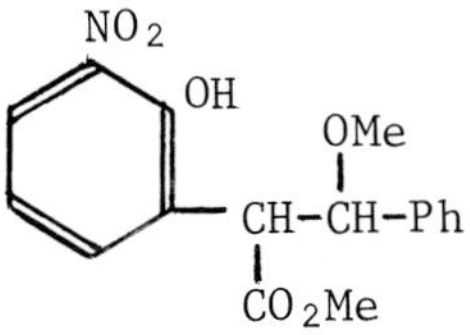

(9)

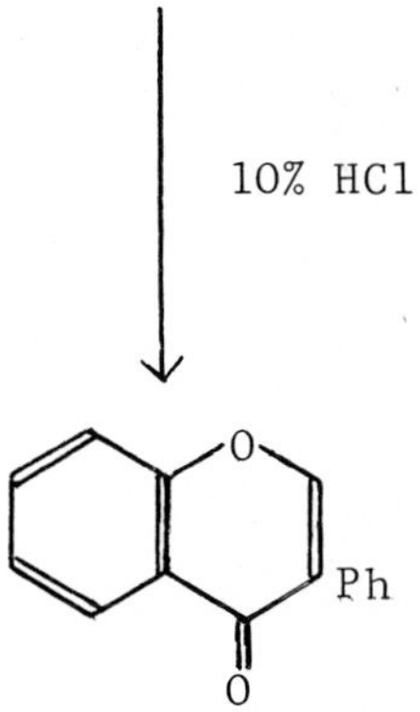

(7)

ω-Substituted 2,4-dihydroxyacetophenones (10) undergo
acetic formic anhydride catalyzed cyclization in the
presence of sodium formate or triethylamine to furnish
isoflavones (11) and 3-hetarylchromones (12). When
R^1 = 2-pyridy, 2-quinolyl, and 2-methyl-4-thiazolyl the
corresponding 7-formate (13) is also obtained
(V.G. Pivovarenko, V.P. Khilya, and F.S. Babichev, Chem.
Abs., 1985, <u>103</u>, 141,792s).

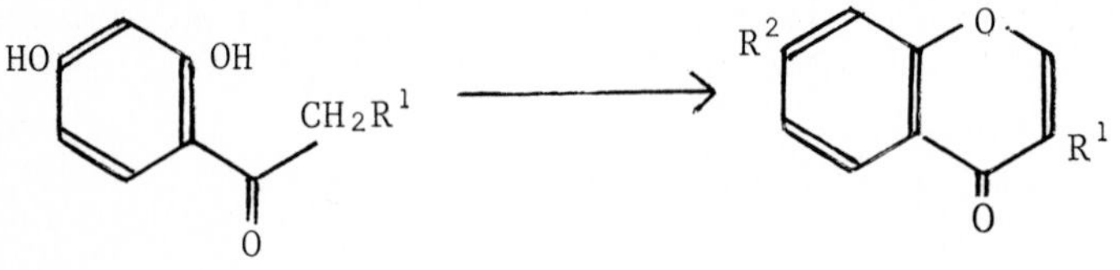

(10)

(11) R^1=Ph, 4-NO$_2$C$_6$H$_4$, 4-MeOC$_6$H$_4$, R^2=OH
Also R^1=Me, R^2=OH
(12) R^1=2-pyridyl,2- and 7-quinolyl,
2-methyl-4-thiazolyl,5(ethoxycarbonyl)-2-furyl, R^2=OH
(13) R^1=2-pyridyl,2-quinolyl,
2-methyl-4-thiazolyl, R^2=HCO$_2$

A number of analogues (15) of natural isoflavones have
been prepared by the cyclocondensation of deoxybenzoins
(14) with ethyl orthoformate, acetic anhydride, trifluoroacetic anhydride, or ethyl oxalyl chloride (Khilya *et al.*,
Ukr. Khim. Zh., 1984, <u>50</u>, 1301).

R^1=H,Et,Pr,Bu,pentyl, R^3=H,Me,CF$_3$,CO$_2$Et
      hexyl

R^2=H,Br,Cl,F,NO$_2$,MeO,Me$_2$CHO

(14)   (15)

Reactions. The rate of hydrogenation of isoflavone over
a palladium/carbon catalyst and the product distribution
are dependent upon the solvent, pH of the solution, and the
activity of the catalyst. Appropriate choice of conditions
yields predominately isoflavanone, 4-isoflavanol, isoflavan,
or in dilute alkali solution 2-hydroxy-α-methyldeoxybenzoin
(V. Szabo and E. Antal, Acta Chim. Acad. Sci. Hung., 1976,
90, 381).
   The reactions of isoflavones with either thionyl chloride or
sulphuryl chloride affords a number of new chloroisoflavones
and isoflavanones, whose structures are supported by spectral
and analytical data. 7-Hydroxyisoflavone (1) with thionyl
chloride yields 7-chloroisoflavone (2), m.p. 144-145°, and
7-methoxyisoflavone (5) gives 6-chloro-7-methoxyisoflavone
(6), m.p. 208-210°. Isoflavone does not react with thionyl
chloride in boiling benzene for 15h, but with sulphuryl
chloride in boiling carbon tetrachloride for 15h 2,3-
-dichloroisoflavanone is obtained. Similar treatment with
sulphuryl chloride in carbon tetrachloride of 7-hydroxy-
isoflavone (1) affords 7-hydroxy-2,3,6,8-tetrachloro-
isoflavanone (3), m.p. 194-196°, and 6,8-dichloro-7-hydroxy-
isoflavone (4), m.p. 215-217°, while 7-methoxyisoflavone (5)
furnishes 7-methoxy-2,3,6,8-tetrachloroisoflavanone (7),
m.p. 165-167°, and 7-methoxy-2,3,8-trichloroisoflavanone (8),
m.p. 175-177° (J.R. Merchant and G. Martyres, J. heterocyclic
Chem., 1980, 17, 1331).

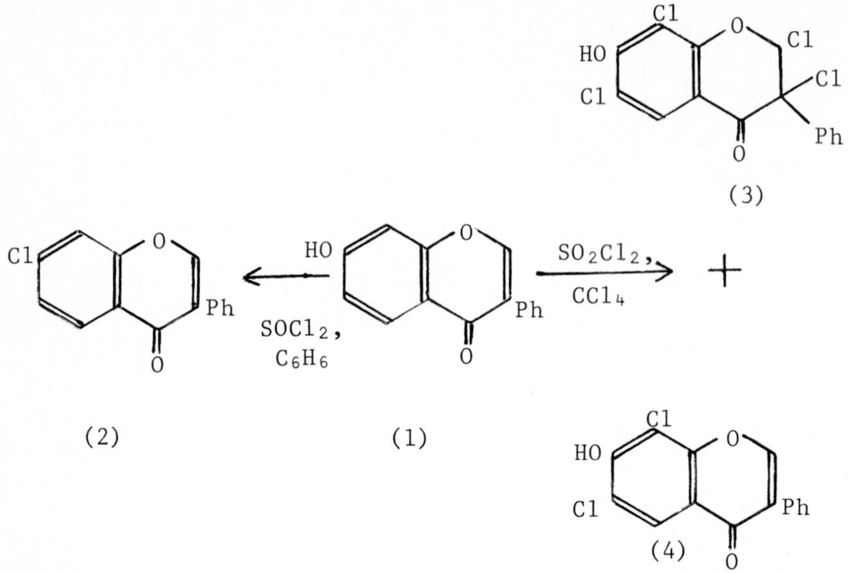

HO
Cl
O
Ph
SOCl2,
C6H6
SO2Cl2,
CCl4

(1) (2) (3) (4)

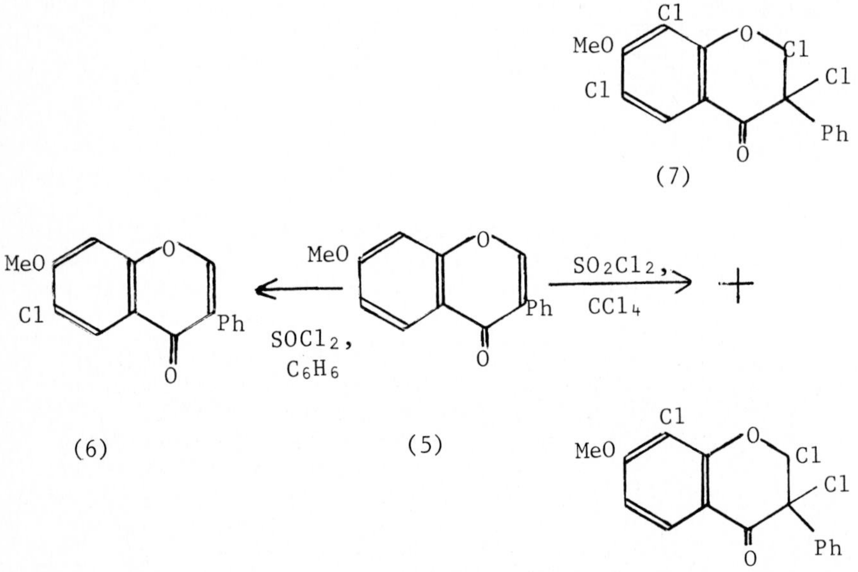

MeO
Cl
O
Ph
SOCl2,
C6H6
SO2Cl2,
CCl4

(5) (6) (7) (8)

An unusual formation of 4'-substituted 6,7-dichloro-
isoflavones (11) occurs during the synthesis of 2-aryl-5,6-
-dichloroisatogens (10) by the base-catalyzed deformylation
of aldehydes (9) (A.B. Sahasrabudhe, B.V. Bapat, and
S.N. Kulkarni, Indian J. Chem., 1981, 20B, 495).

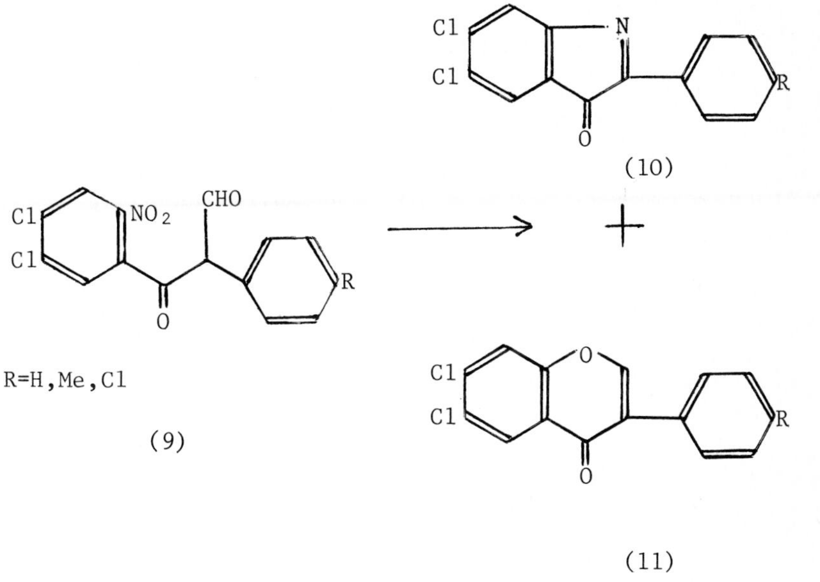

(10)

(9)

R=H,Me,Cl

(11)

Treatment of 5,7-dihydroxyisoflavone at room temperature
with acetic anhydride under either acid- or base- catalyzed
conditions gives the 7-acetoxy derivative. More forcing
conditions yield 5,7-diacetoxyisoflavone. However, it is
found that on acetylation of 5,7-dihydroxy-4'-fluoroisoflavone
(12) using pyridine as catalyst 7-acetoxy-4'-fluoroisoflavone
(13), m.p. 194-197°, is formed, whereas use of perchloric
acid leads to 5-acetoxy-4'-fluoroisoflavone (14), m.p. 195°.
The reaction has been found to be a general one. 5-Acetoxy-
-7-hydroxyisoflavone, needles from ethyl acetate/petroleum
ether, m.p. 195°; 5-acetoxy-7-hydroxy-4'-methoxyisoflavone,

needles, m.p. 188-190°. Reaction of the 5-acetoxy
derivative (14) with dichlorourethane in acetic acid
affords 5-acetoxy-8-chloro-4'-fluoro-7-hydroxyisoflavone
(15), m.p. 185-180°, and the 7-acetoxy derivative (13),
7-acetoxy-6,8-dichloro-4'-fluoro-5-hydroxyisoflavone (16),
m.p. 173-175°. Hydrolysis of the chloro derivatives (15
and 16) with methanolic hydrogen chloride furnishes 5,7-
-dihydroxy-8-chloro-4'-fluoroisoflavone (17), m.p. 224-225°,
and 6,8-dichloro-5,7-dihydroxy-4'-fluoroisoflavone (18),
m.p. 224-226°, respectively (R.J. Bass, Chem. Comm., 1979,
264).

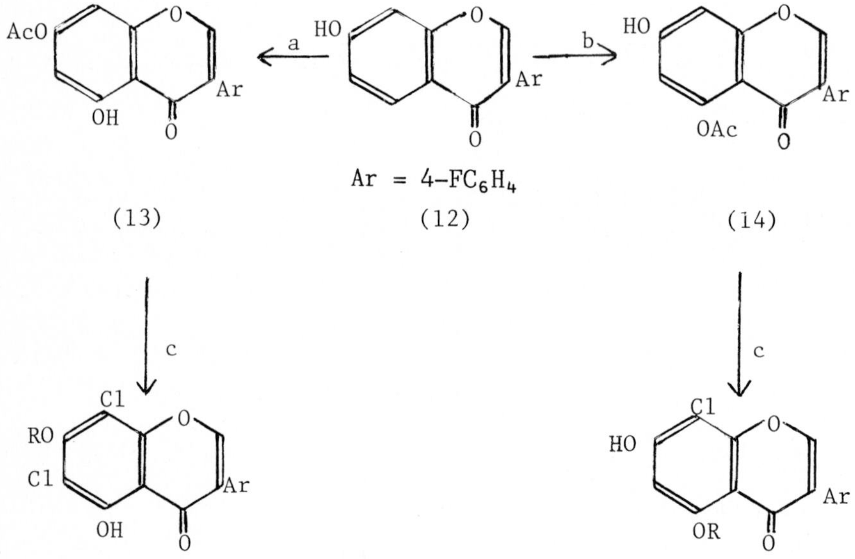

Ar = 4-FC$_6$H$_4$

(13)   (12)   (14)

(16) R=Ac   (15) R=Ac
(18) R=H   (17) R=H

Reagents:-   (a) Ac$_2$O, C$_5$H$_5$N;   (b) AcO$_2$, HClO$_4$;
             (c) Cl$_2$NCO$_2$Et, AcOH

Isoflavone reacts with hydroxylamine between pH4 and 11 to yield the isoxazole (19) and the yield increases with increasing pH. Isoxazole (19) in strongly alkaline solution gives the nitrile (20), which is in ring-chain tautomeric equilibrium with 4-hydroxy-3-phenylcoumarin imine, the equilibrium depends on pH. The imine on boiling in acid solution is converted into 4-hydroxy-3-phenylcoumarin (Szabó, J. Borda, and L. Losonczi, Acta Chim. Acad. Sci. Hung., 1978, 97, 69).

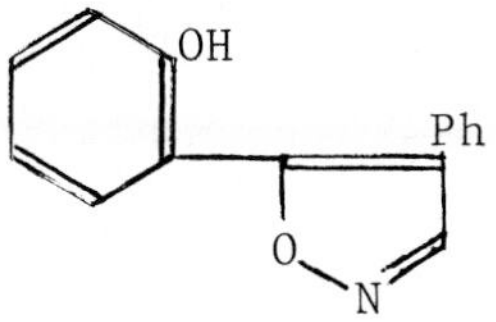

(19) 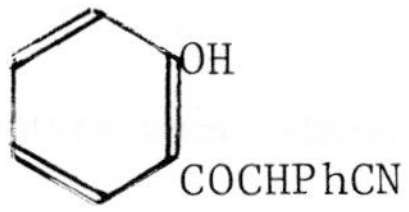   (20)

Several isoflavones have been converted into the corresponding 4-hydroxycoumarins by this method (Szabó, Borda, and E. Theisz, *ibid.*, 1977, 95, 333;   1980, 103, 271).

Isoflavones (21) on treatment with hydroxylamine in aqueous ethanol at pH8 yield isoxazoles (22; R^2 = H), which on treatment with either R^3NCO (R^3 = Me, Et, Bu) or (EtO)$_2$PSCl afford the isoxazole derivatives (22; R^2 = CONHR^3) or [22; R^2 = P(S)(OEt)$_2$], respectively (Szabó *et al.*, Magy. Kem. Foly., 1979, 85, 385).

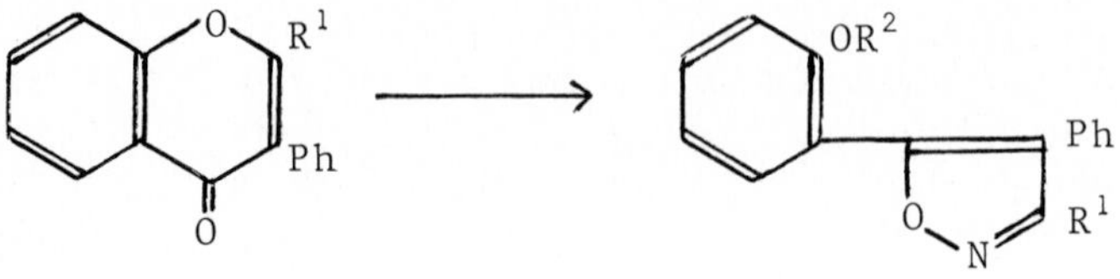

$R^1 = H, Me, CF_3$

(21)   (22)

A number of derivatives of isoxazole have been prepared by
the reaction between isoflavones and hydroxylamine and
their insecticidal activity investigated (Borda *et al.*,
Acta Chim. Acad. Sci. Hung., 1980, **104**, 389). The reaction
of benzofuran analogues of isoflavones and their 4-thioxo
derivatives with hydroxylamine and hydrazine has been
discussed (V.P. Khilya, L.G. Grishko and T.L. Davidkova,
Khim. Geterotsikl, Soedin., 1980, 892).

The reaction between isoflavone and hydrazine, methylhydrazine, and phenylhydrazine under acidic or basic
conditions affords only pyrazole derivatives. With
hydrazine, 5-(2-hydroxyphenyl)-4-phenylpyrazole (23) is
formed and with methylhydrazine a mixture of 5-(2-hydroxy-
phenyl)-1-methyl-4-phenylpyrazole (24) and 5-(2-hydroxyl-
phenyl-2-methyl-4-phenylpyrazole (25). Isoflavone does
not react with 2,4-dinitrophenylhydrazine (Szabó, Borda,
and V. Vegh, Magy. Kem. Foly., 1977, **83**, 393).

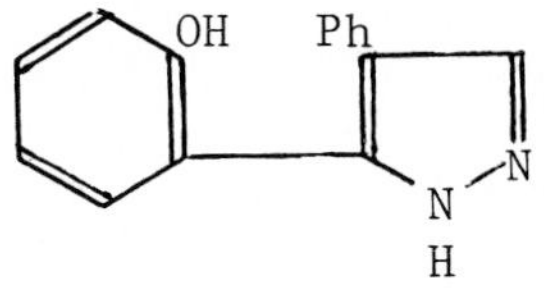

(23) 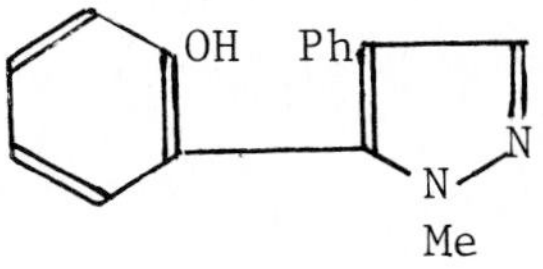   (24)

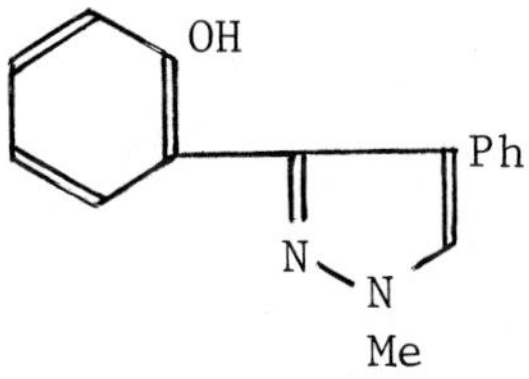

(25)

The reaction between the 4-thioxo derivative of isoflavone
and hydrazine and its derivatives gives the same pyrazoles
[*idem*, Acta Chim. Acad. Sci. Hung., 1978, (Pub. 1979), 98,
457].

The kinetics of the ring cleavage of isoflavones (26)
with HO⁻ at 25° has been measured. Those with R^2 either
2- or 4-MeO are more stable than those with R^1 = R^2 = R^3 = H
and the most stable one is that with R^1 = 7-MeO and
R^2 = R^3 = H, due to the fact that the groups are coplanar
and the interaction is undisturbed (Szabó and M. Zsuga,
ibid., 1978, 97, 451).

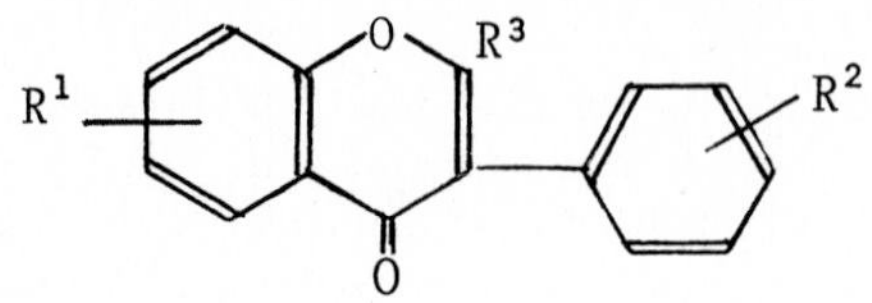

$$R^1 = R^2 = R^3 = H$$
$$R^1 = R^3 = H, \quad R^2 = 2\text{-MeO}, \quad 4\text{-MeO}$$
$$R^2 = R^3 = H, \quad R^1 = 7\text{-MeO}, \quad 6\text{-MeO}$$
$$R^1 = R^2 = H, \quad R^3 = Me, \quad CF_3, \quad Et, \quad CO_2H$$

(26)

The ring cleavage reaction of chromone, flavone, isoflavone, 3-methylchromone, and 3-phenoxychromone in aqueous alkali solution, starts with an isoentropic Ad_N2-type nucleophilic addition, which is the rate-determining step. The carbanion (27) is then ring opened (Zsuga *et al.*, *ibid.*, 1979, **101**, 73).

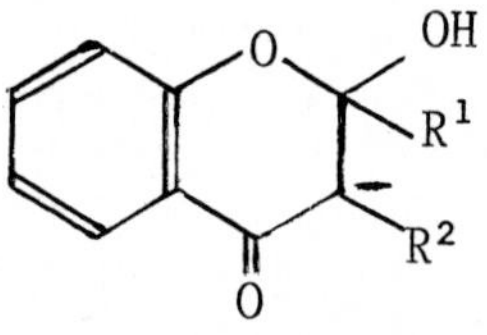

$$R^1 = H, \quad Ph$$
$$R^2 = H, \quad Me, \quad Ph, \quad PhO$$

(27)

The CNDO calculations of the dipole moments of *trans*-
-1-benzoyl-2-phenylethene, flavone, isoflavone, and 2,3-
-dihydro-3-oxo-2-phenylmethylenebenzo[b]furan, indicate
that the first and last compounds are planar and that
flavone is more planar than isoflavone. The calculated
electron density distributions agree with those obtained
from ^{13}C-nmr spectral data (Z. Dinya *et al.*, Flavonoids,
Bioflavonoids, Proc. Hung. Bioflavonoid Symp., 5th, 1977,
247, ed. L. Farkas, M. Gabor, and F. Kallay, Elsevier,
Amsterdam, Neth.). The Claisen rearrangement of 7-(3-
-methylbut-2-enyl)-flavones and -isoflavones has been
discussed (K.V.S. Raju, K. Subha, and G. Srimannarayana,
Indian J. Chem., 1980, **19B**, 866).

(xxi) Naturally occurring isoflavones
The following are some isoflavones, which have been
isolated from natural sources and their structures
elucidated from chemical and spectral data and by reference
to known isoflavones. Petalostetin, *3',4'-methylenedioxy-
-6,7,8-trimethoxyisoflavone* (1) from *Petalostemon candidum*
(S.J. Torrance *et al.*, Phytochem., 1979, **18**, 366); three
new 5-*O*-methylisoflavones, barpisoflavone A, *5-methoxy-
-7,2',4'-trihydroxyisoflavone* (2), 5-*O*-methylupiwighteone,
7,4'-dihydroxy-5-methoxy-8-prenylisoflavone (3) and
barpisoflavone B, *5-methoxy-8-prenyl-7,2',4'-trihydroxy-
isoflavone* (4) and two pyranoisoflavones in addition to 5-
-*O*-methylgenistein from roots of the yellow lupin (*Lupinus
luteus* L. cv. Barpine) (S. Tahara *et al.*, Agric. biol.
Chem., 1986, **50**, 1809); and 3'methoxy-5,6,7,4'-tetrahydroxy-
isoflavone from the rhizomes of *Iris milessii* and
iriskumaonin, iriskumaonin methyl ether, irisflorentin,
junipegenin-A, irigenin, and iridin from those of *Iris
kumaonensis* (V.K. Agarwal *et al.*, Phytochem. 1984, **23**, 2703).
The known isoflavones, orobo and orobo 7-methyl ether have
been isolated along with some flavanones from *Wyethia glabra*
(S. McCormick, K. Robson and B. Bohm, *ibid.*, 1985, **24**, 1614).

(1)

(2) R^1 = OH, $R^2 = R^3$ = H
(3) $R^1 = R^2$ = H, $R^3 = CH_2CH=CMe_2$
(4) R^1 = OH, R^3 = H, $R^3 = CH_2CH=CMe_2$

The heartwood of *Xanthocercis zambesiaca* in addition to
8,4'-dimethoxy-7-hydroxy- and 7,3'-dihydroxy-8,4'dimethoxy-
-isoflavone, also contains 7-hydroxy-8,3',4'-trimethoxy-,
3',4'-dimethoxy-6,7-methylenedioxy-, and 6,7-methylenedioxy-
-8,3',4'-trimethoxy-isoflavone. Also described is a
technique for determining isoflavone hydroxylation patterns
using ^{2}H-labelling (S.H. Harper, D.B. Shirley, and
D.A. Taylor, *ibid.*, 1976, **15**, 1019). The application of
gas-liquid chromatography and mass spectrometry established
that the roots of red clover seedlings grown under sterile
conditions contain formononetin, pseudobaptigenin,
biochanin A, and calicosine (S.A. Popravko, S.A. Sokolova,
and G.P. Kononenko, Bioorg. Khim., 1980, **6**, 1255). Gnonin,
formononetin, genistin, genistein, and 5-methylgenistein
have been obtained from the above ground parts of *Spartium
junceum* (I.I. Ozimina, V.A. Bandyukova, and A.L. Kazakov,
Khim. Prir. Soedin., 1979, 858). The 5-deoxyisoflavones,
afromosin and formononetin have been isolated from healthy
leaves of *Onobrychis viciaefolia* (J.L. Ingham, Z.
Naturforsch., Biosci., 1978, **33C**, 146).

2'-Hydroxygenistein is one of the seven isoflavonoid phytoalexins isolated from the fungi-inoculated hypocotyls of *Lablab nigher* (hyacinth bean) (Ingham, *ibid*., 1977, 32C, 1018). 7,2'-Dihydroxy-4'-methoxy-isoflavone-Me-^{14}C and 7-hydroxy-2',4'-dimethoxyisoflavone-4'-Me-^{14}C have been used in feeding experiments in the investigation of the biosynthesis of pterocarpan and isoflavan phytoalexins in *Medicago sativa* (P.M. Dewick and M. Martin, Phytochem., 1979, 18, 591).

(xxii) Furoisoflavones

Furo[3,2-g]isoflavones (3) have been obtained by the benzylation of the benzofurans (1), followed by conversion into the epoxides (2) and subsequent treatment with hydrochloric acid [O.H. Hishmat and N.M.A. El-Ebrashi, Egypt. J. Chem., 1975 (Pub. 1978), 18, 673].

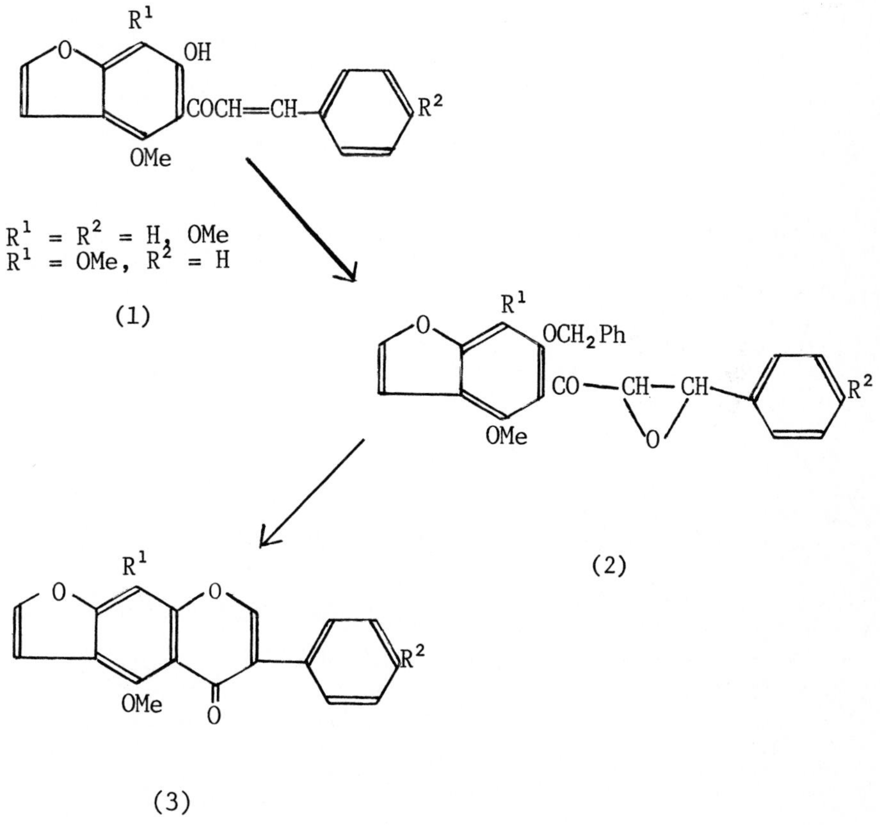

$R^1 = R^2 = H$, OMe
$R^1 = OMe$, $R^2 = H$

(1)

(2)

(3)

4. Chroman, dihydrochromene, 3,4-dihydro-2H-benzo[b]pyran and derivatives

(a) Chromans

Preparations. The dibromo derivative (1) on treatment with butyllithium at -100° undergoes Parham cyclialkylation to furnish chroman (2) (C.K. Bradsher and D.C. Reames, J. org. Chem., 1981, **46**, 1384).

R^1	R^2	(2)%	b.p.°C (torr)
H	H	80	67–70 (2.6–3.0)
H	Me	71	52–53 (0.37–0.38)
H	MeO	72	70–75 (0.25–0.30)
H	Cl	75	89–90 (2.1–2.2)
Br	Me	79	91–95 (0.18–0.22)

6,8-Dimethylchroman (2; R^1 = R^2 = Me) (64%), b.p. 86-89° (2.1-2.2 torr), is obtained by a one-pot double lithiation sequence of 3-(2,6-dibromo-4-methylphenoxy)propyl bromide with butyllithium and then with butyllithium followed by methyl iodide. A solution of sodium 2-allylphenoxide in methanol on irradiation, followed by acidification with acetic acid affords a mixture of products including chroman (T. Kitamura, T. Imagawa, and M. Kawanisi, Tetrahedron, 1978, **34**, 3451).
  3-(2-Fluorophenyl)propan-1-ol is cyclized readily to chroman either through its chromium tricarbonyl complex (3)

or by action of the (η^6-benzene) (η^5-ethyltetramethylcyclo-
pentadienyl)rhodium (III) cation. Complex (3) is prepared
from the parent fluoro alcohol on treatment with trispyridine-
tricarbonylchromium (0) and boron trifluoride-ether. Addition
of potassium *tert*-butoxide to a solution of the complex (3) in
dimethyl sulphoxide at room temperature effects cyclization to
chroman complex (4), which on mild oxidation using iodine in
ether yields chroman. Similarly substituted fluorophenyl-
propanols are converted to substituted chromans.

(3)   (4)

In order to make the chroman formation metal-catalyzed (as
distinct from metal-promoted), the complexed rhodium (III)
cation (5) has been prepared as its hexafluorophosphate (V)
salt, which catalyzed the cyclization of the fluorophenyl-
propanols (6) to the corresponding chromans (7). The
cyclization are also catalyzed by the corresponding tetrafluoroborate salt, but the rates of cyclization are lower
than with the hexafluorophosphate salt. The spiro compound
(9) has been obtained by this procedure from the diol (8)
(R.P. Houghton, M. Voyle, and R. Price, Chem. Comm., 1980,
884).

$$(\eta^5 - \text{Et Me}_4\text{C}_5)\ (\eta^6 - \text{C}_6\text{H}_6)\text{Rh}^{2+}$$

(5)

(6)	R^1	R^2	(7)
	H	H	
	Me	H	
	H	CH_2OH	
	H	$2-FC_6H_4CH_2$	

(8)   (9)

For a review of spirochromans see S. Smolinski [Chem. heterocyclic Compds., John Wiley and Sons, 1981, 36 (Chromans and Tocopherols), 371]. The reaction between 5- -hydroxymethyl-1-oxaspiro[2.5]octa-5,7-dien-4-one (10) and chlorotrimethylsilane yields the spirochroman (11) (P. Cacioli and J.A. Reiss, Austral. J. Chem., 1984, 37, 2599).

(10)   (11)

2,3-Dihydro-2-ethylbenzo[b]furan on treatment with triphenylmethyl perchlorate in acetic or formic acid, isomerizes to give 2-methylchroman (up to 27%), but the yield of dehydrogenation products is increased to 80% on using a mixture of triphenylmethyl perchlorate and tin (IV) chloride (E.A. Karakhanov, E.A. Dem'yanova, and E.A. Viktorova, Doklady Adad. Nauk SSSR, 1972, 204, 879). 2-Methylchroman has also been obtained by the isomerization of 2,3-dihydro-2-ethylbenzo[b]furan at 300-400° in the presence of activated carbon (Karakhanov et al., Khim. Geterotsikl. Soedin., 1975, 321) and 2,3-dihydro-2,3- -dimethylbenzo[b]furan in the presence of aluminosilicate at 433° (Karakhanov, S.V. Lysenko, and Viktorova, Vestn. Mosk. Univ. Khim., 1974, 15, 500). 2-Ethylchroman may be obtained by the hydrogenation of 2-ethyl-2H-benzo[b]pyran in methanol over 10% palladium-carbon (E.E. Schewizer, T. Minami, and S.E. Anderson, J. org. Chem., 1974, 39, 3038).

Crotylbenzaldehydes obtained by the condensation of hydroxybenzaldehydes, for example, 2,3,4-trihydroxy-, 2,4- -dihydroxy-, 2,4-dihydroxy-6-methyl-, and 2,4-dihydroxy-3- -iodo-6-methyl-benzaldehydes with buta-1,3-diene in xylene in the presence of orthophosphoric acid, on cyclization by heating with orthophosphoric acid yield the corresponding 2-methylchroman-6-carboxaldehyde. Thus, crotylbenzaldehyde (12) gives 7,8-dihydroxy-2-methylchroman-6-carboxaldehyde (13) (V.K. Ahluwalia, D. Singh, and R.P. Singh, Monatsh., 1984, 115, 1059).

(12)   (13)

Similarly, 6-benzoyl-2-methylchromans are formed on
cyclization of butenylbenzophenones, prepared by the
condensation of 4-hydroxy-, 2,4-dihydroxy-, 2,4-dihydroxy-
-6-methoxy-, and 4,6-dimethoxy-2-hydroxy-benzophenones with
buta-1,3-diene (Ahluwalia, R. Singh, and R.P. Singh, Gazz.,
1984, 114, 501).

Induced intramolecular cyclization of 2-(3-methylbut-2-
-ene-1-yl)phenol in methanol in the presence of palladium
(II) chloride or acetate gives a mixture of 2,2-dimethyl-
chroman and 2,2-dimethyl-2H-benzo[b]pyran, with trace
amounts of 2-isopropyl- and 2-isopropenyl-benzofuran
(T. Hosokawa et al., Bull. chem. Soc. Japan, 1976, 49,
3662). 2,2-Dimethylchromans are prepared by the direct
condensation of phenols with 2-methylbuta-1,3-diene
(isoprene) in the presence of orthophosphoric acid as
catalyst. Thus resorcinol with isoprene yields 3,4,6,7-
-tetrahydro-2,2,8,8-tetramethyl-2H,8H-benzo[1,2-b;5,4-b']-
dipyran (14), m.p. 97-98°, 2,2-dimethyl-5-hydroxychroman
(15), m.p. 121-121.5°, and 2,2-dimethyl-7-hydroxychroman
(16), m.p. 67-68°, in the ratio 1:2:5, which are separated
by column chromatography over silica gel. Since some
discrepancies have been found from the results obtained by
the condensation of 2-methylbut-3-en-2-ol with phenols in
the presence of aqueous citric acid, the latter method has
been reinvestigated (Ahluwalia, K.K. Arora, and R.S. Jolly,
J. chem. Soc., Perkin I, 1982, 335).

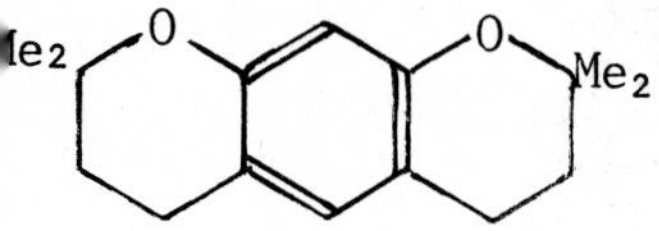

(14) 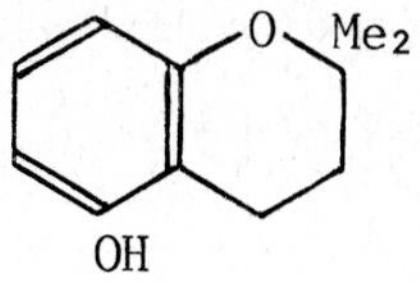   (15) 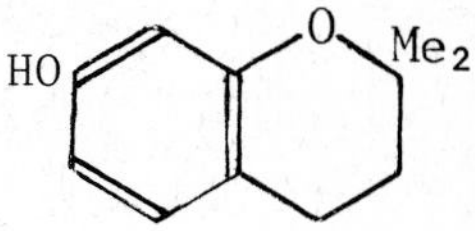   (16)

The cycloaddition of 1,4-dihydroxy-2,3,5-trimethylbenzene
with isoprene in ethyl acetate in the presence of aluminium
chloride affords 6-hydroxy-2,2,5,7,8-pentamethylchroman
(Nippon Petrochemicals Co., Ltd. Chuo Kaseihin Co., Inc.
Japan Kokai Tokkyo Koho JP59 44,376 [84 44,376], 1984).
   Chroman-4-carboxaldehydes (18) have been prepared by the
catalytic hydroformylation of 2*H*-benzo[b]pyrans (17) with
carbon monoxide and hydrogen in the presence of Group VIII B
metals at 80-250° under pressure (A. Widdig *et al.*, Ger.
Offen. DE 3,300,005, 1984).

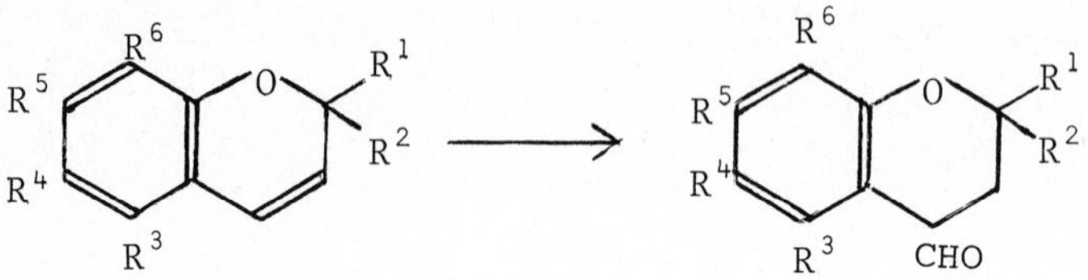

(17)   (18)

$R^1 = R^2 = H$, alkyl, c-alkyl, (un)substituted aryl, aralkyl

$R^1 R^2 =$ carbocycle

$R^3, R^4, R^5, R^6 =$ H, halogeno, alkyl, c-alkyl, (un)substituted aryl, aralkyl, aryloxy, alkoxy

6-Acetylchromans are obtained by condensing appropriate polyhydroxyacetophenones with 2-methylbut-3-en-2-ol in the presence of orthophosphoric acid, for instance, 2,3,4- -trihydroxyacetophenone affords 6-acetyl-7,8-dihydroxy-2,2- -dimethylchroman (19); 2,4-dihydroxyacetophenone gives a 2:2:1 mixture of 6-acetyl-2,2-dimethyl-7-hydroxychroman (20), 6-acetyl-2,2-dimethyl-5-hydroxychroman (21), and 3,4,9,10- -tetrahydro-2,2,8,8-tetramethyl-2H,8H-[1,2-b;5,6-b']dipyran (22); and 2,4-dihydroxy-6-methoxyacetophenone gives equal amounts of 6-acetyl-2,2-dimethyl-7-hydroxy-5-methoxychroman and 6-acetyl-2,2-dimethyl-5-hydroxy-7-methoxychroman (*idem*, Heterocycles, 1981, **16**, 2155).

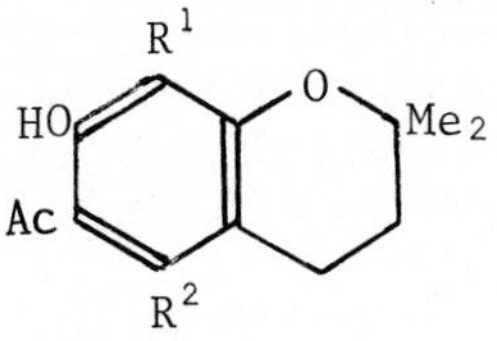

(19) R^1=OH, R^2=H (21)
(20) R^1=R^2=H

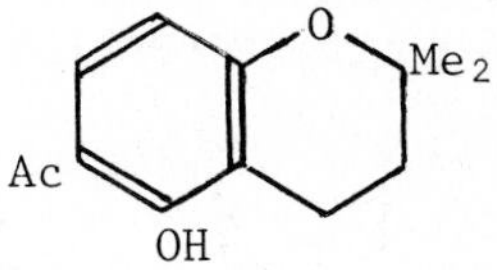

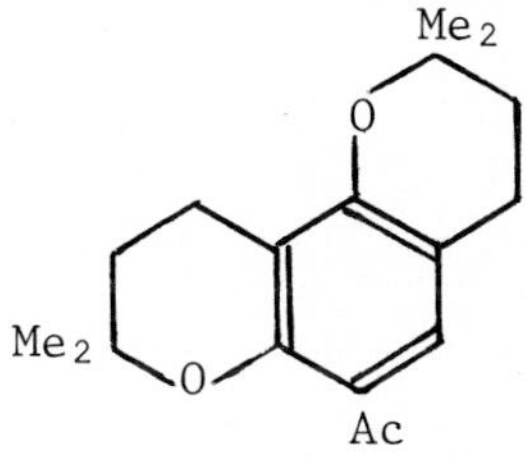

(22)

6-Acetyl-7-hydroxy-2,2,5-trimethylchroman, 6-acetyl-7-
-hydroxy-2,2,8-trimethylchroman, 6-acetyl-5-hydroxy-2,2,7-
-trimethylchroman have been obtained from isoprene and the
necessary hydroxyacetophenone (Ahluwalia and K. Mukherjee,
Indian J. Chem., 1984, 23B, 880).

The Hoesch reaction of 2,2-dimethyl-7-hydroxychroman and
5,7-dihydroxy-2,2-dimethylchroman with acetonitrile/hydrogen
chloride has been reinvestigated and it has been found that
the former chroman yields 6-acetyl and 6,8-diacetyl
derivatives and the latter 8-acetyl and 6,8-diacetyl
derivatives (K.J.R. Prasad and P.R. Iyer, *ibid.*, 1982, 21B,
255).

Condensation of isoprene with 2,4,6-trihydroxybenzophenone in the presence of orthophosphoric acid in xylene at 30-35° furnishes 6-benzoyl-5,7-dihydroxy-2,2-dimethylchroman, 8--benzoyl-5,7-dihydroxy-2,2-dimethylchroman, and 6-benzoyl-5--hydroxy-3,4,9,10-tetrahydro-2,2,8,8-tetramethyl-2H,8H-benzo-[1,2-b;5,6,b']dipyran (Ahluwalia, M. Khanna, and R.P. Singh, Synth., 1983, 404). Similarly 6-butyrylchromans (23) are obtained from 4-butyrylphenols and isoprene (Ahluwalia *et al.*, Gazz., 1984, 114, 359).
Heating the chromans with 2,3-dichloro-5,6-dicyano-1,4--benzoquinone produces the corresponding 2H-benzo[b]pyrans.

$$R^1=OH, \ R^2=H$$
$$R^1=H, \ R^2=OH$$
$$R^3=H, \ Me$$

(23)

2,2-Dimethylchromancarboxylic acids are prepared from appropriate polyhydroxybenzoic acid and isoprene in the presence of phosphoric acid. Hence, 2,5-dihydroxybenzoic acid affords 2,2-dimethyl-6-hydroxychroman-5-carboxylic acid, 2,2-dimethyl-6-hydroxychroman-8-carboxylic acid, and 3,4,8,9-tetrahydro-2,2,7,7-tetramethyl-2H,7H-benzo[1,2-b;-4,5-b']dipyran (24) (Ahluwalia and A.K. Tehim, Monatsh., 1983, 114, 1381).

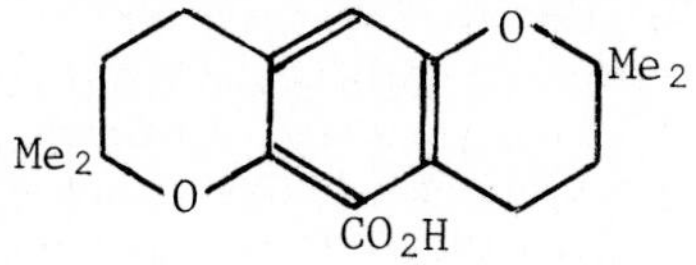

(24)

The esr-spectrum of the cation radical produced by the aerobic oxidation of 2,2,5,7,7,10-hexamethyl-3,4,8,9--tetrahydro-2H,7H-benzo[1,2-b;4,5-b']dipyran in dichloromethane containing trifluoroacetic, dichloroacetic, or trifluoromethanesulphonic acid has been studied (S.A. Fairhurst, L.H. Sutcliffe, and S.M. Taylor, J. chem. Soc., Faraday 1, 1982, 78, 2743).

2-[(un)Substituted allyl]-4-methylphenols with N--iodosuccinimide in dichloromethane cyclize to yield 3--iodochromans, for instance, 2-(γ,γ-dimethylallyl)-4--methylphenol (25) affords 3-iodo-2,2,6-trimethylchroman (26) (90%), which on dehydroiodination with 10% methanolic potassium hydroxide gives 2,2,6-trimethyl-2H-benzo[b]pyran (A. Bongini *et al.*, Tetrahedron Letters, 1979, 2545). Also reported are the preparation of 2,2-dimethyl-3-iodo-7--methoxychroman (67%), 2,6-dimethyl-3-iodo-2-prenylchroman (85%), and that of 3-iodo-6-methyl-2-prenylchroman (45%).

(25)   (26)

A number of 2,2-dimethylchromanyl-3-formates and other chromanyl-3-formates have been synthesized (V.A. Ashwood, Eur. Pat. Appl. EP 126,350, 1984).

2,2-Diethoxychromans (29) are obtained from 2-hydroxybenzyl alcohols (27) with orthoesters (28) in boiling benzene or toluene containing toluene-4-sulphonic acid and removing the water produced as it is formed (R.R. Schmidt and B. Beitzke, Synth., 1982, 750).

$R^1 = R^2 = H$

$R^1 = H, \ R^2 = Ph$

$R^1 = OMe, \ R^2 = H$

$R^3 = H, Me$

(27)   (28)   (29)

The reaction of 2-(allyloxy)bromobenzenes with diethyl copper (I) malonate affords diethyl 2-(allyloxy)phenyl-malonates, which under the reaction conditions cyclize to yield 4,4-bis(ethoxycarbonyl)-3-methylchromans, 4,4-bis-(ethoxycarbonyl)-3-methylenechromans, and 4,4-bis-(ethoxycarbonyl)-3-(3-butenyl)chromans. Thus 2-(allyloxy)-bromobenzene (30) with diethyl copper (I) malonate in boiling dioxane furnishes 4,4-bis(ethoxycarbonyl)-3-methylchroman (31) and 4,4-bis(ethoxycarbonyl)-3-(3-butenyl)chroman (32) (J. Setsune *et al.*, Chem. Letters, 1984, 1931).

R=CH$_2$CH=CH$_2$ (37%)   (11%)

(30)   (31)   (32)

1,2-Bis(chroman-6-yl)ethene has been prepared (F. Eiden and C. Schmiz, Arch. Pharm., 1980, 313, 120). For a review of the preparation, reactions, and physical and spectral properties of alkyl- and aryl-chromans see R. Livingstone [Chem. heterocyclic Compds., John Wiley and Sons, 1981, 36 (Chromans and Tocopherols), 7] and of the total synthesis of chromans and benzopyrans, H.J. Shue (Diss. Abs. Int. B, 1985, 45, 2924).

The following methoxychromans; 6-(G.P. Ellis, Brit. Pat. 1,023,373, 1966; U.S. Pt. 3,322,795, 1967), 7-(G. Graffe, M.-C. Sacquet, and P. Maitte, J. heterocyclic Chem., 1975, 12, 247), and 8-(R.S. Bramwell and A.O. Fitton, J. chem. Soc., 1965, 3882) have been obtained by the Clemmensen reduction of the corresponding methoxychroman-4-one as have 6,7-dimethoxy-2-methyl- (O. Dann, G. Volz, and O. Huber, Ann., 1954, 587, 16) and 7,8-dimethoxy-2,2-dimethyl-chroman (J.S.P. Schwarz et al., Tetrahedron, 1964, 20, 1317; M. Tsukayama, Bull. chem. Soc. Japan, 1975, 48, 80).

2-Methoxyphenol reacts with but-3-enyl diphenyl phosphate and with 3-methylbut-3-enyl diphenyl phosphate in the presence of a Lewis acid to yield 8-methoxy-4-methyl- and 4,4-dimethyl-8-methoxy-chroman, respectively (Y. Butsugan et al., Chem. Letters, 1976, 523).

For a review of the preparation, reactions, and spectral
properties of alkoxychromans see R. Livingstone [Chem.
heterocyclic Compds., John Wiley and Sons, 1981, 36,
(Chromans and Tocopherols), 139].

Diazotization of 6-aminochroman, followed by treatment
with copper (I) chloride affords 6-chlorochroman
(G. Brancaccio, G. Lettieri, and R. Viterbo, J. heterocyclic
Chem., 1973, 10, 623). For the preparation of 6-chloro-2,2-
-dimethylchroman see E.A. Vdovtsova and A.G. Kakharov (Chem.
Abs., 1976, 85, 46313 n); for the methanolysis of a number
of 3,4-dihalogeno- and 3,4-dihalogeno-2,2-dimethyl-chromans,
R. Binns *et al*., (J. chem. Soc., Perkin II, 1974, 732); for
the preparation of 6-iodochroman, A. Goosen and
C.W. McCleland (S. Afr. J. Chem., 1978, 31, 67); and for a
review, which includes a number of halogenochromans not found
elsewhere in the literature, their preparation, reactions,
physical constants and spectral data, R. Livingstone [Chem.
heterocyclic Compds., John Wiley and Sons, 1981, 36 (Chromans
and Tocopherols), 161].

A number of the following chroman derivatives have been
synthesized, 3-amino- and 3-aminomethyl-chromans and
chroman-3-carboxamides (R.C. Gupta *et al*., Indian J. Chem.,
1982, 21B, 344); 8-substituted 6-(substituted amino)- and
6-substituted 8-(substituted amino)-chromans [Lettieri,
Brancaccio, and A. Larizza, Atti Soc. Peloritana Sci. Fis.
Mat. Nat., 1982 (Pub. 1983), 28, 133]; substituted 4-
-(substituted amino)chromans (J.M. Evans and F. Cassidy,
Eur. Pat. Appl. EP 126,367, 1984; Evans, *ibid*., 138,134,
1985); chroman-4-yl-isoindolones and -isoquinolinones
(E.A. Faruk, *ibid*., 93,535, 1983; Evans and V.A. Ashwood,
ibid., 158,923, 1985); chroman-4-yl substituent on the
nitrogen atom contained in the ring of an heterocycle
(*idem, ibid*., 107,423, 1984); and *trans*-4-piperidino-2,2-
-dimethyl-3-hydroxy-7-nitrochroman and its methanesulphonate
(Evans, Brit. Pat. 1,548,222, 1979). The preparations,
reactions, physical properties, and spectral data of nitro-
and amino-chromans have been reviewed [I.M. Lockhart, Chem.
heterocyclic Compds., John Wiley and Sons, 1981, 36
(Chromans and Tocopherols), 189].

Chemical Properties. It has been reported that an
electron-donating substituent in the benzene ring of a 2,2-
-dimethylchroman, facilitates its dehydrogenation by 2,3-
-dichloro-5,6-dicyanobenzoquinone to the related benzopyran
(A.K. Ahluwalia and R.S. Jolly, Synth., 1982, 74).

Treatment of chroman with lithamide in hexametapol in the presence of the electrophilic reagents, water and ethyl bromide, yields 2-(prop-1-enyl)phenol (1) and 1-ethoxy-2--(prop-1-enyl)benzene (2), respectively. Similar treatment with potassium butoxide in dimethyl sulphoxide in presence of water and methyl iodide yields the corresponding products (3 and 4) (E.A. Karakhanov, L.N. Kreindel, and E.A. Runova, Vestn. Mosk. Univ., Ser. 2, Khim., 1983, 24, 286; Karakhanov, S.V. Sharipova, and L.N. Borodina, Chem. Abs., 1980, 92, 180953u.)

(1) R=OH   (43%)
(2) R=OEt (87%)
(3) R=OH   (50%)
(4) R=OMe (72%)

Acid-induced line broadening in the ^{1}H-nmr spectra of 6-hydroxychroman derivatives is produced by acids at least as strong as trichloroacetic acid. The effect is specific for derivatives of hydroquinone, although it is weak in the absence of the heterocyclic rings. It is attributed to the formation of traces of cation radicals and is destroyed by bases or water (I. Al-Khayat et al., J. chem. Soc., Perkin 1, 1985, 1301). The reactivities of chroman, 2,2-dimethyl-chroman and tetralin toward $Me(CH_2)_{10}$, $Ph^\bullet$, and $Me_3CO^\bullet$ in carbon tetrachloride have been investigated (S.S. Zlotskii, D.L. Rakhmankulov, and Borodina, Vestn. Mosk. Univ., Ser. 2, Khim., 1979, 20, 164).

Metastable studies and mass measurements referring to the mass spectrum of chroman show that $[C_6H_6^+]$ (5) from chroman is produced in a two-step process (M.L. Gross *et al.*, J. Amer. chem. Soc., 1977, **99**, 3603). The mass spectra of chroman, homochroman, their bisheterocyclic analogues, and their monoketone derivatives (M.C. Sacquet, B. Graffe, and P. Maitte, Org. mass Spec., 1976, **11**, 1128); and of 5-, 6-, 7-, and 8-nitrochromans (J.H. Bowie and A.C. Ho, Austral. J. Chem., 1977, **30**, 675); and the thermochemical properties of chroman and other polycyclic compounds containing one oxygen atom and one to three rings (R. Shaw, D.M. Golden, and S.W. Benson, J. phys. Chem., 1977, **81**, 1716) have been reported.

(5)

(b) Tocopherols*

A number of compounds related to the tocopherols have been prepared for use as antioxidants and as stabilizers for organic masses, particularly polymers. 6-Hydroxy-2-hydroxy-methyl-2-methylchromans have been prepared as intermediates in the synthesis of tocopherols.

The treatment of oxepinol (1) with trifluoroacetic acid for some hours at 19° causes quantitative conversion into 2-hydroxymethyl-2-methylchroman (2). The reaction is rapid on application of heat. Nmr and ir spectral data indicate that the immediate product is the trifluoroacetate, but

*   R.M. Parkhurst and W.A. Skinner, Chem. heterocyclic Compds., John Wiley and Sons, 1981, **36** (Chromans and Tocopherols), 59.

aqueous workup affords only the alcohol (2). This method (F.M. Dean and M.A. Jones, Tetrahedron Letters, 1983, 24, 2495) provides a shorter route than the previous one (P. Bravo and C. Ticozzi, J. heterocyclic Chem., 1978, 15, 1051) to 2-hydroxymethylchromans, which are of interest in regard to a number of natural products.

Me
Me
HO
Me

O
OH
Me

(1)

$\longrightarrow$

R^3
R^2
HO
R^1

O
CH_2OH
Me

(2) $R^1 = R^2 = R^3 = Me$
(3) $R^1, R^2, R^3 = H$, alkyl

2-Hydroxymethylchromans (3) are also obtained by the following route. Thus the cyclocondensation of methyl hydroxymethoxyacetate with 2-(anilinomethyl)pyrrolidine furnishes the pyrroloimidazole (5), which undergoes a Grignard reaction with 1-(2-bromoethyl)-2,5-dimethoxy--3,4,6-trimethylbenzene (4) to yield the propanoylpyrroloimidazole (6). The latter with methylmagnesium iodide, followed by hydrolysis and hydride reduction affords the diol (7), which on ammonium ceric nitrate oxidation gives the epoxybenzoxepinone (8). Hydrogenation of epoxide (8) yields 6-hydroxy-2-hydroxy-methyl-2,5,7,8-tetramethylchroman (2) (Y. Sakito and G. Suzukamo, Eur. Pat. Appl. EP 65,368, 1982). The 2--hydroxymethylchroman (9), a chiral intermediate for the synthesis of α-tocopherol, has been prepared by a related route involving an asymmetric synthesis of one of the starting materials (*idem*, Tetrahedron Letters, 1982, 23, 4953).

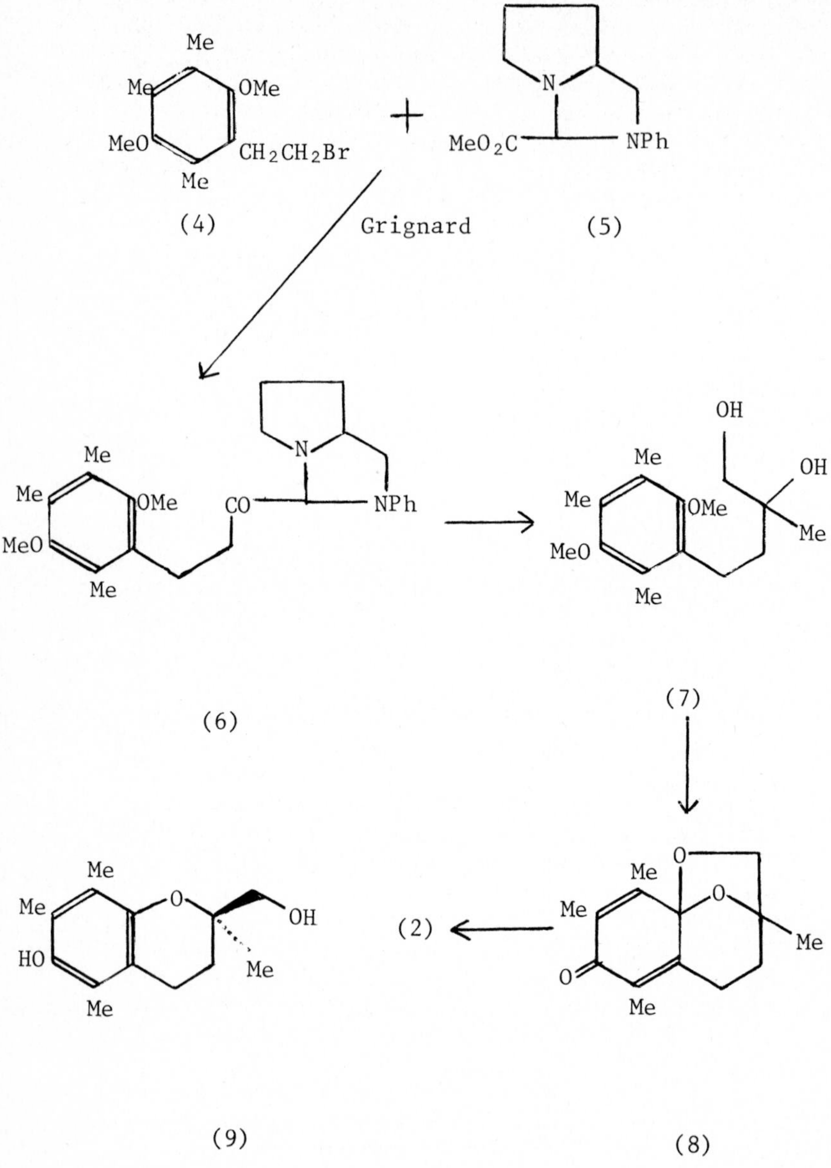
Me
Me
OMe
MeO
CH_2CH_2Br
Me
(4)
+
N
MeO_2C
NPh
(5)
Grignard
Me
Me
OMe
MeO
Me
CO
N
NPh
(6)
OH
OH
Me
Me
OMe
Me
MeO
Me
(7)
O
Me
O
Me
Me
O
Me
(8)
(2)
Me
Me
O
OH
HO
Me
Me
(9)

The following sequence shows the conversion of a 2-(3-
-chloro-3-methylbutyl)phenol (10) into a 2-hydroxymethyl-2-
-methylchroman. 2-(3-Chloro-3-methylbutyl)phenol (10) on
treatment with triphenylmethyllithium yields the terminal
olefin (11), which reacts with 3-chloroperbenzoic acid to
furnish epoxide (12). Base-catalyzed cyclization of the
latter gives a mixture of 2-hydroxymethyl-2-methylchroman
(13) and benzoxepin (14) (K.J. Baird and M.F. Grundon, J.
chem. Soc., Perkin I, 1980, 1820).

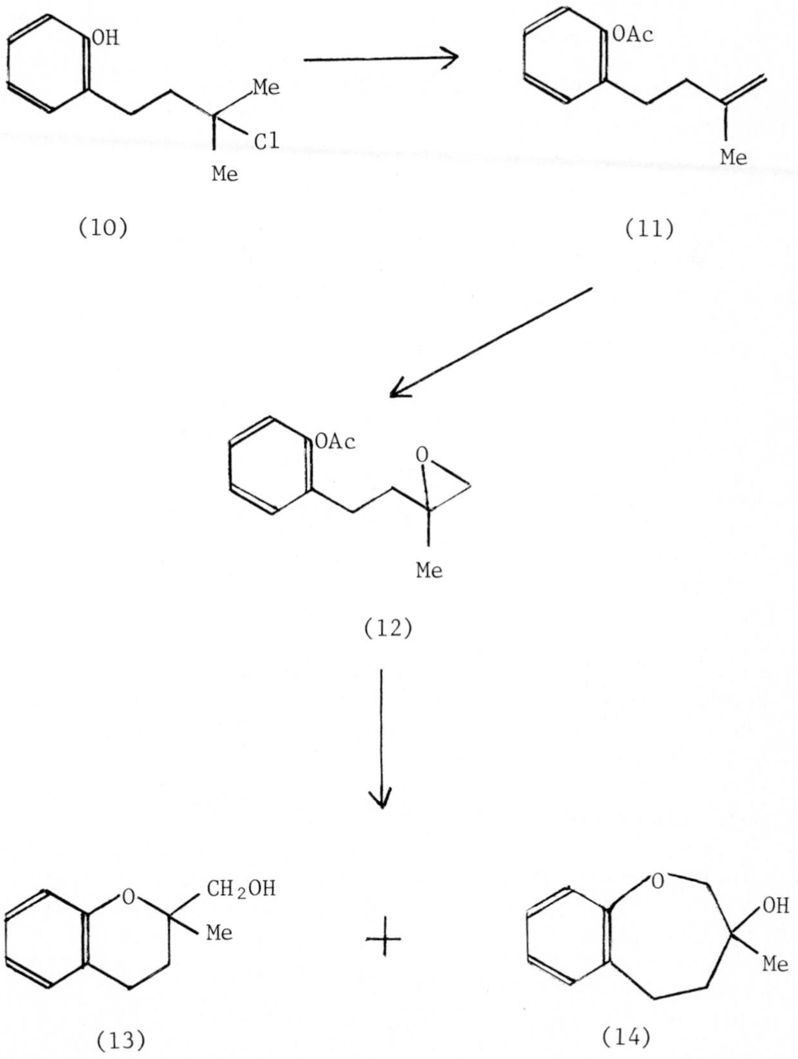

(10)

(11)

(12)

(13)

(14)

2-Hydroxymethyl-2-methylchroman (13; $R^1 = R^2 = H$) (48%),
an oil, *4-toluenesulphonate*, prisms from ethanol, m.p. 76-77°.
A number of 2-hydroxyalkylchromans (15; R^1 = H or group
easily removed by hydrogenolysis; R^{2-5} = H or C_{1-8} alkyl;
n = 1-3), for example, 6-hydroxy-2-(2-hydroxyethyl)-2,5,7,8-
-tetramethylchroman (16) have been reported (M. Horner and
A. Nissen, Ger. Offen. DE 3,010,504, 1981).

(15)

(16) R=H
(17) R=COCH=CH$_2$
(18) R=COCH$_2$CH$_2$SBu

The hydroxychroman (16) reacts with methyl acrylate in
toluene in the presence of toluene-4-sulphonic acid and
hydroquinone to give the acrylate (17), which adds
butanethiol in THF containing sodium methoxide to yield
sulphide (18). A number of related compounds with a variety
of substituents on the benzene ring and modified C-2 side-
-chain have been synthesized (Horner *et al., ibid.,*
3,103,707, 1982).

Reaction between 4-benzoyloxy-2-methylphenol (19) and
γ-vinyl-γ-valerolactone (20), followed by deprotection or
hydrolysis of the resulting chroman derivative (21) yields
2,7-dimethyl-6-hydroxychroman-2-propionic acid (22) (Eisai
Co., Ltd., Japan Kokai Tokkyo Koho JP 57,145,871
[82,145,870], 1982).

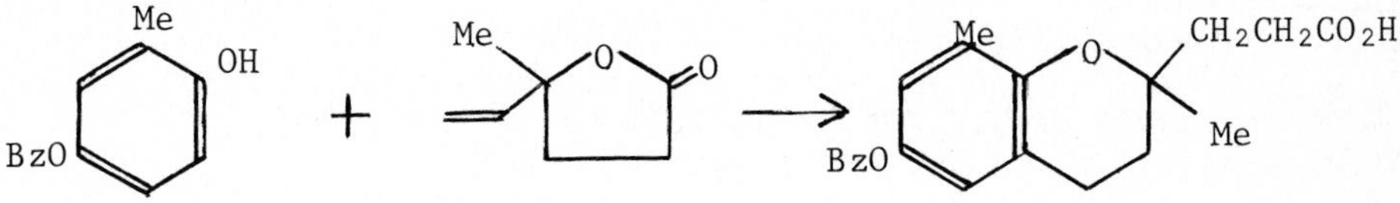

(19)　　　　　　　　　　(20)　　　　　　　　　　(21)

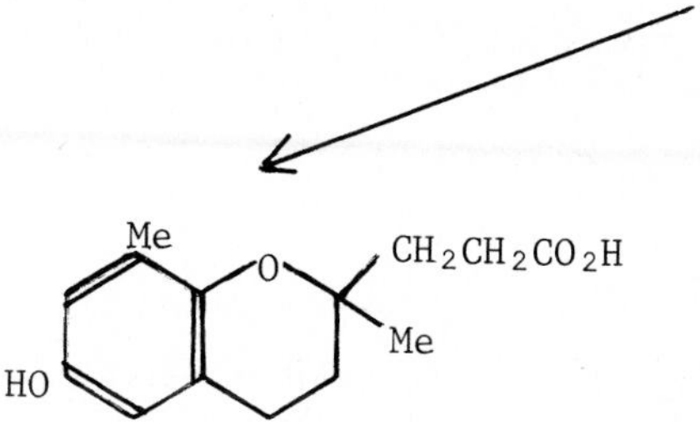

(22)

A number of derivatives related to the tocopherols have been prepared by replacing the hydrogen of the 6-hydroxyl group by a complex group, for instance, the reaction of 6--hydroxy-2,5,7,8-tetramethyl-2-(4,8,12-trimethyltridecyl)-chroman (23) with epichlorohydrin, followed by treatment of the resulting intermediate with isopropylamine furnishes 6--[2-hydroxy-3-(isopropylamino)propoxy]-2,5,7,8-tetramethyl--2-(4,8,12-trimethyltridecyl)chroman (24) (Fujimoto Pharmaceutical Co., Ltd., Japan Kokai Tokkyo Koho JP 57,175,186, 1982).

Me · O · Me · Me · RO · Me · Me · Me · Me · Me · Me

(23) R=H
(24) R=Me$_3$CHNHCH$_2$CH(OH)CH$_2$

A large number of compounds related to the tocopherols have
been synthesized and new syntheses developed, but most of
the effort has been to circumvent a patent position rather
than provide technical advantage. For a review on the
extraction of tocopherols from natural sources and their
synthesis, properties, biological activity, and clinical
application see K. Kostova and V. Lyubenova [Priroda
(Sofia), 1978, <u>27</u>, 60] and of the chemistry of tocopherols
and tocotrienols, S. Kasparek [Clin. Nutr., 1980, <u>1</u> (Vit.
E, Compr. Treatise), 7].

(c) Phenylchromans (flavans and isoflavans) 3,4-dihydro-
   -phenyl-2H-benzo[b]pyrans

  (i) Flavan, 2-phenylchroman, 2,3-dihydro-2-phenylbenzo[b]-
    pyran derivatives
Preparation. 7,2',4'-trihydroxy-2,4,4-trimethylflavan
(1) is obtained by the condensation of resorcinol with
mesityl oxide or acetone in the presence of Lewis acids
(Mitsui Petrochemical Industries Ltd. Japan Kokai Tokkyo
Koho JP 82, 114,585, 1982).

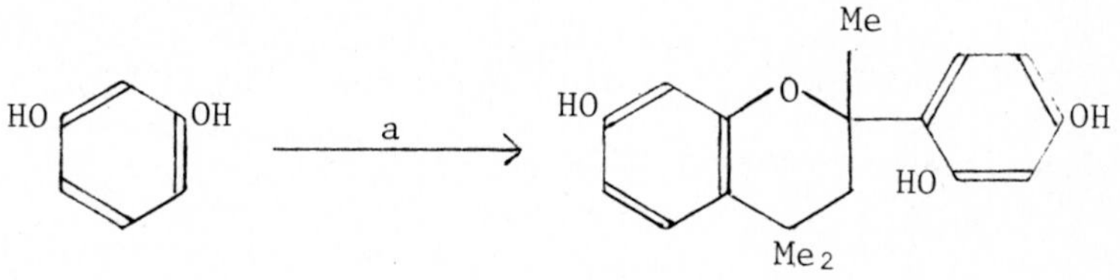

(1)

Reagents:- (a) $Me_2C=CHCOMe$, BF_3-Et_2O,
CS$_2$, boil 5h, yield 77%

2'-Hydroxy-4-methylchalcone on boiling with 2-methoxy-ethanol in the presence of phosphoric acid affords 4'-methyl-flavanone, which on Clemmensen reduction furnishes 4'-methyl-flavan (2; $R^1 = R^2 = H$). A number of derivatives of 4'--methylflavan (2; R^1 can be H or up to four substituents from halogeno, NO_2, CN, CF_3, alkyl, alkyloxy, NH_2, alkylamino, OH; R^2 can be H or up to five substituents as for R^1) have been prepared (J.F. Batchelor *et al.*, Eur. Pat. Appl. 4,579, 1979).

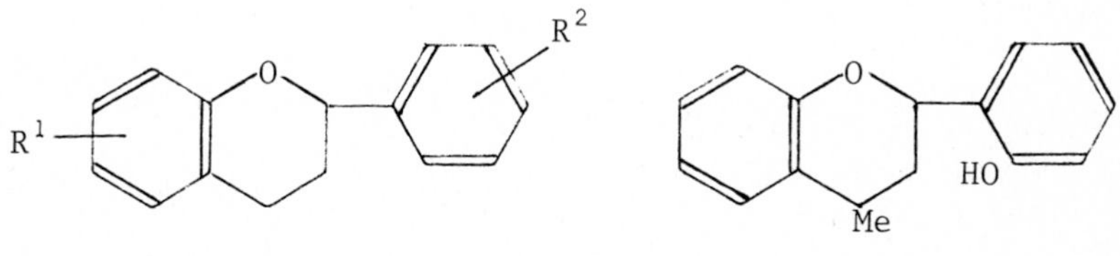

(2)

(3)

Thermal dimerization of 2-vinylphenol affords quantitatively two isomers *cis*- and *trans*-2'-hydroxy-4--methylflavan (3). Analysis of the dimers of of 2--vinylphenol-*O*-d indicates that a quinone methide intermediate is responsible for the formation of these flavans (Y.Y. Chen, R. Oshima, and J. Kumanotani, Bull. chem. Soc., Japan, 1983, <u>56</u>, 2533).

The bromination of flav-2-enes (4) with *N*-bromosuccinimide in methanol affords 2,3-*cis*-3-bromo-2-methoxyflavans (5) in excellent yield. It is evident that addition of bromine to the double bond of flav-2-enes occurs more rapidly than substitution into the aromatic nuclei. 5,7,3',4'-Tetra-methoxyflav-2-ene (4; $R^1 = R^2 = R^3 = R^4$ = OMe) with a highly activated A ring, did not undergo aromatic bromination. Treatment of 2,3-*cis*-3-bromo-2-methoxyflavan (5; $R^1 = R^2 = R^3 = R^4$ = H) with ethanol and a trace of aluminium chloride results in exchange of the 2-alkoxy group and formation of 2,3-*cis*-3-bromo-2-ethoxyflavan (6). Elimination of hydrogen bromide from the methoxybromoflavan (5; $R^1 = R^2 = R^3 = R^4$ = H) yields 2-methoxyflav-3-ene (7), which on treatment with osmium tetraoxide gives a 3,4-*cis*--diol. The diol and its bis-(4-nitrobenzoate) (8) from tentative interpretation of nmr and ir spectral data are proposed to have a 2,3-*cis*-configuration. Also reported are the preparations of number of flavan derivatives (Table 4) (T.G.C. Bird *et al*., J. chem. Soc., Perkin I, 1983, 1831).

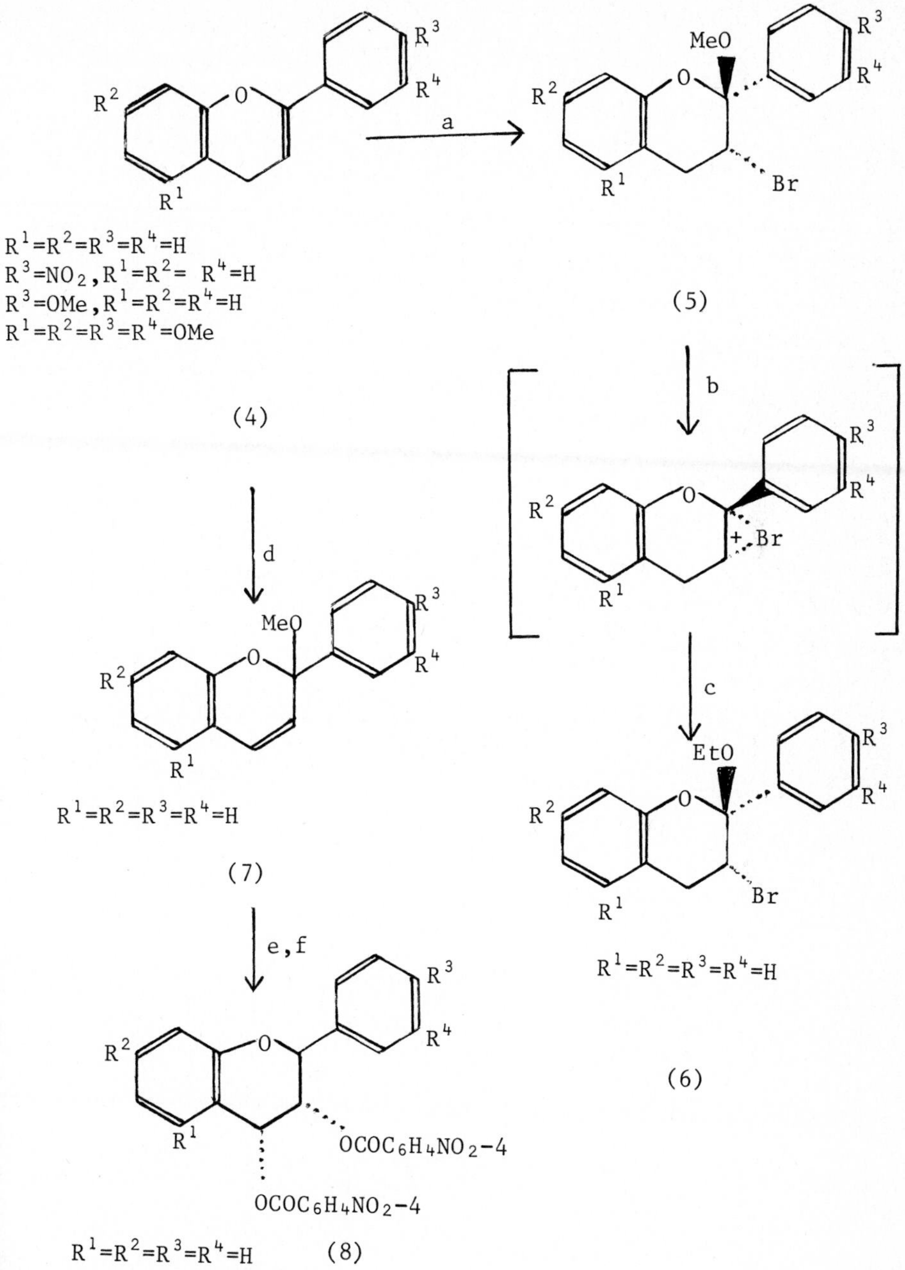

R¹=R²=R³=R⁴=H
R³=NO₂,R¹=R²= R⁴=H
R³=OMe,R¹=R²=R⁴=H
R¹=R²=R³=R⁴=OMe
(4)
a
MeO
Br
(5)
b
+ Br
c
EtO
Br
R¹=R²=R³=R⁴=H
(6)
d
MeO
R¹=R²=R³=R⁴=H
(7)
e,f
OCOC₆H₄NO₂-4
OCOC₆H₄NO₂-4
R¹=R²=R³=R⁴=H
(8)

Reagents:- (a) *N*-Bromosuccinimide-MeOH; (b) $AlCl_3$; (c) EtOH; (d) KOH; (e) OsO_4; (f) $4\text{-}O_2NC_6H_4COCl$

TABLE 4

Some flavan derivatives

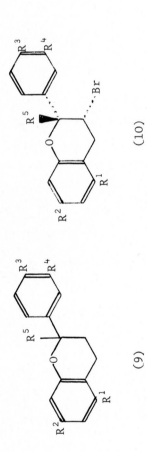

(9)　　　　(10)

Compound		R¹	R²	R³	R⁴	R⁵	M.p. (°C) / B.p. (°C/mm)
		R¹	R²	R³	R⁴	R⁵	
2-Methoxyflavan	(9)	H	H	H	H	OMe	oil
2-Ethoxyflavan	(9)	H	H	H	H	OEt	116/0.35
2,4'-Dimethoxyflavan	(9)	H	H	OMe	H	OMe	oil
2,3-*cis*-3-bromo-2-methoxy-flavan	(10)	H	H	H	H	OMe	134 – 135
2,3-*cis*-3-bromo-2-methoxy--4'-nitroflavan	(10)	H	H	NO₂	H	OMe	158.5 – 160

Compound	Substituents					M.p. (°C) / B.p. (°C/mm)
	R^1	R^2	R^3	R^4	R^5	
2,3-*cis*-3-bromo-2,4'-dimethoxyflavan (10)	H	H	OMe	H	OMe	117 – 118
2,3-*cis*-3-bromo-2,5,7,3',4'-pentamethoxyflavan (10)	OMe	OMe	OMe	OMe	OMe	131 – 132.5 (decomp.)
2,3-*cis*-3-bromo-2-ethoxyflavan (10)	H	H	H	H	OEt	100 – 101
cis-3-Bromoflavan (10)	H	H	H	H	H	110 – 120
cis-3-Bromo-4'-methoxyflavan (10)	H	H	OMe	H	H	113 – 114.5
cis-3-Bromo-5,7,3',4'-tetramethoxyflavan (10)	OMe	OMe	OMe	OMe	H	156 – 157 (decomp.)
2,3-*cis*-2-Acetoxy-3-bromoflavan (10)	H	H	H	H	OAc	119 – 120
2,3-*cis*-2-Acetoxy-3-bromo-4'-nitroflavan (10)	H	H	NO_2	H	OAc	141 – 144
2,3-*cis*-3-Bromo-2-(4-tolyloxy)flavan (10)	H	H	H	H	$4-MeC_6H_4O$	123 – 125

2-Acetoxy-3,3-dibromoflavan, from ether, m.p. 124-125°,
from methanol, m.p. 122-122.5; 3,3-dibromoflavan, m.p.
122.5-124°.

3-Nitroflavans (12) are formed in 51-82% yield by the
sodium tetrahydridoborate reduction of the corresponding
nitrobenzo[b]pyran (11) in methanol/THF (R.S. Varma,
M. Kadkhodayan, and G.W. Kabalka, Heterocycles, 1986, 24,
1647). Similarly 1,2-dihydro-2-nitro-3-phenyl-3H-naphtho-
[2,1-b]pyran (13) (85%) is obtained from 3-nitro-3-phenyl-
-3H-naphtho[2,1-b]pyran.

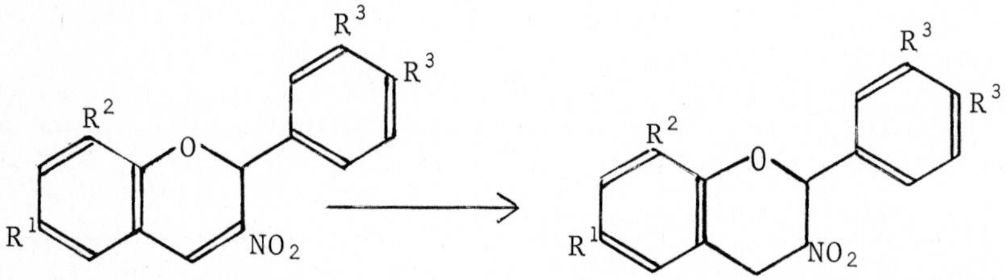

(12)

R^1=H,Cl,NO_2,R^2=R^3=H
R^1=R^3=H,R^2=OMe
R^1=R^2=H,R^3=OEt
R^1=H,R^2=OMe,R^3=OEt

(11)

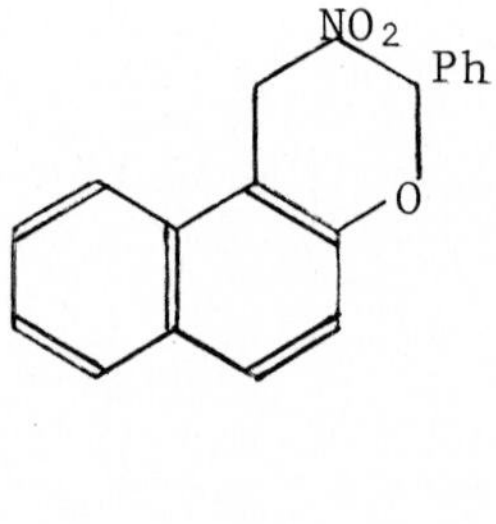

(13)

Naturally occurring flavans. An optically active flavan agylcone, 7-hydroxy-3',4'-methylenedioxyflavan (14) and its 7-glucoside have been isolated from the bulbs of *Zephyranthes flava* collected at flowering. Also for the first time the known flavans, 7,4'-dihydroxy-3-methoxyflavan and 2'-hydroxy-7-methoxy-4',5'-methylenedioxyflavan have been isolated from this species. Spectral data of these flavans and their acetates have been reported (S. Ghosal, S.K. Singh, and R.S. Srivastava, Phytochem., 1985, 24, 15).

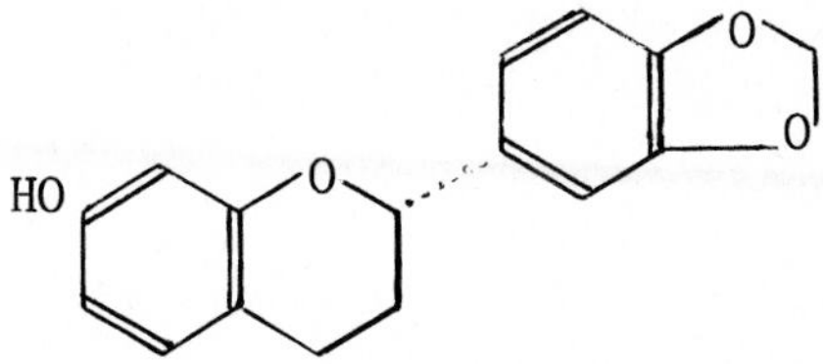

(14)

A prenylated flavan, 5,7-dimethoxy-8-prenylfavan (15), along with sitosterol, stigmasterol, and 5-hydroxy-7-methoxy- -8-prenylflavanone has been isolated from *Tephrosia madrensis* (F. Gomez *et al.*, *ibid.*, 1983, 22, 1305).

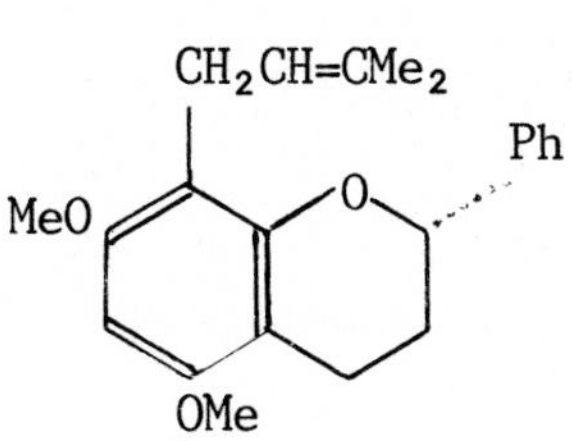

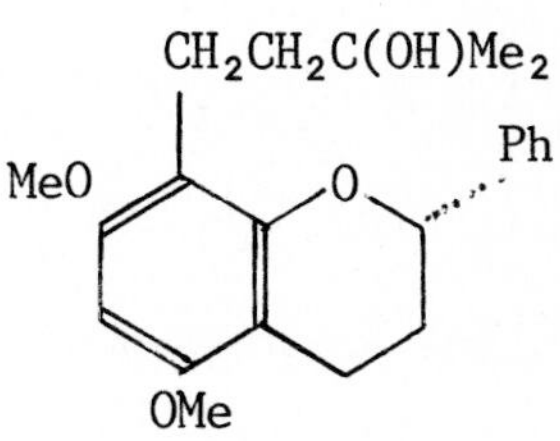

(15)   (16)

A related flavan, nitenin, *5,7-dimethoxy-8-(3-hydroxy-3-* *-methylbutyl)flavan* (16), a colourless viscous oil, $C_{22}H_{28}O_4$ ($M^+/356$), has been isolated from *Tephrosia nitens* Beth. (Leguminosae). Its structure and stereochemistry has been confirmed by dehydration to flavan (15). The configuration at C-2 in flavans (15) and (16) has been shown to be *S*. There are several reports suggesting that extracts of some species of genus *Tephrosia* have, piscicidal, insecticidal and repellent properties (A. Pelter *et al.*, J. chem. Soc., Perkin I, 1981, 2491).

Three phytoalexins, 7-hydroxy-, 7,4'-dihydroxy-, and 7,4'-dihydroxy-8-methyl-flavan, have been isolated from daffodil bulb scales inoculated with *Botrytis cinerea* (D.T. Coxon *et al.*, Phytochem., 1980, 19, 889). For a review of naturally occurring free and glycosylated flavans, unsubstituted in the heterocyclic ring, including distribution, methods of isolation, structure elucidation, and biological properties see K.S. Saini and S. Ghosal (*ibid.*, 1984, 23, 2415).

(ii) Isoflavan, 3-phenylchroman, 2,3-dihydro-3-phenylbenzo[b]- *pyran derivatives*

The key starting materials for the synthesis of isoflav-3- -enes (2) and hence isoflavans (3) are isoflavylium salts (1), which are also used for the synthesis of 3-phenyl-coumarins. The m.p.s of some isoflav-3-enes and isoflavans are given in Table 5 (C. Deschamps-Vallet, J.-B. Ilotse, and M. Meyer-Dayan, Tetrahedron Letters, 1983, 24, 3993).

$$(1) \qquad (2) \qquad (3)$$

Reagents:- (a) KBH_4, THF; (b) H_2, Pd/C

TABLE 5

Isoflav-3-ens and isoflavans

Isoflav-3-ene R (2)	M.p. (°C)	Yield (%)	Isoflavan R (3)	M.p. (°C)	Yield (%)
H	94	80	H	56	85
6 – OMe	74	90	6 – OMe	50	80
7 – OMe	109	65	7 – OMe	99	50
8 – OMe	65	50	8 – OMe	108	60
6 – OH	103	90	6 – OH	85	80
7 – OH	160	70	7 – OH	130	48
6 – Cl	101	98	6 – Cl	liq n_D^{25} 1.594	75

Naturally occurring isoflavans. Catalytic hydrogenation of daidzein, or preferably its *O,O*-diacetyl derivative over palladium/carbon prepared according to the Wessely and Prillinger method (1939) affords (±)-equol in good yield, whereas other palladium/carbon catalysts furnish mixtures of products (J.A. Lamberton, H. Suares, and K.G. Watson, Austral. J. Chem., 1978, **31**, 455).

An isoflavan, tentatively named pendulone, *7-hydroxy--3',4'-dimethoxyisoflavanquinone* (4), has been isolated from the heartwood of *Millettia pendula*, along with equol and two known isoflavonoids (Y. Hayashi *et al.*, Mokuzai Gakkaishi, 1978, **24**, 898).

(4)   (5)

(-)-Duartin, (+)-and (-)-mucronulatol, and the isoflavan-quinone, mucroquinone (5) have been obtained from wood samples of *Machaerium mucronulatum* and *M. villosum* (K. Kurosawa *et al.*, Phytochem., 1978, **17**, 1405). (±)-Mucronulatol and (+)-vestitol have been isolated from the creeper *Dalbergia variabilis* (*idem, ibid.*, p.1417), and (+)-mucronulatol, *(+)-(3R)-7,3'-dihydroxy-2',4'-dimethoxy-isoflavan* from the roots of *Astragalus lusitanicus* Lam. (J. De Pascual Teresa, J.C. Hernandez Aubanell, and M. Grande, An. Quim., 1979, **75**, 1005).

The fungus-inoculated leaflets of *Dalbergia sericea* produce several isoflavonoid phytoalexins including some known pterocarpans and the isoflavan, vestitol. A previously unknown phytoalexin, neovestitol, *2',4'-dihydroxy-7-methoxyisoflavan* (6) has also been isolated (J.L. Ingham, Z. Naturforsch., Biosci., 1979, 34C, 630).

(6)   (7) R = OMe
             (8) R = OH

The fungus-inoculated leaflets of a number of *Trifolium* species have been examined for the presence of isoflavonoid and non flavonoid phytoalexins. Among the compounds isolated were a number of isoflavans, vestitol, isovestitol, sativan, isosativan, and arvensan (*idem,* Biochem. syst. Ecol., 1978, 6, 217). Demethylvestitol, isovestitol, and laxifloran are phytoalexins obtained from the fungi--inoculated hypocotyls of *Lablab niger* (*idem,* Z. Naturforsch., Biosci., 1977, 32C, 1018). Feeding experiments related to the biosynthesis of isoflavan and pterocarpan phytoalexins in *Medicago sativa* show that 7,2'--dihydroxy-4'-methoxy-isoflavone-Me-^{14}C and -isoflavanone--Me-^{14}C are precursorsof sativan (7), vestitol (8), and demethylhomopterocarpin(9). It has also been shown that compound (9) is incorporated into compounds (7) and (8) and (8) into (7) and (9), thus indicating that compounds (8) and (9) are interconvertible (P.M. Dewick and M. Martin, Phytochem., 1979, 18, 591).

(9)

The phenolic content of the heartwood of *Dalbergia nitidula* Welw. *ex* Bak. has been re-examined and has provided the first indication of the natural existence of oligomers among the isoflavonoid metabolites. One isolated in low yield from the methanol extract has been characterized as the dimeric (3S, 4S) -3,4-*trans*-4-[(3S)-7,6'-dihydroxy-4'-methoxyisoflavan-3'-yl]- -7,2'-dihydroxy-4'methoxyisoflavan (12) and its structure has been confirmed by synthesis from the concomitant (3S)- -7,2'-dihydroxy-4'-methoxyisoflavan [(+)-vestitol] (10) and (6aS, 11aS)-3-hydroxy-9-methoxypterocarpan [(+)-medicarpin] (11). Under acidic (2M HCl, 25°) conditions the pterocarpan (11), as prospective electrophile, and the nucleophile isoflavan (10) condense to give the natural [4,3']-bi-isoflavan (12) and its [4,6]-positional isomer (14). The [1]H-nmr spectra of these isomeric ($C_{32}H_{30}O_8$) compounds (12) and (14) support the proposed structures as do their mass spectra. Corresponding results are obtained from the fully methylated ether derivatives (13) and (15) (B.C.B. Bezuidenhoudt, E.V. Brandt, and D.G. Roux, J. chem. Soc., Perkin I, 1984, 2767).

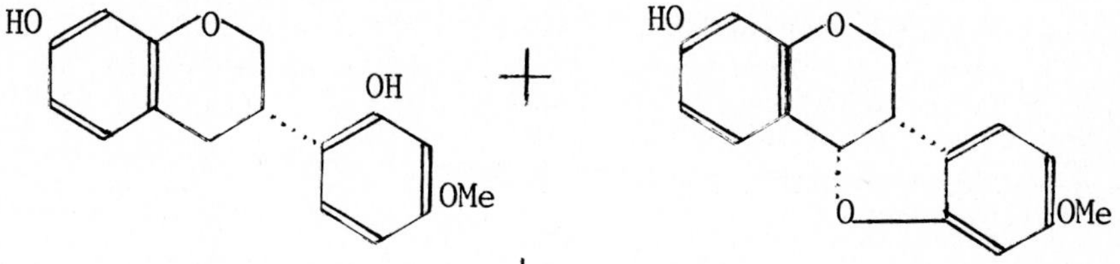

(10)   (11)

H^+ | EtOH – H_2O, 25°

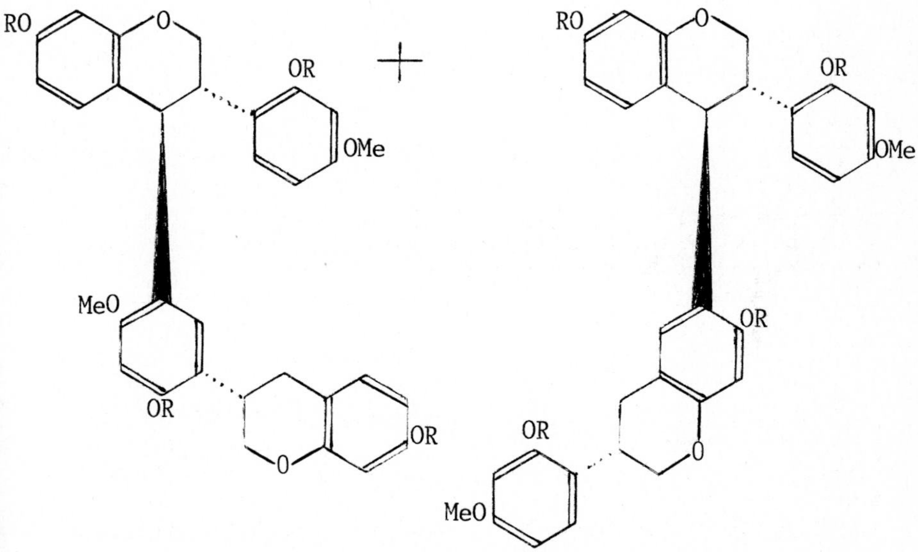

(12) R = H (20%)
(13) R = Me

(14) R = H (15%)
(15) R = Me

(d) Chromanols

Chroman-2-ols (3) are prepared by the catalytic palladium (II) reaction of (2-hydroxyaryl)mercury chlorides (1) with α,β-unsaturated ketones (2) in an acidic two-phase system. Treatment of the chroman-2-ols (3) with trifluoromethanesulphonic acid affords the corresponding 2-ethoxychroman (4) (S. Cacchi, D. Misiti, and G. Palmieri, J. org. Chem., 1982, <u>47</u>, 2995).

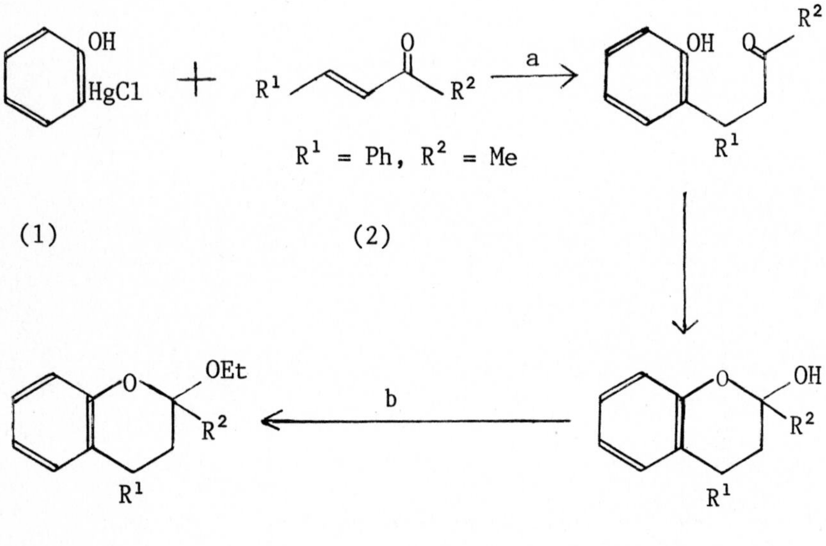

$R^1 = Ph, R^2 = Me$

(1)   (2)

(4)   (3)

Reagents:- (a) $(n-Bu)_4N^+Cl^-$, $PdCl_2$, CH_2Cl_2, 3M HCl, R.T. 4h;
          (b) CF_3SO_3H, EtOH, R.T.

2-Methyl-4-phenylchroman-2-ol (3; R^1 = Ph, R^2 = Me) (86%);
2-benzyl-4-phenylchroman-2-ol (3; R^1 = Ph, R^2 = PhCH$_2$) (91%);
2-ethoxy-2-methyl-4-phenylchroman (98%), m.p. 72-73°;
2-benzyl-2-ethoxy-4-phenylchroman (98%), m.p. 100-101°;
1-ethoxy-3,4-benzo-2-oxabicyclo[3.3.1]nonane [4; R^1R^2 =
(CH$_2$)$_3$] (50%) m.p. 54-55°.
Treatment of flav-2-ene in methanol with aluminium
chloride gives 2-methoxyflavan (5) as an oil; 2-ethoxy-
flavan, b.p. 116°/0.35 mm; 2,4'-dimethoxyflavan, an oil
(T.G.C. Bird *et al.*, J. chem. Soc., Perkin I, 1983, 1831).

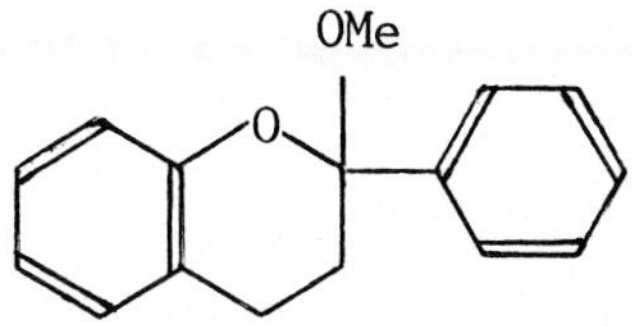

(5)

A number of 4-(substituted amino)-6-cyano-2,2-dimethyl-
chroman-3-ols and 3-alkoxy-(or acyloxy)-4-(substituted
amino)-6-cyano-2,2-dimethylchromans (7) have been prepared
by treating 6-cyano-2,2-dimethyl-3,4-epoxychroman (6) with
the appropriate ω-halogenoalkylamine hydrochloride, for
example, 3-chloropropylamine hydrochloride affords 4-(3-
-chloropropylamino)-6-cyano-2,2-dimethylchroman-3-ol (7;
R^1 = Cl, R^2 = H, n = 3) (J.M. Evans Eur. Pat. Appl. EP
46,652,1982).

NC ... O ... Me$_2$

⟶

NC ... O ... Me$_2$ ... OR^2

NH(CH$_2$)$_n$R^1

R^1 = halogeno, R^2 = H, alkyl acyl, n =2,3,4

(6)   (7)

A number of related derivatives have been prepared (*idem, ibid.*, 139,992, 1985).

*(e) Flavanols, 3-hydroxy-2-phenylchromans, 2,3-dihydro-3-
    -hydroxy-2-phenyl-4H-benzo[b]pyrans, catechins, and
    related condensed tannins*

(i) Flavanols
   A flav-2-ene (e.g. 1) on treatment with perbenzoic acid
and an alcohol yields 2-alkoxyflavan-3-ol, generally as a
mixture of the 2,3-*cis*- (4) and the 2,3-*trans*- (5) isomers
(4 > 5) which are easily separated. The reaction presumably
proceeds through the protonated 2,3-epoxide (2), which is
opened by reaction with an alcohol. Attack of methanol by
an S_N2 mechanism at C-2 of the protonated 2,3-epoxide (2)
would give rise to a *cis*-product (*ie.* a product with the 2-
-methoxy group *trans* to the 3-hydroxy group). Opening of the
epoxide ring by an S_N1 mechanism would give a 2-carbonium ion
(3), which might then be attacked by methanol from either
side, to give rise to either a *cis*- or a *trans*-product,
though the steric effect of the 3-hydroxy group might be
expected to favour formation of the *cis*-product. Removal of
the bromine atom from 2-acetoxy-3-bromoflavan (6) with tri-
-n-butyltin hydride in benzene yields 2,3-*cis*-3-acetoxy-
flavan (7), which on hydrolysis affords 2,3-*cis*-flavan-3-ol
(8). A number of flavan-3-ol derivatives have been
prepared and their m.p.s are given in Table 6 (T.G.C. Bird
et al., J. chem. Soc., Perkin I, 1983, 1831).

TABLE 6

Flavan-3-ol derivatives

Derivative	M.p. (°C)	4-Toluene-sulphonate M.p. (°C)
cis-Flavan-3-ol	81 – 82	
cis-3-Acetoxyflavan	108 – 109 (lit., 110)	
cis-4'-Methoxyflavan-3-ol	131 – 132	
2,3-*cis*-2-Methoxy-4'-nitroflavan-3-ol	oil	182 – 184
2,3-*cis*-2,4'-dimethoxyflavan-3-ol	oil	129.5 – 130.5
2,3-*trans*-2,4'-dimethoxyflavan-3-ol	122 – 124	
2,3-*cis*-2-ethoxy-4'-methoxyflavan-3-ol	oil	134 – 137
2,3-*cis*-2-cyclohexyloxy-4'-methoxy-flavan-3-ol	oil	150 – 154
2,3-*trans*-2-cyclohexyloxy-4'-methoxy-flavan-3-ol	159 – 166.5	

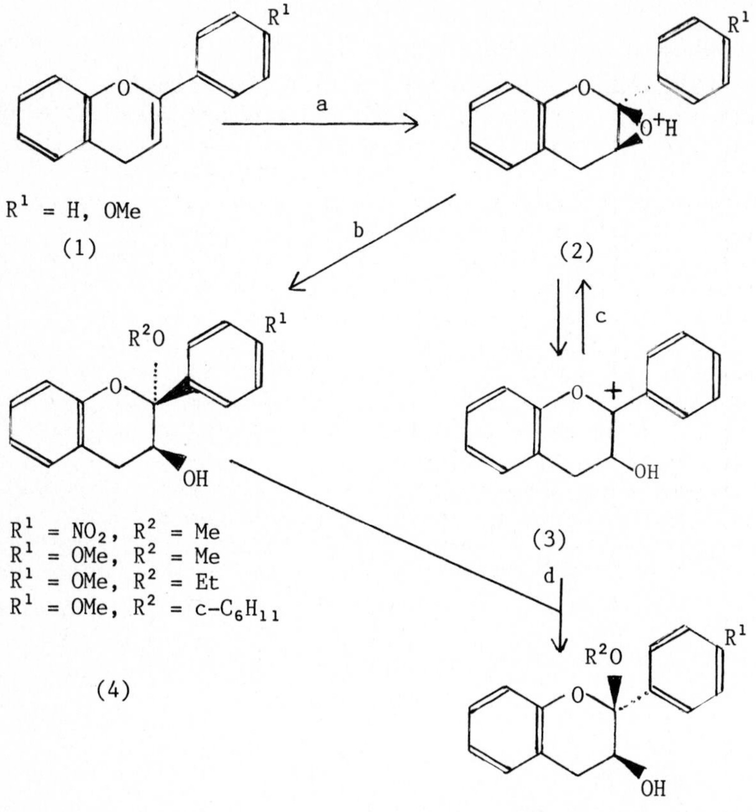
a
b
c
d
R^1 = H, OMe
(1)
(2)
(3)
$R^1 = NO_2$, R^2 = Me
R^1 = OMe, R^2 = Me
R^1 = OMe, R^2 = Et
R^1 = OMe, R^2 = c-C_6H_{11}
(4)
R^1 = OMe, R^2 = Me
R^1 = OMe, R^2 = c-C_6H_{11}
(5)

Reagents:- (a) $ArCO_3H$; (b) R^2OH, SN^2; (c) SN^1; (d) R^2OH

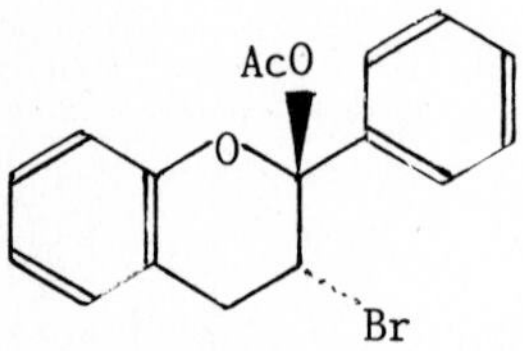

(6)

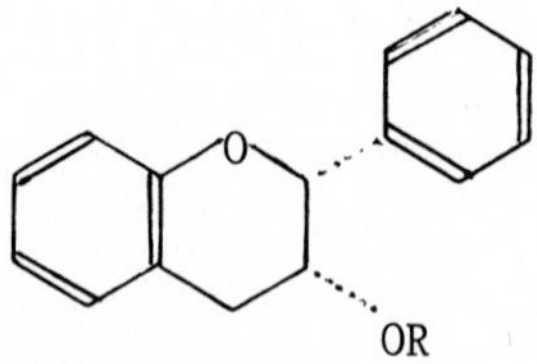

(7) R = Ac
(8) R = H

(±)-7,4'-Dihydroxy-3'-methoxyflavan has been isolated
along with other products from the trunk wood of Brazilian
Iryanthera elliptica (R.B. Filho *et al.*, Phytochem., 1980,
19 455).

(ii) Catechins
The rate of methylolation and that of condensation of
(+)-catechin (1) with formaldehyde in aqueous solution over
pH range 6-9 and temperature range 10-85° have been studied;
(+)-catechin being regarded as an adequate model for
investigating the polymerization of tannins with formaldehyde.
First-order kinetics are observed for the rates of
disappearance of (+)-catechin and formaldehyde as well as for
the increase in molecular weight *via* condensation. At lower
pH, k(methylolation) $\simeq k$(condensation), the reaction mixture
consisting mainly of monomers and polymers. At pH9,
k(methylolation) > k(condensation), the reaction mixture
containing mostly oligomers. Undesirable side reactions of
(+)-catechin and of formaldehyde are much slower than those
of (+)-catechin with formaldehyde. The polymerisation of
formaldehyde with phenol is much slower than that with (+)-
-catechin, suggesting that (+)-catechin cannot directly
replace phenol in resin systems (P. Kiatgrajai *et al.*, J.
org. Chem., 1982, 47, 2913).

(1)

Treatment of (+)-catechin with two equivalents of bromine yields the 6,8-dibromocatechin (2), whereas the reaction with one equivalent affords the 8-bromo derivative (3) and some 6-bromocatechin (4) (E. Kiehlmann, P.J. Van der Merwe, and H.K.L. Hundt, Org. Prep. Proced. Internat., 1983, 15, 341).

(2) $R^1 = R^2 = Br$
(3) $R^1 = H$, $R^2 = Br$
(4) $R^1 = Br$, $R^2 = H$

Catechins have been obtained from new sources, the following being some examples. (+)-Catechin, (±)-
-gallocatechin, and an unidentified dimeric flavan have been isolated from *Betula pendula* bark (G. Chernyaeva *et al.*, Khim. Prir. Soedin., 1983, 531), and (-)-epicatechin along with other products from the fruit of Shan-Li-Hong (hawthorn *Crataegus pinnatifida* var *major*) (Y. Xie *et al.*, Zhiqu Xuebao, 1981, 23, 383).

It is found that in bast fibres and roots of kenaf (*Hibiscus cannabinus*) at all vegetative periods, flavan compounds are present, reaching maximum levels near the flowering period and becoming fairly constant at harvest. (-)-Epicatechin is the predominate monomer and among the products from hydrolysis of the tannins are (-)-epicatechin gallate, (+)-catechin, (-)-epigallocatechin and (-)-
-epicatechin (Phan Van Thien, B. Makhsudova, and O.S. Otroshchenko, Khim. Prir. Soedin., 1981, 558).

Extraction of powdered rhubarb with acetone, followed by
treatment of the concentrated extract with other solvents,
and separation of the products by chromatography affords
[4,8']-bis-3-O-galloyl-(-)-epicatechin (5) and [4,8']-3-O-
-galloyl(-)-epicatechin-(+)-catechin (6), which when left
overnight at room temperature in a 1:1 mixture of acetic
anhydride and pyridine are converted into their respective,
tetradecaacetate (7) and dodecaacetate (8) (Hisamitsu
Pharmaceutical Co. Inc. Japan Kokai Tokkyo Koho JP 81
92,283, 1981).

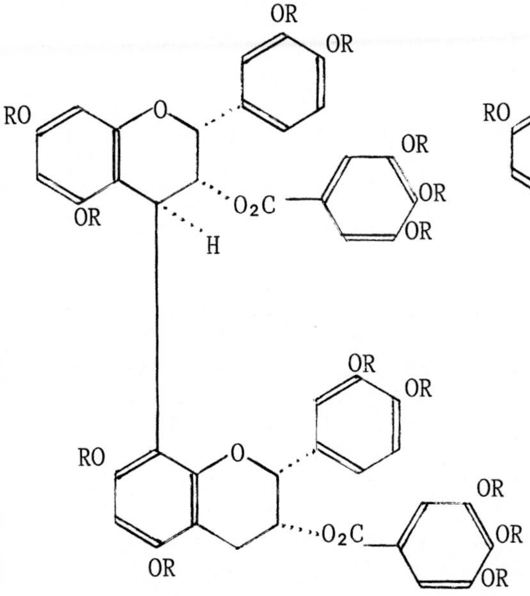

(5) R = H   (6) R = H
(7) R = Ac   (8) R = Ac

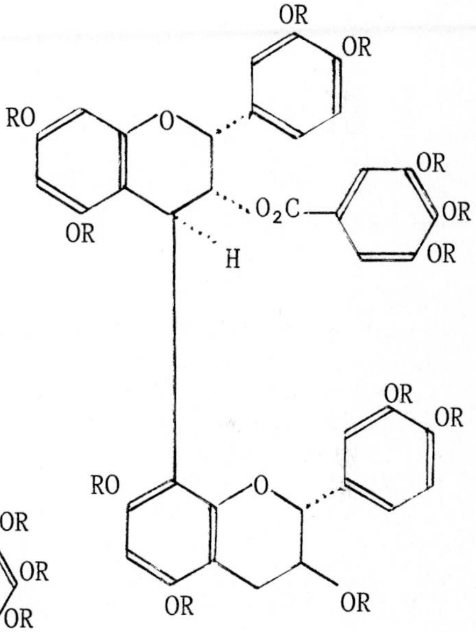

246

(iii) Brazilin and haematoxylin

The chemistry of brazilin (1) and that of haematoxylin
(vegetable products related to catechin) is discussed
separately in Chapter 22, Vol IVE, of the 2nd Edition, but
since the additional chemistry reported has been very
small, they do not warrant a separate chapter and are,
therefore, now included here.

In the present nomenclature brazilin is 7,11b-dihydro-
-3,6a,9,10(H)-tetrahydroxybenz[b]indeno[1,2-d]pyran or
7,11b-dihydrobenz[b]indeno[1,2-d]pyran-3,6a,9,10(6H)-tetrol
(Chem. Abs.).

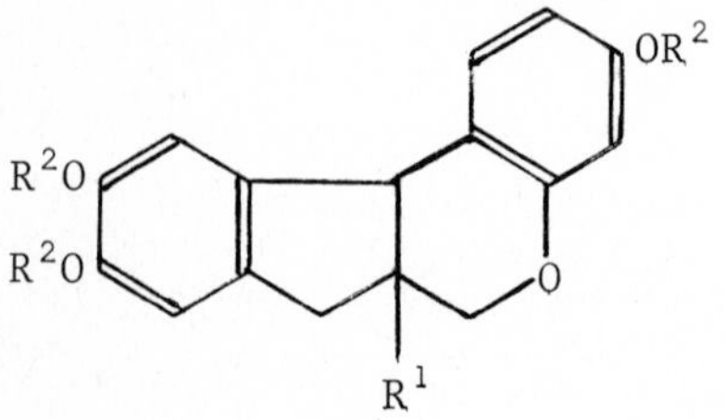

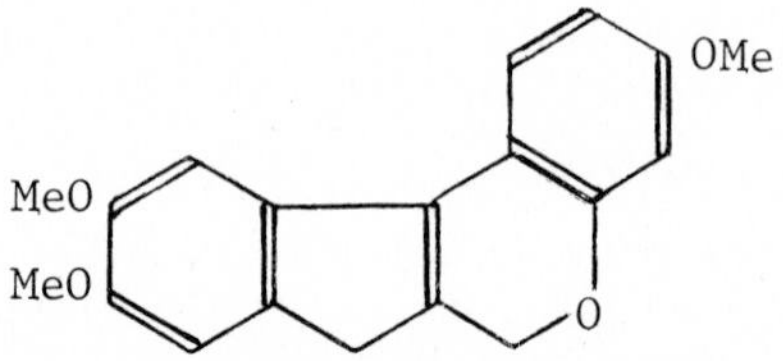

(2)

(1) R^1=OH, R^2=H
(3) R^1=β-OH,R^2=Me
(4) R^1=OMe,R^2=Me
(6) R^1=H,R^2=Me
(7) R^1=α-OH,R^2=Me

Anhydro-O-trimethylbrazilin (2) on reduction with
LiAlH$_4$/BF$_3$·Et$_2$O, followed by hydrogen peroxide oxidation
gives *cis*-O-trimethylbrazilin (3), which on methylation
with methyl iodide affords (±)-O-tetramethylbrazilin (4)
(J.N. Chatterjea, S.C. Shaw, and N.D. Sinha, J. Indian

chem. Soc., 1974, 51, 752). (±)-*O*-Trimethylbrazilin (3)
has also been obtained by the reaction of trimethoxyindenocoumarin (5) in diethylene glycol dimethyl ether and tetrahydrofuran with diborane, followed by treatment with sodium
hydroxyide/hydrogen peroxide. *O*-Trimethylbrazilane (6) and
trans-trimethylbrazilin (7) are also formed
(B.S. Kirkiacharian *et al.*, Tetrahedron Letters, 1974,
3667).

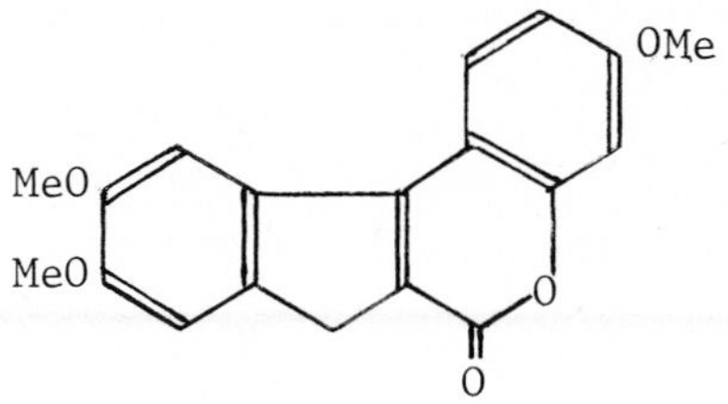

(5)

Reduction of trimethoxyindenocoumarin (5) with lithium
tetrahydridoaluminate give *O*-trimethylanhydrobrazilin (2)
and with sodium in amyl alcohol it affords a phenolic
alcohol (8), which can be converted into *O*-trimethyl-
brazilane (6). The indenocoumarin (5) on hydrogenation
using Adam's catalyst gives the dihydro derivative (9),
which with lithium tetrahydridoaluminate affords a phenolic
alcohol. The latter is converted on treatment with
phosphorus tribromide, followed by aqueous sodium
hydroxide, into *O*-trimethyl*cis*brazilane (10) (Chatterjea,
R. Robinson, M.L. Tomlinson, Tetrahedron, 1974, 30, 507).

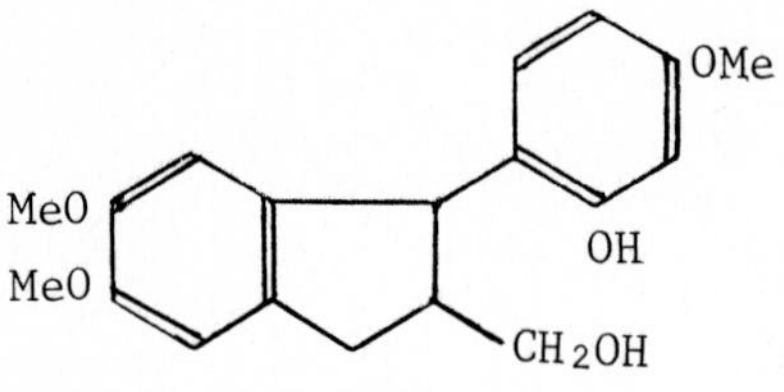

(8)

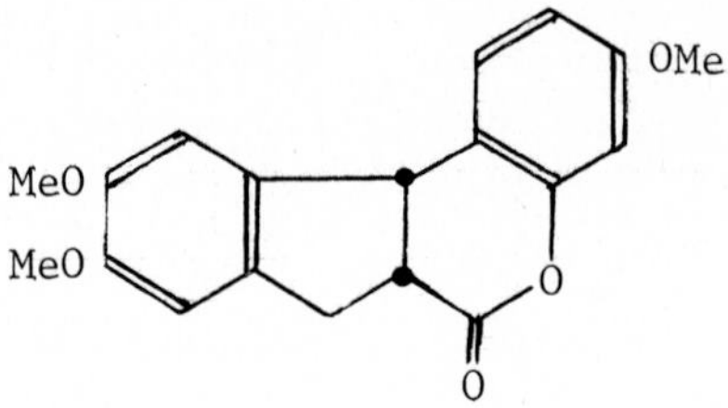

(9)

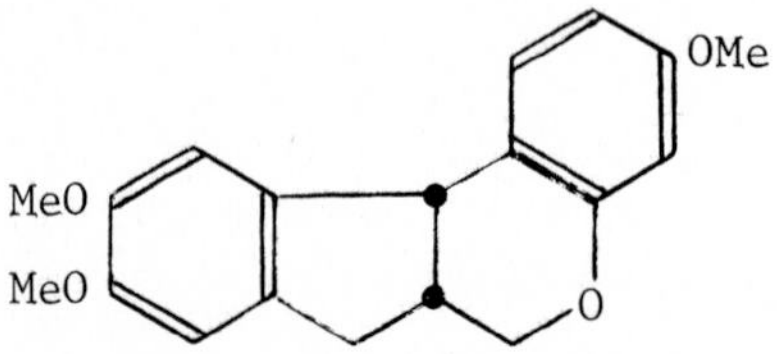

(10)

Reduction of brazilein (11) with zinc dust and acetic acid, followed by acetylation yields a substance, which has hitherto been regarded as a $C_{16}H_8(OAC)_4$ compound. It is now known to be the hexacetyl derivative (12) of a bibrazilanyl formed from two molecules of brazilein (11) by reductive coupling. Also, O-triacetyl-cis-brazilane (13) has been found to be a major, and anhydro-O-tri-acetylbrazilin (2; Me = Ac) a minor reaction product (R.H. Jaeger, P.M.E. Lewis, and Robinson, *ibid.*, p.1295). The structural and conformational details in the dimeric molecule (12), have been defined from data obtained from the X-ray crystal structure analysis of the hexamethyl analogue (14) (M.F. MacKay and N.W. Isaac, *ibid.*, 1979, **35**, 1893).

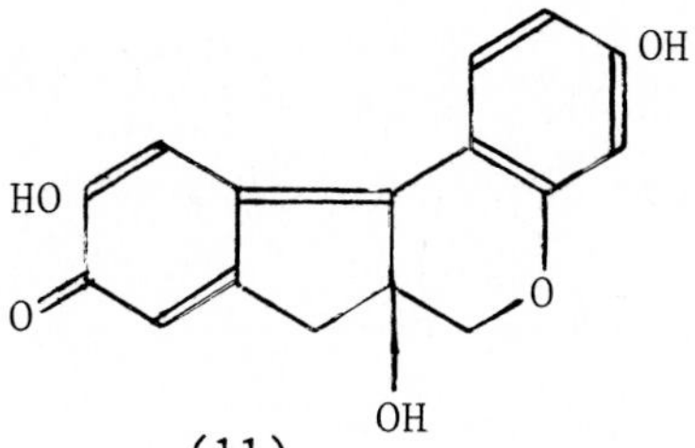

(11)

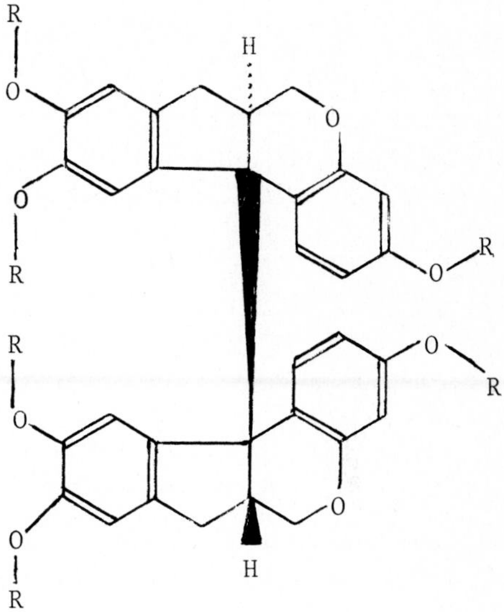

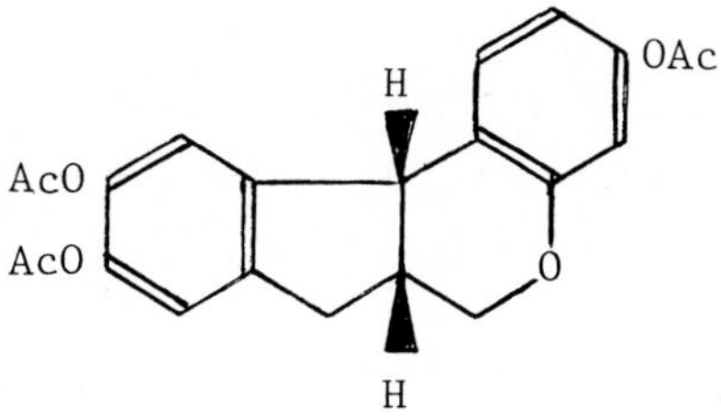

(12) R=Ac
(14) R=Me

(13)

Six aromatic compounds have been isolated from Sappan Lignum, the dried heartwood of *Caesalpinia sappan* and characterized as 3-benzylchroman derivatives. They are closely related biosynthetically to brazilin and two of them may be precusors, while three others may be key intermediates in its biogenesis (T. Saitoh *et al.*, Chem. pharm. Bull., 1986, **34**, 2506).

For a report on the light fastness properties of traditional vegetable dyes see M. Kashiwagi and S. Yamazaki (Kobunkazai no Kagaku, 1982, 27, 54).

(iv) Condensed tannins

Additional information on hydrolysable tannins is found in Vol IIID (2nd Ed., 1976) p.203. Studies have been made of the tannin from Indian wattle (*Acacia mearnii*) bark. Its empirical formula has been shown to be $C_{15}H_{14}O_6$ and its molecular weight as determined by the vapour pressure method is 1025 (S. Gupta, S.P. Singh, and R.C. Gupta, Indian J. For., 1981, 4, 18). The structures of the seven tannin components of vallonea tannin, the mixture of tannins from the acorn cups of *Quercus vallonea* and *Q. macrolepis* have been reviewed and discussed (W. Mayer, Leder, 1977, 28, 17). Certain fundamentals of vegetable tannin chemistry have been reported [K.R.V. Thampuran, A. Doraikannu, and D. Ghosh, Leather Sci. (Madras), 1978, 25, 309] and a review of vegetable tanning agents has appeared (E. Sugano, Hikaku Gijutsu, 1980, 21, 10).

The ir-spectra of tannins have been recorded (J. Kishimoto *et al.*, Tottori Daigaku Nogakubu Enshurin Hokoku, 1979, 11, 129) and wood species have been classified according to the ir-spectra of their tannin extracts. It has been found that the heartwood and sapwood of gymnosperms do not contain hydrolyzable tannin, as indicated by an absence of an absorption band at 1720 cm^{-1} (*idem, ibid.*, 1981, 33, 65).

(v) Flavan-4-ols

Reactions. Flavan-4α- or -4β-ols (1) unsubstituted in ring A react with phosphorus or similar halides to yield 4α-halogenoflavans (2), which react with a variety of nucleophiles (phenolate ions, thiophenolate ions, amines, and alcohols) furnishing the corresponding 4-substituted flavans whose stereochemistries depend upon the ratio of S_N1 to S_N2 reaction occurring. The lability of the halogen atom at C-4 is illustrated by the fact that 'recrystallisation' of 4α-chloro-3',4'-dimethoxyflavan (2; $R^1 = R^2 = $ OMe, X = Cl) from methanol affords 4α,3',4'-trimethoxy-flavan (3). 4α-Chloro-4'-methoxyflavan (2; $R^1 = $ OMe, $R^2 = $ H, X = Cl) does not react at room temperature with potassium phenolate but at 100° reaction occurs yielding a mixture of 4α- and the hitherto unknown 4β-phenoxy-4'-methoxyflavan together with 4β-(2-hydroxyphenyl)-4'-methoxyflavan. 4β-Aryloxyflavans

(4) free from 4-arylflavans are obtained by use of a phase transfer catalyst. Benzylamine reacts with 4α-chloroflavan to give 4β-benzylaminoflavan and none of the 4α-isomer, whereas aniline, being less nucleophilic than benzylamine, affords a mixture of 4α- and 4β-phenylaminoflavan. The m.ps. and yields of some 4-substituted flavans are presented in Table 7 and some 4-aminoflavans in Table 8.

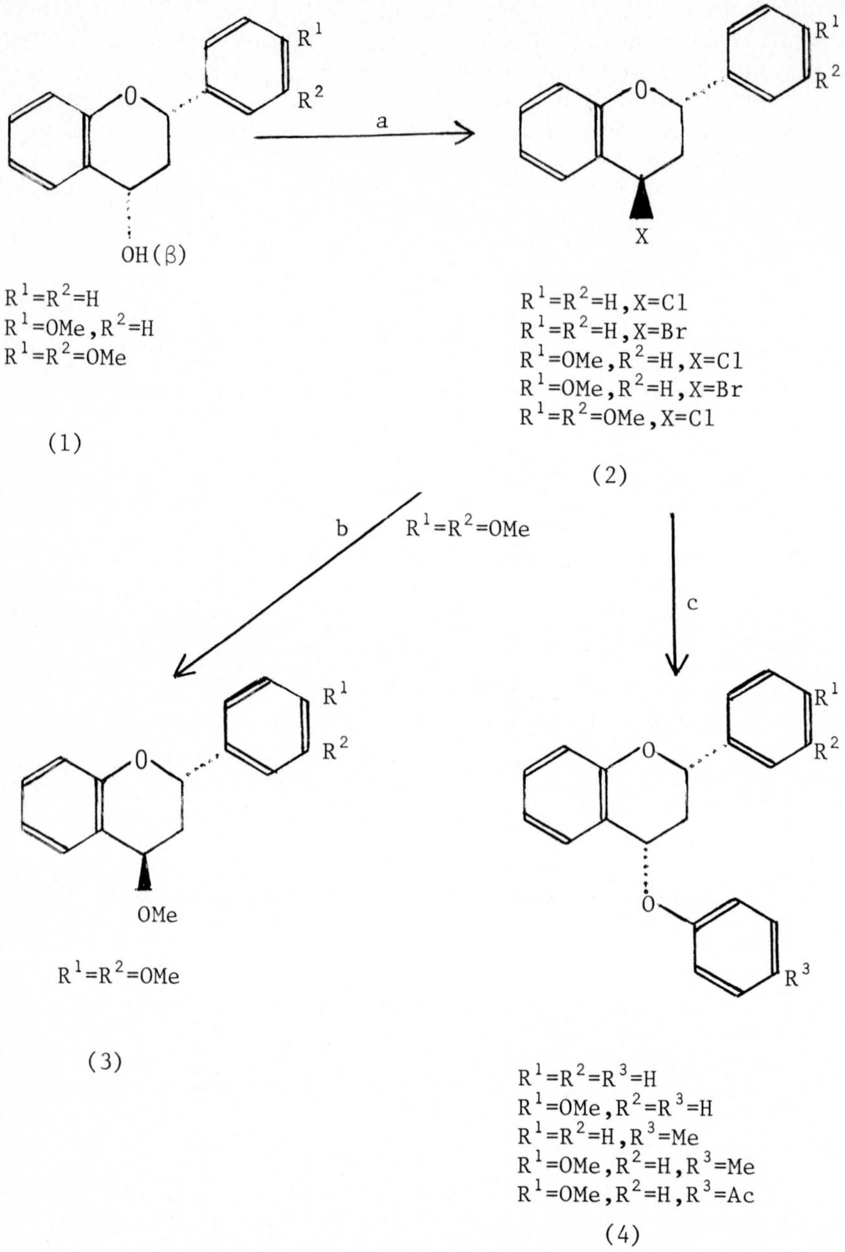

$R^1=R^2=H$
$R^1=OMe,R^2=H$
$R^1=R^2=OMe$

(1)

$R^1=R^2=H,X=Cl$
$R^1=R^2=H,X=Br$
$R^1=OMe,R^2=H,X=Cl$
$R^1=OMe,R^2=H,X=Br$
$R^1=R^2=OMe,X=Cl$

(2)

b $R^1=R^2=OMe$

c

$R^1=R^2=OMe$

(3)

$R^1=R^2=R^3=H$
$R^1=OMe,R^2=R^3=H$
$R^1=R^2=H,R^3=Me$
$R^1=OMe,R^2=H,R^3=Me$
$R^1=OMe,R^2=H,R^3=Ac$

(4)

Reagents:- (a) $PX_3(X=Cl,Br)$,Et_2O; (b) MeOH; (c) phase transfer, e.g., 4-cresol,CH_2Cl_2,H_2O,NaOH, $PhCH_2(n-Bu)_3N^+Br^-$,20^0,8h

TABLE 7

4 -Halogenoflavans and 4-substituted flavans

Compound	Method of preparation (D = Direct) (PT = Phase transfer)	Substituents			M.p. (°C)	Yield %
		R^1	R^2	R^3		
4α-Chloroflavan	From 4α-OH	H	H		56– 57	60
4α-Chloroflavan	From 4β-OH	H	H			50
4α-Bromoflavan	From 4α-OH	H	H		85– 86	75
4α-Bromoflavan	From 4β-OH	H	H			82
4α-Chloro-4'-methoxy-flavan	From 4β-OH	OMe	H		94– 95	77
4α-Bromo-4'-methoxy-flavan	From 4β-OH	OMe	H		102–103	34
4α-Chloro-3',4'-di-methoxyflavan	From 4β-OH	OMe	OMe		70– 71	75

TABLE 7   (*Contd.*)

Compound	Method of preparation (D = Direct) (PT = Phase transfer)	Substituents			M.p. (°C)	Yield %
		R^1	R^2	R^3		
Derivatives from 4α-halogenoflavans						
4α,3',4'-Trimethoxy-flavan (3)	D	OMe	OMe	OMe	70- 71	quantitative
4β-Phenoxyflavan (4)	D	H	H	H	158-159	23
4β-Phenoxyflavan (4)	PT	H	H	H	159-159	47
4'-Methoxy-4β-phenoxyflavan (4)	D	OMe	H	H	137-138	30
4β-(4-methylphenoxy)-flavan (4)	D	H	H	Me	99-100	25
4'-Methoxy-4β-(4-methyl-phenoxy)flavan (4)	D	OMe	H	Me	118.5-119.5	24
4'-Methoxy-4β-(4-methyl-phenoxy)flavan (4)	PT	OMe	H	Me	119-120	42
4β-(4-acetylphenoxy)-4'-methoxyflavan (4)	PT	OMe	H	Ac	146-147	52

4β-(4-Acetyl-3-hydroxyphenoxy)-4'-methoxyflavan (45%), m.p. 112-113° (PT);
4β-(4-acetyl-3-hydroxyphenoxy)-3',4'-dimethoxyflavan (10%), m.p. 149-150° (PT);
4'-methoxy-4β-(2-naphthyloxy)flavan (39%), m.p. 124-125° (PT).

TABLE 8

4-Aminoflavans from 4α-halogenoflavans

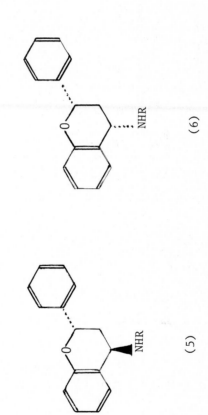

(5)　　　　　　　(6)

Compound	R	M.p. (°C)	Yield (%)
4α-Phenylaminoflavan (5)	Ph	146.5-148	6
4β-Phenylaminoflavan (6)	Ph	162 -163	16
4β-Benzylaminoflavan (6)	CH_2Ph	92 - 92.5	24
4α-Ethoxycarbonylaminoflavan (5)	CH_2CO_2Et	154 -156	8
4β-Ethoxycarbonylaminoflavan (6)	CH_2CO_2Et	99 -100	30

Biflavonoids (9) are obtained by the reaction of 4α-halogenoflavans with flavonols under conditions of phase transfer catalysis, for example, the reaction of 4α-chloro-4'-methoxyflavan (7) with flavonol (8) (B.R. Brown *et al.*, J. chem. Soc., Perkin I, 1983, 1825).

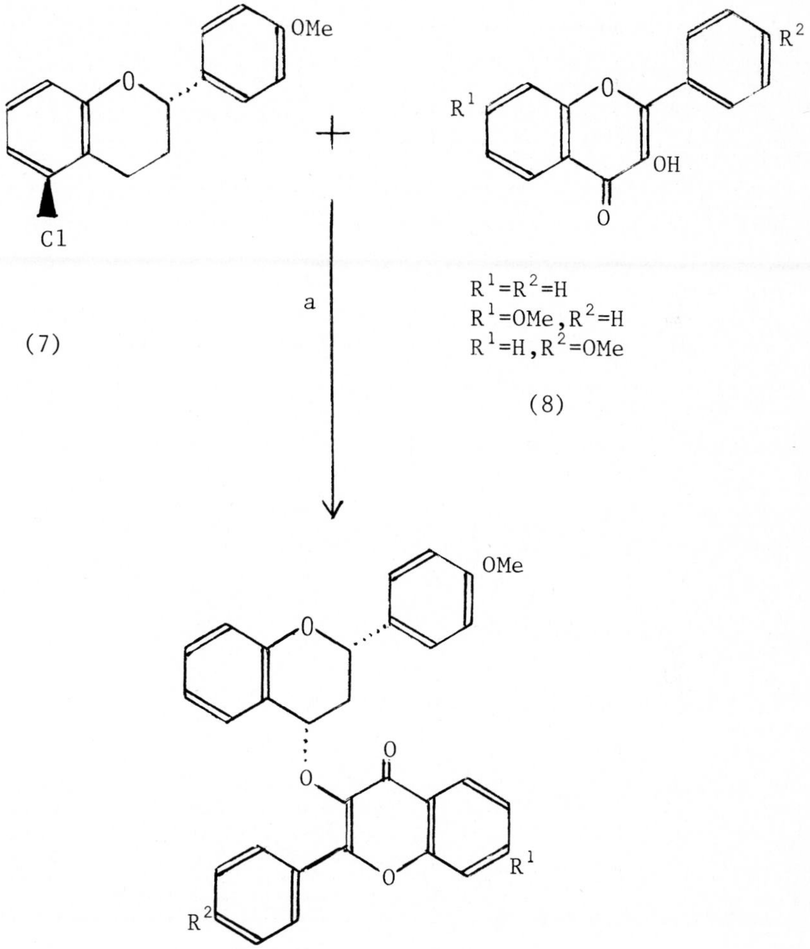
OMe
O
Cl
a
R¹
O
OH
O
R²
R¹=R²=H
R¹=OMe,R²=H
R¹=H,R²=OMe
OMe
O
O
O
O
R¹
R²

(7)

(8)

(9)

Reagents:- (a) $PhCH_2(n\text{-}Bu)_3N^+Br^-$, 2M NaOH, H_2O, CH_2Cl_2, 12h, stir, R.T.

3-(4'-Methoxyflavan-4-yloxy)flavone, m.p. 168-170°; 7-
-methoxy-3-(4'-methoxyflavan-4-yloxy)flavone, m.p. 188-190°;
4'-methoxy-3-(4'-methoxyflavan-4-yloxy)flavone, m.p. 151.5-
152.5°.

(f) Chromanones, dihydrobenzo[b]pyranones

(i) Chroman-3-ones
(±)-2,2-Dimethyl-3,4-epoxychroman (2) [readily obtained
from 2,2-dimethyl-2H-benzo[b]pyran (1)], on treatment with
a catalytic amount of boron trifluoride etherate in dry
benzene yields 2,2-dimethylchroman-3-one, as an oil.

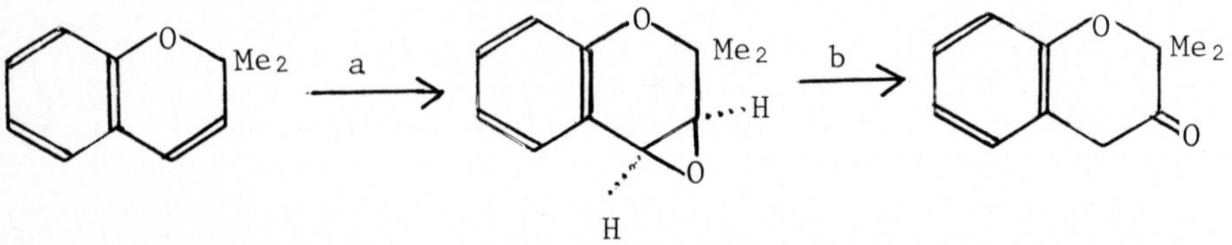

(1)   (2)   (3)

Reagents:- (a) 3-ClC$_6$H$_4$CO$_3$H,CH$_2$Cl$_2$,0.5M aq.NaHCO$_3$,
            0°→R.T.;   (b) BF$_3$-Et$_2$O,C$_6$H$_6$,R.T., 0.33h

2,2-Dimethylchroman-3-one (3) undergoes ethoxycarbonylation
on boiling with sodium hydride in diethyl carbonate to give
2,2-dimethyl-4-ethoxycarbonyl-3-hydroxy-2H-benzo[b]pyran (4),
which on hydrogenation in ethanol in the presence of platinum
yields (±)-2,2-dimethyl-4-ethoxycarbonylchroman-3-ol (5).

(3) 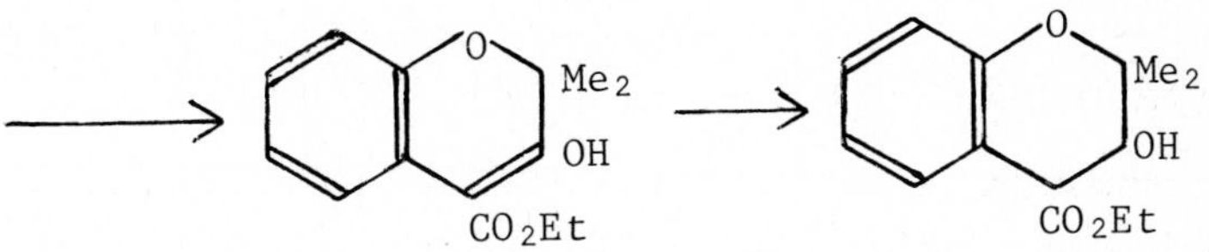 $\longrightarrow$ (4) $\longrightarrow$ (5)

    2,2-Dimethyl-7-methoxychroman-3-one is unstable and
cannot be ethoxycarbonylated, but 2,2-dimethyl-4-ethoxy-
carbonyl-7-methoxychroman-3-one (6) is obtained by a
standard synthesis illustrated below, involving a Dieckmann
ring closure. The chroman-3-one (6) is found to exist
entirely in the enol form (7) (P. Anastasis and P.E. Brown,
J. chem. Soc., Perkin I, 1983, 1431).

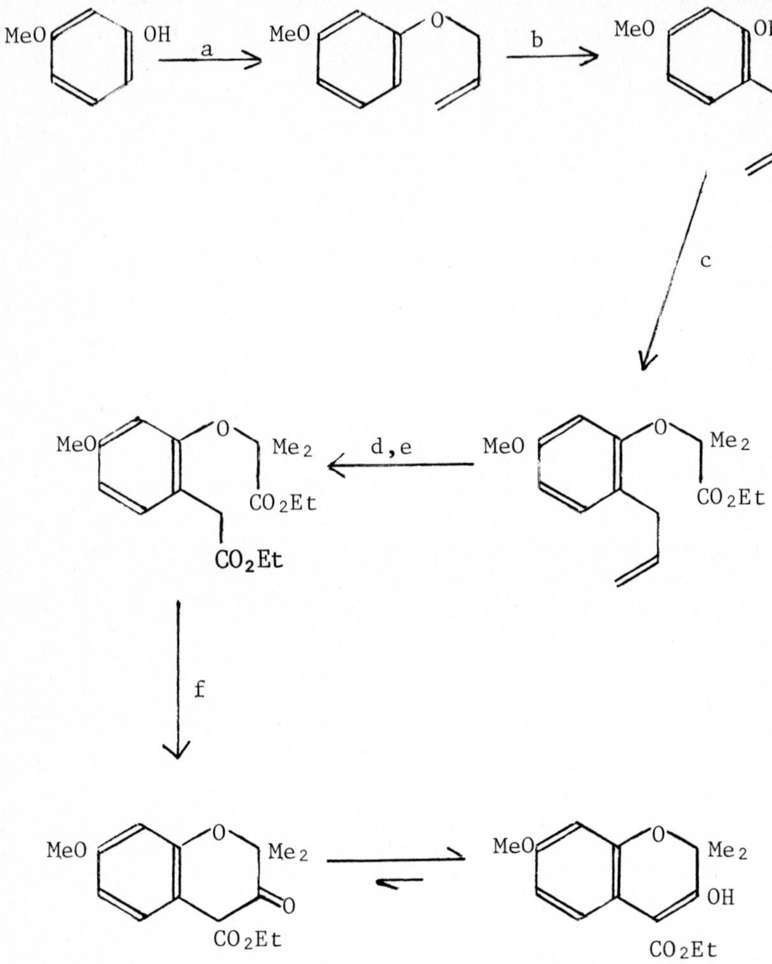

Reagents:- (a) $CH_2=CHCH_2Br$, K_2CO_3, Me_2CO; (b) $PhNME_2$, Δ;
(c) EtO^-, $EtOH$, $BrCMe_2CO_2Et$; (d) O_3, $MeOH$,
H_2O_2, $AcOH$; (e) $SOCl_2-EtOH$; (f) EtO^-, $EtOH$

1-Aryloxy-3-diazoacetones (8), formed from the appropriate aryloxyacetyl chloride and diazomethane, on treatment with bis[hexafluoroacetoacetonato]copper (II) yield the chroman-3-ones (9). 3,4-Dihydro-2H-naphtho[1,2-b]pyran-3-one (10) and 2,3-dihydro-1H-naphtho[2,1-b]pyrane-2-one (11) are prepared in a similar manner (A. Saba, Synth., 1984, 268).

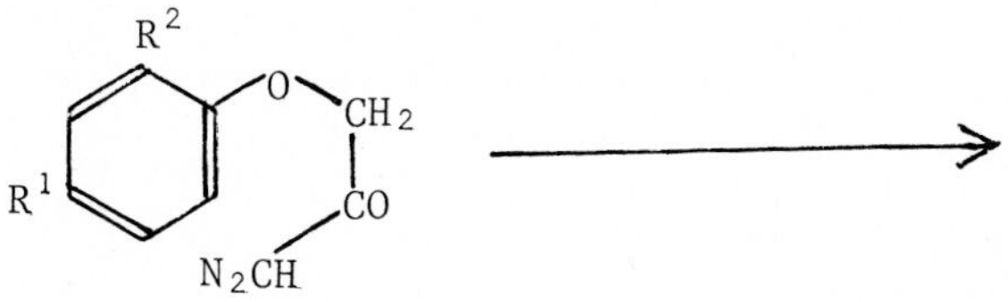

R^1=H,alkyl, R^2=H,alkyl

(8)　　　　　　　　　　　　　　　　　(9)

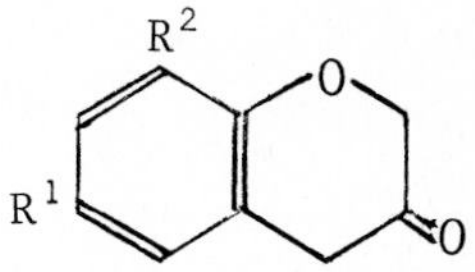

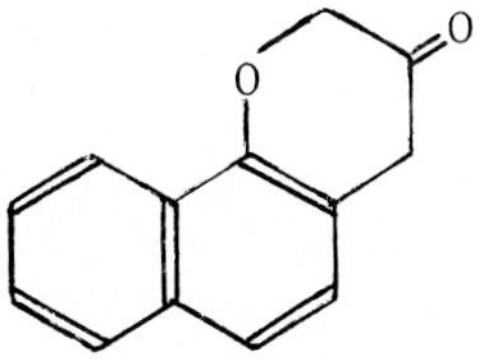

(10) 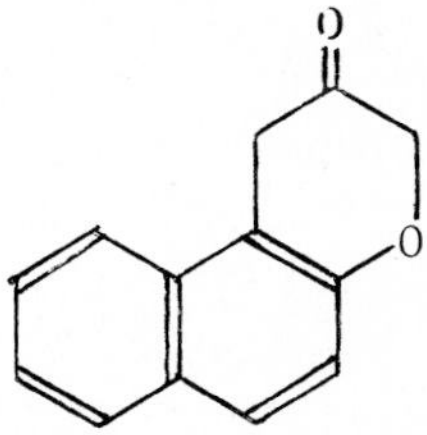　　　　　　　　　　　　　　　　(11)

(ii) Chroman-4-ones
Preparations. Chromones (1) activated by carbonyl substituents at position C-3 are transformed on treatment with lithium dimethylcuprate into the corresponding 2-methyl-chroman-4-ones, which exist as the keto (2) and/or enol (3) form (Table 9).

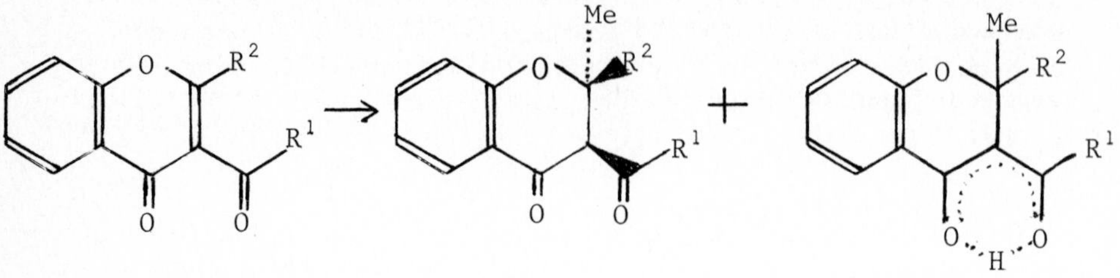

(1)   (2)   (3)

TABLE 9
3-Acyl- and 3-Acyloxy-chroman-4-ones

Substituents		Tautomer formed	Yield (%)
R^1	R^2		
Ph	H	(2) : (3) 9:1	77
OMe	H	(2) : (3) 4:1	78
Me	H	(2) : (3) 1:4	70
H	H	(3)	50
Me	Me	(2)	42
OMe	Me	(2)	74

Less activated substrates, such as, chromone, 2-methyl-chromone, 3-bromochromone, and 7-methoxy-3-phenylchromone yield complex reaction mixtures under the same reaction conditions. The chroman-4-ones produced in the above reaction (Table 9) arise *via* conjugated addition, but the nature of the intermediate remains unclear (T.W. Wallace, Tetrahedron Letters, 1984, 25, 4299).

Chroman-4-ones (5) are obtained by the photolysis of 4--methoxyphenyl crotonates (4) in benzene/10% aqueous sodium hydroxide mixture. The chroman-4-ones (5) have been reduced and dehydrated to provide 2H-benzo[b]pyrans (J. Primo, R. Tormo, and M.A. Miranda, Heterocycles, 1982, 19, 1819).

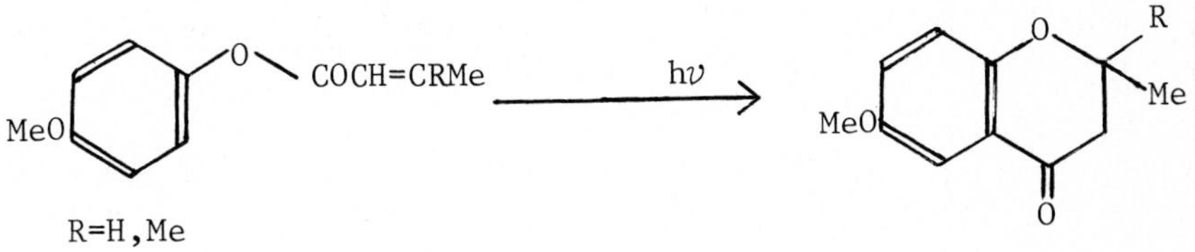

R=H,Me

(4)   (5)

2-Hydroxyacetophenones (6) react with aliphatic aldehydes and ketones in the presence of pyrrolidine to afford chroman-4-ones (7) (Table 10). The syntheses and reactions of chroman-4-ones have been reviewed (H.-J. Kabbe and A. Widdig, Angew. Chem. intern. Edn., 1982, 21, 247).

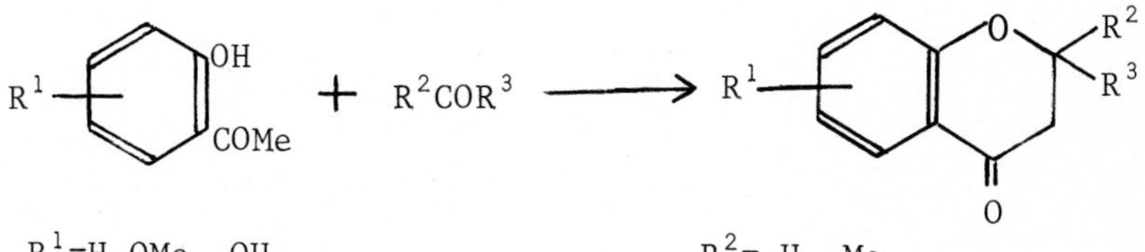

R^1=H,OMe, OH, R^2= H, Me
      Cl, Me, CO_2H R^3= alkyl chain with
                                 substituents
                          R^2R^3= $(CH_2)_5$, $(CH_2)_6$

(6)   (7)

TABLE 10
Chroman-4-ones (7)

Substituents			M.p. (^{o}C) B.p. (^{o}C/torr)	Yield (%)
R^1	R^2	R^3		
H	H	CHMe$_2$	125/0.07	93
H	H	CMe$_3$	54- 56	78
H	H	n-C$_9$H$_{19}$	175/0.1	79
H	H	(CH$_2$)$_2$Ph	185/0.05	30
H	Me	Me	88- 90	65
6-OMe	H	n-C$_6$H$_{13}$	53- 55	62
7-OH	— (CH$_2$)$_3$ —		170-171	73
H	— (CH$_2$)$_4$ —		38- 39	85
H	Me	CH(OMe)$_2$	130/0.01	55
6-OH	Me	(CH$_2$)$_2$CH=CMe$_2$	79- 81	74
6-Cl	Me	CO$_2$H	92- 94	61
6-OH	Me	(CH$_2$)$_2$CO$_2$H	164-166	66
5,6,7-Me$_3$, 6-OMe	Me	(CH$_2$)$_4$CO$_2$H	76- 78	61
H	Me	(CH$_2$)$_3$N Et$_2$	185/0.01	75
6-Cl	Me	(CH$_2$)$_3$N Et$_2$	180/0.01	81
6-OMe	Me	(CH$_2$)$_3$N Et$_2$	175/0.01	66
6-CO$_2$H		(CH$_2$)$_4$	229-231	73

3-Arylmethylenechroman-4-ones (8) on treatment with borane
followed by sodium dichromate and sulphuric acid give 3-
-arylmethylchroman-4-ones (9) and 3-arylmethyl-3-hydroxy-
chroman-4-ones (10). From (8; R = Ph) the yield of (9) is
52% and (10) (24%) (B.S. Kirkiacharian and M. Gomis, Compt.
rend., 1982, 295, 27).

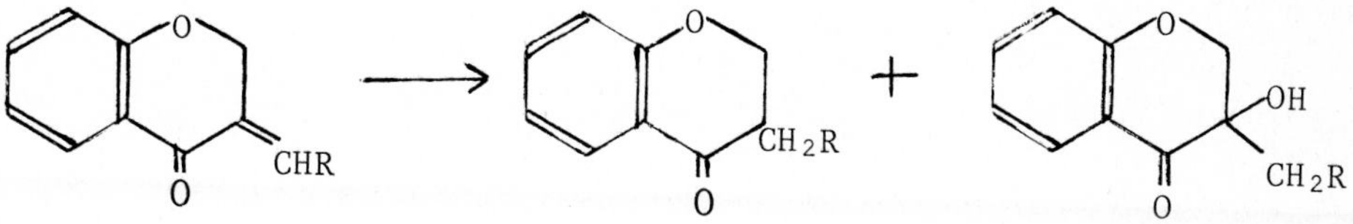

R=Ph, 4-MeOC$_6$H$_4$,
  (MeO)$_2$C$_6$H$_3$, naphthyl,

(8)   (9)   (10)

Acylating the benzenediols (11) with β,β-dimethylacrylic
acid or its acid chloride in the presence of aluminium
chloride and phosphoryl chloride gives the 2,2-dimethyl-
chroman-4-ones (12). Similar acylation of benzene-1,3-diol
yields 2,4-dihydroxyphenyl 2,2-dimethylvinyl ketone, which
is cyclized to 2,2-dimethyl-7-hydroxychroman-4-one on
treatment with acid (D. Sowmithran and K.J.R. Prasad, Synth.,
1985, 545).

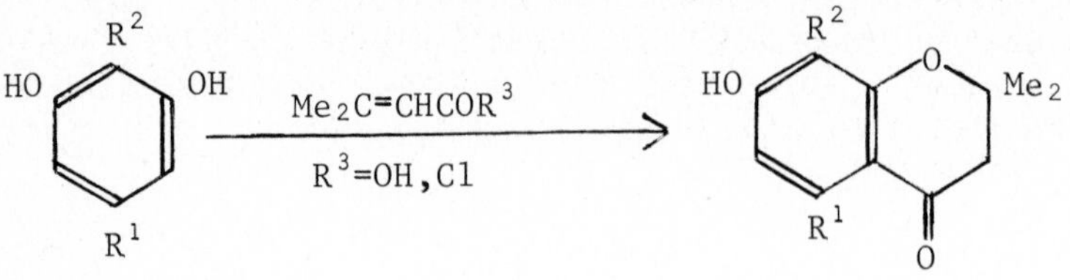

HO—(R²)—OH →[Me₂C=CHCOR³][R³=OH,Cl]→ (12)

$R^1 = OH, Me, R^2 = H$
$R^1 = H, R^2 = OH$

(11)   (12)

Some 3,3-bis(dialkylaminomethyl)-6,8-dihalogenochroman-4-
-ones (15) are obtained *via* a Mannich reaction between a
3,5-dichloro-2-hydroxyacetophenone (13), formaldehyde and
a secondary amine (14). The chroman-4-ones (15) on
hydrogenation furnish the corresponding chroman-4-ols and
3,3-bis(morpholinoamino)- and 3,3-bis(piperidino)-6,8-
-dichlorochroman-4-ones exhibit bactericidal activity
(A. Cascaval, E. Miscovici, and V. Chiorean, Rom. Pat. RO
79,907, 1982; Cascaval, Synth., 1983, 579).

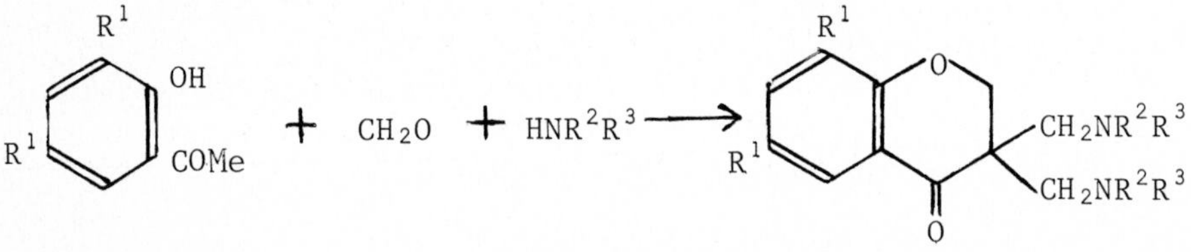

R^1 ... OH ... COMe + CH_2O + HNR^2R^3 → $CH_2NR^2R^3$ / $CH_2NR^2R^3$

$R^1 = Cl, Br$

$R^2 = R^3 = Me, Et$
$R^2R^3 = (CH_2)_4, (CH_2)_5,$
morpholino

(13)   (14)   (15)

The photolysis of 4-acetoxy-2H-benzo[b]pyran (16) gives
chroman-4-one (17), 2-acetoxyphenyl vinyl ketone (18),
chromone (19), 3-acetoxy- (20) and 3-hydroxy- (21) chroman-
-4-one and the cyclobutane dimer (22) (M.J. Climent *et al.*,
Tetrahedron, 1987, **43**, 999).

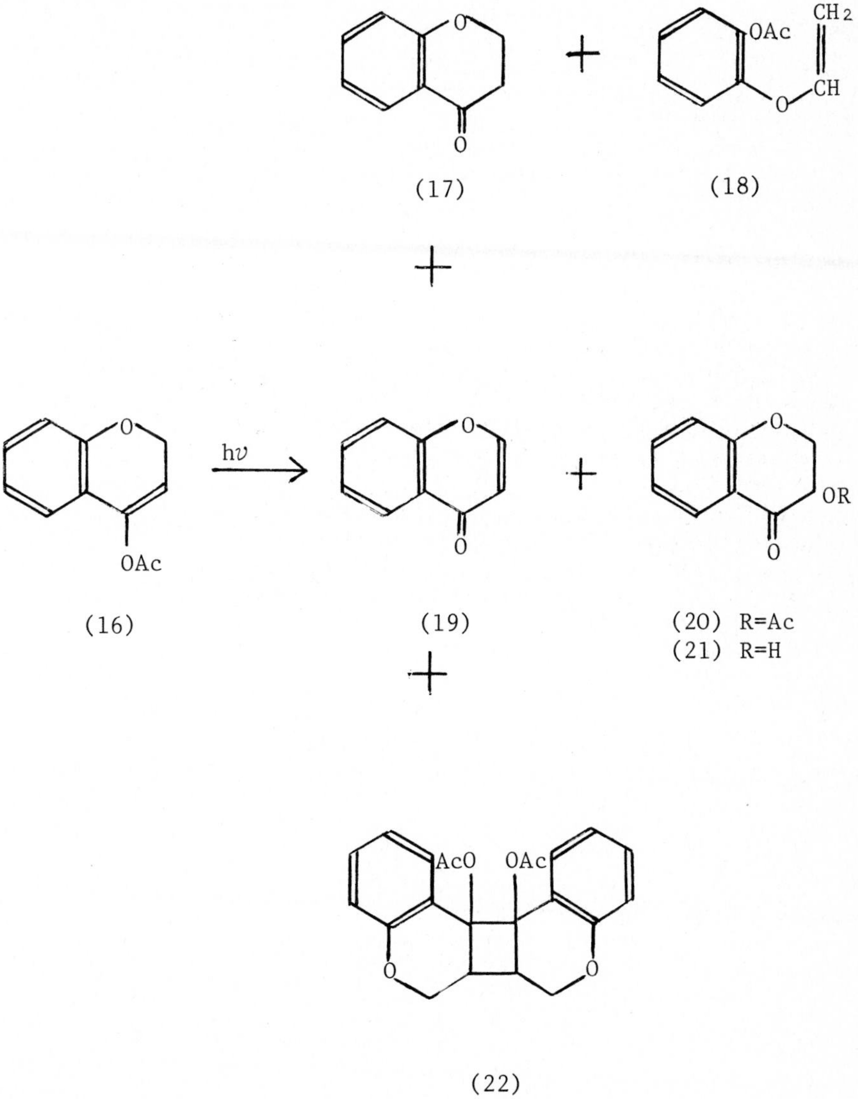

(17)   (18)

(16)   (19)   (20) R=Ac
                        (21) R=H

(22)

6-Fluorochroman-4-one has been recovered from by-products during the synthesis of sorbinil (B.W. Cue Jr., P.D. Hammen, and S.S. Massett, U.S. Pt. US 4,431,828, 1984). 2,2--Dimethyl-5-hydroxy-7-methoxy-6-octyl- and -8-octyl-chroman--4-ones have been prepared and their properties as uv--absorbers discussed (A. Ninagawa, T. Iseki, and H. Matsuda, Makromol. Chem., 1983, <u>184</u>, 71).

Reactions. Treatment of chroman-4-ones (23) with thallium (III) nitrate in methanol, containing a small amount of perchloric acid, results mainly in dehydrogenation to yield the corresponding chromone (24), along with small amounts ($\leq$ 10%) of the related 3-methoxy-chroman-4-one (25). Chroman-4-ones (23; R = H and R = NO_2) again undergo dehydrogenation and α-methoxylation when treated with thallium (III) nitrate in trimethyl orthoformate, but the latter reaction now predominates. In the reaction with chromanone (23; R = H) a substantial amount of methyl benzofuran-3-carboxylate (26) is formed. A tentative explanation is that it results from two successive methoxythallation, according to sequence proposed by Taylor and McKillop, to account for the formation of methyl α-methoxyphenylacetate from acetophenone in trimethyl orthoformate. The reaction between a number of chroman-4-ones and thallium (III) nitrate has been investigated (P.G. Ciattini, E. Morera, and O. Giorgio, J. heterocyclic Chem., 1982, <u>19</u>, 395).

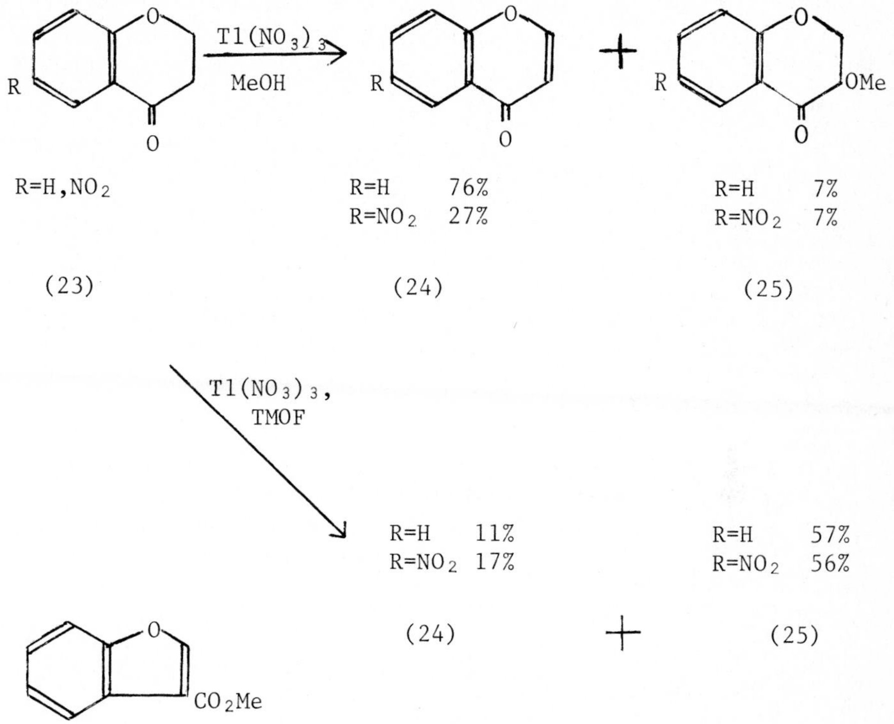

$$\text{(23)} \xrightarrow[\text{MeOH}]{\text{Tl(NO}_3)_3} \text{(24)} + \text{(25)}$$

R=H,NO$_2$

R=H   76%
R=NO$_2$ 27%

R=H   7%
R=NO$_2$ 7%

(23)   (24)   (25)

Tl(NO$_3$)$_3$, TMOF

R=H   11%
R=NO$_2$ 17%

R=H   57%
R=NO$_2$ 56%

(24)   +   (25)

CO_2Me

(26)

The action of Vilsmeier reagent on 2,2-dimethyl-7-methoxy-chroman-4-one (27; R = H) and its 6-bromo- and 6-methoxy analogues affords low yields of the 4-chloro-2,2-dimethyl--2*H*-benzo[b]pyran-3-carboxaldehydes (29), but high yields of the 4-chloro-2,2-dimethyl-2*H*-benzo[b]pyrans (28), which are not precusors of the carboxaldehydes. 4-Chloro-2,2-dimethyl--7-methoxy-6-nitrochroman is too labile to allow a dehydrogenation to the 6-nitro-2*H*-benzo[b]pyran (28; R = NO$_2$), but the latter is obtained by regioselective nitration of 4-chloro-2,2-dimethyl-7-methoxy-2*H*-benzo[b]pyran. 6-Amino--2,2-dimethyl-7-methoxychroman-4-one, m.p. 103-104°, protected as the *N*-ethoxycarbonyl derivative gives 6-amino--2,2-dimethyl-7-methoxy-2*H*-benzo[b]pyran.

MeO ... Me$_2$ →(a) MeO ... Me$_2$ (Cl) + MeO ... Me$_2$ (CHO, Cl)

R=H,Br,OMe

(27)   (28)   (29)

R(27)	yield(28)(%)	Yield(29)(%)
H	92	8
Br	88	12
OMe	88	12
Prolonged reaction H	83	9

7-Methoxy-2-methylchroman-4-one affords 4-chloro-7-methoxy-
-2-methyl-2H-benzo[b]pyran (95%).

Reagent:- (a) POCl$_3$, HCONMe$_2$, dry solvent CH$_2$Cl$_2$, CCl$_4$,
or trichloroethene.

2,2-Dimethyl-7-methoxy-3,3,6-tribromochroman-4-one, m.p.
161-162°, on debromination with zinc yields 6-bromo-2,2-
-dimethyl-7-methoxychroman-4-one, m.p. 97°. 2,2-Dimethyl-
-7-methoxy-3,3-6-trinitrochroman-4-one, m.p. 147-148°
(P.E. Brown, W.Y. Marcus, and P. Anastasis, J. chem. Soc.,
Perkin I, 1985, 1127). 3-(4-Acetylamino-4-carboxy-2-
-thiabutyl)chroman-4-one has been synthesized from chroman-
-4-one and shows antiinflammatory properties (R.Y. Mauvernay
et al., Fr. Demande 2,285,858, 1976).
    The Michael addition of chroman-4-one (30) to methyl
vinyl ketone at room temperature gives cannabichromanone
(31) along with spiro compound (32) (L. Chiodini,
M.D. Ciommo, and L. Merlini, Heterocycles, 1981, __16__, 1899).

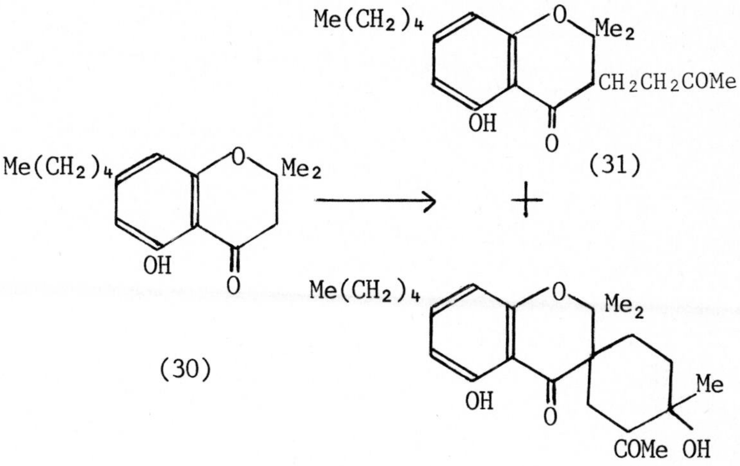
Me(CH2)4
O
Me2
OH
O
Me(CH2)4
O
Me2
CH2CH2COMe
OH
O
Me(CH2)4
O
Me2
Me
OH
O
COMe OH

(30) (31) (32)

3-Benzylidenechroman-4-ones (33) in the presence of alkali undergo isomerization to yield both the 3-benzyl-chromones (34) and the 3-methylflavones (35). It is found that both types of isomerization are competitive and always occur, although at different rates (A.C. Jain, A. Sharma, and R. Srivastava, Indian J. Chem., 1983, 22B, 1119).

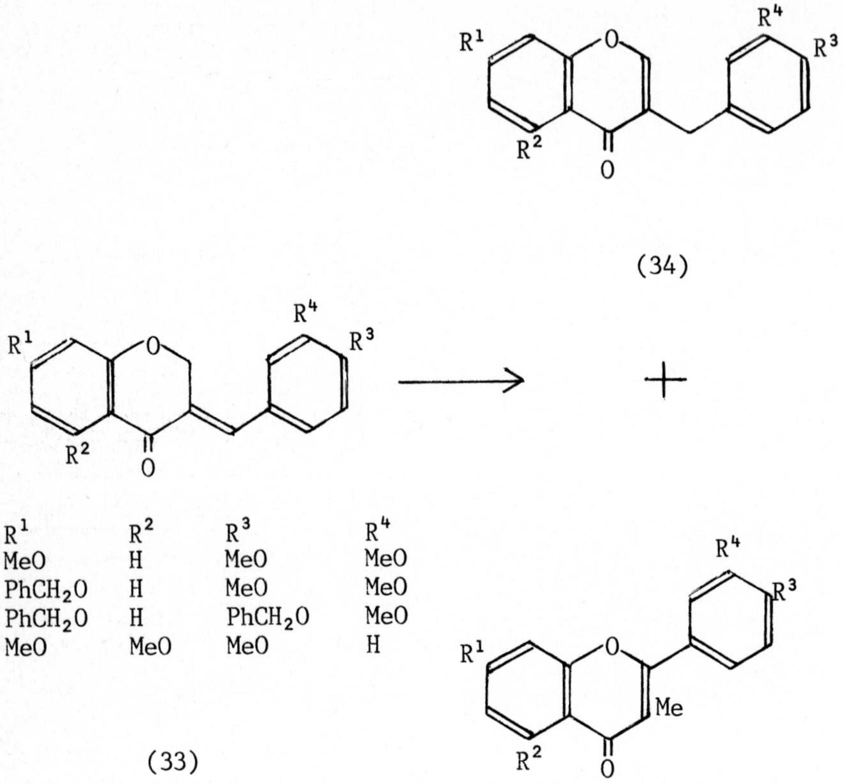

(34)

R^1	R^2	R^3	R^4
MeO	H	MeO	MeO
$PhCH_2O$	H	MeO	MeO
$PhCH_2O$	H	$PhCH_2O$	MeO
MeO	MeO	MeO	H

(33)

(35)

2-Hydroxychroman-4-ones (36) react with excess hydroxyl-amine hydrochloride in aqueous ethanol at pH<6 to give oximes (37) (V. Szabó *et al.*, Acta Chim., Hung., 1984, __115__, 331).

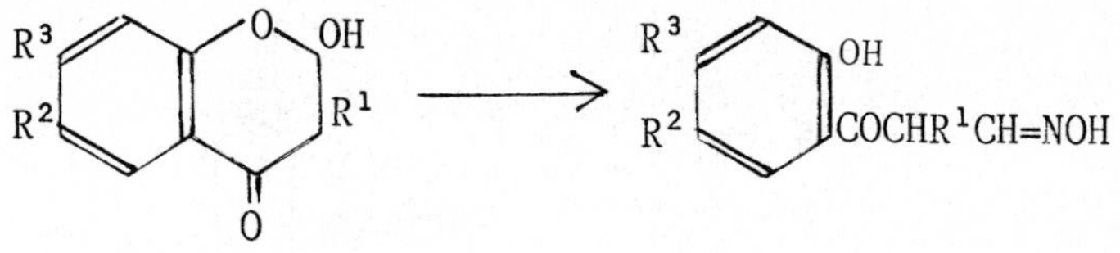

R^1 = H, Me, Ph

R^2, R^3 = H, Cl, OMe, NO_2

(36)   (37)

The ring-chain isomerism of 2-hydroxychroman-4-one has been examined using ^{1}H-nmr spectroscopy (J. Borbély and Szabó, Tetrahedron Letters, 1984, 25, 5813).

The addition of iodobenzene diacetate to chroman-4-one and potassium hydroxide in methanol at 0° gives the dimethoxy derivative (38), which on treatment with hydrochloric acid in aqueous ethanol affords 3-hydroxychroman-4-one (39). This a convenient method for its preparation (R.M. Moriarty, O. Prakash, and C.T. Thachet, Synth. Comm., 1984, 14, 1373).

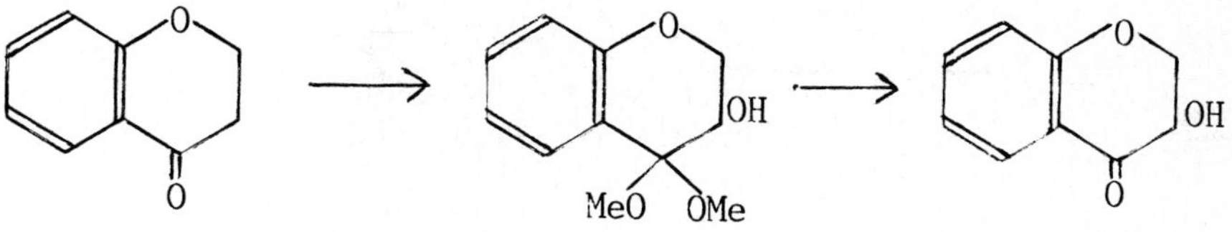

(38)   (39)

Treatment of chroman-4-ones (40) with carbon disulphide in the presence of sodium *tert*-butoxide, followed by methylation gives the dithioacetals (41) [H. Ila and H. Junjappa, J. chem. Res., (S), 1979, 268].

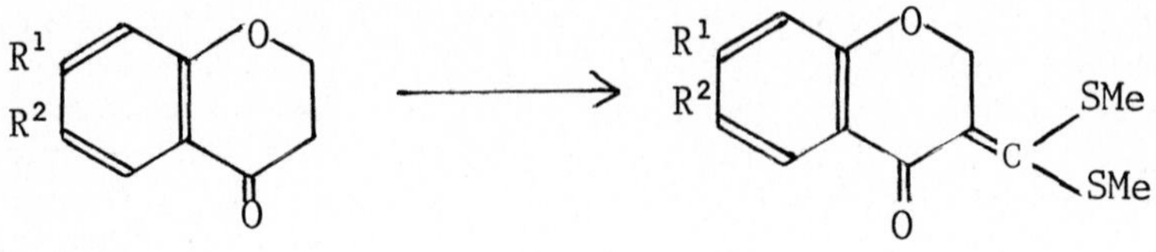

R^1 = H, Me

R^2 = H, Me, OMe

(40)   (41)

Chroman-4-ones have been converted into pyranobenzo-
pyranones and benzothiopyranopyranones (A.H. Philipp and
I.L. Jirkovsky, Canad. J. Chem., 1979, 57, 3292; U.S. Pt.
4,177,284, 1979).

The ^{13}C-nmr spectral data of chroman-4-one and some of its
methyl and phenyl derivatives have been discussed (Y. Senda
et al., Bull. chem. Soc., Japan, 1977, 50, 2789) and the
chemical shifts and carbon-proton coupling constants for
chromone, some methyl- and phenyl-chromones, chroman-4-one,
and chroman-4-ol have been examined (T.N. Huckerby and
G. Sunman, J. mol. Struct., 1979, 56, 87). Studies have been
made of the mass spectra of chroman-4-ones, chroman-4-ols and
related compounds (M.C. Sacquet, B. Graffe, and P. Maitte,
Org. mass Spec., 1976, 11, 1128; H. Schwarz et al., ibid.,
1977, 12, 39; S. Eguchi, ibid., 1978, 13, 653).

(iii) Chromanochromanones, rotenone and related substances

Interconversion between rotenone, *8,9-dimethoxy-2α-*
-isopropenyl-1,2,12,12aα-tetrahydrobenzo[b]pyrano[3,4-b]-
furo[2,3-h]benzo[b]pyran-6-(aαH)-6-one or *8,9-dimethyl-2α-*
-isopropenyl-1,2,12,12aα-tetrahydro[1]benzopyrano[3,4-b]-
furo[2,3-h][1]benzopyran-6-(aαH)-one (1) and (-)-dalpanol
(3) is achieved by treatment of rotenone with osmium
tetroxide ⸺ sodium periodate to give rotenone norketone
(2), which on treatment with methylmagnesium iodide affords
(-)-dalpanol. Alcohol (3) on dehydration with phosphorus
tribromide/pyridine furnishes rotenone (1) (N. Nakatani and
M. Matsui, Agric. biol. Chem., 1977, **41**, 601).

(1)

(2) R = MeCO
(3) R = HOCMe$_2$

The boron tribromide cleavage product of (-)-rotenone (1),
on reduction with sodium cyanoborohydride in hexamethyl-
phosphoramide gives (-)-rot-2'-enonic acid (4), which has
been converted into rotenoids (-)-deguelin (5), (-)-dalpanol
(3), and (-)-elliptone (6), previously prepared only in the
racemic form (P.B. Anzeveno, J. heterocyclic Chem., 1979, **16**,
1643).

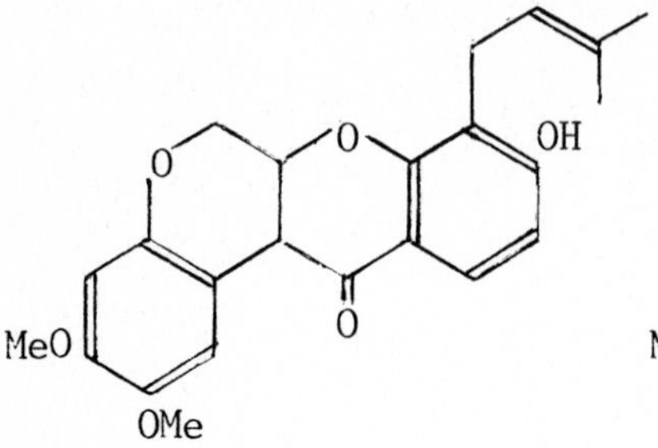

(4)

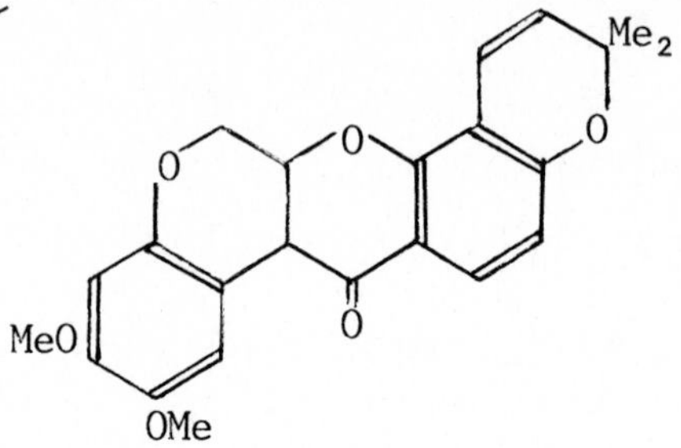

(5)

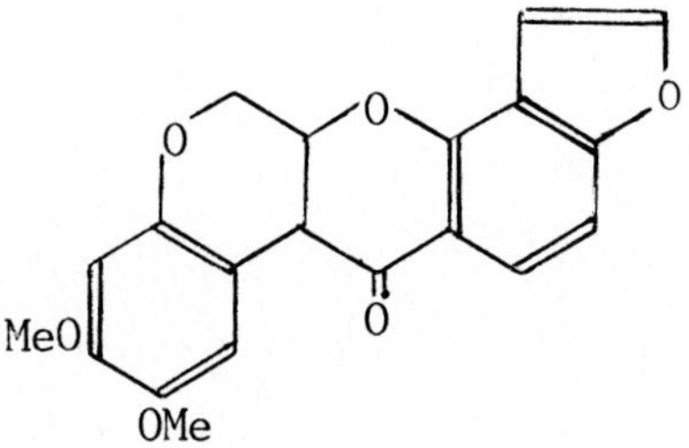

(6)

Also the conversion of (-)-rotenone into rotenone
norketone (2), followed by Baeyer-Villiger oxidation and an
elimination reaction yields (-)-elliptone (6). Treatment
of rotenone with perbenzoic acid affords 12a-hydroxyrotenone
(A.K. Singhal *et al.*, Chem. and Ind., 1982, 549).

Rotenoids in the tissue cultures of *Crotalaria burhia*
have been identified as rotenone, elliptone, deguelin,
toxicarol, sumatrol, and tephrosin (A. Uddin and P. Khanna,
Planta Med., 1979, 36, 181). A high-efficiency resolution
of isomeric rotenone compounds (S.L. Abidi, J Chromatog.,
1984, 317, 383) and the determination of rotenone and
rotenonone in fresh water by high-performance liquid
chromatography (S.M. McCown, LC, liq. Chromatogr. HPLC
Mag., 1984, 2, 318) have been reported. The selective two

dimensional INEPT (insensitive nuclei enhanced by polarization transfer) nmr technique has been used to determine the ^{1}H-^{13}C coupling constants of rotenone (T. Jippo, O. Kamo, and K. Nagayama, J. magn. Reson., 1986, 66, 344).

The rotenoids, villosin (7), $C_{23}H_{22}O_8$, m.p. 133°, *diacetate*, m.p. 147°, villol (8), $C_{23}H_{22}O_9$, m.p. 223°, *triacetate*, m.p. 124°, villinol (9), $C_{24}H_{22}O_8$, m.p. 200°, *monoacetate*, m.p. 170°, and villosone (10), $C_{23}H_{18}O_8$, m.p. 268° have been obtained from *Tephrosia villosa*. A comparison of their nmr, optical rotary dispersion, and CD data suggests that villosin and villol have 6aS, 12aS, 12S, and 2R, villinol has 12S and 2R, and villosone has 2R absolute configuration (G.L.D. Krupadanam *et al.*, Tetrahedron Letters, 1977, 2125).

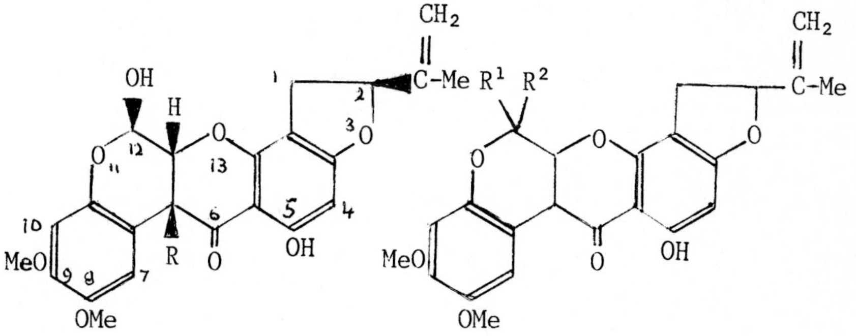

(7) R = H
(8) R = OH

(9) R^1 = H, R^2 = OMe
(10) R^1R^2 = O

Villosin, villol, and rotenone are stable to uv-light, but the photolysis of villosol (11), dehydroisosumatrol (12), and dehydrorotenone (13) yields derivatives (14), (15), and (16), respectively. Compound (17) is converted nearly quantitatively into (14). (Krupadanam, G. Srimannarayana, and N.V.S. Rao, Indian J. Chem., 1978, 16B, 770).

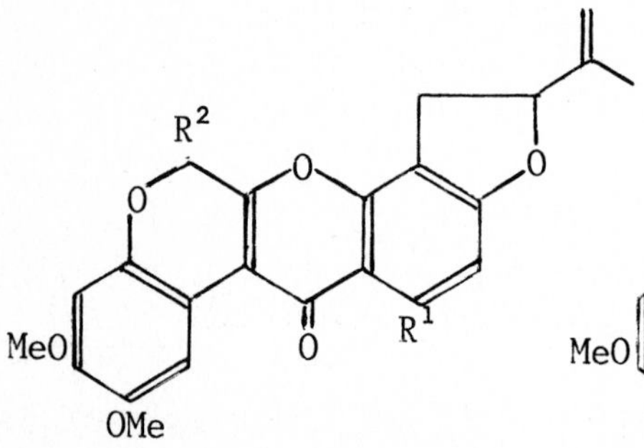

(11) R^1 = OH, R^2 = H
(13) R^1 = R^2 = H
(17) R^1 = R^2 = OH

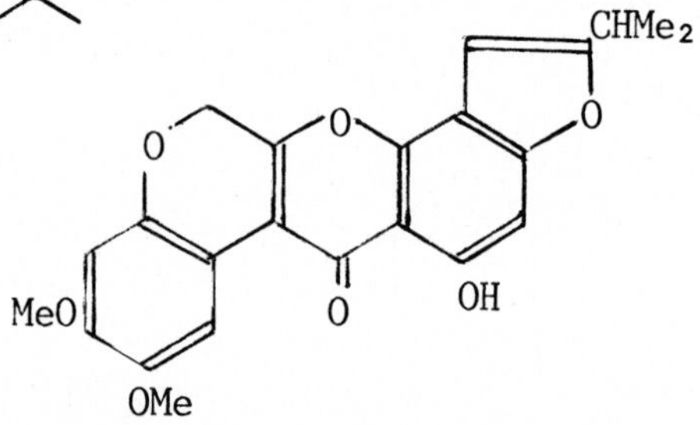

(12)

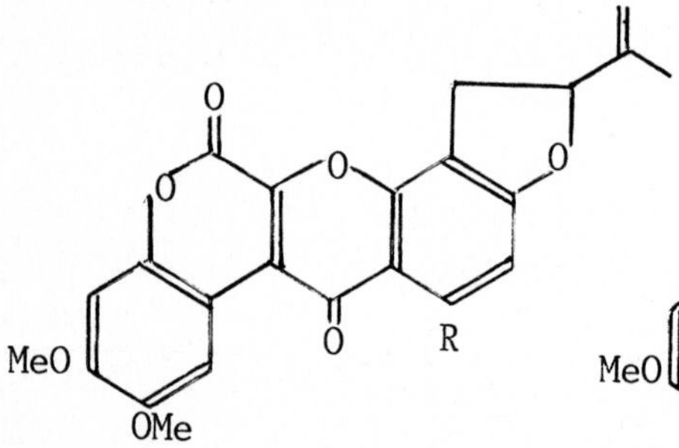

(14) R = OH
(16) R = H

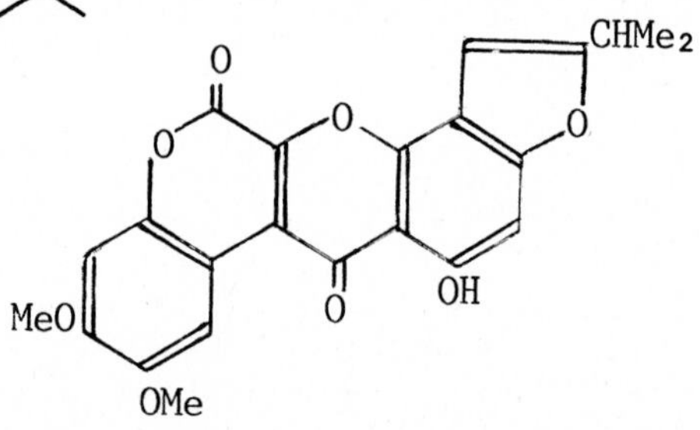

(15)

Dalbinol (18) and dalbin (19), a 6a-hydroxy rotenoid β-D-
-glucopyranosyl, along with 7,2'-dimethoxy-6-hydroxy-4',5'-
-methylenedioxyisoflavone have been isolated from the seed
of *Dalbergia assamica* [S.S. Chibber and R.P. Sharma, Natl.
Acad. Sci. Letters (India), 1978, 1, 253].

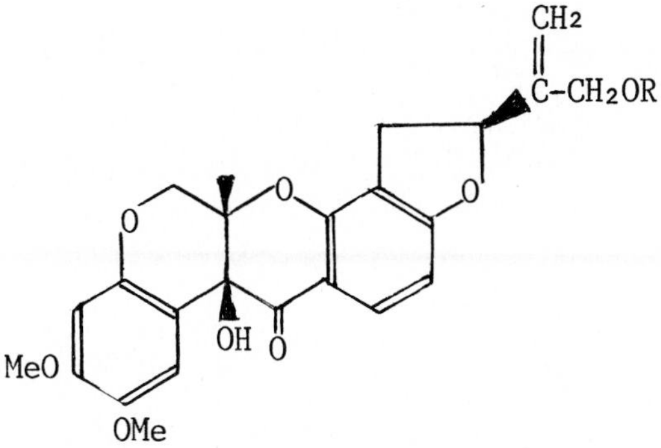

(18) R = H

(19) R =

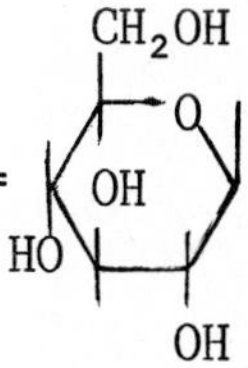

Dalbin has also been obtained from *Dalbergia nitidula*
(F.R. Van Heerden, E.V. Brandt, and D.G. Roux, J. chem.
Soc., Perkin I, 1980, 2463) and *D. latifolia*. Its structure
has been determined from its chemical and spectral
properties (Chibber and U. Khera, Phytochem., 1979, 18,
188). The first 6-dihydrorotenone, 6-dihydrodalbinol, and
its β-D-glucoside have been isolated from *D. monetaria*
(F. Abe *et al.*, *ibid.*, 1985, 24, 1071). Volubinol (20) has
been isolated from the non-green branches of *D. volubilis*
(H.M. Chawla, R.S. Mittal, and D.K. Rastogi, Indian J.
Chem., 1985, 23B, 680).

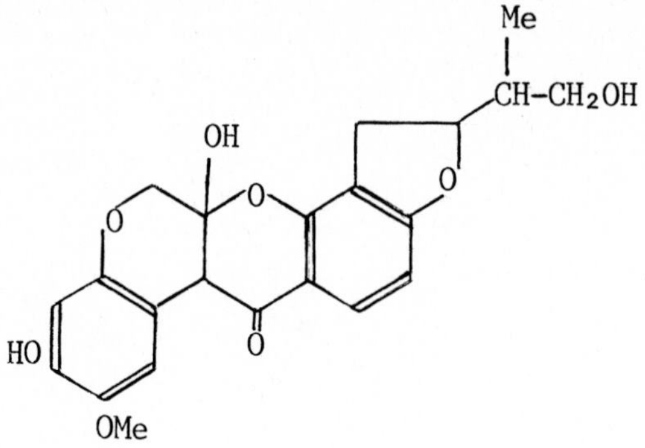

(20)

(-)-7a,13a-Dehydro-13-hydroxy-α-toxicarol [*(-)-6a,12a-
-dehydro-6-hydroxy-α-toxicarol* if numbering based on rotexen
nucleus] and 10-*O*-demethylamorphigenin (3-O-demethylamorphigen
if based on rotexen nucleus) along with tephrosin and 7,4'-
-dimethoxyisoflavone have been obtained from the fruits of
Amorpha fruticosa [G. Ganchev *et al.*, Farmatsiya (Sofia),
1984, <u>34</u>, 18]. Neobanone, rotenonone, neobanol, and a
hydroxyisomillettone have been isolated from the roots of
Neorautaneia amboensis (M.E. Oberholzer, G.J.H. Rall, and
D.G. Roux, Phytochem., 1976, <u>15</u>, 1283).

The biosynthetic path leading to formation of the E ring
in the *Amorpha fruticosa* rotenoid, amorphigenin (21) has
been determined by isotope incorporation studies
(L. Crombie *et al.*, J. chem. Soc., Perkin I, 1982, 789).

(21)

Despite low incorporations of doubly labelled [^{13}C]acetate during the biosynthesis by *A. fruticosa* seedlings, it has been possible to solve the ring-D folding by an INADEQUATE nmr experiment. New proposals have been made for the mechanism of the enzyme catalyzed flavanone-isoflavone transformation step (*idem*, Chem. Comm., 1986, 1063).

The configuration at C-6a and C-12a of the rotenoids (22) has been determined from the ^{1}H-nmr spectral data of their deuteriobenzene solutions (I. Kostova and I. Ognyanov, Monatsh., 1986, 117, 689), and mass spectral fragmentation behaviour of a number of rotenoids functionalized in the alicyclic portions of the molecule have been examined (Kostova, Ognyanov, and H. Budzikiewicz, Org. mass Spec., 1983, 18, 234). Also the mass spectra of some rotenones containing an oxo group at C-6 (C-12 if based on rotexen nucleus), with different stereochemistry at the B/C ring junction have been compared (Kostova and N. Mollova, *ibid*., 1985, 20, 765). The chemical reactivity of the 6a-hydroxy-methyl derivatives of rotenone and amorphigenin have been compared and discussed (Kostova and Ognyanov, Doklady Bolg. Akad. Nauk, 1985, 38, 1509).

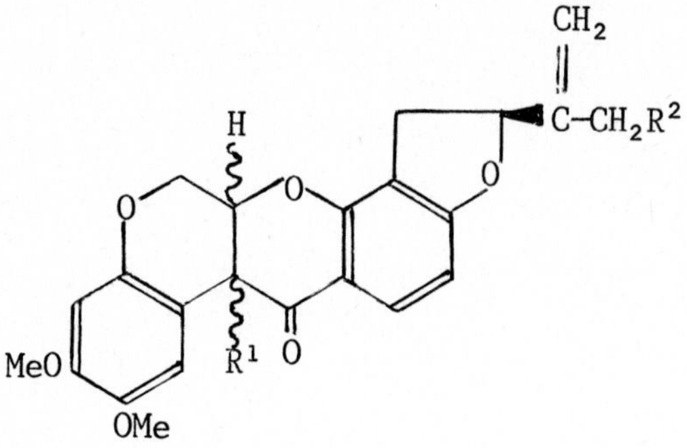

R^1 = H, OH, CH_2O_2CH, CH_2Cl, CH_2OH, CH_2OAc
R^2 = H, OH, Cl

(22)

(g) *Flavanones, 2-phenylchroman-4-ones, 2,3-dihydro-2-*
    -phenyl-4H-benzo[b]pyran-4-ones

Preparations. Manganese dioxide oxidation of 4-oximino-
flavans (1) gives flavanones (2) (T.C. Sharma and V. Saksena,
Indian J. Chem., 1977, 15B, 748).

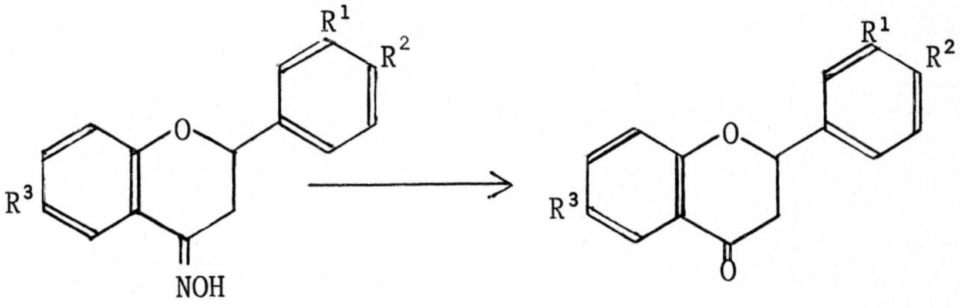

R^1	R^2	R^3
H	H	H
H	MeO	H
H	MeO	Me
Me	MeO	Me

(1)

(2)

On visible irradiation in polar aprotic solvents 4-
-substituted 2'-hydroxychalcones (3) undergo photocyclization
to afford the corresponding flavanones (4), but in low
quantum yields (R. Matsushima and H. Kageyama, J. chem. Soc.,
Perkin II, 1985, 743).

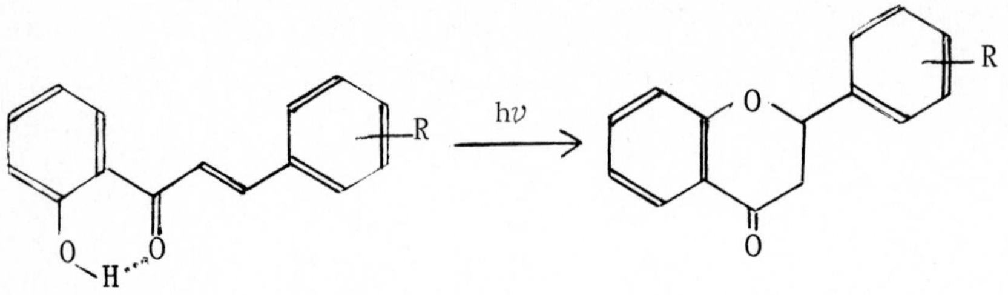

R = H, 4-Me, 4-MeO, 4-F
4-Cl, 4-Br, 4-NO$_2$, 2,4(MeO)$_2$

(3)   (4)

A a one-step synthesis of 3-nitroflavanones (6) is
accomplished by the cyclocondensation of the 5-substituted
ω-nitro-2-hydroxyacetophenones (5) with the appropriate
benzaldehyde in the presence of ammonium acetate in acetic
acid. The flavanones on bromination followed by dehydrobromination furnish the related 3-nitroflavones (C. Paparao,
K.V. Rao, and V. Sundaramurthy, Synth., 1981, 236).

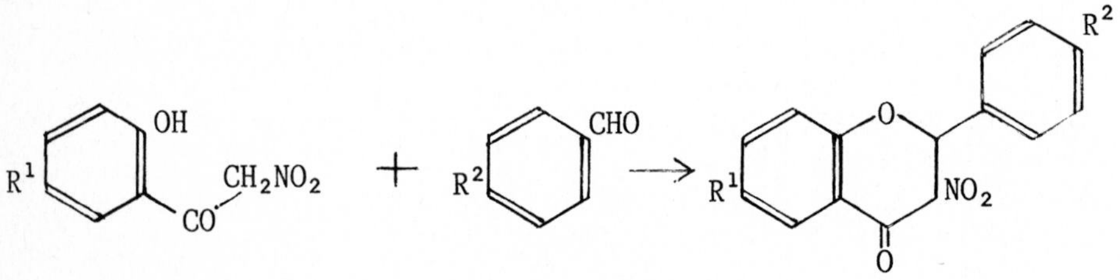

R^1 = H, Me   R^2 = H, OMe

(5)   (6)

2'-Hydroxy-4'-n-propoxy-5'-nitrochalcones (7) with 10% sulphuric acid yields 6-nitro-7-propoxyflavanones (8), whereas boiling with selenium dioxide in pentanol affords flavones and treatment with alkaline hydrogen peroxide in methanol gives flavonols (N.J. Shah, L.C. Jhaveri, and H.B. Naik, Chem. Era, 1980, 16, 108).

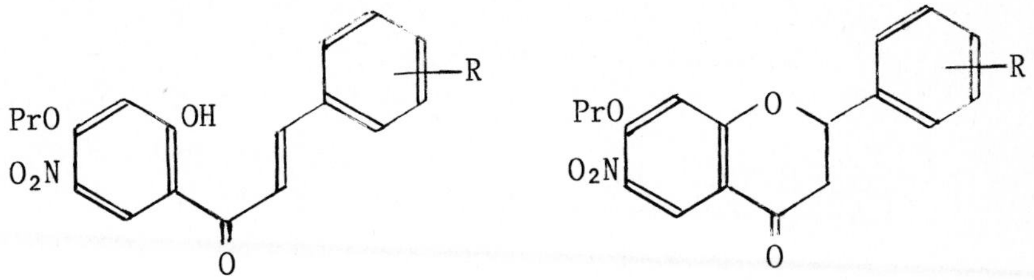

R = H, 4-MeO, 4-NO$_2$, 3-HO

(7)   (8)

Reactions. Treatment of flavanone with sodium cyanoborohydride affords 4-hydroxyflavan (9), while dihydroquercetin yields catechin (C.A. Elliger, Synth. Comm., 1985, 15, 1315).

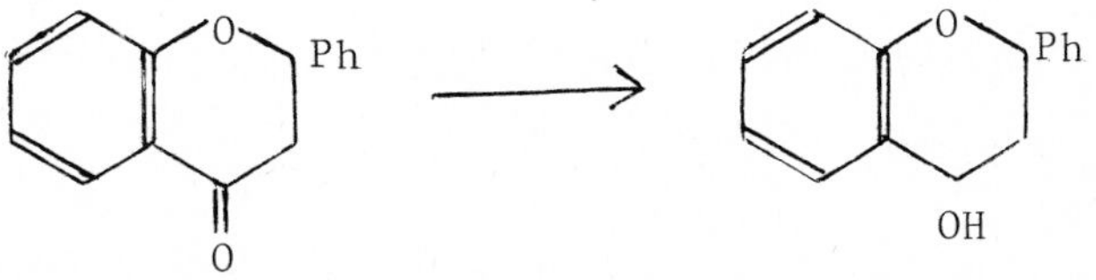

(9)

Flavanones (10) with copper (II) chloride furnish the 3-
-chloro- (11) and/or the 3,3-dichloro-flavanones (12)
depending on the substituents. Some of the dichloro
derivatives have been dehydrochlorinated to yield the
related 3-chloroflavones. Reduction of 3-chloro- and 3,3-
-dichloro-flavanones with sodium tetrahydridoborate gives
the corresponding 3-chloro- and 3,3-dichloro-4-hydroxy-
flavans, with inversion of configuration at the C1-bearing
carbon atom for monochlorides (F.G. Weber, E. Birkner, and
H. Koeppel, Pharmazie, 1980, $\underline{35}$, 328).

R^1=H,Cl,OMe
R^2=H,Me,OMe
R^3=H,F,Cl,Br,OMe
R^4=H,OMe

(10)

(11) R^5=H,R^6=Cl
(12) R^5=R^6=Cl

Flavanones (13) are brominated with pyridinium tribromide
in acetic acid or by (2-carboxyethyl)triphenylphosphonium
perbromide to afford the corresponding 3-bromoflavanones
(14), which are converted back into the flavanones on
boiling with thiourea in ethanol or by dimethyl sulphoxide
at room temperature (N.J. Reddy and T.C. Sharma, Bull.
chem. Soc., Japan, 1979, $\underline{52}$, 647; Sharma et $al.$, Indian J.
Chem., 1981, $\underline{20B}$, 80).

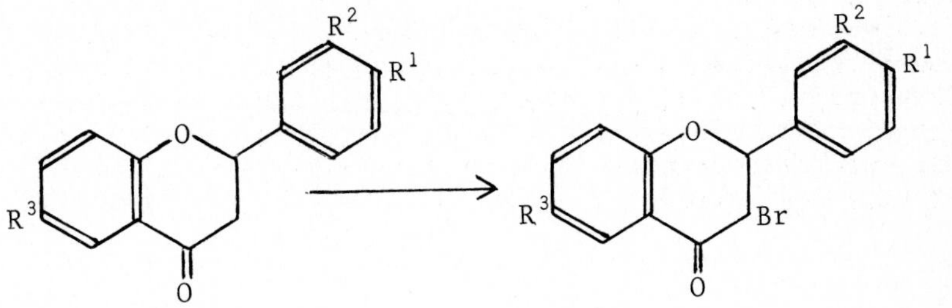

R^1	R^2	R^3
H	H	H
H	H	Me
OMe	H	Me
OMe	OMe	Me

(14)

(13)

The ^{13}C-nmr spectra of chroman-4-one and its methyl and phenyl substituted homologues, including flavanone have been examined (Y. Senda *et al.*, Bull. chem. Soc., Japan, 1977, <u>50</u>, 2789).

288

(h) Hydroxyflavanones

Preparations. Flavone and 4'-methoxyflavone react with
N-bromosuccinimide in methanol to give the corresponding
3-bromo-2-methoxyflavanones (1) and (2), which can be
converted into the related 3-bromo-4-hydroxy-2-methoxy-
flavans. Flavonol (3) and 4'-methoxyflavonol (4) react
rapidly with N-bromosuccinimide in methanol to yield 2,3-
-dimethoxyflavanones (5) and (6), but no bromo derivatives
(T.G.C. Bird *et al.*, J. chem. Soc., Perkin I, 1983, 1831).

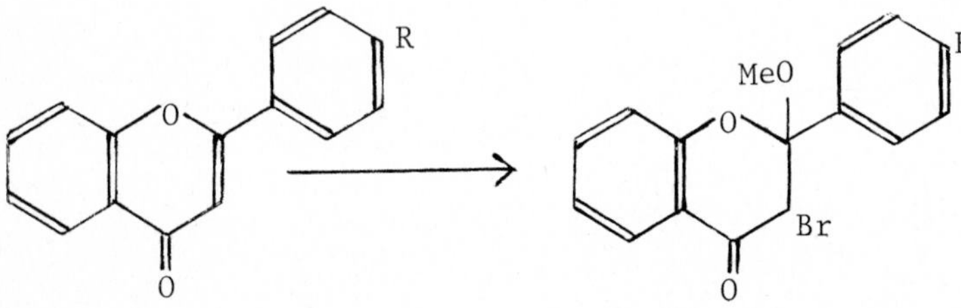

R=H,OMe

(1) R=H
(2) R=OMe

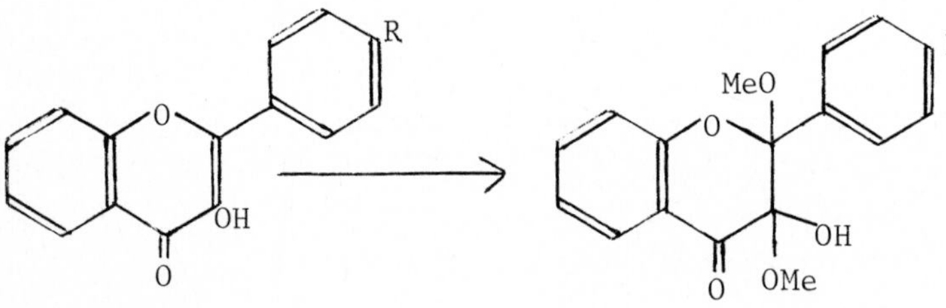

(3) R=H
(4) R=OMe

(5) R=H
(6) R=OMe

The conversion of 2'-hydroxy-, 2',4'-dihydroxy-, 3,2'--dihydroxy-, and 4,2'-dihydroxy-chalcone (7) in aqueous alkali, and of 2'-hydroxy- and 2'-hydroxy-4-nitro-chalcone (8) in alkaline methanolic medium into the corresponding substituted flavanones (9) has been studied kinetically and spectroscopically. It has been found that results from both methods are consistent with the existence of chalcones in the *trans-s-cis* conformation and indicate a mechanism that involves general acid attack to the ionized form of the 2'-hydroxychalcones, rotation through the $CO-C_\alpha$ bond and annelation to the flavanone (J.J.P. Furlong and N.S. Nudelman, J. chem. Soc., Perkin II, 1985, 633).

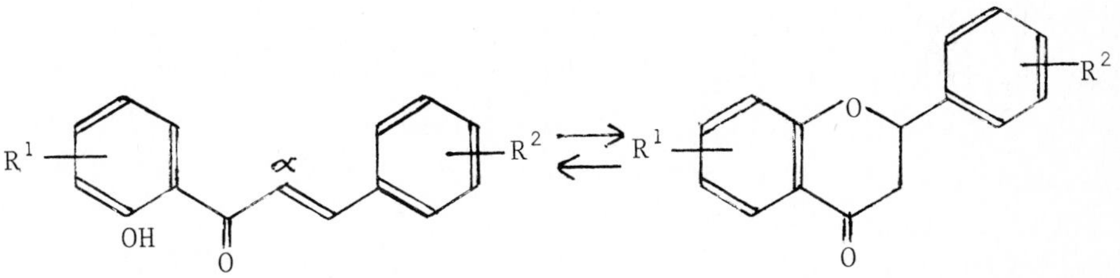

$R^1=R^2=H$
$R^1=4'-OH, R^2=H$
$R^1=H, R^2=3-OH$
$R^1=H, R^2=4-OH$

(7)

(9)

(8) $R^1=R^2=H$; $R^1=H, R^2=4-NO_2$

Rate coefficients for the chalcone ⟶ flavanone equilibration reaction have been established for some 2'--hydroxychalcones over the pH range (*ca*. 8-11) (K.B. Old and L. Main, *ibid*., 1982, 1309).

The hypervalent iodine oxidation of flavanone in methanol gives the dimethyl acetal (10), which on acid-catalyzed hydrolysis affords *cis*-3-hydroxyflavanone (11), m.p. 75-77°, *acetate*, m.p. 96-98°. Under more vigorous hydrolytic conditions isomerization accompanies hydrolysis of the dimethyl acetal (10), resulting in the formation of *trans*-3--hydroxyflavanone (12), m.p. 185-187°. Alkaline hydrogen peroxide oxidation of flavanone yields *trans*-3-hydroxy-flavanone (12) together with flavonol, whereas lead tetraacetate oxidation gives *trans*-3-acetoxyflavanone, m.p. 96-97°, along with flavone and isoflavone (R.M. Moriarty and O. Prakash, J. org. Chem., 1985, 50, 151).

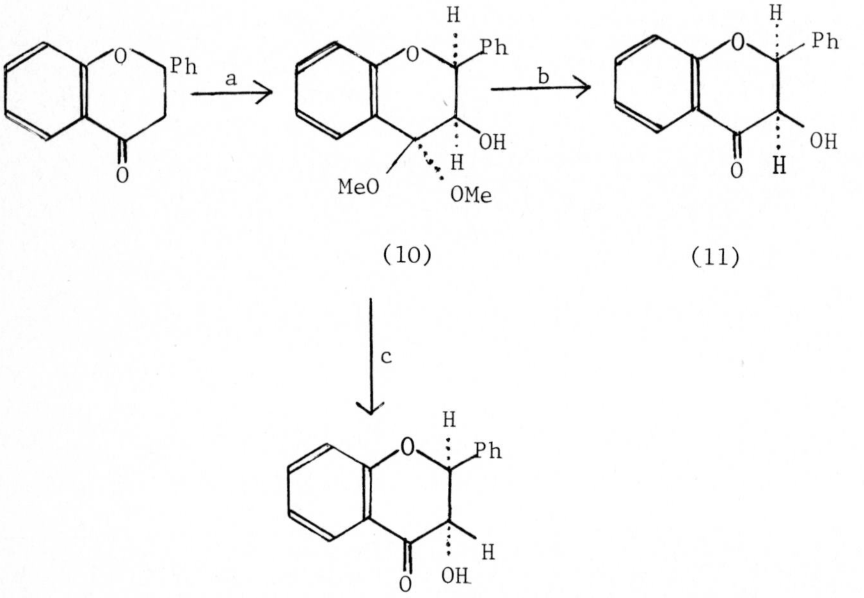

(10)   (11)

(12)

Reagents:-   (a) $C_6H_5I(OAc)_2$,KOH,MeOH,0°;   (b) 3M HCl, EtOH, 0.5h or 50% AcOH,50-70°,6h, or 4-TsOH,Me_2CO,H_2O,3 days; (c) conc.HCl,Me_2CO,2h

Properties. 5,7,4'-Trimethoxyflavanone (13) on heating
for *ca.* 0.5h with dimethyl sulphoxide in the presence of a
catalytic amount of iodine and concentrated sulphuric acid
gives 5,7,4'-trimethoxyflavone (14) in almost quantitative
yield. The method is also effective with flavanones having
free hydroxy groups and is, therefore, a general and
convenient one for the dehydrogenation of flavanoids
[W. Fatma *et al.*, J. chem. Res. (S), 1984, 298].

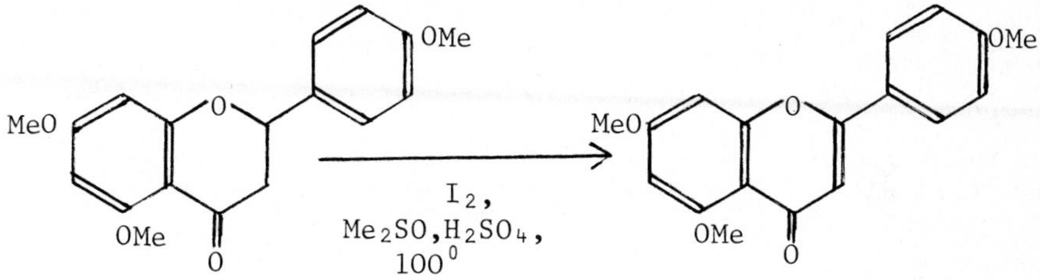

(13)   (14)

Reduction of flavanonol (15) with sodium tetrahydridoborate gives flavan-3,4-diol (16) (88%), whereas 3,5,4'-
-triacetoxyflavanone (17) affords a mixture of the flavans
(18) and (19) (H. Takahashi *et al.*, Chem. pharm. Bull.,
1985, **33**, 3134).

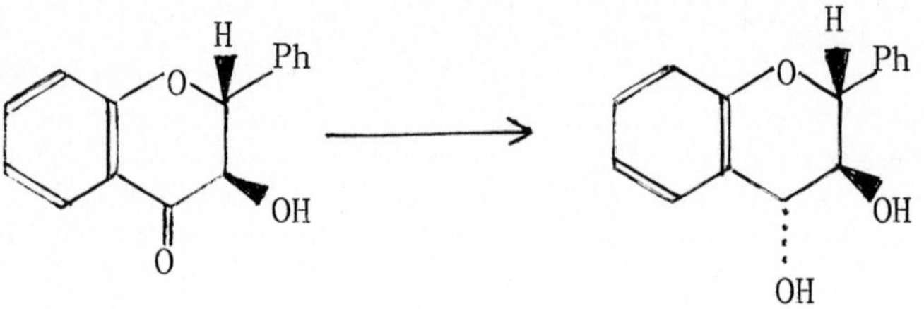

(15)   (16)

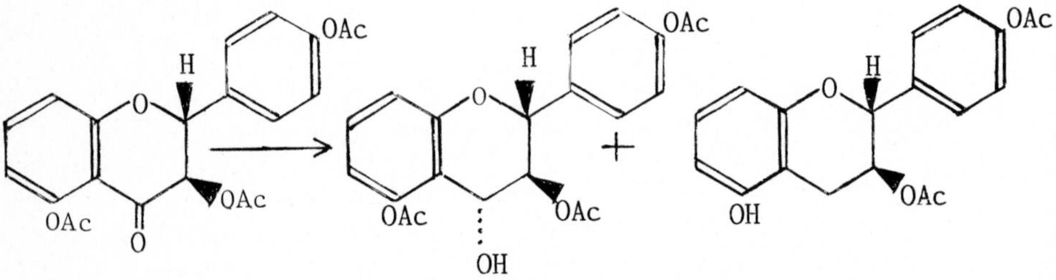

(17)   (18)   (19)

Treatment of 5,7,4'-trimethoxyflavanone with vanadyl chloride in benzene under nitrogen at room temperature furnishes 8-chloro-5,7,4'-trimethoxyflavanone, m.p. 158-160° (H.I. Siddiqui and W.A. Shaida, *ibid.*, 1985, 168). Bromination of 7-hydroxyflavanone with excess bromine (8 mole) in acetic acid gives 7-hydroxy-3,6,7,4'-tetrabromo-flavanone, which on heating with dimethyl sulphate and anhydrous potassium carbonate yields 7-methoxy-6,8,4'--tribromoflavanone. Heating the tetrabromo derivative with potassium acetate in acetic acid affords 7-hydroxy-6,8,4'--tribromoflavanone (Y.B. Vibhute and U.K. Jagwani, Chem. Abs., 1978, 89, 43027u).

The ^{13}C-nmr spectral data of a number of chalcones, flavanones, and related compounds (A. Pelter, R.S. Ward, and T.I. Gray, J. chem. Soc., Perkin I, 1976, 2475) and of some C-benzylated flavanones (C.D. Hufford and W.L. Lasswell Jr., Lloydia, 1978, 41, 151) have been discussed.

Naturally occurring hydroxyflavanones. The following are
some of the naturally occurring hydroxyflavanones, which
have been isolated. Didymocarpin A, *5,8-dihydroxy-6,7-*
-dimethoxyflavanone, $C_{17}H_{16}O_6$, m.p. 213-215°, has been
obtained from *Didymocarpus pedicellata* and on methylation
with diazomethane it gave 5-hydroxy-6,7,8-trimethoxy-
flavanone, m.p. 99-100° (A. Bhattacharyya, A. Choudhuri, and
N. Adityachaudhury, Chem. and Ind., 1979, 348). It has been
synthesized (*idem*, Indian J. Chem., 1980, <u>19B</u>, 428).
Aurapin, *7,4'-dimethoxy-5-hydroxy-3-0-rhamnosylfavanone*, a
new glycoside has been isolated from *Acinos alpinus* along
with naringenin, taxifolin, and neoponcirin (P. Venturella
et al., Heterocycles, 1980, <u>14</u>, 1979). 5,7,2',5'-Tetra-
-hydroxy- and 5'-methoxy-3,5,7,2'-tetrahydroxy-flavanone
have been isolated from the aerial parts of *Inula cappa*
(N.C. Baruah *et al.*, Phytochem., 1979, <u>18</u>, 2003).
     The white farinose exudate on the fronds of *Cheilanthes*
argentea contains 5,4'-dihydroxy-6,7-dimethoxyflavanone and
its 4'-methyl ether, 5,4'-dihydroxy-7,8-dimethoxyflavanone
and its 4'-methyl ether, 5,6-dihydroxy-7,8,4'-trimethoxy-
flavanone, and 5,4'dihydroxy-6,7,8-trimethoxyflavanone and
its 4'-methyl ether as minor constituents. The farina of
Notholaena limitanea var *mexicana* contains eriodictyol
7,4'-dimethyl ether and eriodictyol 7,3',4'-trimethyl ether
as minor components, also found together with naringenin 7-
-methyl ether in the exudate of *N. fendleri*. Trace amounts
of 5-hydroxy-7,3',4',5'-tetramethoxyflavanone are detected
in *N. lemmonii* var *lemmonii* (E. Wollenweber *et al.*,
Z. Naturforsch., Biosci., 1980, <u>35C</u>, 685).
     Cedeodarin, *6-methyl-3,5,6,3',4'-pentahydroxyflavanone*
(6-methyltaxifolin), dihydromyricetin, cedrin (6-methyl-
dihydromyricetin), and cedrinoside along with the known
taxifolin have been isolated from *Cedrus deodara* (cedar
wood) (P.K. Agrawal, S.K. Agarwal, and R.P. Rastogi,
Phytochem., 1980, <u>19</u>, 893); and 2'-hydroxydemethoxy-
matteucinol, *6,8-dimethyl-5,7,2'-trihydroxyflavanone* (1),
$C_{17}H_{16}O_5$, m.p. 198-200°, $[\alpha]_D^{25}$ - 97.8° (MeOH), and the
known demethoxymatteucinol from *Uvaria afzelii* (C.D. Hufford,
B.O. Oguntimein, and J.K. Baker, J. org. Chem., 1981, <u>46</u>,
3073).

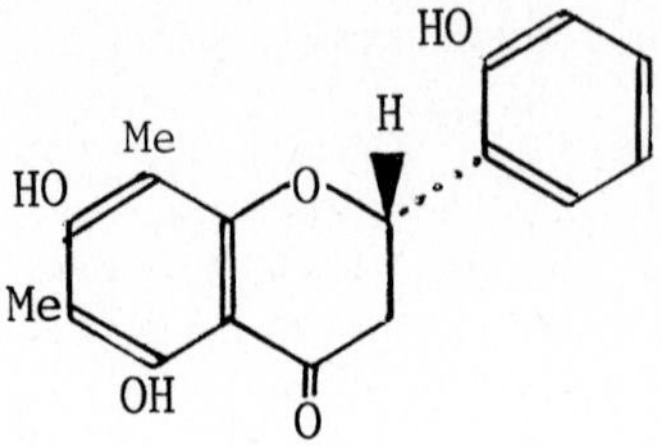

(1)

Flavanones containing prenyl substituents include
sophoraflavanone B, m.p. 193-195°, isolated from the aerial
parts of *Sophora tomentosa* L. and originally given the
structure of 6-prenylnaringenin, *6-prenyl-5,7,4'-trihydroxy-*
flavanone (M. Komatsu, I. Yokoe, and Y. Shirataki, Chem.
pharm. Bull., 1978, **26**, 3863), but later revised to 8-prenyl-
naringenin (2). 6-Prenylnaringenin (3) and isoxanthohumole
(4) have been isolated from hard resins of hops (*Humulus
lupulus* L.) and found to have antifungal properties
(S. Mizobuchi and Y. Sato, Chem. Abs., 1987, **106**, 135069g).

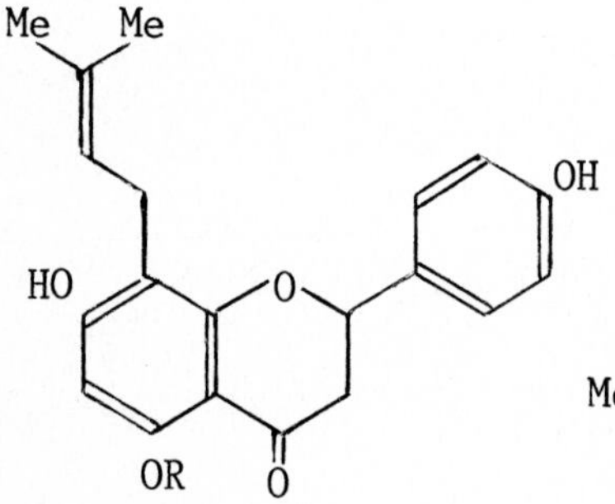

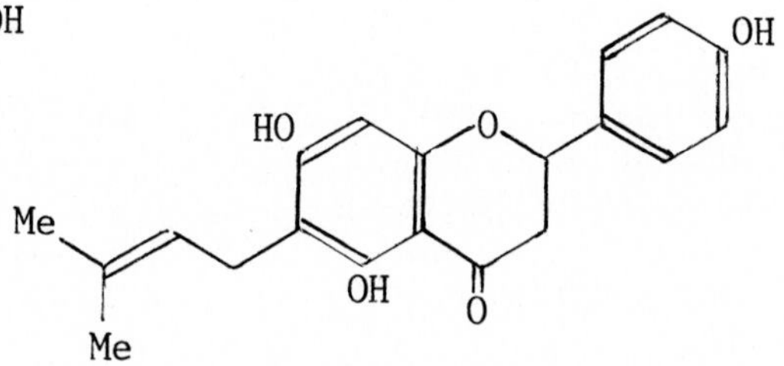

(2) R = H
(4) R = Me

(3)

Tirumalin, *7,4'-dimethoxy-8-prenyl-3,5,3'-trihydroxy-flavanone* [(+)-(2*R*,3*R*)-8-*C*-prenyltaxifolin 7,4'-dimethyl ether] (5) has been isolated from the leaves of *Rhynchosia cyanosperma* (D. Adinarayana *et al.*, Phytochem., 1980, <u>19</u>, 478), and flemiflavanone A, *6,8-diprenyl-4'-methoxy-5,7,2'-trihydroxyflavanone* (6), flemiflavanone B (7), and flemiflavanone C (8), from the roots of *Flemingia stricta* (S. Sivarambabu, J.M. Rao, and K.V.J. Rao, Indian J. Chem., 1979, <u>17B</u>, 85).

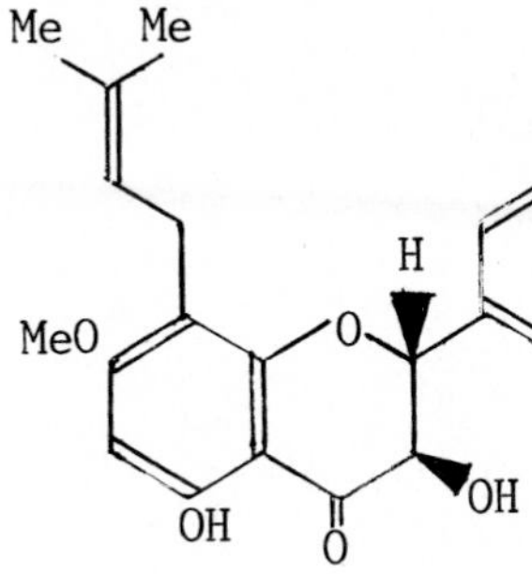

(5) 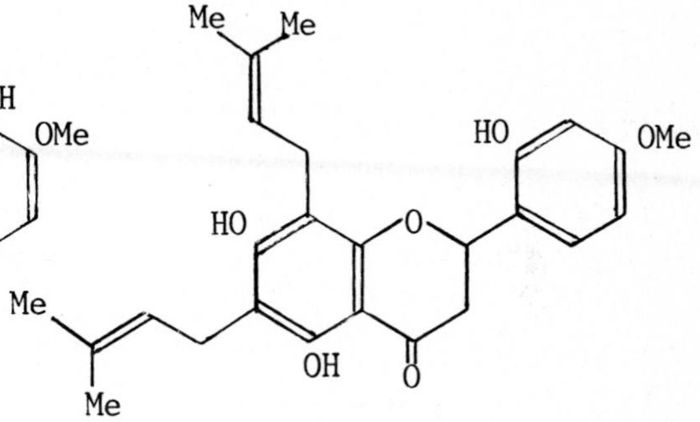   (6)

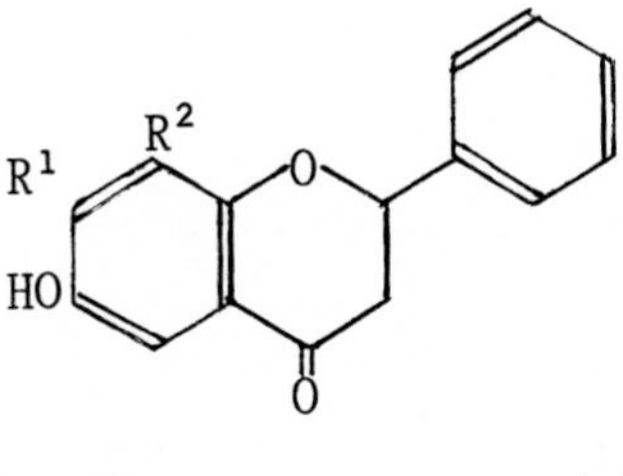

(7) R¹ = , R² = OH

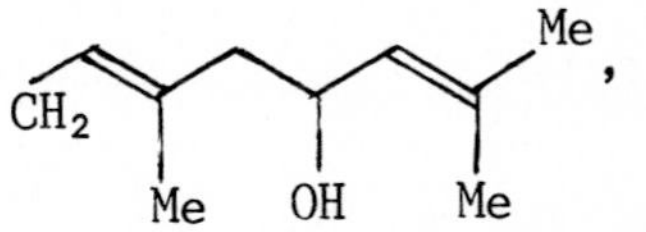

(8) R¹ = OH, R² =

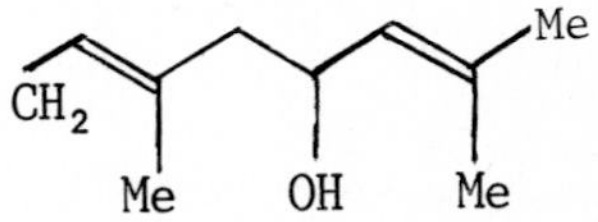

8-*C*-Prenyldihydroisorhamnetin, 7,8-dihydrooxepinoeriodictyol, and 7,8-dihydrooxepinodihydroquercetin, along with the known flavanones, 6-*C*-prenyleriodictyol, 8-*C*-prenyleriodictyol, 6-*C*-prenylnaringenin, 8-*C*-prenylnaringenin, naringenin 4'- -methyl ether, eriodictyol, and naringenin and other flavonoids have been isolated from *Wyethia angustifolia* (S. McCormick, K. Robson, and B. Bohm, Phytochem., 1986, 25, 1723).

The cytotoxic *C*-benzylated flavanones, chamanetin, *5,7- -dihydroxy-8-(2-hydroxybenzyl)flavanone*, $C_{22}H_{18}O_5$, m.p. 210- -211°, $[\alpha]_D^{25}$-52.5° (MeOH), isochamanetin, *5,7-dihydroxy-6- -(2-hydroxybenzyl)flavanone*, $C_{22}H_{18}O_5$, m.p. 215-217°, $[\alpha]_D^{25}$ -10.5° (Me₂CO), and dichamanetin, *5,7-dihydroxy-6,8-di(2- -hydroxybenzyl)flavanone*, $C_{29}H_{24}O_6$, m.p. 118-120° (EtOH), $[\alpha]_D^{25}$-9.75 (Me₂CO) have been isolated from *Uvaria chamae* along with the known flavanones, pinocembrin and pinostrobin (Lasswell Jr. and Hufford, J. org. Chem., 1977, 42, 1295). Chamanetin 5-methyl ether and dichamanetin 5-methyl ether have been obtained from the root bark of *U. chamae* and the ^{13}C-nmr spectral data of a number of methyl ether derivatives of chamanetin and dichamanetin have been discussed (H.N. El-Sohly and Lasswell Jr., J. nat. Prod., 1979, 42, 264). Uvarinol (9), $C_{36}H_{30}O_7$, m.p. 152-154°, $[\alpha]_D^{20}$-16.5° (Me₂CO), a novel cytoxic tribenzylated flavanone has also been obtained from *U. chamae* (Hufford *et al*., J. org. Chem., 1979, 44, 4709).

(9)

Structure (10) has been confirmed for silylchristin by
[13]C-nmr spectroscopy. It had previously been assigned
the alternative structure (11). Based on structure (10),
structure (12) has been proposed for anhydrosilychristin
(A. Pelter, R. Haensel, and M. Kaloga, Tetrahedron Letters,
1977, 4547).

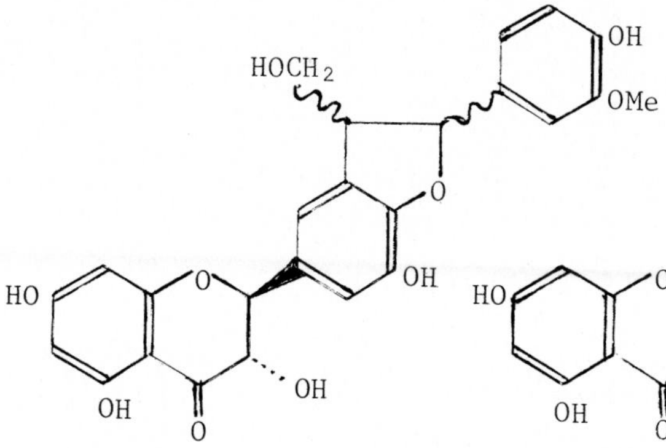

(10) 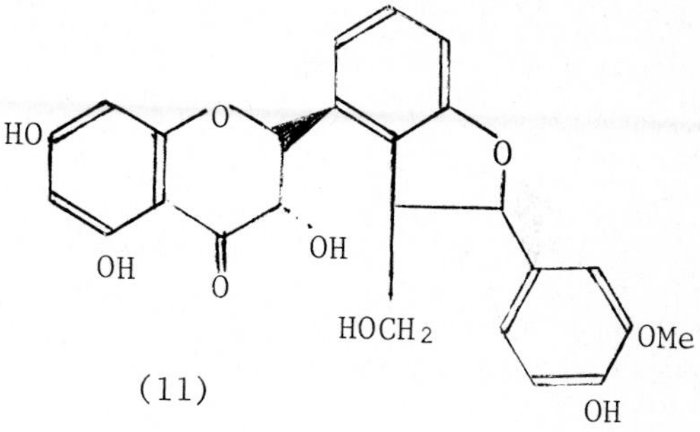   (11)

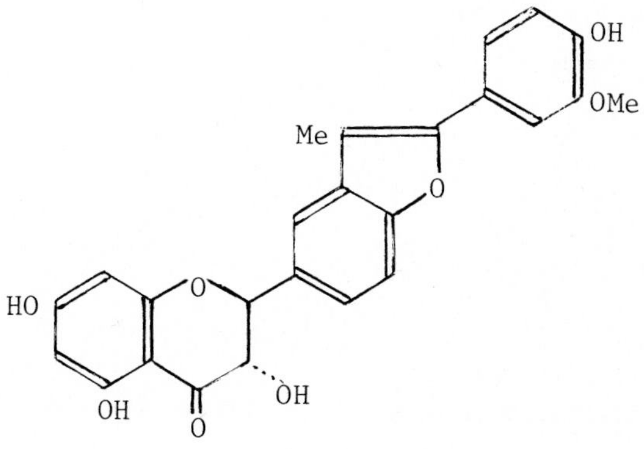

(12)

Four flavanonols, kushenol J(13), K(14), L(15), and M(16) have been isolated from the dry roots of *Sophora flavescens* (L.J. Wu *et al.*, Yakugaku Zasshi, 1985, <u>105</u>, 1034).

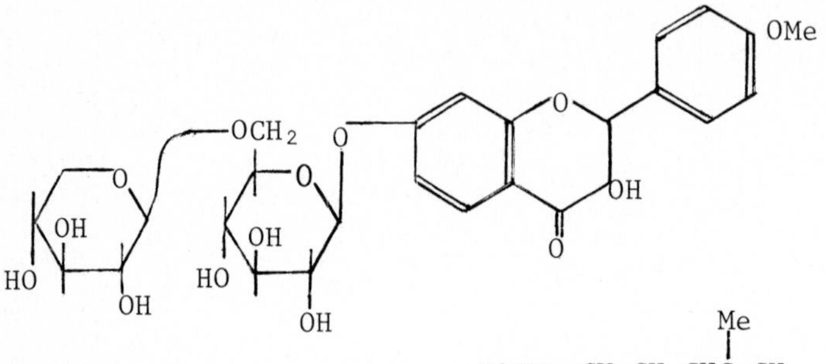

(13)

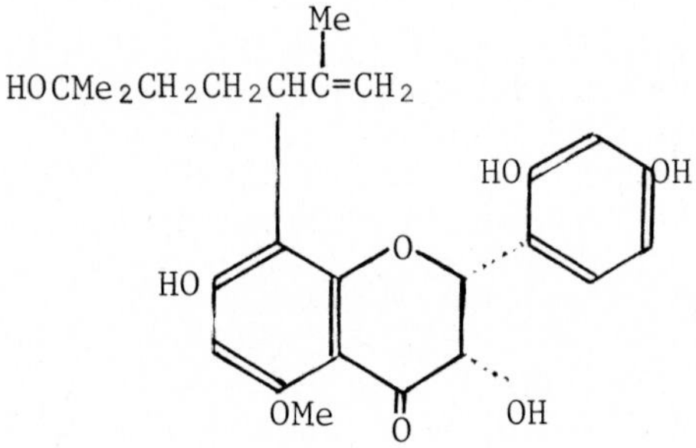

(14)

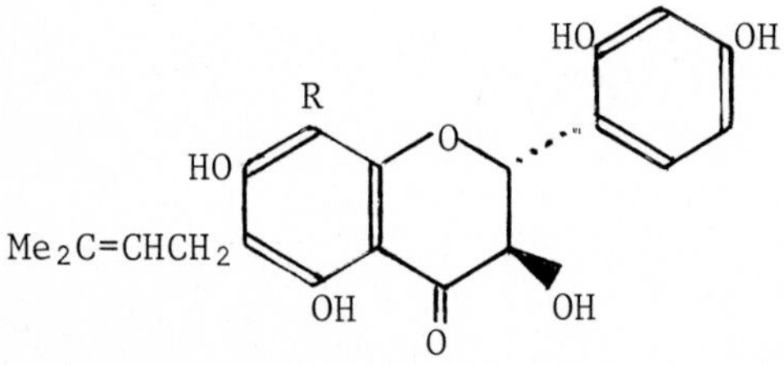

(15) R=CH$_2$CH=CMe$_2$
(16) R=CH$_2$CH(CMe=CH$_2$)CH$_2$CH=CMe$_2$

Some known naturally occurring flavanones have been isolated from additional sources, for instance, pinocembrin along with other flavonoids from *Chenopodium graveolens* (R. Mata *et al.*, Phytochem., 1986, <u>26</u>, 191); 7-hydroxy-flavanone, 8-methylpinocembrin, and 6,8-dimethylpinocembrin from the ground aerial parts of *Ceratiola ericoides*

(N. Tanrisever *et al.*, *ibid.*, 1986, <u>26</u>, 175); (±)-
-pinostrobin, (-)-demethoxymatteucinol, and (-)-cryptostrobin
from *Agonis spathalata* (Myrtaceae) (J.R. Cannon and
P.F. Martin, Austral. J. Chem., 1977, <u>30</u>, 2099);
strobopinin, desmethoxymatteucinol, and cryptostrobin from
the white farinose exudate on the fronds of *Pityrogramma
pallida* (E. Wollenweber *et al.*, Z. Pflanzenphysiol., 1979,
<u>94</u>, 241); and dihydroquercetin 7,3'-dimethyl ether and
other flavonoids from the aerial tissues of *Jasonia
tuberosa* (A.G. Gonzalez *et al.*, An. Quim., 1977, <u>73</u>, 460).
Prenylation of 7-hydroxyflavanone with 2-methylbut-3-en-2-ol
in the presence of boron trifluoride etherate gives
ovaliflavanone A, *6,8-diprenyl-7-hydroxyflavanone* (17),
ovaliflavanone B, *7-hydroxy-8-prenylflavanone* (18), 7-
-hydroxy-6-prenylflavanone, and 7-prenyloxy-flavanone.
Ovalichromene B (19) is obtained on treating 3',4'-methylene-
dioxy-7-hydroxy-8-prenylflavanone with dichlorodicyanobenzoquinone (A. Islam, R.K. Gupta, and M. Krishnamurti, Indian
J. Chem., 1981, <u>20B</u>, 21).

(17)

(18)

(19)

5,7,4'-Trihydroxyflavanone on prenylation affords a
mixture of products, the major one being 6,8-diprenyl-
-5,7,4'-trihydroxyflavanone, which undergoes oxidative
cyclization on treatment with dichlorodicyanobenzoquinone
to yield lupinifolin (20) (A. Nagar, V.K. Gujral, and
S.R. Gupta, Tetrahedron Letters, 1978, 2031).

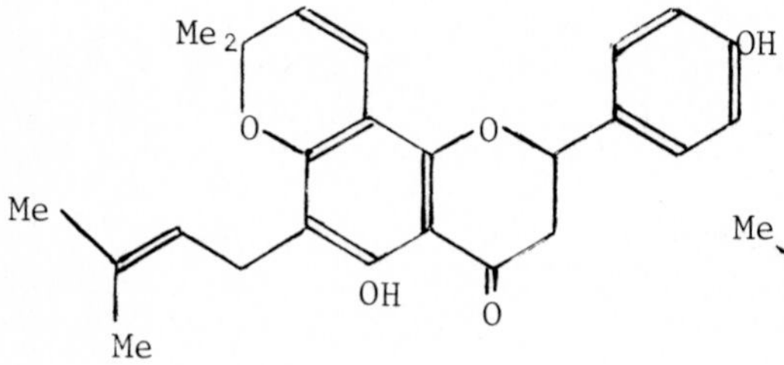

(20)

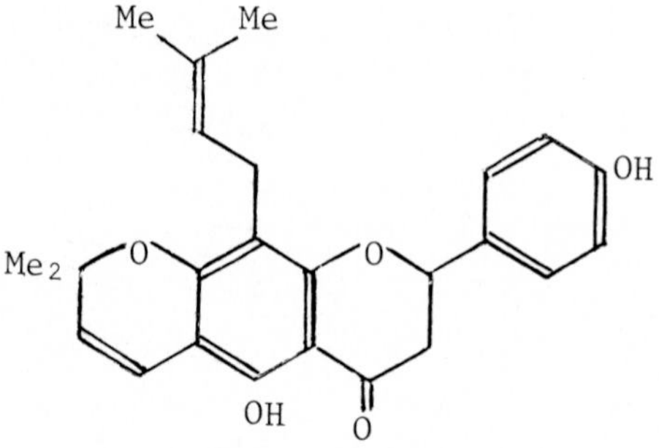

(21)

Racemic 8,8-dimethyl-10-C-prenyl-8H-pyrano[3,2-g]naringenin
(21), the structure originally assigned to lupinifolin, has
been synthesized (A.C. Jain, R.C. Gupta, and P.D. Sarpal,
Chem. Letters, 1978, 995). 7-Hydroxy-5,6,8-trimethoxy-
flavanone has been prepared and shown not to be didymocarpin
(S.K. Garg, S.R. Gupta, and N.D. Sharma, Indian J. Chem.,
1979, 17B, 394). Abyssinone I (22) and other flavonoids
have been isolated from an East African medical plant
Erythrina abyssinica (V.S. Kamat *et al*., Heterocycles 1981,
15, 1163).

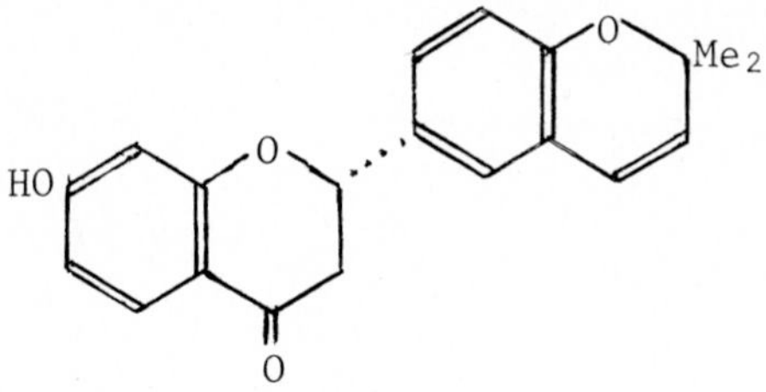

(22)

The ^{13}C-nmr spectral data of some flavanones, flavonoids, and isoflavanoids have been reported (E. Wenkert and H.E. Gottlieb, Phytochem., 1977, 16, 1811).

(i) Biflavanones

A mixture of triterpenes and four 3,8-linked biflavanones, kolaflavanone, *3-(4'-methoxy-3,5,6,3'-tetrahydroxy-8-
-flavanonyl)-5,7,4'-trihydroxyflavanone* (1), and the known biflavanones GB-1, GB-1a, and GB-2 have been obtained from the nuts of *Garcinia kola* Heckel. It has been characterized as the II-4' methyl ether of GB-2, as part of a ^{13}C-nmr study of flavonoids (P.J. Cotterill, F. Scheinmann, and I.A. Stenhouse, J. chem. Soc., 1978, 532).

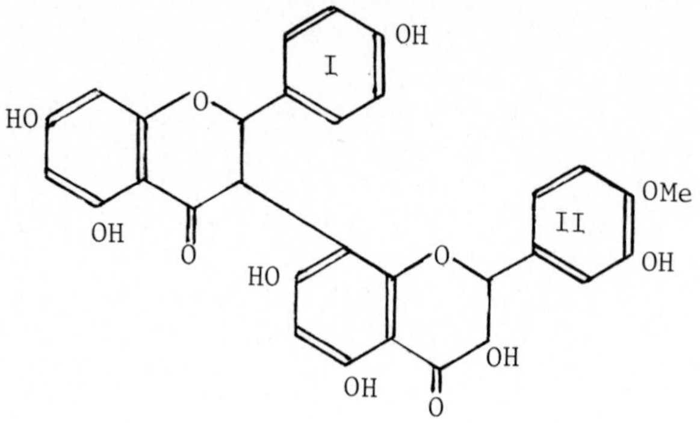

(1)

Semecarpuflavanone, *8-(7,4'-dihydroxy-3'-flavanonyl)-
-7,3',4',5'-tetrahydroxyflavanone* (2), a 8,3'-linked
biflavanone has been isolated from the nut shells of
Semecarpus anacardium Linn. On dehydrogenation with iodine
and potassium acetate in acetic acid it gives the
relatively more stable biflavone SA-4 (S.S.N. Murthy, Proc.
Indian Acad. Sci., chem. Sci., 1986, <u>97</u>, 63).

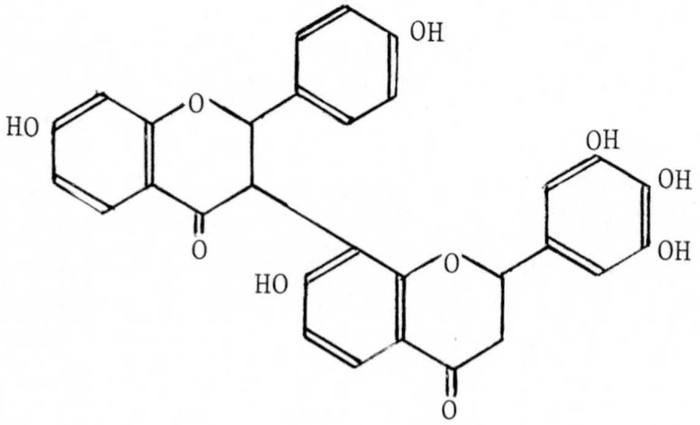

(2)

*(j) Isoflavanones, 3-phenylchroman-4-ones, 2,3-dihydro-3-
-phenyl-4H-benzo[b]pyran-4-ones*

The structures of a number of new naturally occurring
isoflavanones have been elucidated, including the following.
Sophoraisoflavanone A, *2'-methoxy-3'-prenyl-5,7,4'-trihydroxy-
isoflavanone* (1), m.p. 178-180°, together with other
flavonoids has been isolated from the aerial parts of *Sophora
tomentosa* L. It exhibits antifungal activity (M. Komatsu,
I. Yokoe, and Y. Shirataki, Chem. pharm. Bull., 1978, 26,
3863).

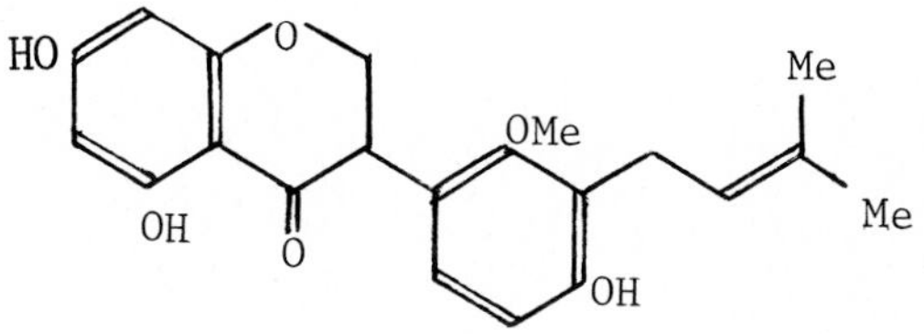

(1)

A diprenylated isoflavanone, isosophoranone, *6,3'-diprenyl-
-2'-methoxy-5,7,4'-trihydroxyisoflavanone* (2) has also been
obtained from *S. tomentosa* along with a minor component
isosophoronol (3) (G. Delle Monache *et al.*, Gazz., 1977,
107, 189).

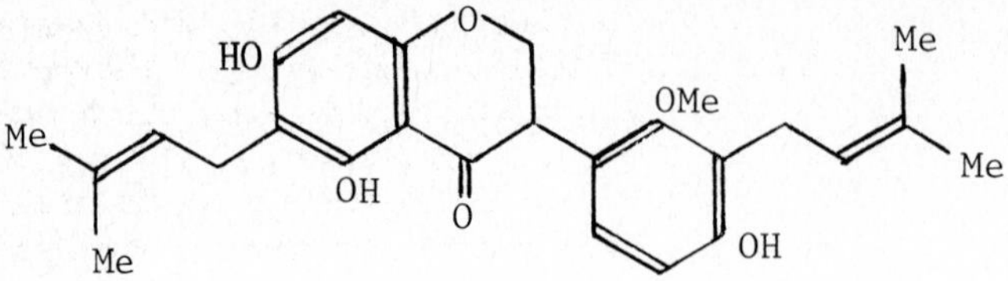

(2)

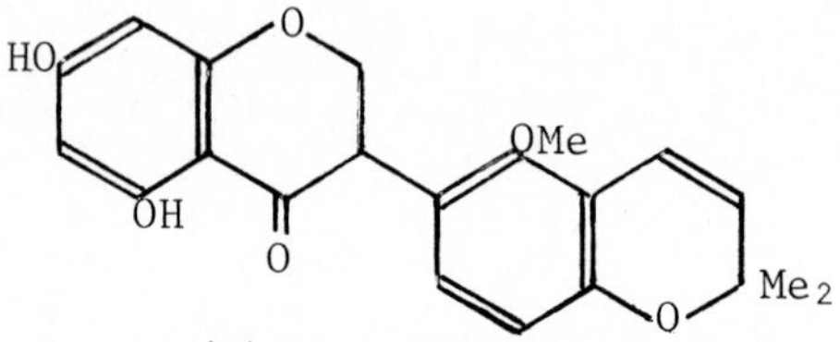

(3)

Cajanol, a major phytoalexin from *Cajanus cajan*, has been shown to be 5,4'-dihydroxy-7,2'-dimethoxyisoflavanone (4). The peroxide oxidation of cajanol ethyl ether yields the B--ring derived product, 4-ethoxy-2-methoxybenzoic acid, thus providing the evidence for the revised structure (J.L. Ingham, Z. Naturforsch., Biosci., 1979, 34C, 159).

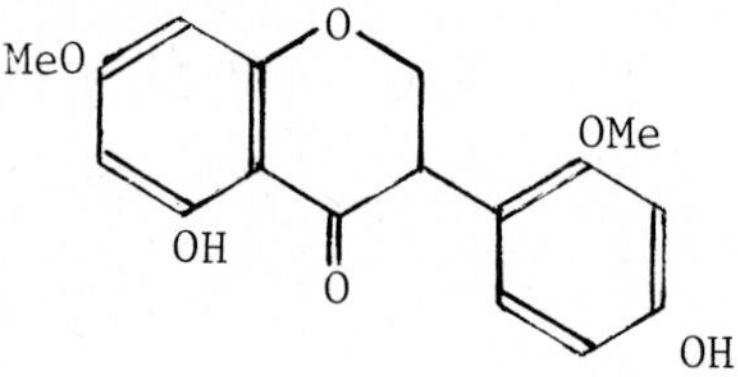

(4)

Vestitone, *7,2'-dihydroxy-4'-methoxyisoflavanone*, along with other flavonoids and isoflavonoids has been found in leaves of sainfoin *(Onobrychis viciifolia)*, inoculated with the fungus *Helminthosporium carbonum* (*idem*, *ibid.*, 1978, 33C, 146).

Isoferreirin, *2'-methoxy-5,7,4'-trihydroxyisoflavanone*, together with dalbergioidin and other products accumulate in the leaves and stems of *Dolichos biflorus* (horsegram) following inoculation with the nonpathogens *Pseudomonas pisi* and *Phytophthora megasperma glycinea*, respectively (N.T. Keen and Ingham, *ibid.*, 1980, **35C**, 923).

Structurally related to the isoflavanones are the following naturally occurring compounds. (+)-Variabilin (5) isolated together with a number of known compounds from the creeper *Dalibergia variabilis* (K. Kurosawa *et al.*, Phytochem., 1978, **17**, 1417) and mucronucarpan (6) from wood samples of *Machaerium mucronulatum* and *M. villosum* (*idem*, *ibid.*, p.1405).

(5)   (6)

Two other new pterocarpans, erythrabyssin I (7) and erythrabyssin II, phytoalexins, phaseolin (8) and phaseollidin, and a number of new flavones, have been isolated from the East African medicinal plant *Erythrina abyssinia*. Compounds (7) and (8) have been noted for their antiyeast and antifungal activities (V.S. Kamat *et al.*, Heterocycles, 1981, **15**, 1163).

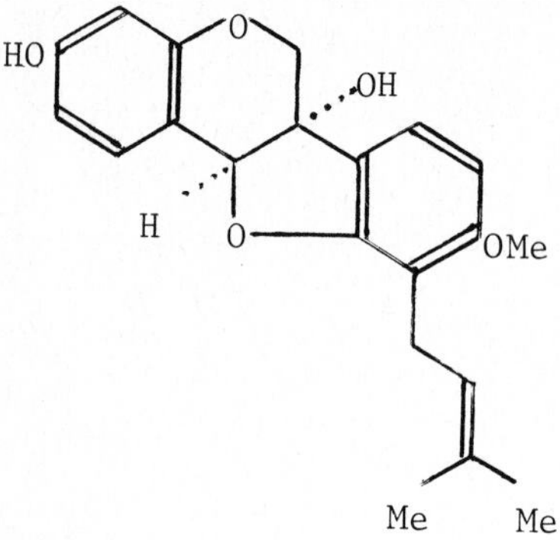

(7)

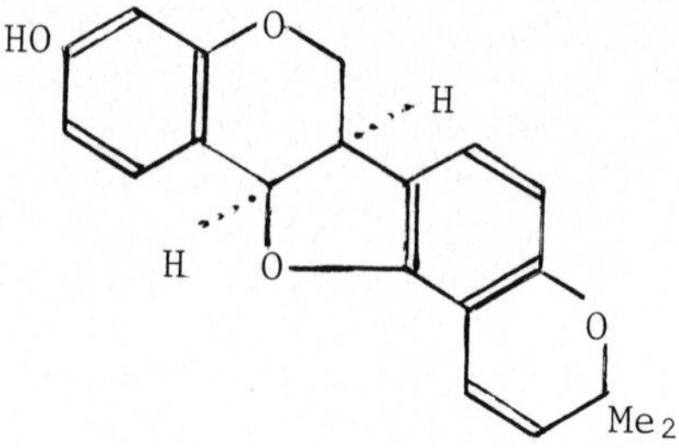

(8)

Sophorapterocarpan A (9), m.p. 149°, sophoracoumestan A (10), m.p. > 300°, and sophoraisoflavanone B, *6,5'-diprenyl--4'-methoxy-5,7,2'-isoflavanone*, m.p. 171-172°, have been isolated from the roots of *Sophora franchetiana* (Komatsu, Yokoe, and Shirataki, Chem. pharm. Bull., 1981, 29, 532).

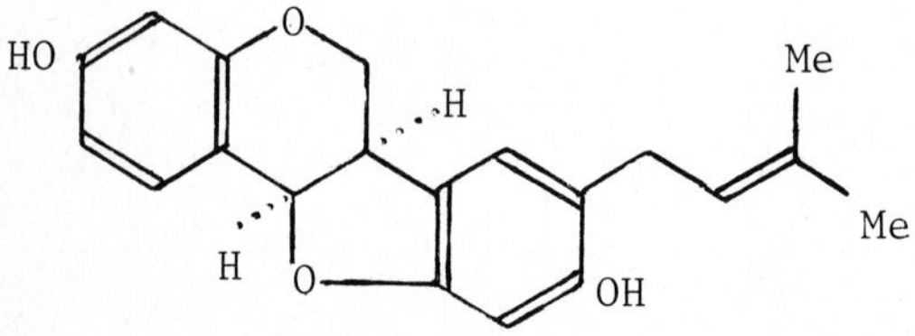

(9)

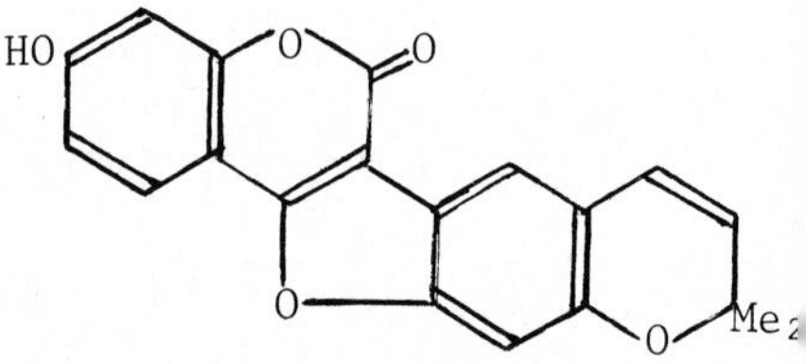

(10)

Two phytoalexins isolated from the fungus-inoculated leaflets of *Erythrina sandwicensis* have been identified as (-)-6aS;11aS-10-prenyl-3,6a,9-trihydroxypterocarpan (sandwicarpin) and (-)-6aR;11aR-3-hydroxy-9-methoxy-10- -prenylpterocarpan (sandwicensin). Also present were the known pterocarpans, demethylmedicarpin (see p. 308), 3,6a,9- -trihydroxypterocarpan, phaseollidin, and cristacarpin (Ingham, Z. Naturforsch., Biosci., 1980, **35C**, 384). Phytoalexins isolated from the fungi-inoculated hypocotyls of *Lablab niger*, include the isoflavanones, dalbergioidin and kievitone and the pterocarpan, phaseollidin (*idem, ibid.*, 1977, **32C**, 1018), and inoculated leaflets of *Dalbergia sericea* afford the pterocarpans, medicarpin and maackiain (*idem, ibid.*, 1979, **34C**, 630).

Feeding experiments relating to the biosynthesis and biochemical interconversion of sativan, *2',4'-dimethoxy-7- -hydroxyisoflavan*; vestitol, *7,2'-dihydroxy-4'-methoxy- isoflavan*; and demethylhomoptero-carpin (11) in copper (II) chloride treated lucerne seedlings have been investigated. The incorporation of vestitol - [14]C into demethylhomopterocarpin (11), but not into maackiain, pterocarpin phytoalexins of red clover has also been demonstrated (P.M. Dewick and M. Martin, Phytochem., 1978, **18**, 591).

(11)

Feeding experiments with ^{13}C- and ^{2}H-labelled precursors in copper (II) chloride-treated *Trifolium pratense* (red clover) seedlings have demonstrated that the isoflavone, formonetin (12), and the pterocarpan phytoalexins, medicarpin (13) and maackiain (14) are biosynthesized from 4,2',4'-trihydroxychalcone (H.A.M. Al-Ani and Dewick, J. chem. Soc., Perkin I, 1984, 2831).

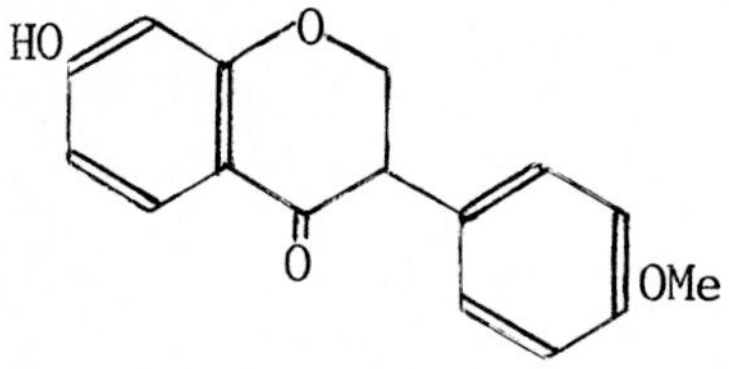

(12)

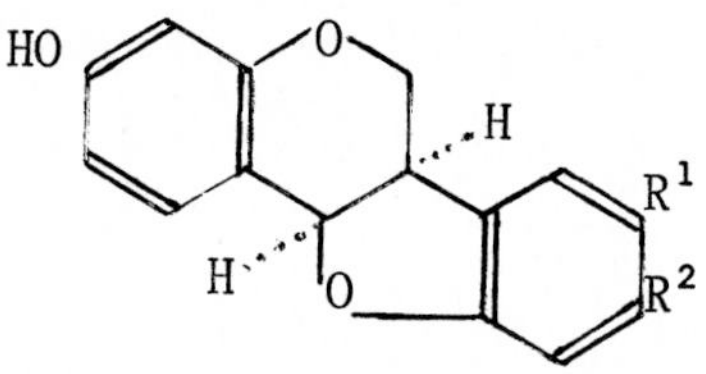

(13) R^1 = H, R^2 = OMe
(14) R^1R^2 = OCH$_2$O

(k) Naturally occurring homoisoflavanones

Additional homoisoflavanones (previously referred to as dihydrohomoisoflavones in Vol. IVE, p.218), (E)-7-O--methyleucomin (1), (-)-7-O-methyleucomol (2), (+)-3,9--dihydroeucomin (3), and 7-O-methyl-3,9-dihydroeucomin (4) have been isolated from the waxy material between the bulb leaves of *Eucomis bicolor*. Only the previously isolated (Z)-isomer of eucomin is genuine the (E)-isomer is only an artifact and possibly (1) is not genuine but an artifact (W. Heller *et al.*, Helv., 1976, 59, 2048).

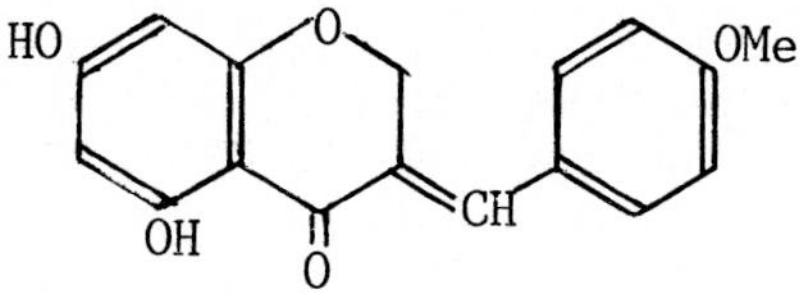

(1)

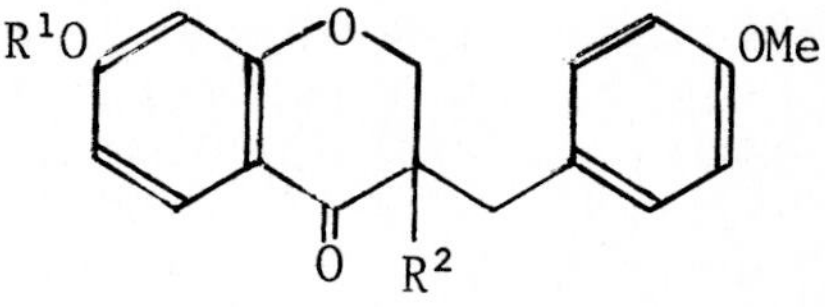

(2) R^1 = Me, R^2 = OH
(3) R^1=R^2 = H
(4) R^1 = Me, R^2 = H

The absolute configuration of (-)-7-O-(4-bromophenacyl)-eucomol has been determined by X-ray crystal analysis (H.P. Weber, Heller, and C. Tamm, Helv., 1977, 60, 1388).

A number of homoisoflavanones (5 and 6), for example, 5,7-dihydroxy-3-(3,4-dihydroxybenzy)isoflavanone (5; R^1= R^3 = R^5 = R^6 = OH, R^2 = R^4 = H), have been isolated from the bulbs of *Muscari armeniacum* and *M. botryoides* (M. Adinolfi *et al.*, Phytochem., 1986, 26, 285).

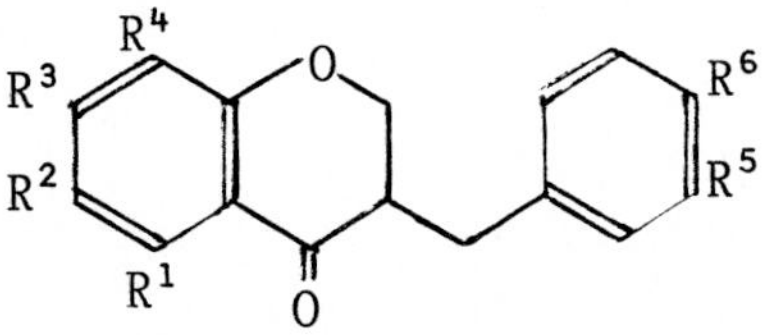

R^1 = R^3 = R^5 = R^6 = OH, OMe;
R^2 = H, OMe; R^4 = H, OH

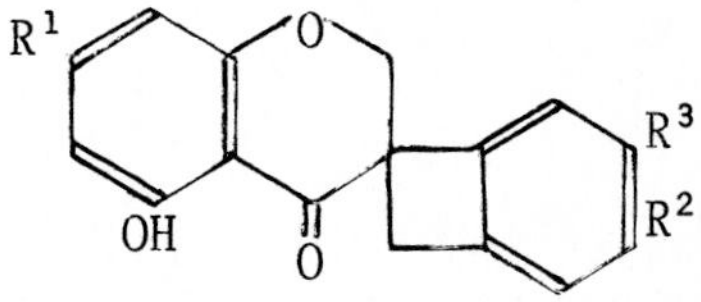

R^1 = R^2 = R^3 = OH, OMe

(5)

(6)

The ^{13}C-nmr chemical shifts of the homoisoflavanones (7) and some related compounds have been assigned, mainly on the basis of long-range heteronuclear [2]D nmr shift correlation experiments and of long-range H–C coupling constants. Also [1]H-nmr chemical shift have been reported and general relationships between substitution patterns and nuclear shielding, useful for the identification of naturally occurring homoisoflavanones, have been discussed (Adinolfi *et al.*, Mag. resn. Chem., 1986, 24, 663).

$$R^1 = H, \ OMe$$
$$R^2 = OH, \ OMe$$
$$R^3 = H, \ OH, \ OMe$$

(7)

5. 1H-Benzo[c]pyran, 1H-2-benzopyran, isobenzopyran, 3,4- -benzopyran, isochromene and its derivatives

The names 1H-benzo[c]pyran and 1H-2-benzopyran are generally used, but in some publications the name isochromene is still used. The names isocoumarin, isochroman, and isochromanone are still generally used and will be adopted in this section along with 1H-benzo[c]pyran.

(a) 1H-Benzo[c]pyrans

1H-Benzo[c]pyran (4) is obtained by the dehydration of 3- -hydroxyisochroman (2) when it is distilled in the presence of potassium hydrogen sulphate. Esterification of homophthalic acid (1), followed by selective reduction with diisobutylaluminium hydride affords 3-hydroxyisochroman (2), also obtained by similar reduction of isochroman-3-one (3), prepared by the Baeyer-Villiger oxidation of indan-2- -one (5) with *m*-chloroperbenzoic acid (F. Cottet, L. Cottier, and G. Descotes, Synth., 1987, 497).

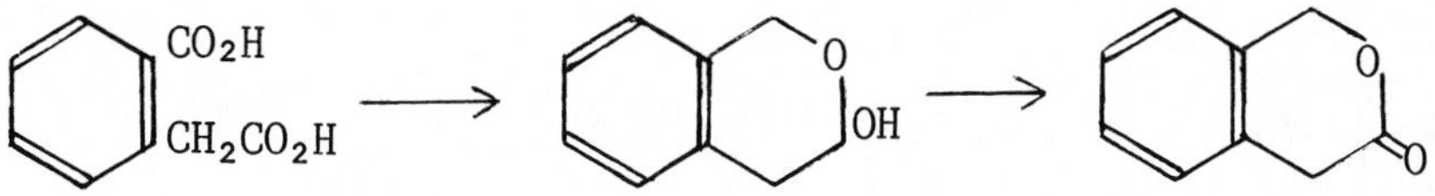

(1)   (2)   (3)

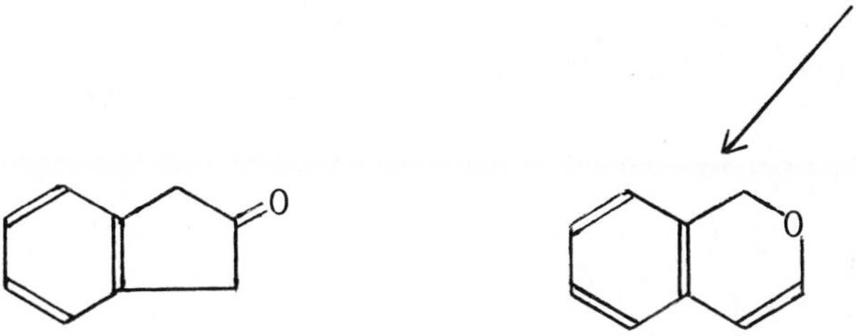

(5)   (4)

Benzo[c]pyrans have been prepared by the cyclization of appropriate aralkylphosphonium salts (intramolecular Wittig reaction), for example, 3-methyl- and 3-phenylbenzo[c]pyran [M. Le Corre *et al*., Actes Congr. Int. Composes Phosphores, 1st. 1977 (Pub. 1978), 607, Inst. Mond. Phosphate, Paris; Le Corre, A. Hercouet, and B. Begasse, Ger. Offen. 2,837,736, 1979].

Benzo[c]pyrylium-4-olate (2-benzopyrylium-4-olate) (7) is readily liberated from 1-acetoxybenzo[c]pyran-4-one (6) under thermal or base-catalyzed conditions. It undergoes cycloadditions with a wide range of dipolarophiles and the regioselectivity of these processes follows frontier MO predictions, for example, base-generated (7) undergoes cycloaddition with dimethyl acetylenedicarboxylate to yield dimethyl epoxybenzocycloheptenedicarboxylate (8) (88%) (P.G. Sammes and R.J. Whitby, J. chem. Soc., Perkin I, 1987, 195).

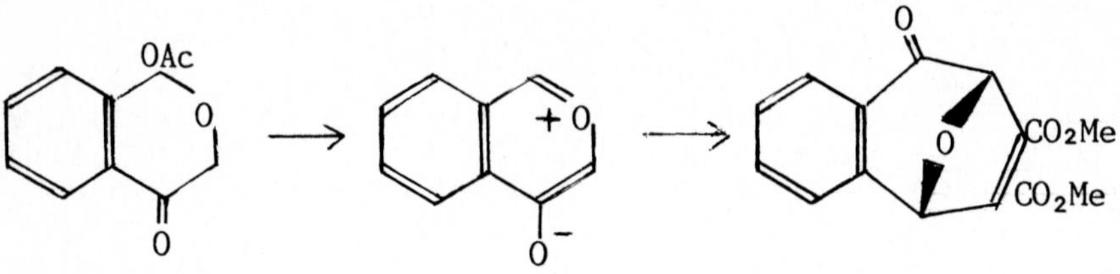

(6)   (7)   (8)

Treatment of benzo[c]pyran (9) or 1,1'-bisbenzo[c]pyranyl (10) with 2,2,6,6-tetramethyl-1-oxopiperidinium perchlorate in anhydrous acetonitrile in an inert atmosphere affords 6,7-dimethoxy-3-methylbenzo[c]pyrylium perchlorate (11) (I.M. Sosonkin *et al.*, Khim. Geterotsikl. Soedin., 1982, 553).

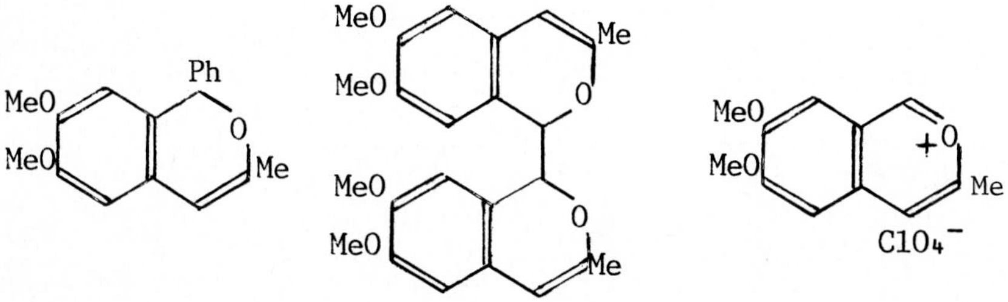

(9)   (10)   (11)

(b) Isocoumarin, 1H-benzo[c]pyran-1-one

Preparations. The thermolysis of phenylacetaldehyde at
800° and 0.01 - 0.15 torr affords isocoumarin (61%)
(P. De Champlain *et al*., Canad. J. Chem., 1976, **54**, 3749).
   Isocoumarins are prepared by cyclization of homophthalic
acid derivatives, 2-vinylbenzoic acid derivatives, and 2-
-carboxybenzyl ketones, by the ortholithiation of *N*-methyl-
benzamides, and by oxidation of isochromans. These methods
frequently suffer from requiring starting materials which
are difficult to prepare and/or severe conditions for
cyclization. The following methods involve the use of π-
-allylnickel halide complexes for the facile introduction of
the acetonyl group into aromatic substrates and the
palladium assisted cyclization to isocoumarins. 2-
-Acetonylbenzoic esters (3) obtained on treating 2-bromo-
benzoic esters (1) with π-(2-methoxyallyl)nickel bromide
(2) are cyclized to 3-methylisocoumarins (4) by treatment
with sodium hydride in *tert*-butanol. 2-Allylbenzoic acids
(5) produced by the reaction of sodium 2-bromobenzoates
with a variety of π-allylnickel halide complexes, furnish
isocoumarins (6) on treatment with palladium chloride. The
yields of some isocoumarins are given in Table 11
(D.E. Korte, L.S. Hegedus, and R.K. Wirth, J. org. Chem.,
1977, **42**, 1329).

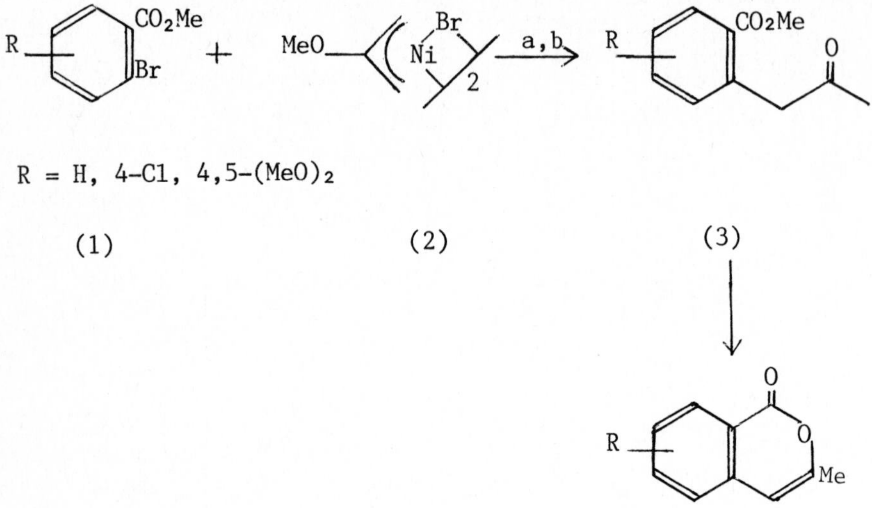

R = H, 4-Cl, 4,5-(MeO)$_2$

(1) (2) (3)

(4)

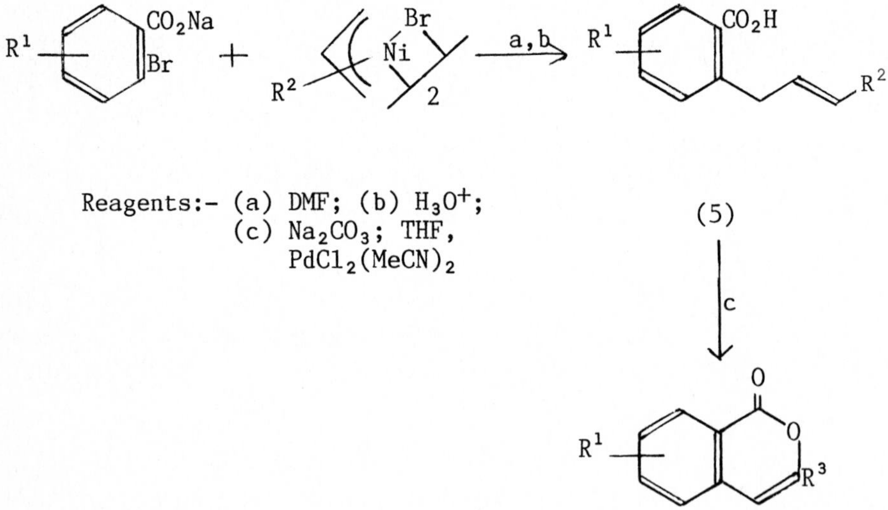

(5)

Reagents:- (a) DMF; (b) H$_3$O$^+$;
(c) Na$_2$CO$_3$; THF,
PdCl$_2$(MeCN)$_2$

(6)

TABLE 11
Isocoumarins (6)

Substituents	R^1	R^3	Yields (%)	R^2 (5)
3-Methyl	H	Me	86	H
3-Ethyl	H	Et	86	Me
3-isopropyl	H	i-Pr	96	Me_2
6-Chloro-3-methyl	6-Cl	Me	63	H
6-Chloro-3-ethyl	6-Cl	Et	54	Me
7-Methoxy-3-methyl	7-OMe	Me	65	H

4-Ethyl- and 3,4-diethyl-isocoumarins have been obtained by the bromination and subsequent dehydrobromination of the related 3,4-dihydroisocoumarins (R.P. Singh and J.N. Srivastava, J. Indian. chem. Soc., 1983, 60, 964).
Benzoic acid and substituted benzoic acids are readily thallated by thallium (III) trifluoroacetate to afford the intermediates (7), which react with palladium chloride and simple alkenes, allylic halides, vinyl halides, or vinyl esters to yield isocoumarins (8). The yields of some isocoumarins from alkenes are given in Table 12 (R.C. Larock et al., J. Amer. chem. Soc., 1984, 106, 5274).

(7)   (8)

Reagents: - (a) Alkene, $PdCl_2$, MeCN, 25°, 16-20h;
          (b) Na_2CO_3, Et_3N, Δ, 5h

TABLE 12

Isocoumarins from some alkenes

Alkene	R¹ (7)	Product (8)	(8) R¹	(8) R²	Yield (%)
t-BuCH=CH$_2$	H	3-*tert*-Butylisocoumarin	H	Bu^t	79
PhCH=CH$_2$	H	3-Phenylisocoumarin	H	Ph	73[1]
t-BuCH=CH$_2$	5-Cl	7-Chloro-3-*tert*-butylisocoumarin and 5-Chloro-3-*tert*-butylisocoumarin 9:1	7-Cl / 5-Cl	Bu^t / Bu^t	65[2]
t-BuCH=CH$_2$	5-Me	7-Methyl-3-*tert*-butylisocoumarin and 5-Methyl-3-*tert*-butylisocoumarin 6:1	7-Me / 5-Me	Bu^t / Bu^t	41[2,3]
t-BuCH=CH$_2$	3-OMe	5-Methoxy-3-*tert*-butylisocoumarin and 7-Methoxy-3-*tert*-butylisocoumarin 10:1	5-OMe / 7-OMe	Bu^t / Bu^t	66[2]

1 Also see A. Kasahara (Japan Kokai 76,108,060, 1976)

2 The reaction was run at room temperature for 2 days
with 2 equiv. of lithium chloride added (no bases added)

3 Yield determined by gas chromatography 50%

The condensation of substituted benzoic acids with benzoin
in the presence of polyphosphoric acid at 130° affords, for
example, 7-methyl-, 7-methoxy-, 6,7-dimethoxy-, 6,7-
-methylenedioxy-, 7,8-dimethoxy-, and 5,6,7-trimethoxy-3,4-
-diphenylisocoumarins (D.L. Brahmbhatt and B.H. Bhide, Indian
J. Chem., 1984, 23B, 889).
4-Bromo-3-phenylisocoumarin (9), m.p. 130-130.5°, is
prepared by adding bromine in acetic acid to a mixture of
methyl 2-(2-phenylethynyl)benzoate and lithium bromide in
acetic acid, and by the base-catalyzed cyclization of methyl
2-(1-bromo-2-oxo-2-phenylethyl)benzene (10). It is also
isolated in the bromination of 3-phenylisocoumarin. A
previous reported synthesis of (9) by the reaction between
3-bromo-2-phenylinden-1-one and trifluoroperacetic acid in
the presence of sodium diphosphate is incorrect. It is
speculated that the product is the epoxide of 3-bromo-2-
-phenylinden-1-one (M.A. Oliver and R.D. Gandour, J. org.
Chem., 1984, 49, 558).

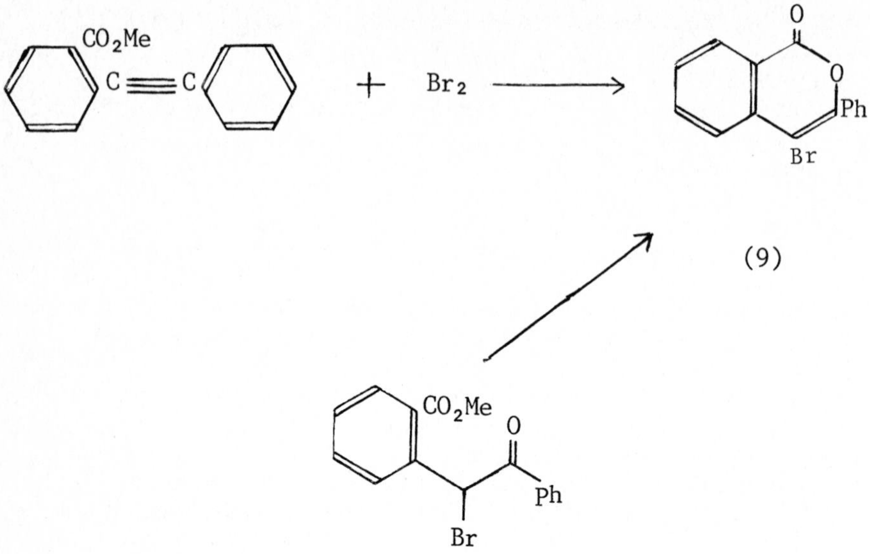

(9)

(10)

A number of chlorinated isocoumarins have been synthesized by a route involving carboxylation of chlorinated 2,4--dimethoxy-6-methylbenzoic acids to give chlorohomophthalic acids, which are converted into the corresponding anhydrides, acetylated and decarboxylated to produce chloroisocoumarins. The naturally occurring 7-chloro-8--hydroxy-6-methoxy-3-methylisocoumarin (11), $C_{11}H_9ClO_4$, m.p. 208-215° (decomp.) (163-165°, R.B. Filho, M.P.L. de Mornes and O.R. Gottlieb, Phytochem., 1980, 19, 2003) has been synthesized for the first time by this method.

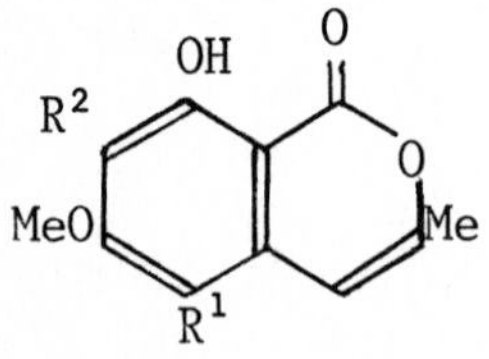

(11) R^1 = H, R^2 = Cl
(12) R^1 = Cl, R^2 = H

5-Chloro-8-hydroxy-6-methoxy-3-methylisocoumarin (12), m.p. 156-158°; 4-carboxy-7-chloro-6,8-dimethoxy-3-methyl-isocoumarin, m.p. 164° (decomp.); 5-chloro-6,8-dimethoxy--3-methylisocoumarin, m.p. 204°; 5,7-dichloro-6,8-dimethoxy--3-methylisocoumarin, m.p. 140-144°; and 7-chloro-6,8--dimethoxy-3-methylisocoumarin, m.p. 187-200° (decomp.). The chlorinated isocoumarins (11 and 12) have been isolated from the trunkwood of *Swartzia laevicarpa*. 4-Carboxy-6,8--dimethoxy-3-methylisocoumarin, m.p. 145-147°; 6,8--dimethoxy-3-methylisocoumarin, m.p. 157-158° (G.B. Henderson and R.A. Hill, J. chem. Soc., Perkin I, 1982, 1111).

A number of 2-substituted arylacetylenes, for example, acetylene (13), undergo facile intramolecular solvomercuration with mercury (II) acetate in acetic acid to yield 4-(chloromercurio)isocoumarins (14) after aqueous sodium chloride workup. This method may be adopted to give benzofuran, benzothiophene, and chromone organomercuric chlorides. 4-(Chloromercurio)-3-n-propylisocoumarin (14), m.p. 195-197°, may be converted into 3-n-propylisocoumarin (15) and 4-iodo-3-n-propylisocoumarin (16) (R.C. Larock and L.W. Harrison, J. Amer. chem. Soc., 1984, 106, 4218).

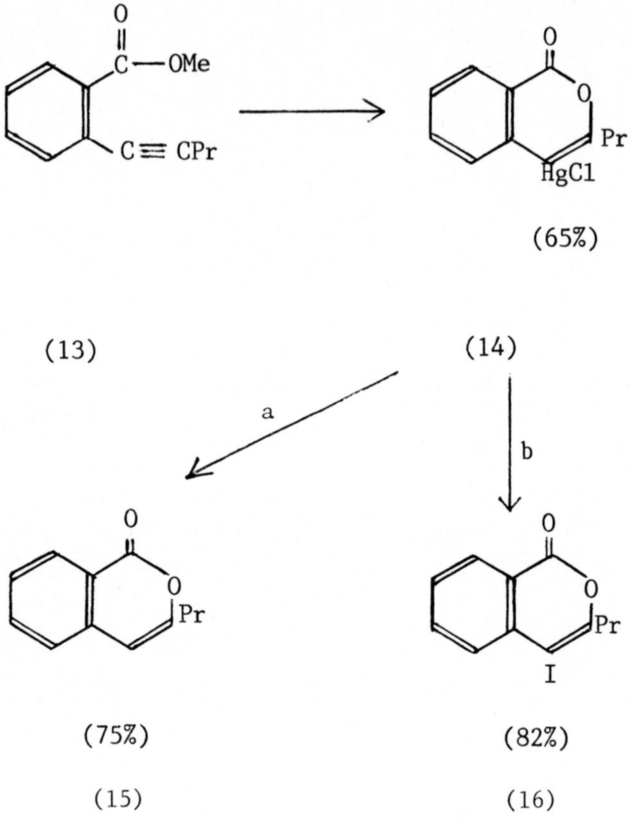

(13)

(14)

(65%)

a

b

(75%)

(82%)

(15)

(16)

I

Reagents:- (a) THF-H_2O(1:1), 3M NaOH, add $NaBH_4$ in
3M NaOH, NaCl added to separate H_2O layer;
(b) I_2, $CHCl_3$

5,6,7-Trimethoxyisocoumarin-3-carboxylic acid (18) has
been prepared by the hydrolysis of the product, from the
reaction between phthaldehydic ester (17) and benzoylamino-
acetic acid [J.N. Chatterjea, K.R.R.P. Singh, and
C. Bhakta, Natl. Acad. Sci. Letters (India), 1981, **4**, 83].

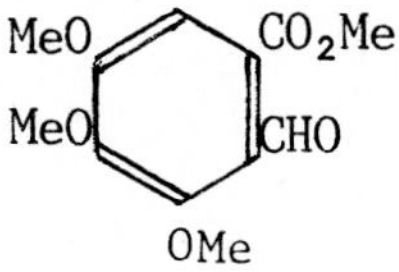

(17) 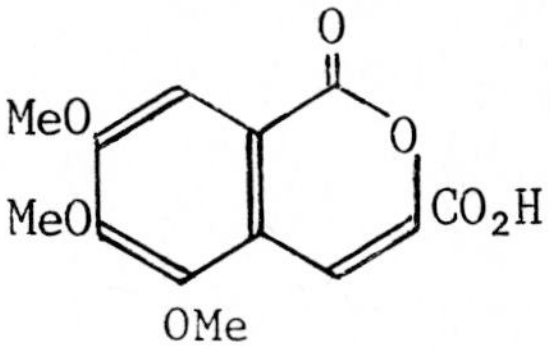   (18)

2-Arylcarbonylbenzoyl chlorides (19) on reacting with the
phosphorane, $Ph_3P=CHCO_2Et$, yield the intermediates (20),
which may be thermally decomposed to ethyl 4-arylisocoumarin-
-3-carboxylates (21), also obtained by oxidation of the
expected indenones (22) (P. Babin and J. Dunoguès,
Tetrahedron Letters, 1984, 25, 4389).

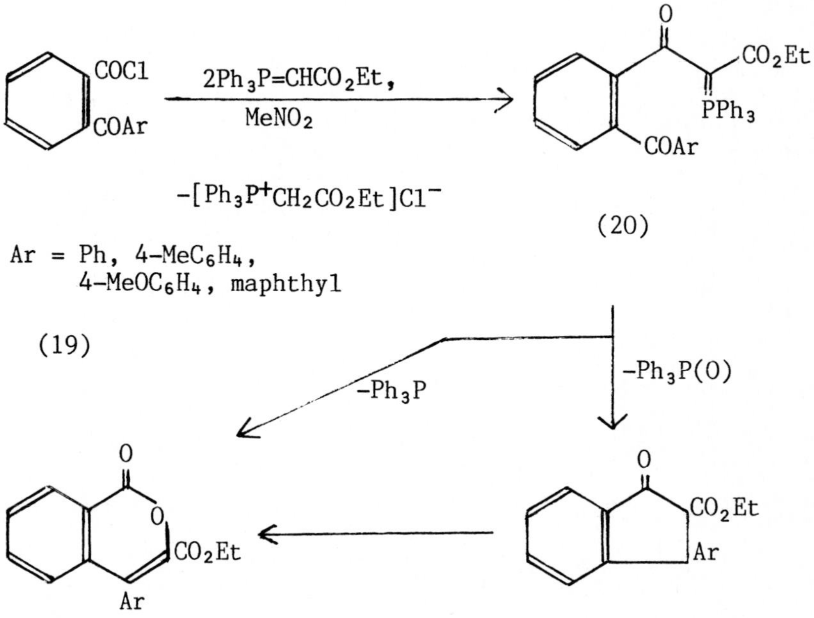

$Ar = Ph, 4-MeC_6H_4,$
    $4-MeOC_6H_4$, maphthyl

(19)

(21)   (22)

322

Isocoumarin-4-carboxylic acid is obtained by rearrangement of 4-(methoxymethylene)isochroman-1,3-dione (23) in hydrochloric acid (O.S. Wolfbeis, Ann., 1981, 819).

(23)

The preparations of ethyl 5- and 7-methoxyisocoumarin-3--carboxylate (N.S. Narasimhan and S.S. Kusurkar, Indian J. Chem., 1983, 22B, 349) and of 8-hydroxy-3-phenyl-, 8--acetoxy-3-phenyl-, 8-hydroxy-3-(4-methoxyphenyl)-, 3-(4--hydroxyphenyl)-, 8-hydroxy-3-(4-hydroxyphenyl)-, and 5--chloro-8-hydroxy-6-methoxy-3-phenyl-isocoumarin (K. Nozawa et al., Chem. pharm. Bull., 1981, 29, 2491) have been described.
Properties. Of a number of substituted isocoumarins and phthalides investigated it is reported that 3,4-dichloro-isocoumarin is the most biologically potent, in that it inactivated all the serine proteases tested, but did not inhibit papain, leucine aminopeptidase, or β-lactamase. It is suggested that 3,4-dichloroisocoumarin should find wide application as a general serine protease inhibitor (J.W. Harper, K. Hemmi, and J.C. Powers, Biochem., 1985, 24, 1831).
Lithio ethyl acetate is a superior reagent for the transformation of isocoumarins (1) into naphthoates (2). It is often used as a replacement for the Reformatsky reagent (F.H. Hauser and S.A. Pogany, J. heterocyclic Chem., 1978, 15, 1535).

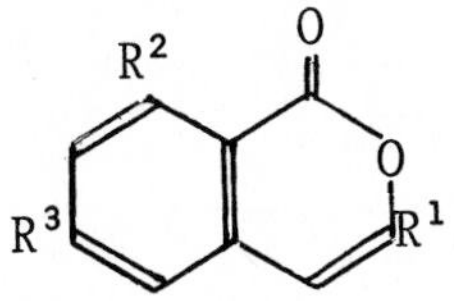

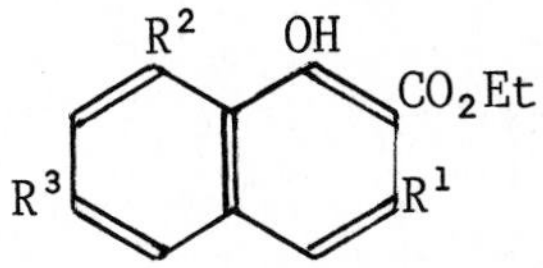

$R^1 = R^2 = R^3 = H$
$R^1 = Me, R^2 = R^3 = H$
$R^1 = Me, R^2 = OMe, R^3 = H, OMe$ (2)

(1)

Naturally occurring isocoumarins. A number of substituted isocoumarins, including the naturally occurring isocoumarins, tetrahydrocapillarine, O-methylglomellin, and oospolactone have been prepared by pyridine-catalyzed acylation of homophthalic acids with alkanoic anhydrides. Homophthalic anhydride on reacting with valeric anhydride in the presence of pyridine affords the dione (1), which on treatment with 80% sulphuric acid yields tetrahydrocapillarine (2). 4,6- -Dimethoxyhomophthalic acid on acylation with butyric anhydride at 100° gives the isocoumarincarboxylic acid (3), which is decarboxylated to furnish O-methylglomellin (4). Acetylation of 3-methoxy-α-methylhomophthalic acid at 100° with simultaneous decarboxylation yields O-methyloospolactone (5), which on demethylation gives oospolactone (6). Isocoumarins (7) are prepared in a similar manner (J.N. Chatterjea *et al.*, Ber., 1980, **113**, 3927).

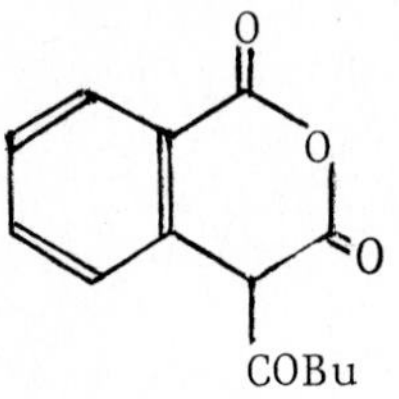

(1)

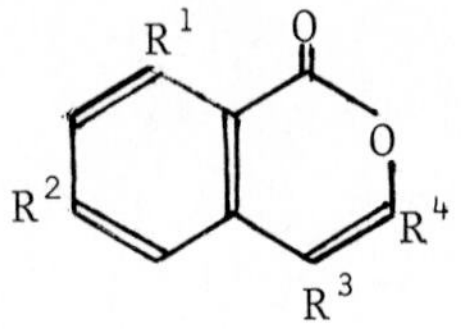

(2) $R^1=R^2=R^3=H,R^4=Bu$
(3) $R^1=R^2=OMe,R^3=CO_2H,R^4=Pr$
(4) $R^1=R^2=OMe,R^3=H,R^4=Pr$
(5) $R^1=OMe,R^2=H,R^3=R^4=Me$
(6) $R^1=OH,R^2=H,R^3=R^4=Me$
(7) $R^1=Me,H; R^2=H,OMe;$
    $R^3=Me,Ph; R^4=Me,Et,Pr$

1,4-Diphenyl-3H-benzo[c]pyran-3-one (9) has been obtained in small amounts along with diketone (10) by oxidation of the indane derivative (8) (A. Arcoleo, G. Fontana, and G. Giammona, J. chem. Eng. Data, 1984, <u>29</u>, 354).

(8)

(9)

(10)

6. Isochroman, 3,4-dihydro-1H-benzo[c]pyran, 3,4-dihydro-2-
    -benzopyran and derivatives

(a) Isochromans

Preparations. Appropriate 1-benzyltetrahydroisoquino-
lines (e.g. 1) containing a tertiary alcohol group as
nucleophile, react with ethyl chloroformate to give 3-
-phenylisochromans (2) (R. Ambros, S. Prior, and
W. Wiegrebe, Sci. Pharm., 1982, 51, 179).

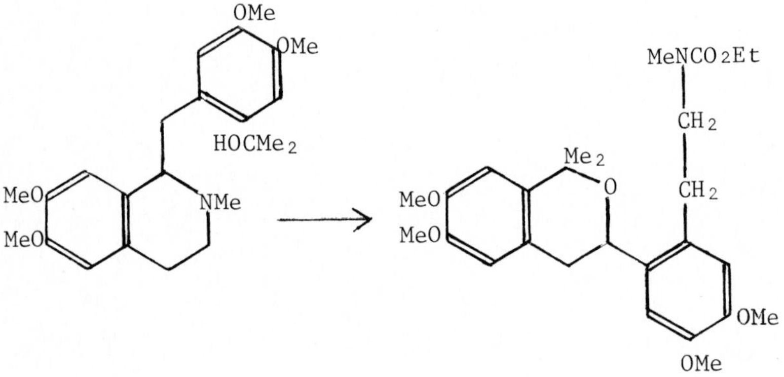

(1)   (2)

Acid chloride (3) undergoes a Grignard reaction with 1-
-bromopentane to yield ketone (4), which on demethylation,
formylation with ethyl orthoformate, and reduction by
sodium tetrahydridoborate affords alcohol (5), cyclizes in
acetone containing sulphuric acid to 5-butyl-6,8-dihydroxy-
-3-pentylisochroman (6). Methylation of (6) by methyl
iodide in the presence of potassium carbonate gives (±)-5-
-butyl-6,8-dimethoxy-3-pentylisochroman. [S.M. Afzal and
W.B. Whalley, Pak. J. sci. Res., 1981 (Pub. 1984), 33,
52.]

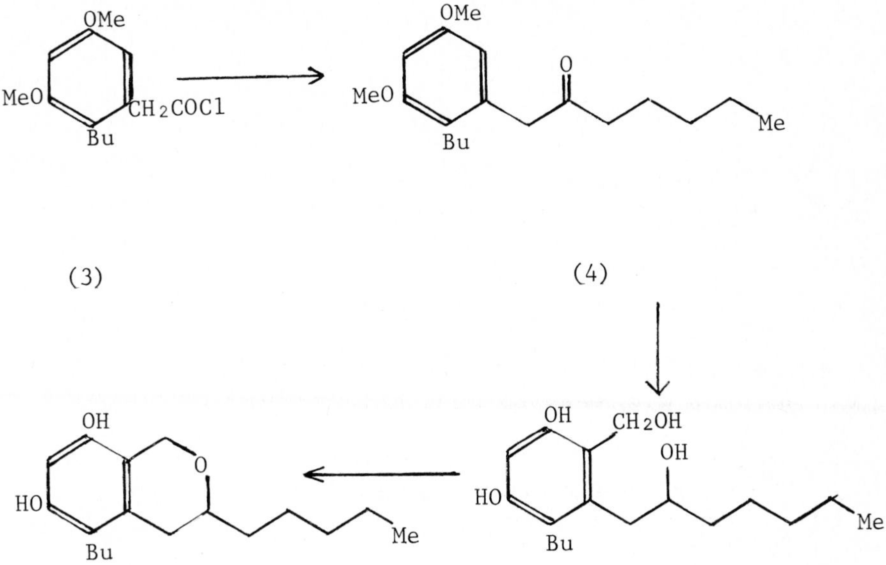

(3)

(4)

(6)

(5)

A number of appropriate aminoisochroman-3-ones have been reduced with diborane to furnish 4-(substituted amino)-1-
-phenyl/methylisochromans, a new class of antihistaminics (S. Ram *et al.*, Indian J. Chem., 1984, 23B, 1261). 1-
-Substituted isochroman dihydrochloride derivatives [7; R = Me$_2$N, Et$_2$N, pyrrolidino, morpholino, piperidino, PhCH$_2$NH, (2-furylmethyl)amino, (1-isochromanylmethyl)amino], and their dimethiodides have been prepared. Some of the compounds possessed sympatholytic and parasympatholytic activity, two have some antiinflammatory activity, but none have antipyretic activity (A.G. Samodurova *et al.*, Arm. Khim. Zh., 1982, 35, 391).

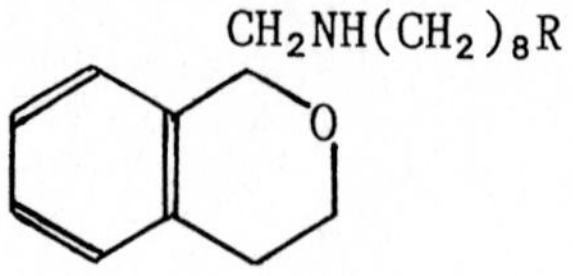

$$CH_2NH(CH_2)_8R$$

(7)

Treatment of polyene (8) with tin (IV) chloride in dichloromethane causes cyclization to the isochroman-1-yl derivative (9). The primary OH group participates as an internal nucleophile delivered intramolecularly to capture the intermediary tetracyclic benzylic cation (M.E. Garst, Y.-F. Cheung, and W.S. Johnson, J. Amer. chem. Soc., 1979, 101, 4404).

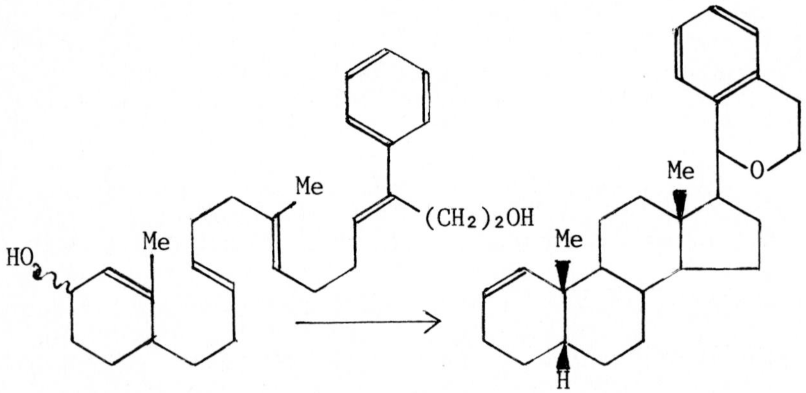

(8)   (9)

Reactions. The reaction of 1-ethoxyisochroman (1; R = OEt) with anilines affords 1-arylaminoisochromans [1; R = PhNH, 2-MeO$_2$CC$_6$H$_4$NH, 4-MeO$_2$CC$_6$H$_4$NH, 2-MeOC$_6$H$_4$NH, 4-MeOC$_6$H$_4$NH, 4-H$_2$NC$_6$H$_4$NH, 4-Et$_2$NC$_6$H$_4$NH, 3,4-MeO(H$_2$N)C$_6$H$_3$NH]. It also reacts with a number of amides and heterocycles (M. Yamato, T. Ishikawa, and S. Yamada, Chem. pharm. Bull., 1982, 30, 843).

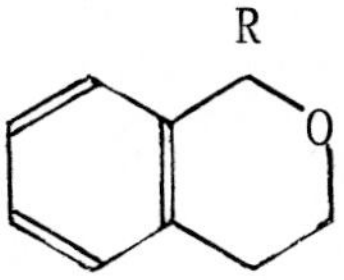

(1)

Treatment of 1-ethoxyisochroman with nucleophilic reagents such as PhCH$_2$OH, PhCHMeOH, HOCHMeCO$_2$Et, and Me$_2$NCH$_2$CH$_2$OH yields derivatives (1; R = PhCH$_2$O, PhCHMeO, OCHMeCO$_2$Et and Me$_2$NCH$_2$CH$_2$O), respectively, whereas, with phenol it gives 1-(2-hydroxyphenyl)isochroman, 1-(4-hydroxy-phenyl)isochroman, and 2,4-di(1-isochromanyl)phenol (2) and with 1,3-dimethoxybenzene, 2,4-di(1-isochromanyl)-1,5--dimethoxybenzene (3) (Yamato, Ishikawa, and T. Kobayashi, *ibid*., 1980, 28, 2967).

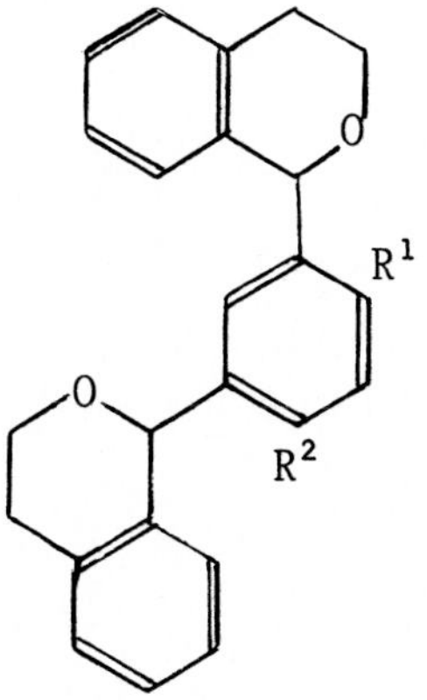

(2) R^1 = OH, R^2 = H
(3) R^1 = R^2 = OMe

Bromination of isochroman in the presence of uv-light gives 1-bromoisochroman (85-88%), which on treatment with copper (I) cyanide affords 1-cyanoisochroman [A.G. Samodurova and E.A. Markaryan, (USSR), Sint. Geterotsikl. Soedin., 1981, 12, 63].

Treatment of 1-isochromanylmethyl ketones (4) with potassium *tert*-butoxide gives the naphthalene derivatives (5) and (6) as major products along with other compounds. Cyclic ketone (7) also yields the spironaphthalenones (8) and (9) (M. Yamato *et al*., Heterocycles, 1985, 23, 1741).

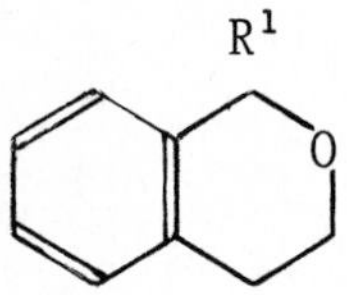

$R^1 = R^2COCH_2$, $PhCOCHR^3$ $R^2 = Ph$, Bu^t $R^3 = Me$, Ph

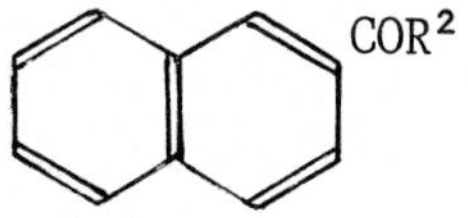

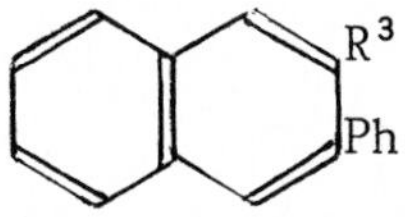

(4)   (5)   (6)

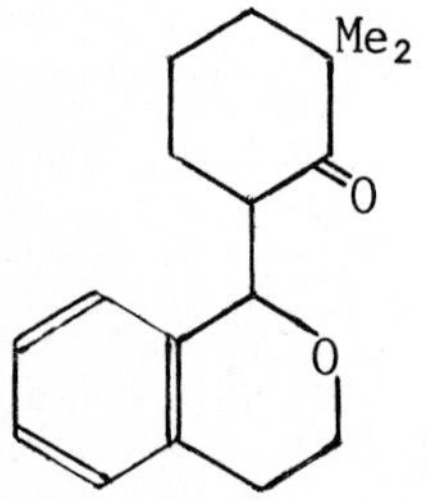

(7) 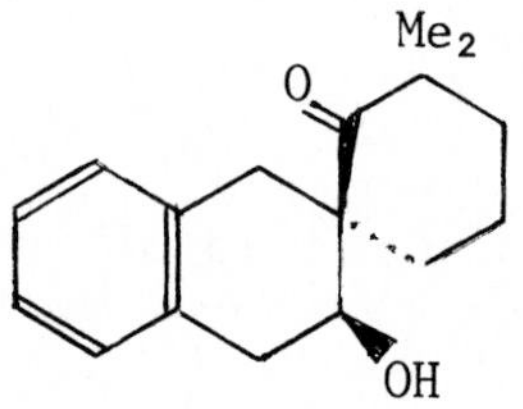   (8) 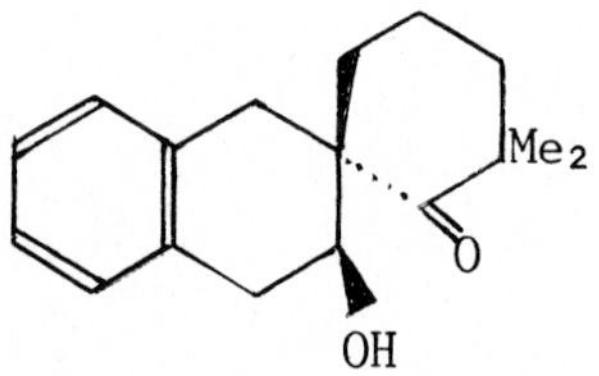   (9)

1-Hydroxyisochroman (acyclic hemiacetal) (10) exists in equilibrium with 2-(β-hydroxyethyl)benzaldehyde (11). The equilibrium is displaced towards the hemiacetal in strongly basic solutions because of its ionization. The kinetics of the equilibration have been studied in the pH range 1-8 in carboxylic and alkylphosphonic acid buffers. Several acetals related to 1-hydroxyisochroman (10) have been prepared and the kinetics and mechanism of their hydrolysis discussed (J. Harron *et al.*, J. org. Chem., 1981, **46**, 903).

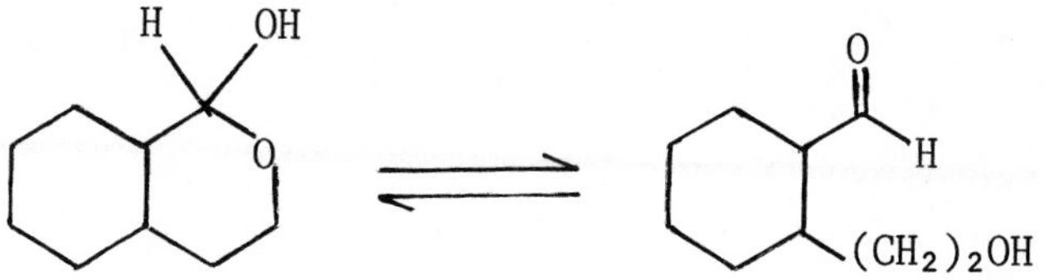

(10)   (11)

1,6-Dihydroxy-3-methylisochroman-5-carboxylic acid, m.p. 135°; 6-hydroxy-1-methoxy-3-methylisochroman-5-carboxylic acid, m.p. 149-150° (J.R. Anderson, R.L. Edwards, and A.J.S. Whalley, J. chem. Soc., Perkin I, 1983, 2185). 1--Aryl-6,7-dimethoxyisochromans prepared by condensation of 3,4-dimethoxyphenylalkanols with aromatic aldehydes, on oxidation with chromium trioxide afford substituted benzophenones (12) (F. Gatta *et al.*, Farmaco, Ed. Sci., 1985, **40**, 942).

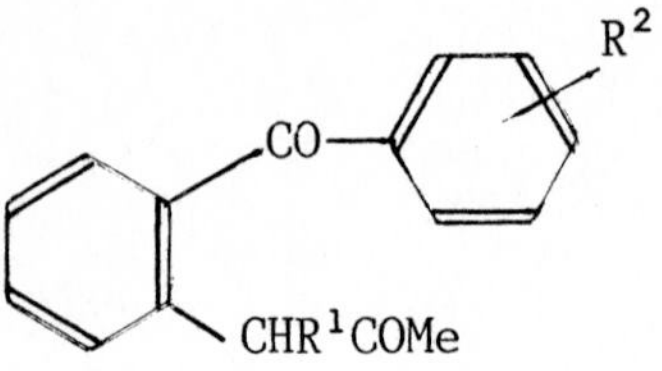

332

R^1 = H, Me, Et
R^2 = H, Cl, F, 3,4-(MeO)$_2$

(12)

The uv-spectra of isochroman, tetralin, and some
β-substituted derivatives of tetralin have been discussed
and compared with those of indane and benzyl compounds
(B. Vidal, G. Bastaert, and J. Brocard, Compt. rend., 1978,
<u>286C</u>, 163). Tricyclic isochromans (13) are useful as
flavours and aromas for various consumable materials, such
as foodstuffs, perfumes, toothpastes, *etc.* (W.J. Wiegers
et al., U.S. Pt. US 4,308,412, 1981).

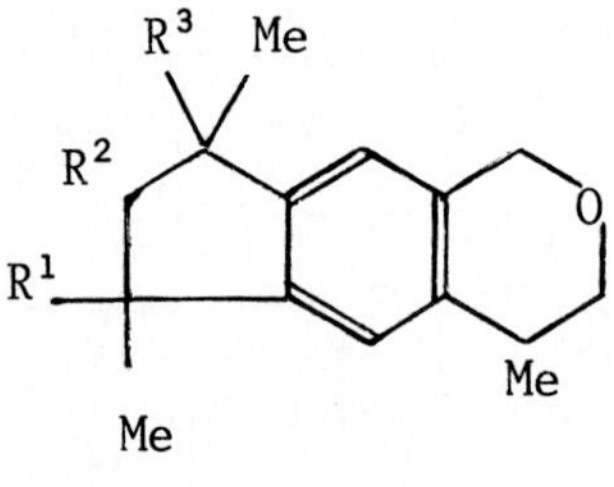

R^1 = Et, R^2 = H, R^3 = Me
R^1 = Me, R^2 = H, R^3 = Et
R^1 = R^2 = R^3 = Me

(13)

The thermochemical properties of some oxygenated six-
-membered cyclic and polycyclic compounds including
isochroman and benzo[b]pyran have been reported (R. Shaw,
D.M. Golden, and S.W. Benson, J. phys. Chem., 1977, 81,
1716).

(b) Isochromanones

*Isochroman-1-ones, 3,4-dihydroisocoumarins and naturally
occurring derivatives.* Methyl 2-acetonylbenzoates (1)
(p 318) on treatment with sodium tetrahydridoborate in
methanol give 3-methylisochroman-1-ones (2) (D.E. Korte,
L.S. Hegedus, and R.K. Wirth, J. org. Chem., 1977, 42,
1329).

R = H, 4-Cl, 4,5-(MeO)$_2$

(1)   (2)

The uv-irradiation of 2-carboxystyrene in methanol yields
isochroman-1-one and derivatives of 2-carboxystyrene, for
example, its β-(4-methoxyphenyl), β-methyl, β,β-dimethyl,
and β,β-pentamethylene derivatives give the corresponding
cyclized compounds. Some of these reactions are reversible
since styrene derivatives are obtained from isochroman-1-ones
and in the ring closure of β-alkyl-substituted 2-carboxy-
styrenes rearrangement products are also formed (M. Yamato
et al., Chem. pharm. Bull., 1978, 26, 1990).

o-Lithiated benzamides (3) undergo transmetallation with
MgBr$_2$·2Et$_2$O to yield intermediates, which in contrast to the
lithiated intermediates react with allyl bromide to afford
2-allylbenzamides (5). On treatment with acid the 2-allyl-
benzamides (5) are converted into isochroman-1-ones (6),
including the natural products mellein (6; R = 8-OH) and
kigelin (6; R = 6,7-(OMe)$_2$-8-OH) (M.P. Sibi, M.A. Jalil Miah,
and V. Snieckus, J. org. Chem., 1984, **49**, 737), previously
isolated from *Kigelia pinnata* DC by T.R. Govindachari,
S.J. Patankar, and N. Viswanathan (Phytochem., 1971, **10**,
1603) and shown to be 6,7-dimethoxy-8-hydroxy-3-methyl-
isochroman-1-one. Elemicin (2nd. Edn., Vol. III E, p.172)
has been transformed into the methyl ether of kigelin, *3-*
-methyl-6,7,8-trimethoxyisochroman-1-one, m.p. 104°
(N.S. Narasimhan and C.P. Bapat. J. chem. Soc., Perkin I,
1982, 2099). 5,8-Dihydroxy-3-methylisochroman-1-one has
been converted into (±)-mellein, *8-hydroxy-3-methylisochroman-*
-1-one (L.M. Harwood, Chem. Comm., 1982, 120).

R = H, 2-OMe, 3-OMe,
   2.5-(OMe)$_2$, 2,3,4(OMe)$_3$

(3)   (4)

R = H, 8-OH, 5-OMe, 5-OH,   R = H, 6-OMe, 3-OMe,
   6,7-(OMe)$_2$ -8-OH   3,6-(OMe)$_2$, 4,5,6-(OMe)

(6)   (5)

Reagents:- (a) s-BuLi; (b) MgBr$_2$.2Et$_2$O; (c) 6N aqHCl

A systematic survey of the metabolites of the Xylariaceous fungi has resulted in the isolation of mellein (6; R = 8-OH) and its 5-methyl-(7), m.p. 131-133°, $[\alpha]_D^{20}$-115°(CHCl$_3$)[1], 5-formyl-(8), m.p.127°, $[\alpha]_D^{20}$-180°(CHCl$_3$), 5-carboxy-(9), m.p. 245°, 5-carbomethoxy-(10), m.p.66°, 5-hydroxymethyl-(11), m.p. 163-164°, and 6-methoxy-5-methyl-(12), m.p. 121° derivatives.

(7) R^1 = Me, R^2 = H
(8) R^1 = CHO, R^2 = H
(9) R^1 = CO$_2$H, R^2 = H
(10) R^1 = CO$_2$Me, R^2 = H
(11) R^1 = CH$_2$OH, R^2 = H
(12) R^1 = Me, R^2 = OMe

5-Carboxymellein (9) and 5-carboxy-6-hydroxy-3-methyl-isochroman-1-one have been prepared from 5-formylmellein (8). Mellein, ramulosin, *8-hydroxy-3-methyl-4a,5,6,7-tetra-hydroisochroman-1-one* (13), C$_{10}$H$_{14}$O$_3$, m.p. 120°, the dihydrobenzo[c]furan-1-ones, isoochracein and 4-hydroxy-isoochracein, have been obtained from *Hypoxylon howieanum* (J.R. Anderson, R.L. Edwards, and A.J.S. Whalley, J. chem. Soc., Perkin I, 1983, 2185; [1]M.A. de Alvarenga *et al.*, Phytochem., 1978, **17**, 511).

(13)

A number of *cis*- and *trans*-3-aryl-4-(carboxy or methoxy-carbonyl)-8-(hydroxy or methoxy)isochroman-1-ones (14) and diazo and acetyl (15) derivatives have been prepared. Some of them showed *in vitro* antifungal activity, but this disappeared with the introduction of a carboxyl, carbomethoxy, acetyl, diazoacetyl, or hydroxyacetyl group at the 4-position (K. Nozawa *et al.*, Chem. pharm. Bull., 1981, <u>29</u>, 3486).

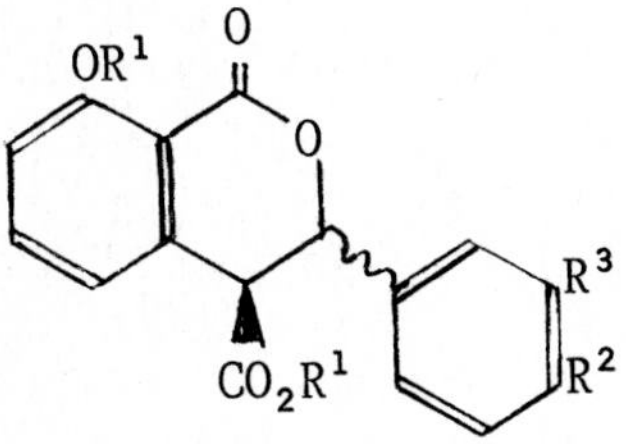

R^1 = H, Me

R^2 = H, MeO, PhCH$_2$O

R^3 = H, HO, MeO, AcO, PhCH$_2$O

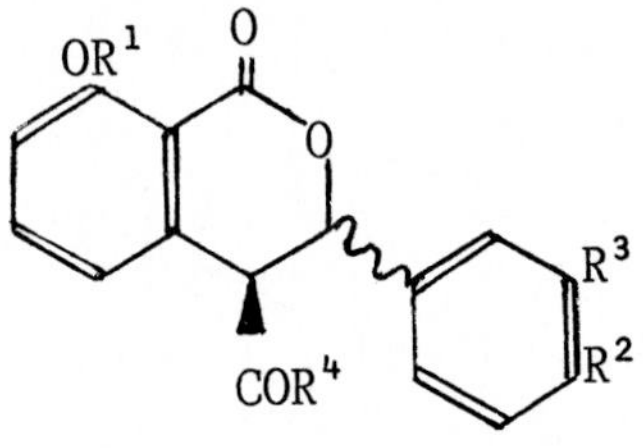

R^4 = N$_2$CH, Me, CH$_2$OH, AcCH$_2$

(14)   (15)

1'-Benzylspiro[isochroman-1-one-piperidines] (16) and (17) with varying substituents on the isochroman-1-one moiety have been synthesized (Yamato *et al.*, *ibid.*, p.3494).

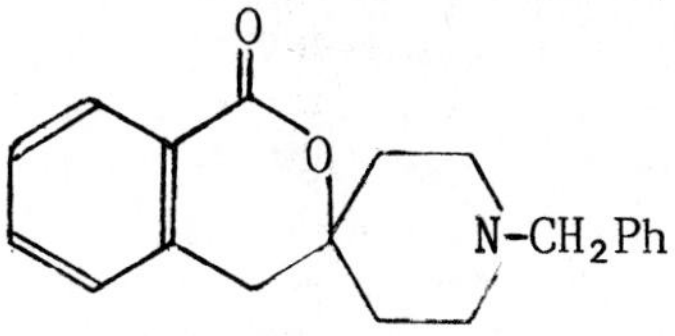

(16) 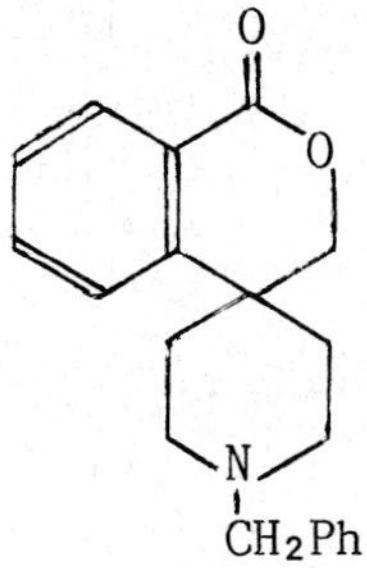   (17)

The chlorinated isochroman-1-ones, 5,7-dichloro-8-hydroxy-
-6-methoxy-3-methylisochroman-1-one (18), m.p. 121-125°,
and 7-chloro-8-hydroxy-6-methoxy-3-methylisochroman-1-one
(19), m.p. 168-171°, together with cryptosporiopsin and
its dechloro-derivative have been isolated from *Sporormia
affinis*; and 5-chloro-8-hydroxy-6-methoxy-3-methylisochroman-
-1-one (20), m.p. 119-122° along with a chlorocyclopentenol
from *Periconia macrospinosa*.

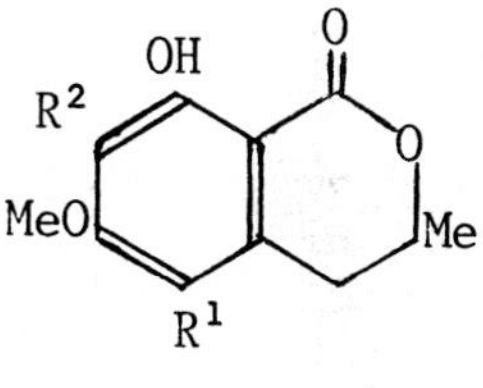

(18) $R^1 = R^2 = Cl$
(19) $R^1 = H$, $R^2 = Cl$
(20) $R^1 = Cl$, $R^2 = H$

6,8-Dimethoxy-3-methylisochroman-1-one, m.p. 125-126° [1];
7-chloro-6,8-dimethoxy-3-methylisochroman-1-one, m.p. 154-157°; 5-chloro-6,8-dimethoxy-3-methylisochroman-1-one, m.p. 134-136° [2]; 5,7-dichloro-6,8-dimethoxy-3-methyl-isochroman-1-one, m.p. 76-80° (decomp.); 8-hydroxy--6-methoxy-3-methylisochroman-1-one, m.p. 96° [3]; 6,8-di-hydroxy-3-methylisochroman-1-one, m.p. 214-215° [1]; 7-chloro--6,8-dihydroxy-3-methylisochroman-1-one, m.p. 193-196°; 5--chloro-6,8-dihydroxy-3-methylisochroman-1-one, m.p. 179-181°; 5,7-dichloro-6,8-dihydroxy-3-methylisochroman-1--one, m.p. 230-234° (G.B. Henderson and R.A. Hill, J. chem. Soc., Perkin I, 1982, 1111; [1] R.H. Carter *et al.*, *ibid.*, 1976, 1438; [2] Nozawa *et al.*, Chem. pharm. Bull., 1980, **24**, 1622; [3] H.L. Slates, S. Weber, and N.L. Wendler, Chimia, 1967, **21**, 468; E. Sondheimer, J. Amer. chem. Soc., 1957, **79**, 5036).

Isochroman-3-one. Preparations. A convenient preparation of 7-methoxyisochroman-3-one (3) involves the chloromethylation of 3-methoxybenzyl methyl ether (1), which yields a mixture of isomeric cyano compounds (2). Acid hydrolysis of the mixture furnishes 7-methoxyisochroman--3-one (3) (46%) and 5-methoxyisochroman-3-one (4) (8%), separated by chromatography (S.P. Khanapure, B.M. Bhawal, and B.G. Hazra, Indian J. Chem., 1982, **21B**, 889).

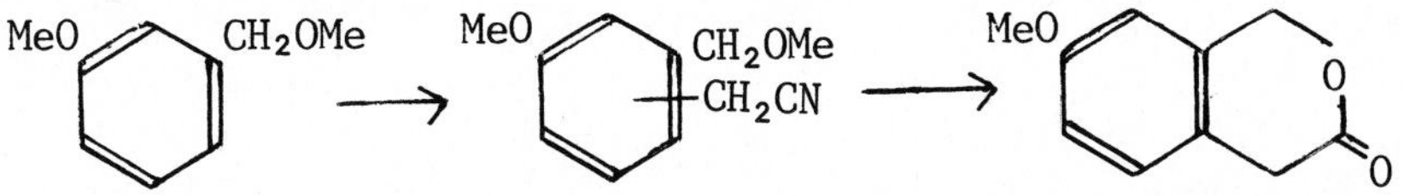

MeO CH$_2$OMe $\longrightarrow$ MeO CH$_2$OMe CH$_2$CN $\longrightarrow$ MeO O O

(1)   (2)   (3)

O O OMe

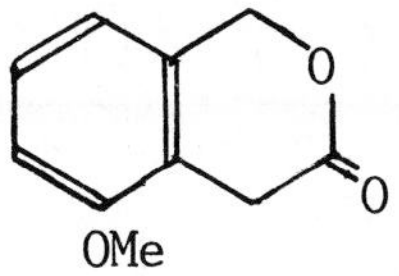

(4)

An improved synthesis of 7-methoxyisochroman-3-one (3) is
from 2-bromo-5-methoxybenzyl methyl ether, which on
treatment with magnesium and dimethylformamide is converted
into 2-formyl-5-methoxybenzyl methyl ether. Reduction of
the formyl group to a hydroxymethyl group, followed by
replacement of the OH group by Cl and then CN, on
subsequent hydrolysis affords 7-methoxyisochroman-3-one (3)
(Kanapure, K.G. Das, and Bhawal, Synth. Comm., 1984, 14,
1205).
Prolonged treatment of 2-bromo-4,5-dimethoxyphenylacetic
acid (5) with 37% formaldehyde and hydrochloric acid in
acetic acid, results in chloromethylation, with loss of the
bromo substituent and formation of 6,7-dimethoxyisochroman-
-3-one (6). This is reported as an improved synthesis of
(6). A previously reported formation of 5,6-dibromo-7,8-
-dimethoxyisochroman-3-one in the above reaction is incorrect
(S. Nimgirawath and Y. Srikirin, Indian J. Chem., 1983, 22B,
272).

(5)   (6)

6-Methoxyisochroman-3-one, b.p. 125-145°/0.04 mm, m.p.
74-78°; 5,6-dimethoxyisochroman-3-one, m.p. 59.5-62°;
7-benzyloxy-6-methoxyisochroman-3-one, m.p. 137-138.5°;
7-hydroxy-6-methoxyisochroman-3-one, m.p. 174-177°; 8-
-acetoxy-7-methoxyisochroman-3-one, m.p. 133-135°; 6,7-
-methylenedioxyisochroman-3-one, m.p. 130.5-132°
(R.J. Spangler, B.G. Beckmann, and J.H. Kim, J. org. Chem.,
1977, <u>42</u>, 2989).

Reactions. 6,7-Dimethoxyisochroman-3-one (6) on heating
with PhCH=C(CN)$_2$ yields 2,2-dicyano-6,7-dimethoxy-3-
-phenyl-1,2,3,4-tetrahydronaphthalene (7) and 4-benzylidene-
-6,7-dimethoxyisochroman-3-one (8). Compound (8) is formed
via elimination of CH$_2$(CN)$_2$ from the intermediate Michael
adduct (9), which can be isolated by carrying out the
original reaction at 160°. Similar reactions have been
investigated with related compounds (J. Afzal *et al.*,
Synth. Comm., 1980, <u>10</u>, 843).

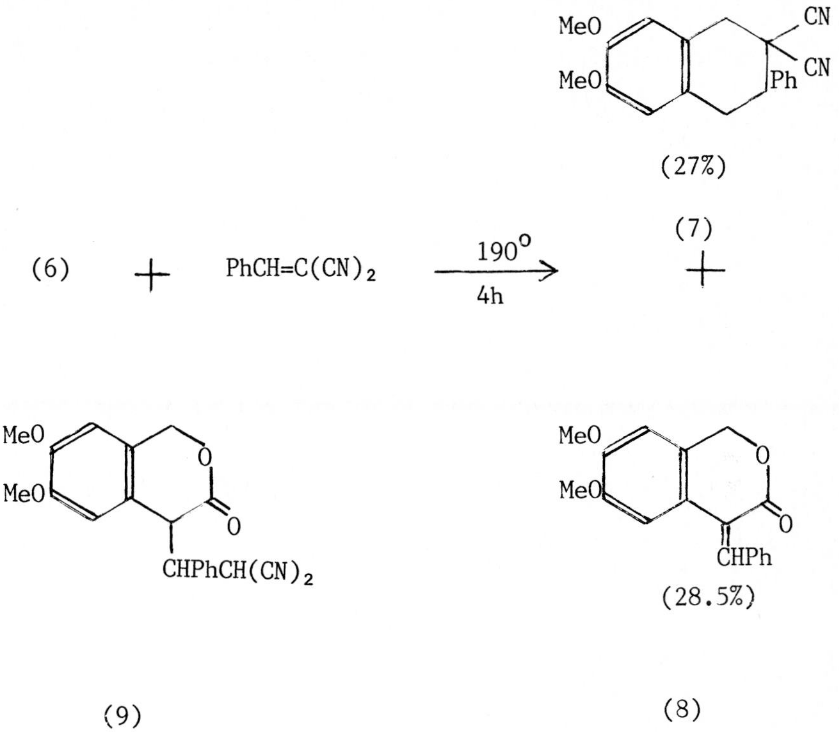
MeO
MeO
CN
CN
Ph
(27%)
(7)
(6)
PhCH=C(CN)2
190°
4h
MeO
MeO
O
O
CHPhCH(CN)2
MeO
MeO
O
O
CHPh
(28.5%)
(9)
(8)

The reaction between 6,7-dimethoxyisochroman-3-one (6) and phenol ethers (10) yields homoveratic acids (11), ketones (12), or the corresponding dibenzo[a,d]tropylium salts, depending on the reaction conditions. For instance, the reaction of (6) with (10; $R^1 = H$, $R^2 = OMe$) in polyphosphoric acid affords (12; $R^1 = H$, $R^2 = OMe$), whereas in the presence of formic acid they yield (11; $R^1 = H$, $R^2 = OMe$) (A.V. Bicherov, G.N. Dorofeenko, and E.V. Kuznetsov, Zh. org. Khim., 1979, 15, 588).

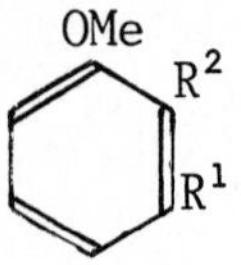

$R^1 = R^2 = H$
$R^1 = H, R^2 = OMe$
$R^1 = R^2 = OMe$

(10)

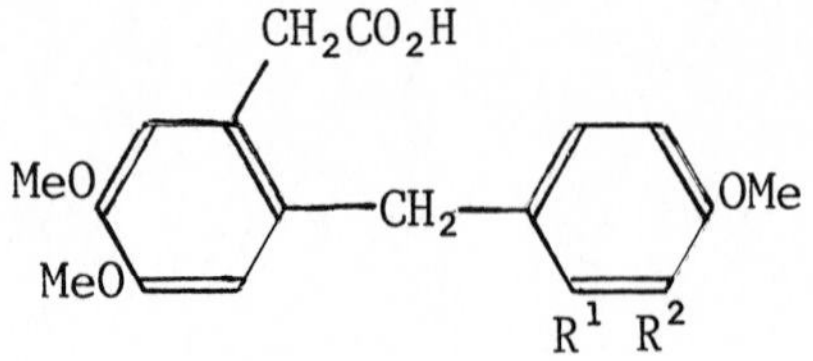

(11)

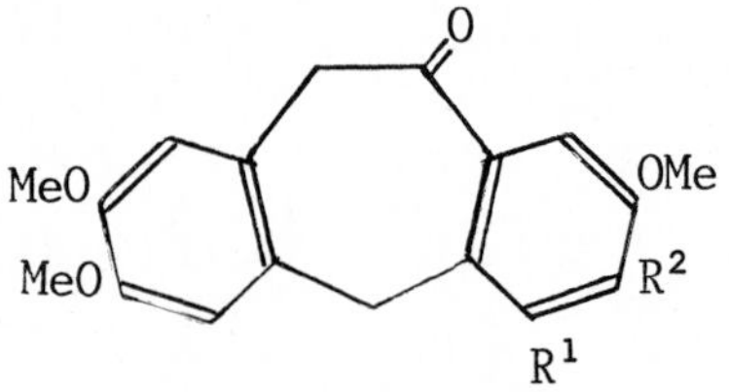

(12)

Gas phase pyrolysis of isochroman-3-ones, including iso-chroman-3-one, furnishes a new synthesis of benzocyclobutenes. A two-step sequence provides a simple and efficient synthesis of benzocyclobutene (14) from commercially available indan-2--one (13) in an overall yield of 60-65%. Fulveneallene (17) is obtained with benzocyclobuten-1-one (16) from isochroman--1,3-dione (15). The relationship between the thermal and electron impact induced decomposition of the isochroman-3--ones has been discussed. (Spangler, Beckmann, and Kim, J. org. Chem., 1977, 42, 2989).

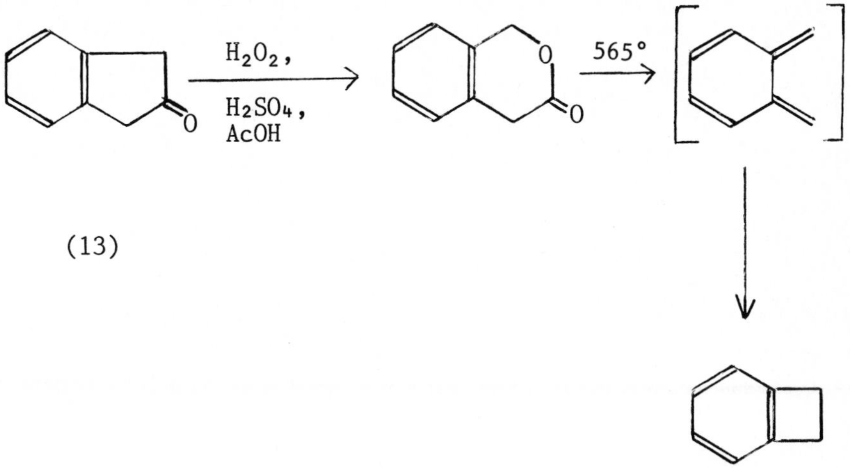

(13)

(14)

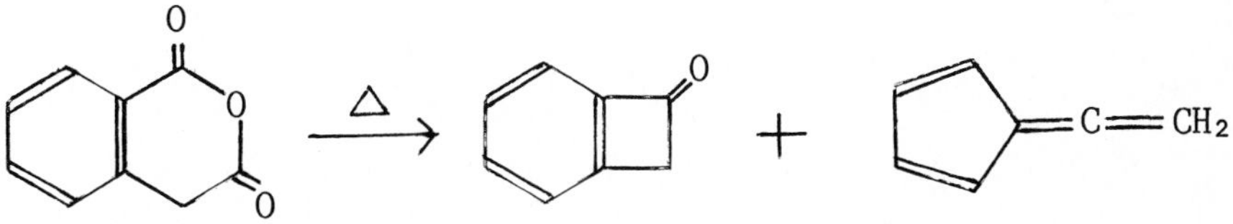

(15)   (16)   (17)

The influence of the solvent upon the coupling reactions of 4-(3',4'-dimethoxybenzyl)-6,7-dimethoxyisochroman-3-one (18), during electrochemical oxidation, has been studied and it has been shown that formation of the product, 7a,8--dihydro-3,10,11-trimethoxy-2*H*-phenanthro[9,8a-c]furan--2,7(5H)-dione (19) proceeds through a six-membered ring transition state (A.J. Majeed, M. Sainsbury, and S.A. Hall, J. chem. Soc., Perkin I, 1984, 833; M.P. Carmody, Sainsbury, R.F. Newton, *ibid.*, 1980, 2013).

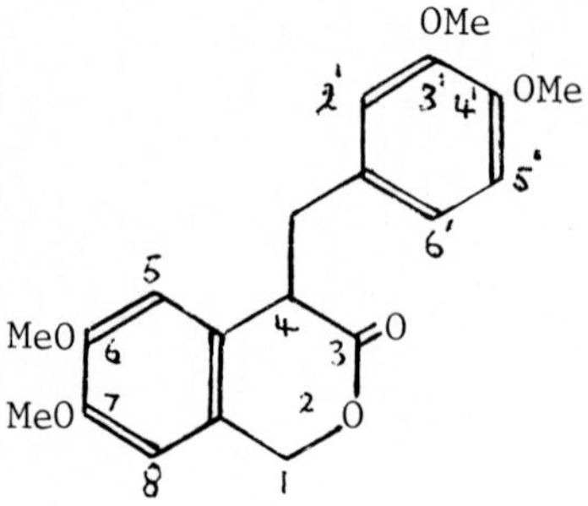

(18)

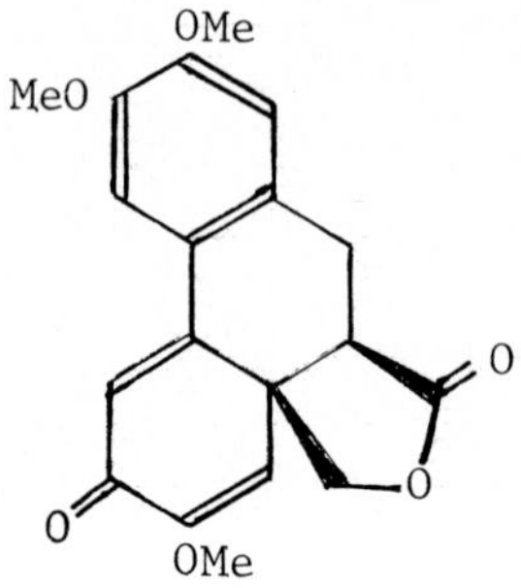

(19)

The anodic coupling reactions of 4-benzylisochroman-3-
-ones and 1- and 4-benzyl-1,2,3,4-tetrahydroisoquinolines
have been compared and discussed (Majeed, P.J. Patel, and
Sainsbury, *ibid*., 1985, 1195).

O-Methylcorypalline (21) has been prepared by reduction
of the product obtained from 6,7-dimethoxyisochroman-3-one
(20), following its cleavage with hydrogen bromide and
cyclization of the resulting product with methylamine
(G.D. Pandey and K.P. Tiwari, Pol. J. Chem., 1979, 53,
2159).

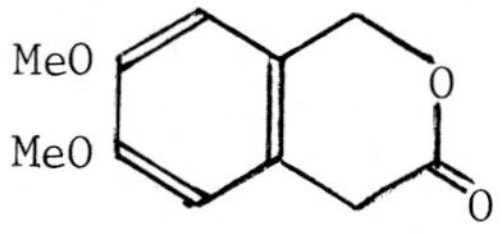

(20)

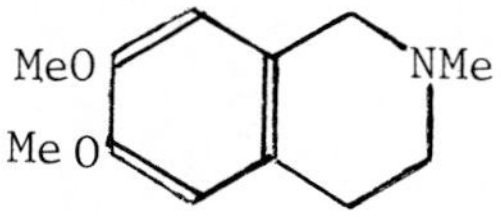

(21)

3-Hydroxy-9,10-dimethoxyberberine has been prepared from
7,8-dimethoxyisochroman-3-one (*idem*, Curr. Sci., 1979, __48__,
1032). 7,8-Dimethoxyisochroman-1,3-diones has been
utilized in the synthesis of the protoberberine alkaloids
(±)-thalictricavine, berlambine, and (±)-canadine
(M. Cushman and F.W. Dekow, J. org. Chem., 1979, __44__, 407).
  5,8-Dimethoxyisochromans (22) are oxidatively demethylated
with silver (I) oxide to give the isochroman-5,8-diones (23)
(J.I. Retamal *et al*., Synth. Comm., 1982, __12__, 279).

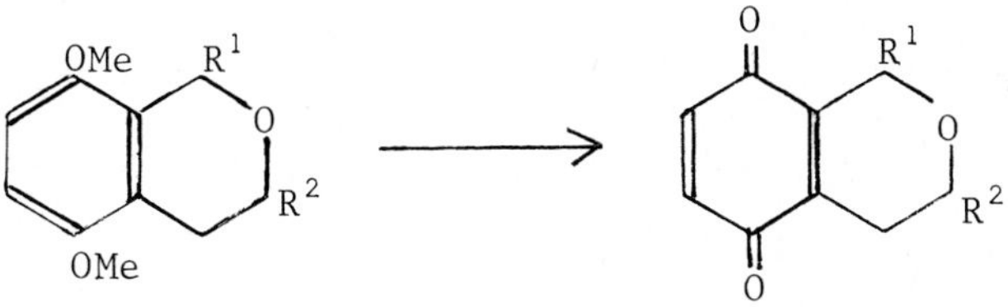

R^1=H,Me,Ph,R^2=H
R^1=H,Ph,4-MeC$_6$H$_4$,4-MeOC$_6$H$_4$,R^2=Me

(22)   (23)

The antibiotics, (±)-nanaomycin (24) and (±)-frenolicin (25) have been prepared *via* the hemiacetal (26) obtained from juglone by sequential Diels-Alder reaction with 1--acetoxybuta-1,3-diene (A. Ichihara *et al.*, Tetrahedron Letters, 1980, 21, 4469).

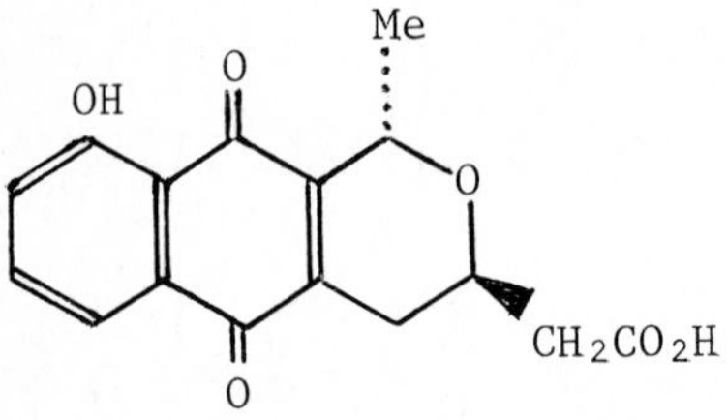

(24)

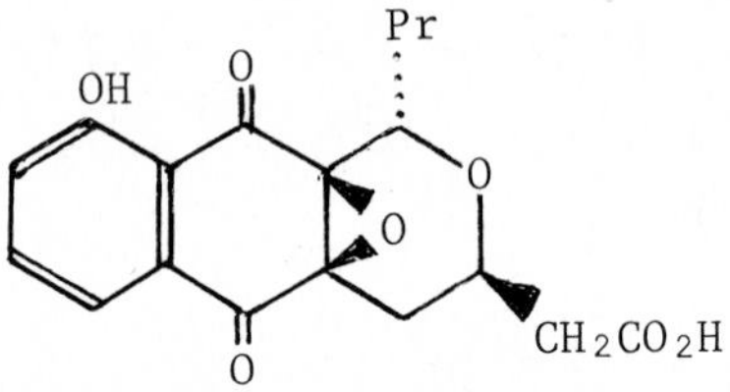

(25)

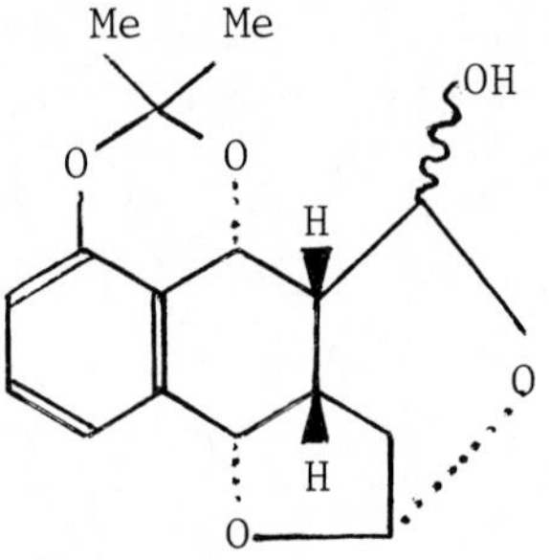

(26)

7. Xanthene, 2,3:5,6-dibenzopyran, dibenzo[b,e]pyran and
    its derivatives

(a) Xanthene derivatives

Preparations. 3-Methoxyphenylmagnesium iodide reacts
with xanthone to yield the 9-ethoxy derivative (1),
following crystallization of the product from ethanol.
Heating the derivative (1) with sodium carbonate and formic
acid gives 9-(3-methoxyphenyl)xanthene (2), which on
heating with trimethylsilyl iodide and quinoline at 180°
affords 9-(3-hydroxyphenyl)xanthene (3) (J.L. Hinds,
S.N. Rajadhyaksha, and N. Shirish, Heterocycles, 1983, 20,
481).

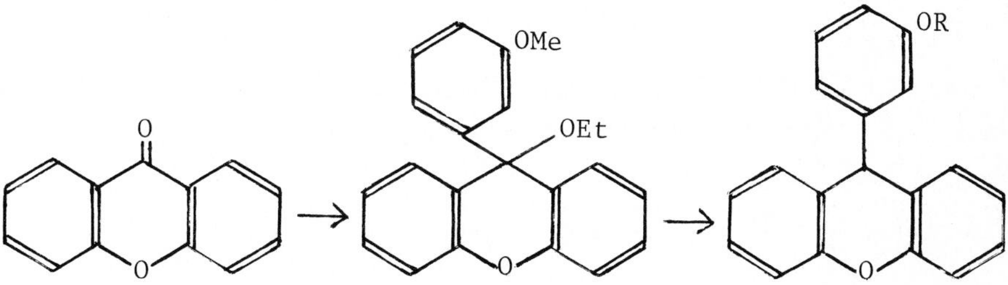

(1)

(2) R=Me
(3) R=H

9-Arylxanthenes, including 3-chloro-9-(4-chlorophenyl)-
xanthene and benzo[b]naphtho[2,3-e]pyran have been prepared
from appropriately substituted triarylmethanols *via*
cyclization of their triarylmethyl chlorides with a
tertiary amine, for example, triethylamine (M. Soucek and
M. Pisova, Czech. Pat. CS 200,381, 1983).

Intramolecular cycloaddition of 3-chloro-6-(2-allyloxy-phenoxy)pyridazine (4) on heating in diethylaniline gives 4-hydroxyxanthene (5). Various hydroxyxanthenes have been obtained from appropriate 3-chloro-6-(2-allyloxyphenoxy)-pyridazines and 3-substituted-6-(2-allylphenoxy)pyridazines (T. Jojima, H. Takeshiba, and T. Kinoto, Chem. pharm. Bull., 1980, 28, 198). Other derivatives have also been prepared by the thermal cyclization of pyridazines (Sankyo Co. Ltd., Japan Tokkyo Koho 80 15,478, 1980).

(4)   (5)

2,5-Bis(4-methoxycinnamyl)-1,3,4,6,7,8- and 4,5-bis(4--methoxycinnamyl)1,2,3,6,7,8-hexamethoxyxanthene have been synthesized and compared with carthamin derivatives with reference to their structure (H. Obara *et al*., Bull. chem. Soc., Japan, 1981, 54, 3225). A number of 2-mercaptomethyl-xanthones have been prepared (M. Eckstein and M. Henryk, Pol. J. Chem., 1980, 54, 1281).

The condensation of 2,3-diacetoxybenzaldehyde (6) with the lithium enolate of 2-(isopropoxymethylene)-cyclo-hexanone (7) gives the dihydro derivative (8), which on dehydrogenation with palladium/carbon yields 5-hydroxy-xanthene-4-carboxaldehyde (9). Other alternative reagents (dichlorodicyanoquinone, sulphur fusion, Br_2, NiO_2, Rh/Al_2O_3, CrO_3) fail to achieve satisfactory dehydrogenation of (8) (D.S. Kemp *et al*., J. org. Chem., 1981, 46, 490).

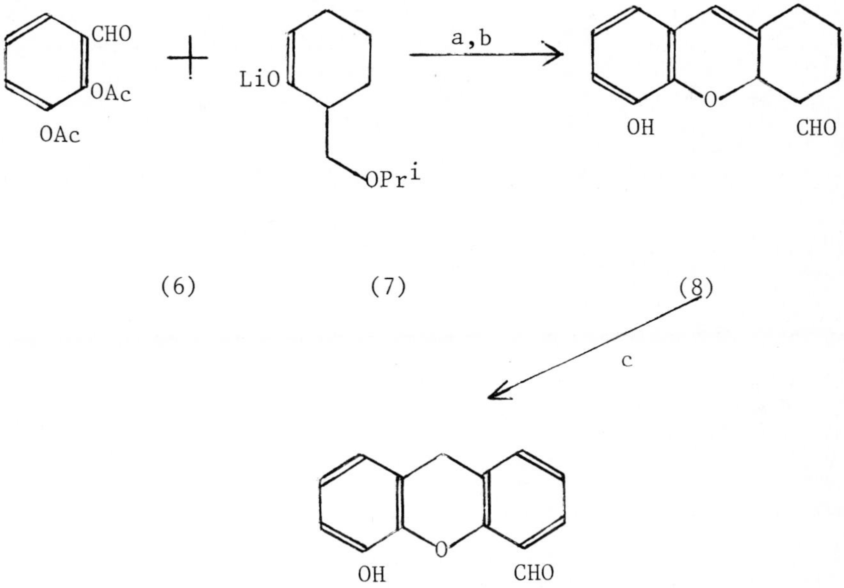

(6)   (7)   (8)

(9)

Reagents:- (a) THF,N_2,-78°,0.5h,0°,1h; (b)M HCl;
     (c) Pd/C,PhMe,Δ

A number of 7-substituted xanthene-3-carboxylic acids and 3-(5-tetrazolyl)xanthenes have been synthesized (J.F. Batchelor et al., Eur. Pat. Appl. EP 93,381, 1983).

Properties. Ozone attacks the methylene group of xanthene *via* a 1,3-dipolar insertion reaction to give a hydrotrioxide intermediate, which loses singlet oxygen to yield xanthydrol (1). The ozonization of xanthydrol (1) gives singlet oxygen, water, and xanthone accompanied by autoxidation (M. Matsui *et al.*, Bull. chem. Soc., Japan, 1984, **57**, 603).

(1)

The benzophenone-sensitized oxidation of 9-phenylxanthene with oxygen has been investigated (S.A. Glover *et al.*, J. chem. Soc., Perkin II, 1985, 1205).

The autoxidation of xanthene carried out in the presence of catalytic amounts of diacetyldiethylammonium chloride in a benzene/50% aqueous sodium hydroxide double phase, with air or oxygen as oxidant, at atmospheric pressure and at nearly room temperature or slightly higher furnishes high yields of xanthone (E. Alneri, G. Bottaccio, and V. Carletti, Tetrahedron Letters, 1977, 2117).

Xanthyl acetate is obtained on treating xanthydrol (1) with acetic anhydride in pyridine. Dixanthyl ether has been prepared (G.E. Ivanov and B.T. Kaminskii, Ukr. Khim. Zh., 1979, **45**, 1211).

Xanthene on alkylation with 2-chloroethyl vinyl ether at room temperature with dimsyl sodium as base surprisingly, affords mainly the vinyl migration product, 9-(2-hydroxy-ethyl)-9-vinylxanthene (2), 15% of the expected product, 9,9-di(2-vinyloxyethyl)xanthene (3), and a minor component, xanthene-9-spirocyclopropane (4). At lower temperatures (0-5°) little rearrangement occurs and (3) is obtained in high yield (P. Doyle *et al.*, *ibid.*, 1976, 3729).

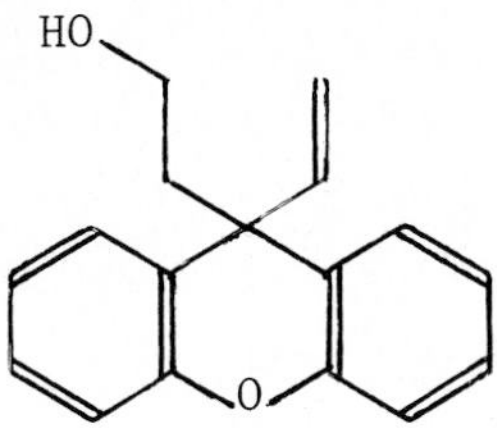

(2)

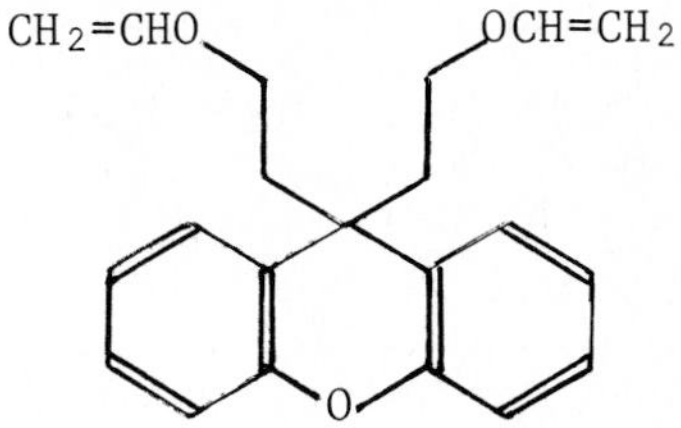

(3)

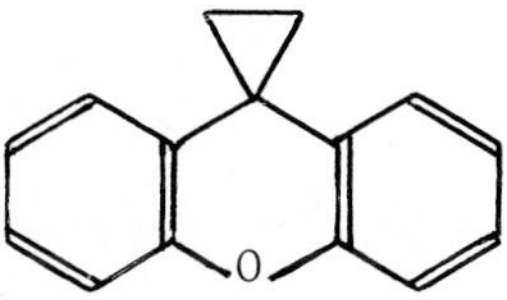

(4)

Friedel-Crafts reaction between xanthene and methylsuccinic anhydride and phthalic anhydride affords the 2-acylxanthene derivatives (5 and 6) and (7), respectively. They can be oxidized to xanthones or to xanthone-2-carboxylic acid and derivatives (5) and (7) can be reduced to acids (8) and (9) and cyclized to compounds (10) and (11) [V.B. Baghos, A.A. Alhawathari, and M. Gindy, Egypt. J. Chem., 1980 (Pub. 1981), 23, 423].

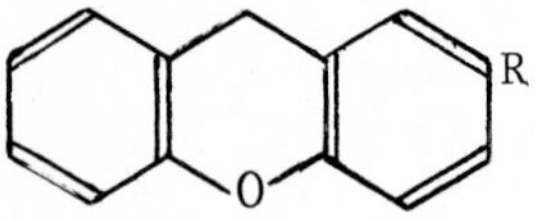

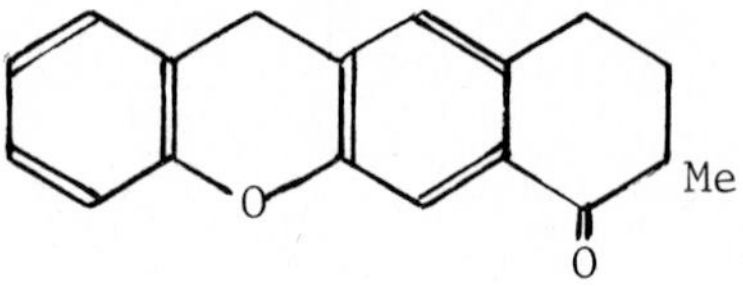

(10)

(5) R=COCH$_2$CHMeCO$_2$H
(6) R=COCHMeCH$_2$CO$_2$H
(7) R=COC$_6$H$_4$CO$_2$H-2
(8) R=CH$_2$CH$_2$CHMeCO$_2$H
(9) R=CH$_2$C$_6$H$_4$CO$_2$H-2

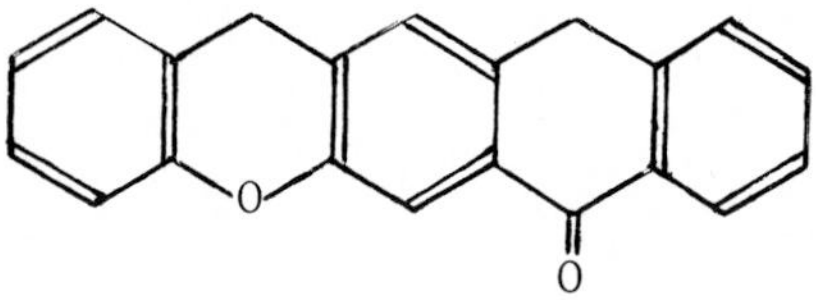

(11)

Xanthene-9-carboxylic acid has been obtained by treating xanthene with *tert*-butylsodium in the presence of initiators (e.g. chlorobenzene and amyl alcohol, *tert*-butyl chloride and amyl alcohol, or 2-chloroethanol) and an activator (e.g. Me$_2$NCH$_2$CH$_2$NMe$_2$) and then with solid carbon dioxide (I. Dory *et al.*, Teljes 11,886, 1976). Xanthene-9-carbonyl chloride has been treated with 4-H$_2$NCH$_2$CH$_2$C$_6$H$_4$CH$_2$CH$_2$CO$_2$Et and the resulting ester hydrolysed to give acid (12). A number of related derivatives have been synthesized and tested for anti-diabetic and anti-cholesteremic properties (M. Huebner *et al.*, Ger. Offen. 2,629,752, 1978).

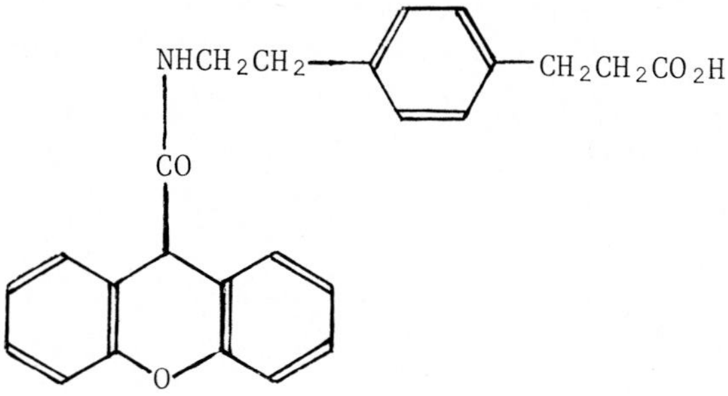

(12)

1,3-Dimethoxyxanthene on nitration with urea nitrate in
polyphosphoric acid affords 1,3-dimethoxy-4-nitroxanthene
(V.B. Nabar and N.A. Kudav, Indian J. Chem., 1977, 15B,
89).

Cyclization of xanthene derivative (13) with phenylhydrazine yields 5-methyl-2-phenyl-4-(9-xanthenyl)-3H-
-pyrazol-3-one (14; 32%), also obtained in 91% yield by
condensing xanthydrol (1) with 5-methyl-2-phenyl-3H-pyrazol-
-3-one (15). Condensation of (1) with 5-amino-2-phenyl-3H-
-pyrazol-3-one (16) affords 5-amino-2-phenyl-4-(9-xanthenyl)-
-3H-pyrazol-3-one (17; 84%) (I. Okabayashi, J. heterocyclic
Chem., 1980, 17, 1339).

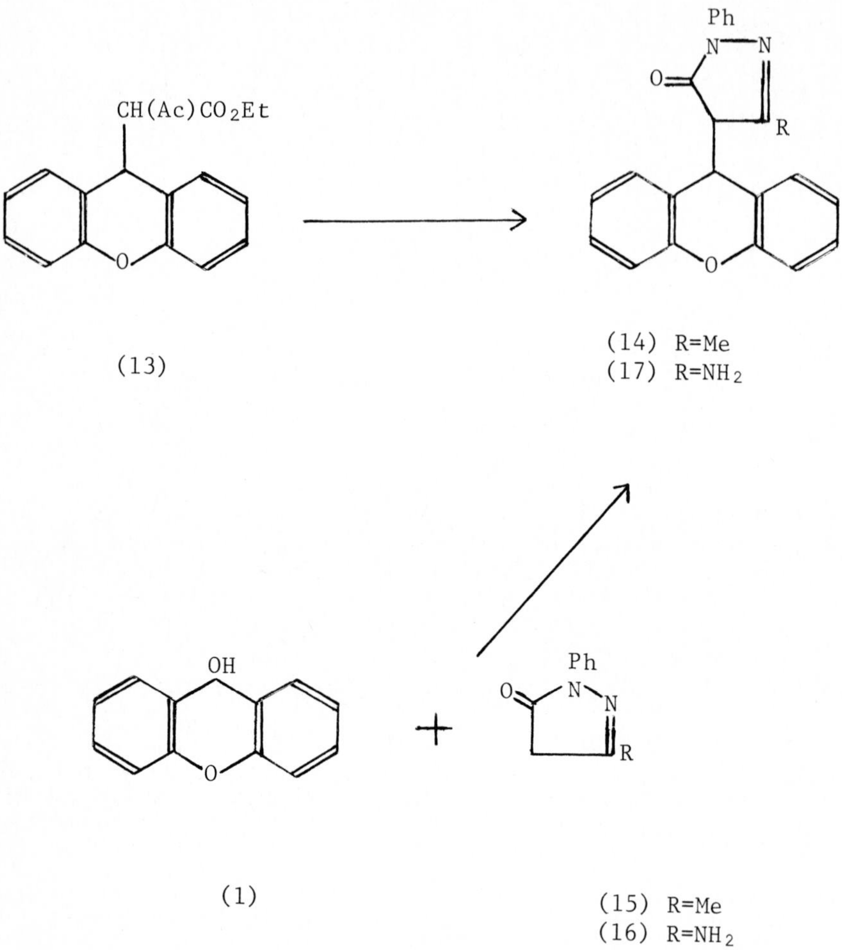
CH(Ac)CO2Et
O
(13)
Ph
N
N
O
R
O
(14) R=Me
(17) R=NH2
OH
O
+
Ph
O
N
N
R
(1)
(15) R=Me
(16) R=NH2

9-Diazoxanthene (18) and 9-xanthylidene (19) are of interest from a synthesis and theoretical point of view. The latter is generated from the former by photolysis. As a dipolar reagent 9-diazoxanthene (18) undergoes addition to 1,4-quinones to yield pyrazolines and is reported to be inert to styrene (G.W. Jones, K.T. Chang, and H. Shechter, J. Amer. chem. Soc., 1979, 101, 3906).

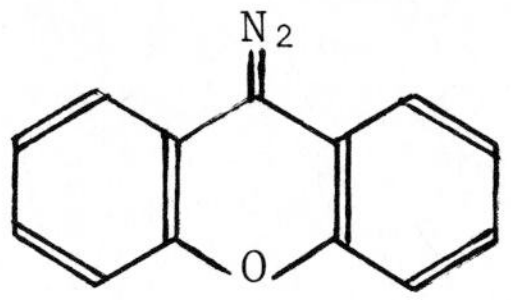

$$\text{(18)} \hspace{6cm} \text{(19)}$$

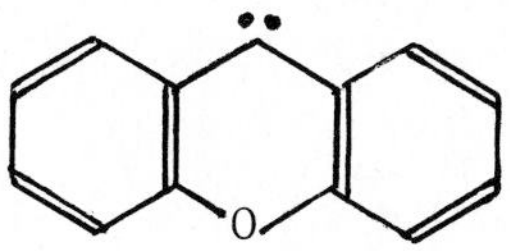

Photolysis of 9-diazoxanthene and dimethyl maleate affords the spiropyrazole (20), also prepared by 1,3-dipolar reactions (H. Duerr, S. Froehlich, and M. Kausch, Tetrahedron Letters, 1977, 1767).

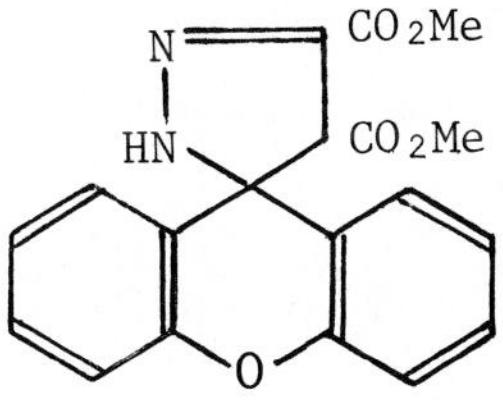

$$\text{(20)}$$

Xanthen-9-ylidenecyanoacetates react with organomagnesium halides to yield 1,4-adducts regardless of the nature of the Grignard reagent and its steric requirements (N. Latif, N. Mishriky, and M. Hammad, Austral. J. Chem., 1977, $\underline{30}$, 2263).
$\Delta^{4,9'}$-4-(9'-Xanthenyl)piperidine derivatives (Smith Kline Corp. Fr. Demande 2,290,202, 1976) and $\Delta^{4,9'}$-2,6-diphenyl--4-(9'-xanthenyl)-4H-pyran, m.p. 245-246° (G.A. Reynolds and C.H. Chen, J. org. Chem., 1980, $\underline{45}$, 2456) have been synthesized.

356

Reductive cyclization of 1,5-diketones (21) by Raney
nickel/hydrogen furnishes the 1,2,3,4,4a,5,6,7,8,9a-
-decahydroxanthenes (22; 96-98%), which on further
hydrogenation over Rhodium/carbon affords the perhydroxanthenes (23; 98-99%). Similar hydrogenation of
decahydroxanthenes (24; R = H, 76%, R = Me, 69.5%),
obtained on treating (21) with sodium tetrahydridoborate,
yields the perhydroxanthenes (25; 98% and 99%)
(V.G. Kharchenko *et al.*, Zh. org. Khim., 1987, 23, 576).

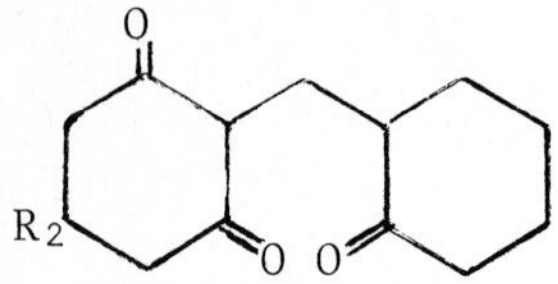

R=H,Me

(21)

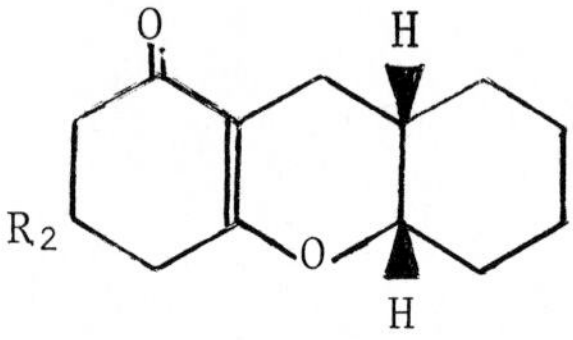

(22)

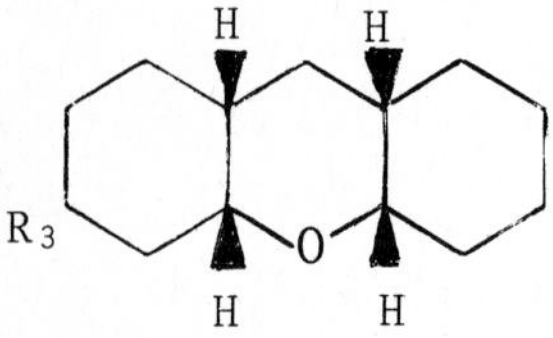

(23)

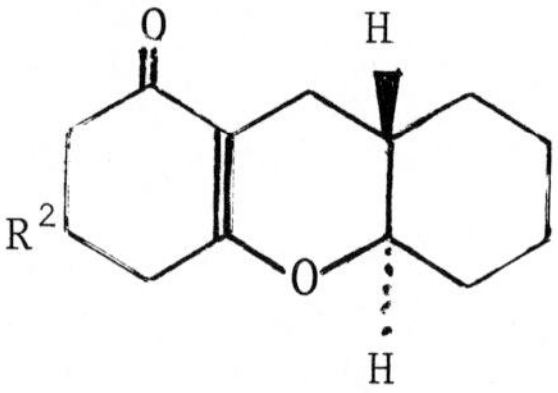

(24)

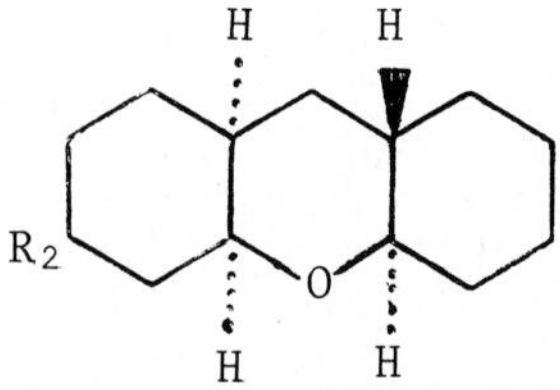

(25)

Perhydroxanthenes have been prepared by the hydrogenation of substituted bis(2-oxocyclohexanyl)methane (26) in trifluoroacetic acid in the presence of $(Ph_3P)_2PtCl_2$, $(Ph_3P)_2Pt(CF_3CO_2)_2$, or $(Ph_3P)_2PtClH$, at 30-70° and 50-200 atmospheres (Z.H. Parnes *et al*., U.S.S.R. Pat. 534,454, 1976).

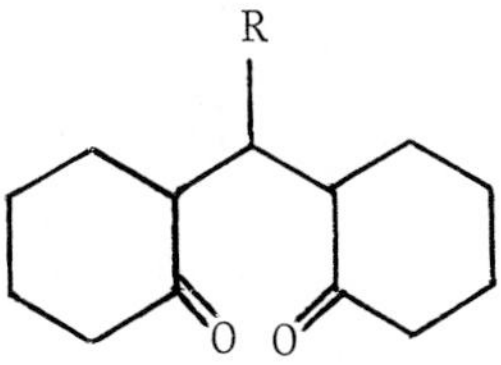

R=H,alkyl

(26)

(b) *Xanthylium salts and xanthene colouring matters*

Xanthylium salts (1) react with trimethyl phosphite in acetonitrile in the presence of sodium iodide to yield phosphonate (2), which on treatment with butyllithium in tetrahydrofuran at -78° is deprotonated and the resulting carbanion reacts with 4-tolualdehyde, cinnamaldehyde and 4,4'-dichlorobenzophenone to give the corresponding exomethylene compound (eg. 3) (K. Akiba, K. Ishikawa, and N. Inamoto, Synth., 1977, 862; Bull. chem. Soc., Japan, 1978, *51*, 2684).

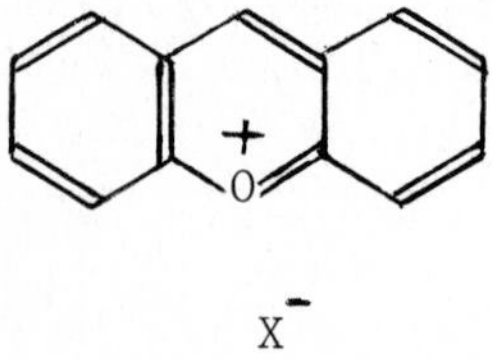

X=Br,ClO₄,MeOSO₃,EtOSO₃

$X = Br, ClO_4, MeOSO_3, EtOSO_3$

(1)

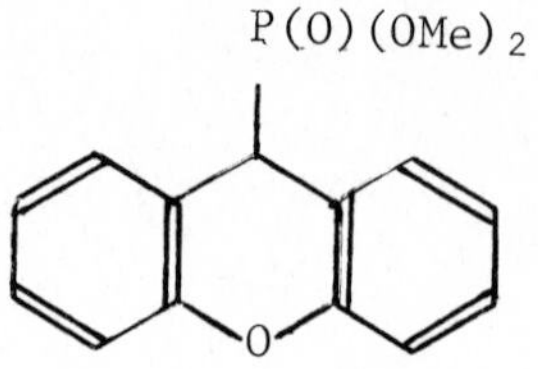

(2)

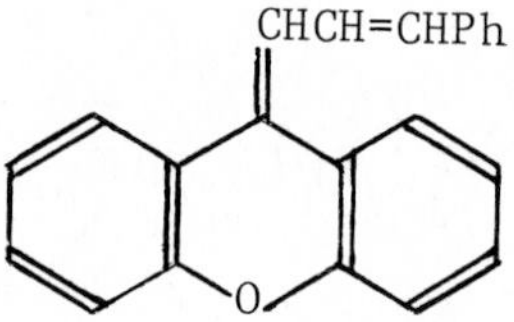

(3)

9-Vinylidenexanthenes (4) react with perchloric acid, hydrochloric acid, and acetic acid to afford the corresponding highly coloured xanthylium salts (5), which on heating in acid media rearrange to yield the spiro compounds (6) (N.F. Abdul-Malik, S.B. Awad, and A.B. Sakla, *ibid.*, 1979, **52**, 3431).

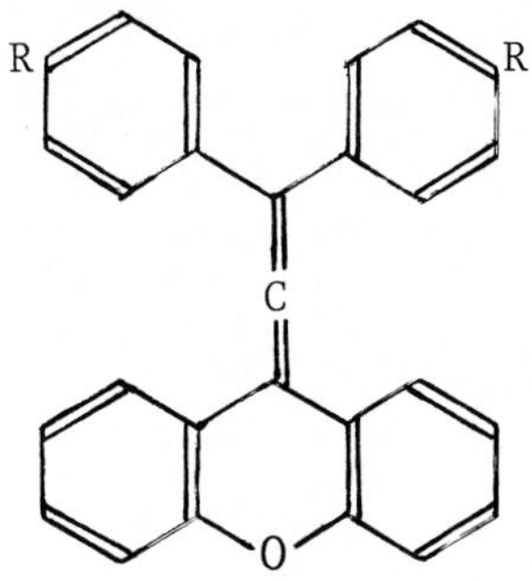

R=H,OMe,OEt,OPr^i
(4)

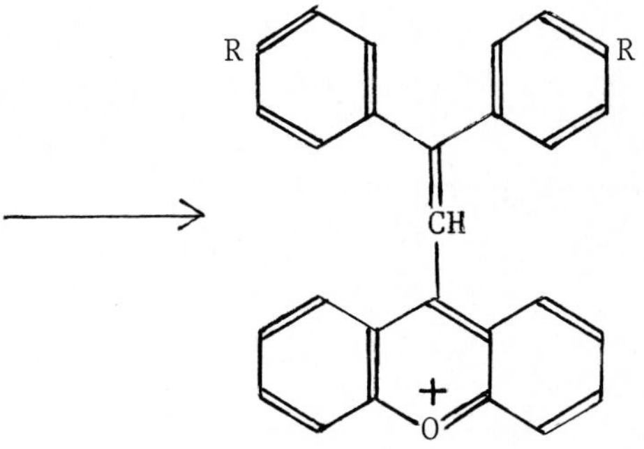

X=ClO$_4$,Cl,OAc

(5)

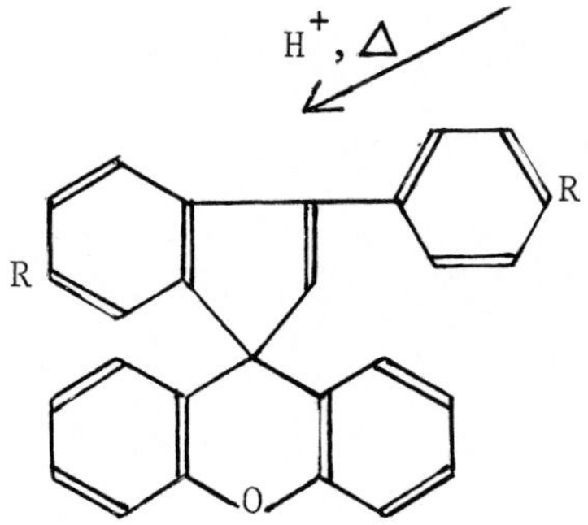

H$^+$,$\triangle$

(6)

Xanthydrol (7) on treatment with 4-toluidine in ethanol and acetic acid gives xanthyltoluidine (8) and treating xanthylium perchlorate (9) with dimethylaniline in pyridine over 24h yields 4-xanthyl-N,N-dimethylaniline (10) (G.E. Ivanov *et al.*, Ukr. Khim. Zh., 1985, **51**, 655).

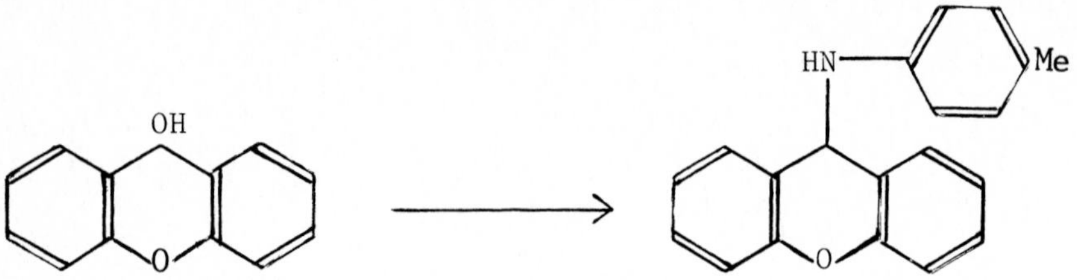

OH

(7)

HN—⟨ ⟩—Me

(8)

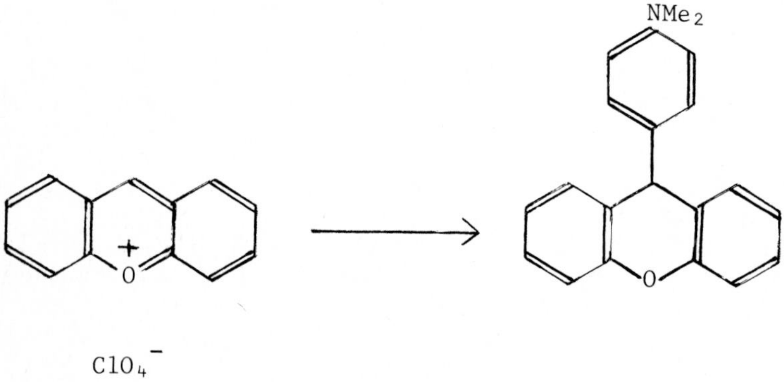

ClO₄⁻

(9)

NMe₂

(10)

9-Acridylamino-, -morpholino-, -(4-methylphenylamino)-, -(2-amino-5-methylphenyl)-, -ureylene-, -succinimido-, -(4--toluenesulphonamido)-, and -ethoxyxanthenes are obtained by treating 9-pyridylxanthylium perchlorate in pyridine with the necessary amine or ethanol (*idem*, Zh. org. Khim., 1986, 22, 842).

The products from the reaction between xanthylium perchlorate and an amine depends on the structure of the reacting amine, for instance, with aniline, 3-toluidine, and 4-chloroaniline xanthenes (11), (12) and (13), respectively, are obtained. The amination of 9-ethyl- and -phenyl- xanthylium perchlorates is also reported. [H.M. El-Namaky and M.A. Salama, Egypt. J. Chem., 1977 (Pub. 1979), 20, 125.]

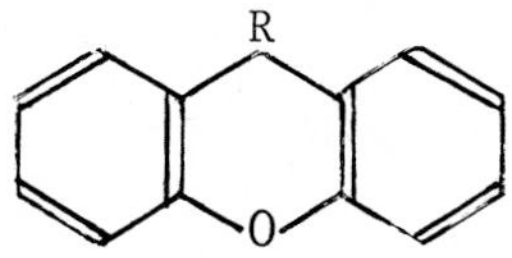

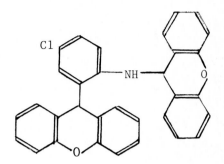

(11) R=4-H₂NC₆H₄

(11) $R=4-H_2NC_6H_4$
(12) $R=3-MeC_6H_4NH$

(13)

(c) Xanthones

2-Phenoxybenzaldehyde and copper (II) chloride on boiling
in nitrobenzene under nitrogen yields xanthone, m.p. 173-175°
(J.I. Okogun, J. chem. Soc., Perkin I, 1976, 2241).

The irradiation of a 1:1 mixture of xanthene and xanthone
in benzene with a mercury high pressure lamp affords a
mixture of dixanthenes (1) (M. Takacs *et al.*, Arch. Pharm.,
1976, 309, 735).

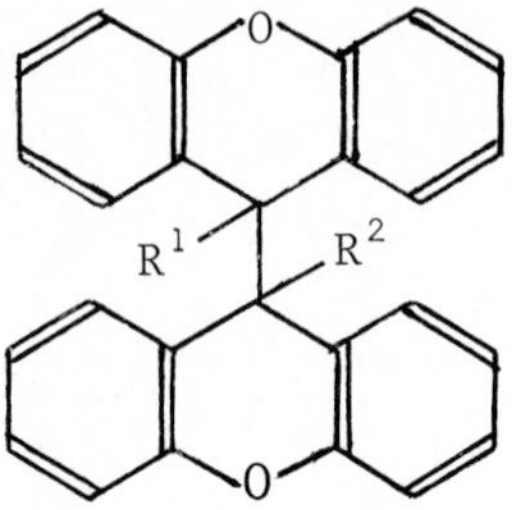

R^1	R^2
H	H
OH	H
OH	OH

(1)

Solvent effects have been studied and it has been found that the hydrogen-bonding properties of the media play a dominant role in the photochemistry of xanthone (J.C. Scaiano, J. Amer. chem. Soc., 1980, 102, 7747).

(i) Halogenoxanthones

1-Chloroxanthone (3) has been obtained by the cyclization of 2-chloro-6-phenoxybenzoic acid (2). Similarly, 1,5- and 1,7-dichloroxanthenes have been prepared from appropriate benzoic acids and 2,6-dichloroxanthone by cyclization of 4-chloro-2-(4-chlorophenoxy)benzoic acid (I. Okabayashi and N. Iwata, Chem. pharm. Bull., 1980, 28, 2831).

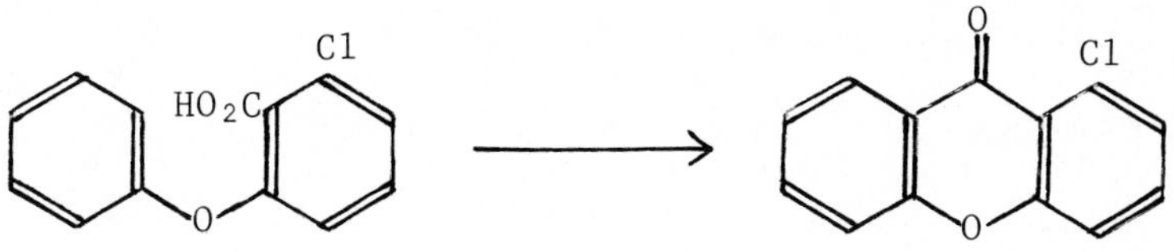

(2)

(3)

Iodination of 2-hydroxyxanthone with iodine-iodic acid or iodine-ammonium hydroxide gives 2-hydroxy-1-iodoxanthone, whereas 3,6-dihydroxyxanthone affords 3,6-dihydroxy-4-
-iodoxanthone, 3,6-dihydroxy-4,5-diiodoxanthone, 3,6-
-dihydroxy-2,4,5-triiodoxanthone and 3,6-dihydroxy-2,4,5,7-
-tetraiodoxanthone. The methyl ethers of the hydroxy-
iodoxanthone undergo Rosenmund-von Braun cyanation to give nitriles, for example, 1-iodo-2-methoxyxanthone furnishes 1-cyano-2-methoxyxanthone. 1-Iodo-2-methoxyxanthone undergoes the Ullmann coupling reaction to yield the bixanthonyl (4) (Y.G. Gaekwad and S. Sethna, J. Indian chem. Soc., 1978, 55, 794).

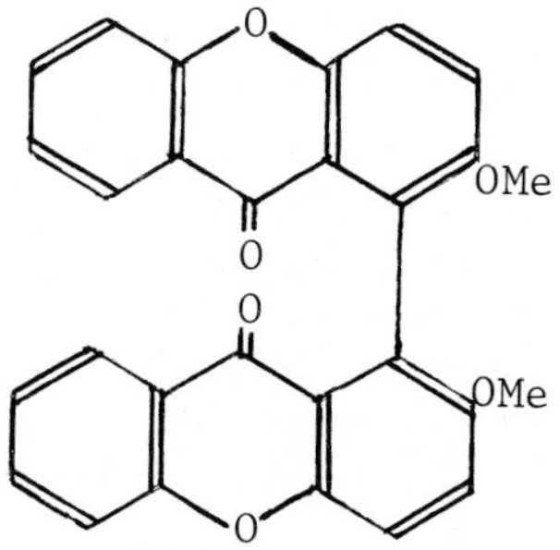

(4)

(ii) Nitro- and amino-xanthones

4-Dimethylamino-1,3-dimethoxyxanthone, 4-dimethylamino-
-1,3-dimethoxy-5-methylxanthone, and 4-dimethylamino-1,3-
-dimethoxy-6-methylxanthone have been prepared in one step
from the respective nitroxanthones (N.A. Kudav,
N.A. Kulkarni, and A.B. Kulkarni, Indian J. Chem., 1976,
14B,1 484).

(iii) Hydroxyxanthones

Ethyl salicylate condenses with phenols (e.g. resorcinol,
hydroquinone, catechol, and 3,4-xylenol) in boiling
diphenyl ether to yield hydroxyxanthones and also in some
instances the corresponding phenyl salicylate. Thus, ethyl
salicylate with 3,5-dihydroxytoluene yields 1-hydroxy-3-
-methyl- and 3-hydroxy-1-methyl-xanthone (R.J. Patolia and
K.N. Trivedi, *ibid.*, 1983, 22B, 444; Chem. and Ind., 1978,
235).

Various 2-hydroxybenzoic acids have been condensed with
reactive phenols, in a mixture of polymeric phosphoric acid
[prepared by mixing equimolar quantities of phosphoryl
chloride and orthophosphoric acid (85%) and heating the
mixture at 50° for 1 hour] and anhydrous zinc chloride
to yield the following 1-hydroxyxanthones; 1-hydroxy-,
1,3-dihydroxy-, 1-hydroxy-3-methyl-, 1-hydroxy-5-methyl-,
1,3-dihydroxy-5-methyl-, 3,5-dimethyl-1-hydroxy-, 1-hydroxy-
-6-methyl-, 1,3-dihydroxy-6-methyl-, and 3,6-dimethyl-1-
-hydroxy-xanthone. The reaction medium has distinct
advantage over the earlier methods of preparation, as the
condensation reactions can be carried out under homogeneous
conditions and even at elevated temperatures. Furthermore,
the reaction medium leads to the preparation of 1-hydroxy-
xanthones and not the 3-hydroxyxanthones or the respective
benzophenones when resorcinol is used as the phenolic
substrate (N.B. Nevrekar *et al.*, Chem. and Ind., 1983,
479).

One new procedure for the preparation of 1-hydroxy-
xanthones, in good yield, involves the condensation of 2-
-hydroxybenzoyl ureas with reactive phenols, including
resorcinol, phloroglucinol, and orcinol in a mixture of
phosphoryl chloride, orthophosphoric acid, and anhydrous
zinc chloride at 150°. The hitherto unknown 2-hydroxy-
benzoyl ureas are prepared from the corresponding acid
chlorides and urea. The following known 1-hydroxyxanthones
have been obtained by this method; 1-hydroxy-, 1,3-
-dihydroxy-, 1-hydroxy-3-methyl, 1-hydroxy-5-methyl-, 1,3-

-dihydroxy-5-methyl-, 3,5-dimethyl-1-hydroxy-, 1-hydroxy-
-6-methyl-, 1,3-dihydroxy-6-methyl-, 3,6-dimethyl-1-hydroxy-,
1,6-dihydroxy-, 1,3,6,-trihydroxy-, 1,6-dihydroxy-3-methyl-,
1,7-dihydroxy-, and 1,8-dihydroxyxanthone (M.V.R. Mucheli
and N.A. Kudav, *ibid*., 1985, 31). For the preparation of
4,6-dimethoxy-1-hydroxy-8-methylxanthone see T. Kato,
N. Katagiri, and J. Nakano (Heterocycles, 1976, 5, 167),
and for 1-methoxy-3-hydroxyxanthone, V. Eswaren *et al*.,
(Indian J. Chem., 1979, 17B, 517).

Benzoyl peroxide reacts with 1-hydroxyxanthone to give
2,2-bisbenzoyloxyxanth-1,9-dione, but 1-hydroxyxanthones
(5) yield the corresponding 2-benzoyloxy-1-hydroxyxanthones
(6) (Y.S. Chauhan and K.B.L. Mathur, *ibid*., 1977, 15B, 51).

R=Me,MeO

(5)   (6)

2-Aryloxybenzaldehydes with copper (II) chloride or
bromide yield the corresponding xanthones including 2-
-methoxyxanthone, m.p. 129-131°; 4-methoxyxanthone, m.p.
174-175°; 2-methylxanthone, m.p. 119-122°; and 2-nitro-
xanthone, m.p. 204-206°, along with bis-4-nitrophenyl ether,
m.p. 138-140°, and an unidentified compound, m.p. 125-127°
(J.I. Okogun, J. chem. Soc., Perkin I, 1976, 2241).

Procedures have been presented for selective demethylation
of eight polymethoxyxanthones, for example, boiling 1,3-
-dimethoxyxanthone in aqueous piperidine for 2½ days affords
1-hydroxy-3-methoxyxanthone (44%), 1,3-dihydroxyxanthone
(11%) and some polymeric material (R.K. Chaudhuri,
F. Zymalkowski, and S. Ghosal, J. pharm. Sci., 1978, 67,
1321). The regioselective cleavage of the methylenedioxy
ring in mono- and bis-methylenedioxypolymethoxyxanthones has

been discussed (S. Ghosal *et al.*, Indian J. Chem., 1984, <u>23B</u>, 1226). The [1]H-nmr chemical shifts for methoxy groups (1-4 MeO groups) in methoxyxanthones have been determined in chloroform, benzene, trifluoroacetic acid, and 3% trifluoro-acetic acid in benzene (R.A. Finnegan and K.E. Merkel, J. pharm. Sci., 1977, <u>66</u>, 884).

In the synthesis of xanthone *O*-glucosides, it has been found that the highest product yield is obtained with 4--hydroxylated xanthones followed by progressively lower yields from 3-, 2-, and 1-hydroxylated xanthones (T.J. Nagem and O.R. Gottlieb, Rev. Latinoam. Quim., 1981, <u>12</u>, 50).

The hexacyanoferrate (III) oxidation of 2,4,6,3'-tetra-hydroxybenzophenone yields 1,3,7-trihydroxyxanthone by a *para*-intramolecular coupling. Similar oxidation of 2,4,5'--trihydroxy-2'-methoxy-3-methylbenzophenone affords xanthenedione (7), which with hydrogen chloride is aromatized to give 4-chloro-2,6-dihydroxy-5-methylxanthone (8) and with zinc-acetic acid to give 2,6-dihydroxy-5--methylxanthone. 5-Methyl-1,4,6-trihydroxyxanthone is obtained from 3-methyl-2,4,2',5'-tetrahydroxybenzophenone (R.C. Ellis, W.B. Whalley, and K. Ball, J. chem. Soc., Perkin I, 1976, 1377).

(7)   (8)

(iv) Acylxanthone, xanthonecarboxylic acids and derivatives

Xanthonecarboxaldehydes (9) have been converted into ethynylxanthones (11) by the application of the Wittig reaction, using bromomethylenetriphenylphosphorane (10), prepared from bromomethyltriphenylphosphonium bromide (M. Matsumoto and K. Kuroda, Tetrahedron Letters, 1980, 4021).

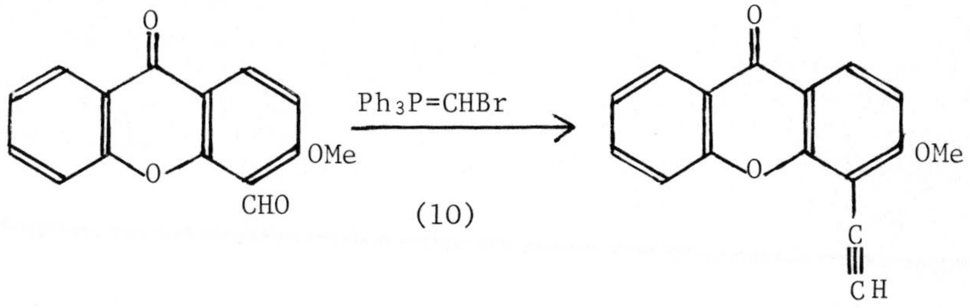

(9)   (11)

2-Acylxanthones (12) have been obtained from the reaction of the appropriately 4-substituted 2-chlorobenzoic acid with the necessary 4-hydroxyacylbenzene in the presence of sodium methoxide and cyclizing the resulting 4-substituted 2-(4-acylphenoxy)benzoic acid under acidic conditions (H. Marona, Pol. J. Chem., 1979, **53**, 1645).

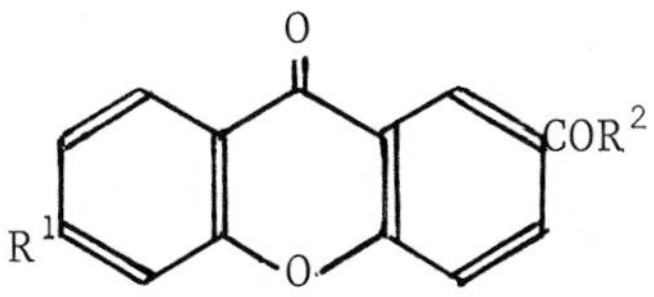

R^1=H,Cl
R^2=Me,Et

(12)

The preparations of a number of 2-acylxanthones have been reported (I. Iijima *et al.*, J. chem. Soc., Perkin I, 1979, 3190).

The reaction of either 3,5-di-isopropyl- or 3,5-di-*tert*-butyl-salicylic acid (13) with methyl 4-hydroxybenzoate in the presence of dicyclohexylcarbodiimide gives the depside type esters (14), which undergo a Smiles rearrangement on treatment with base to yield the diphenyl ethers (15). Hydrolysis of ethers (15) give dicarboxylic acids (16), cyclized by sulphuric acid to the 5,7-disubstituted xanthone-2-carboxylic acids (17), which are not readily accessible by other routes. 5,7-di-isopropylxanthone-2-carboxylic acid, m.p. 248-249.5°, 5,7-di-*tert*-butyl-xanthone-2-carboxylic acid, m.p. 256-258° (J.R. Pfister, J. heterocyclic Chem., 1982, <u>19</u>, 1255).

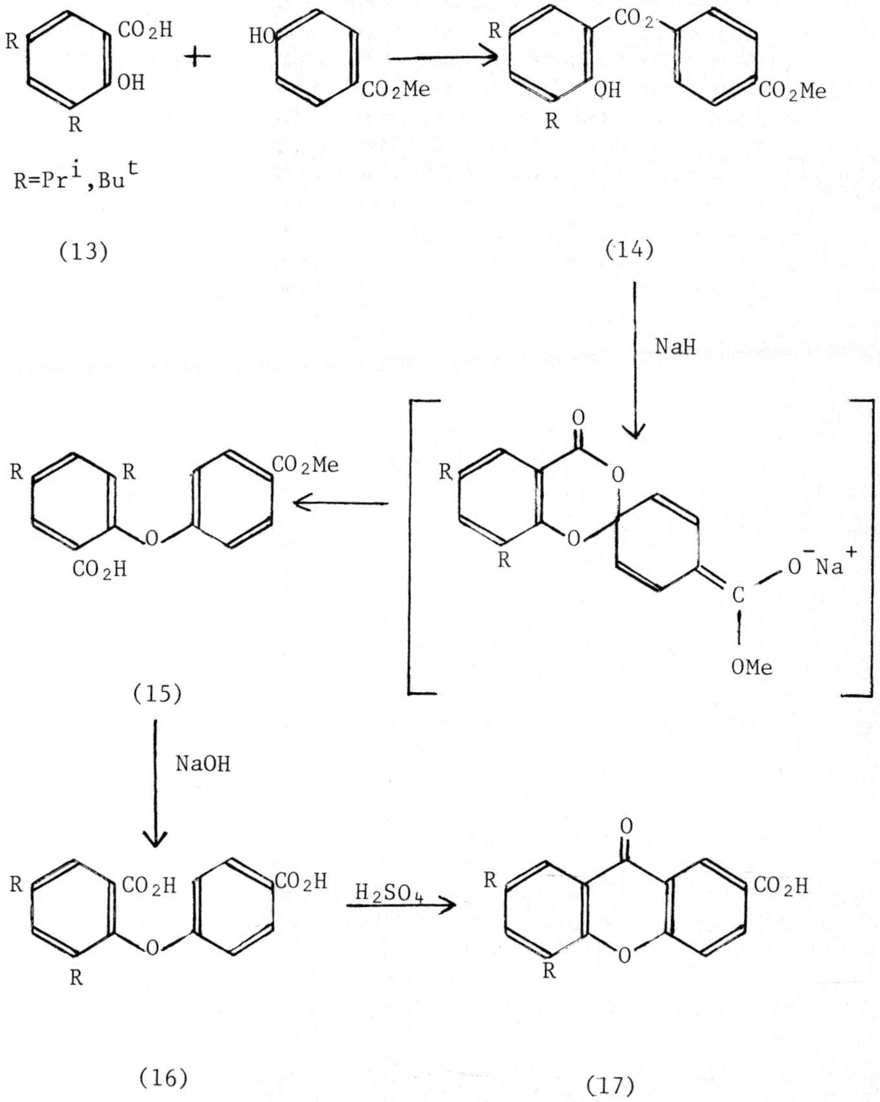

CO2H
OH
R
HO
CO2Me
CO2
R=Pri,But
(13)
(14)
NaH
O
O
O
O⁻Na⁺
C
OMe
CO2Me
CO2H
(15)
NaOH
CO2H
H2SO4
(16)
(17)

Some chloro-, bromo-, nitro-, and alkoxy-substituted
xanthone-2-carboxylic acids have been prepared and their
biological properties have been investigated (M. Eckstein,
Marona, and J. Mazur, Pol. J. pharmacol. Pharm., 1983, 35,
159), some alkanolamide derivatives of xanthone-2-carboxylic
acids (Marona and M. Cegla, Acta Pol. Pharm., 1986, 43, 399)
and some xanthone-2-alkylcarboxylic acids have been
synthesized (Marona, Pol. J. Chem., 1980, 54, 2059).

Xanthones on oxidation with manganese (III) acetate give
carboxy-, dicarboxy-, (acetomethyl)(carboxy)-,
(carboxy)(hydroxymethyl)-, and acetoxymethyl-xanthones, and
small amounts of (carboxymethyl)(carboxy)-, carboxymethyl-,
(acetoxymethyl)(carboxymethyl)-, bis(acetoxymethyl)-, and
diacetoxy-xanthones. The carboxyl group is preferentially
introduced into the *peri*-positions and the acetoxymethyl
groups are located at positions *ortho* to the methoxy or
methyl group (H. Nishino and K. Kurosawa, Bull. chem. Soc.,
Japan, 1983, 56, 474). The oxidation of xanthones with
lead (IV) acetate has also been investigated (*idem*, *ibid*.,
p.2847).

The ^{13}C-nmr spectra for mangiferin and eleven hydroxy-
and two hydroxymethoxy-xanthones have been reported and
indicate that some previously reported ^{13}C-nmr signal
assignments should be revised (A.W. Frahm and R.K. Chaudhuri,
Tetrahedron, 1979, 35, 2035). Ir-vibration frequencies of
xanthone have been interpreted according to form and symmetry
determined in the Raman scattering spectra (V.G. Klimenko,
E.A. Gastilovich, and D.N. Shigorin, Zh. Fiz. Khim., 1979,
53, 580). The absorption spectra of xanthene, xanthone, and
five different symmetrical compounds in several solvents
(D. Grasso *et al.*, Tetrahedron, 1976, 32, 2105) have been
studied, and the phosphorescence spectra and lifetimes have
been measured for xanthone, xanthone-d$_8$, 2,6-, 2,7-, and
3,6-dimethylxanthone at various temperatures and found to be
temperature dependent (R.E. Connors and W.R. Christian, J.
phys. Chem., 1982, 86, 1524). For the absorption and
emission properties of xanthone and xanthione see
D.A. Capitanio (Diss. Abs. Int. B, 1976, 37, 2874).

7-Bromo-1,2,3,4-tetrahydroxanthone (18) has been obtained
by treating 2-acetyloxy-5-bromobenzoic acid successively with
ethyl chloroformate and 1-pyrrolidinocyclohexene. It has been
converted into 2-cyanoxanthone by treatment with copper (I)
cyanide and then to either xanthone-2-carboxylic acid or 2-(5-
-tetrazolyl)xanthone. A number of related derivatives have
been prepared (K. Onogi *et al.*, U.S. Pt. US 4,290,954, 1981).

(18)

The thermal condensation of cyclohexanone and ethyl 2-
-oxocyclohexanecarboxylate affords octahydroxanthone
(B.S. Kirkiacharian, Comp. rend., 1981, __293__, 149).

(v) Naturally occurring hydroxyxanthones

A number of new naturally occurring xanthones have been
isolated in the last ten years and some of them are named
below.

Laxanthone I, *1,3-dihydroxy-6,7-dimethoxyxanthone*, and
laxanthone II, *3,6-diacetoxy-1-hydroxy-7-methoxyxanthone*,
have been isolated from *Lawsonia inermis* (D.K. Bhardwaj,
T.R. Seshadri, and R. Singh, Phytochem., 1977, __16__, 1616)
along with laxanthone III, *6-acetoxy-3,7-dimethoxy-1-
-hydroxyxanthone* (Bhardwaj *et al.*, *ibid.*, 1978, __17__, 1440;
Bhardwaj, R.K. Jain, and C.K. Mehta, Curr. Sci., 1979, __48__,
614). The structures of laxanthone II, and isolaxanthone
II have been confirmed by synthesis of their diethyl ethers
(Bhardwaj, M.S. Bisht, and Jain, Proc. Indian Natl. Sci.
Acad., 1980, __46A__, 381) and laxanthone III has been synthe-
sized (Bhardwaj *et al.*, Indian J. Chem., 1979, __17B__, 288).

Veratrilogenin, *1,7-dihydroxy-3,4-dimethoxyxanthone*,
$C_{15}H_{12}O_6$, m.p. 285 ~ 288°, and veratriloside, *3,4-
-dimethoxy-7-O-β-D-glucopyranosyl-1-hydroxyxanthone*,
$C_{21}H_{22}O_{11}$, m.p. 272 ~ 275°, along with the known 1-
-hydroxy-2,3,4,7-tetramethoxyxanthone, have been obtained
from the roots of *Veratrilla baillonii* Franch, a folk
medicine (Y-B. Yang and J. Zhou, Yao Hsueh Hsueh Pao, 1980,
__15__, 625). Also isolated from same source, 1-hydroxy-2,3,5-
-trimethoxy-, 1-hydroxy-2,3,7-trimethoxy-, 1-hydroxy-2,3,4,5-
-tetramethoxy-, 1,4-dihydroxy-2,3,7-trimethoxy-xanthones
and the new xanthone 1,3-dihydroxy-2,7-dimethoxyxanthone
(*idem*, Yun-nan Chih Wu Yen Chiu, 1980, __2__, 468).

Bellidifoloside, *1,8-dihydroxy-5-O-β-glucopyranosyl-3-*
-methoxyxanthone, $C_{20}H_{20}O_{11}$, m.p. 165-168°, the known
compound mangiferin, and 1-*O*-β-primeverosyl-3,7,8-trimethoxy-
xanthone have been isolated from the above ground parts of
Swertia iberica (O.A. Denisova *et al.*, Khim. Prir. Soedin.,
1980, 724). Bellidifoloside is also contained in the aerial
parts of *Swertia connata* along with mangiferin, and
gentiacaulein 1-*O*-glucoside (E.V. Solov'eva *et al.*, *ibid.*,
p.840).

Zeyloxanthonone, *6,8-dihydroxy-1,2,3,4-tetrahydro-1,1,7-*
tris(3-methylbut-2-enyl)xanthone-2,9-dione (1), $C_{28}H_{34}O_5$,
m.p. 137°, and other compounds have been isolated from the
bark of *Calophyllum zeylanicum* Kosterm (Guttiferae) (renamed
Calophyllum lankaensis Kosterm). Its timber yielded
jacareubin, 6-deoxyjacareubin, 2-hydroxy-, 1,5-dihydroxy-,
1,7-dihydroxy-, and 1,6-dihydroxy-5-methoxy-xanthones.
Zeyloxanthonone has the structure (1) and has been shown to
be identical to wightianone. The structure of the latter
has therefore been revised (S.P. Gunasekera,
S. Sotheeswaran, and M.U.S. Sultanbawa, J. chem. Soc.,
Perkin I, 1981, 1831).

R=CH₂CH=CMe₂

(1)

The timber extract of *Calophyllum cuneifolium* Thw. affords
2-(3-methylbut-2-enyl)-1,3,5-trihydroxyxanthone (2) and the
known xanthones, calabaxanthone, trapezifolixanthone,
jacareubin, 6-deoxyjacareubin, guanandin, euxanthone,
buchanaxanthone, and scriblitifolic acid, and that of
Calophyllum soulattri Burmf. yields xanthone (2), 6-deoxy-
jacareubin, buchanaxanthone, euxanthone, and 1-hydroxy-5-
-methoxyxanthone (Gunasekera *et al.*, *ibid.*, 1977, 1505).

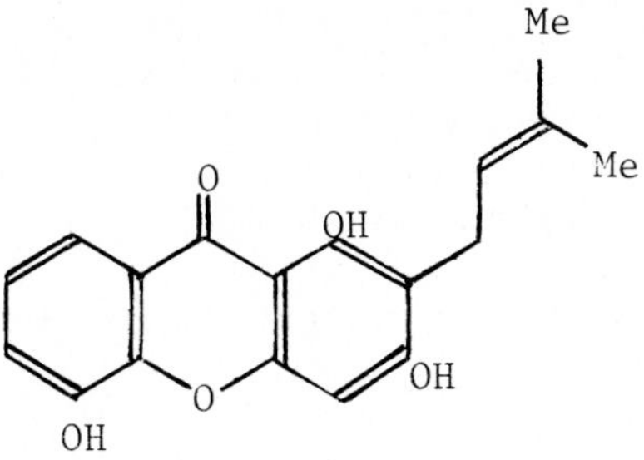

(2)

Isogentiakochianin, *7-methoxy-1,3,8-trihydroxyxanthone*
and swertiaiberin, *7,8-dimethoxy-1,2,3-trihydroxyxanthone*,
along with swertiaperennin, decussatin, gentiakochianin and
norswertianin have been obtained from the roots *Swertia
iberica* (Denisova *et al.*, Khim. Prir. Soedin., 1980, 190).
   Eustomin, *1-hydroxy-3,5,6,7,8-pentamethoxyxanthone* (3)
has been isolated along with five other polyoxygenated
xanthones from the roots of *Eustoma grandiflorum* collected
in Texas. Four of these xanthones have been identified as
3,5-dimethoxy-1-hydroxy-, 3,7-dimethoxy-1-hydroxy-1, 1-
-hydroxy-3,7,8-trimethoxy-, and 1-hydroxy-3,5,6,7-tetra-
methoxy-xanthone. It has been reported that eustomin and
the five xanthones probably occur in the plant as
glycosides (G. Sullivan, F.D. Stiles, and K.H.A. Rosler, J.
pharm. Sci., 1977, **66**, 828).

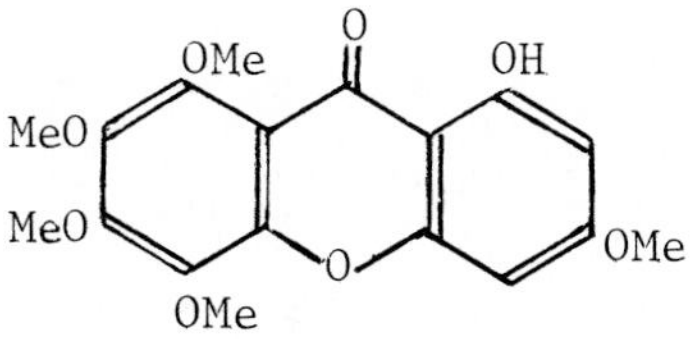

(3)

Two new *O*-glucosides of mangiferin, mangiferin 7-*O*-β-D-
-glucoside and mangiferin 6-*O*-β-D-glucoside have been
isolated from the leaves of *Gentiana asclepiadea* L.
(M. Goetz and A. Jacot-Guillarmod, Helv., 1977, <u>60</u>, 2104).

2-Chlorolichexanthone, *2-chloro-3,6-dimethoxy-1-hydroxy-*
-8-methylxanthone, and other new lichen xanthones, 3,6-
-dimethoxy-1-hydroxy-, 2,4-dichloro-1,3-dihydroxy-6-methoxy-,
2-chloro-1,3-dihydroxy-6-methoxy-, and 4-chloro-1,3-
-dihydroxy-6-methoxy-xanthones have been isolated from the
lichen *Pertusaria sulphurata* and each structure confirmed
by unambiguous synthesis. Some chloroxanthones and the
chlorination of 1,3-dihydroxy-6-methoxy-8-methylxanthone
have been reported (J.A. Elix *et al.*, Austral. J. Chem.,
1978, <u>31</u>, 145). 2,4-Dichloro-, 2,5-dichloro-, and 2,4,5-
-trichloro-xanthones have been obtained from *Pertusaria*
aleiant [S. Huneck, IUPAC Int. Symp. Chem. nat. Prod., 11th,
1978, 4 (Part 1), 197. ed. N. Marekov, I. Ognyanov, and
A. Orahovats, Izd. BAN: Sofia., Bulg.]. Lichexanthone,
along with a phthalide derivative (djalonensin) and two
triterpenes, has been obtained from the stem bark of
Anthocleista djalonensis and *A. vogelli* (D.A. Okorie,
Phytochem., 1976, <u>15</u>, 1799) and the relationship of some
model syntheses to its biosynthesis have been discussed
(T.M. Harris and J.V. Hay, J. Amer. chem. Soc., 1977, <u>99</u>,
1631).

Several chlorinated derivatives of norlichexanthone (4)
have been prepared by unambiguous methods and their [1]H-nmr
spectra discussed, resulting in the structures previously
assigned to several lichen xanthones being questioned.
Suggested revised structures are given in Table 13
(E.G. Sundholm, Tetrahedron, 1978, <u>34</u>, 577).

(4)

TABLE 13

Revised structures of some norlichexanthone derivatives

Previously assigned positions		New assigned positions	
Cl	OMe	Cl	OMe
2		4 and 5	
2	6	5	3
2, 5	3	2, 7	3
2, 5	3, 6	2, 7	3, 6
2, 7		4, 5	
2, 7	3, 6	4, 5	3, 6
2, 4, 7		2, 4, 5	
2, 4, 7	3	2, 4, 5	3
2, 5, 7		4, 5, 7	

The synthesis of further lichen xanthones and several other
norlichexanthone derivatives have confirmed the suggested
revisions. However, the components originally described as
1,6-dihydroxy-3-methoxy-8-methyl-2,5,7-trichloro- and 8-
-methyl-2,5,7-trichloro-1,3,6-trihydroxy-xanthones have been
found to have the assigned structure (Sundholm, Acta Chem.
Scand., 1979, B33, 475).

2,5-Dichloro-3,6-dimethoxy-1-hydroxy-8-methyl-, 2,4-
-dichloro-3,6-dimethoxy-1-hydroxy-8-methyl-, and 3,6-
-dimethoxy-1-hydroxy-8-methyl-2,4,5-trichloro-xanthones
have been isolated from *Pertusaria* species and their
structures determined from chemical and spectral data
(including ^{13}C-nmr). Revised structures, 1,6-dihydroxy-
-3-methoxy-8-methyl-2,4,5-trichloro-, 8-methyl-2,4,5-
-trichloro-1,3,6-trihydroxy-, and 5-chloro-1,6-dihydroxy-3-
-methoxy-8-methyl- xanthone have been reported for thuringione

arthothelin, and vinetorin, respectively. 7-Chloronorliche-
xanthone has been prepared and erythrommone isolated from
Haematomma erythromma shown to be 3,6-diacetoxy-2,4,5-
-trichloronorlichexanthone (5) (Huneck and G. Hoefle,
Tetrahedron, 1978, 34, 2491).

(5)

The ^{13}C-nmr spectral data of norlichexanthone, including
substituent effects of chlorination and *O*-methylation, have
allowed the interpretation of the spectra of a number of
lichen xanthones (Sundholm, Acta Chem. Scand., 1978, B32,
177).

The total synthesis of eight chlorine-containing lichen
xanthones by ring closure of appropriately substituted
benzophenones, 5-chloro-8-methyl-1,3,6-trihydroxy-, m.p.
304-305°, 5-chloro-1,6-dihydroxy-3-methoxy-8-methyl-
-(vinetorin), m.p. 254-255°, 2,5-dichloro-3,6-dimethoxy-
-1-hydroxy-8-methyl-, m.p. 314-315°, *acetate*, m.p. 248-
250°, 4,5-dichloro-1,6-dihydroxy-3-methoxy-8-methyl-,
m.p. 255-256°, *di-O-methyl ether*, m.p. 254-256°, 4,5-
-dichloro-8-methyl-1,3,6-trihydroxy-, m.p. 292-294°
(slight decomp.), 4,5-dichloro-3,6-dimethoxy-1-hydroxy-8-
-methyl-, m.p. 285-286.5°, *acetate*, m.p. 235-237°, 1,6-
-dihydroxy-3-methoxy-8-methyl-2,4,5-trichloro-(thuringione),
m.p. 278-280°, and 8-methyl-2,4,5-trichloro-1,3,6-
-trihydroxy-(arthothelin)xanthones, m.p. 283-285°, and
related compounds, 5-chloro-1,3-dimethoxy-6-hydroxy-8-
-methyl-, m.p. 325-327° (decomp.), *acetate*, m.p. 240-241°
with sublimation from 220°, 2,5-dichloro-8-methyl-1,3,6-
-trimethoxy-, m.p. 208-209°, 8-methyl-2,4,5-trichloro-
-1,3,6-trimethoxy-, m.p. 212-214°, 3,6-dimethoxy-1-hydroxy-
-8-methyl-2,4,5-trichloro-, m.p. 225-227°, *acetate*, m.p.

200-202°, and 6-acetoxy-1,3-dimethoxy-8-methyl-2,4,5-
-trichloro-xanthones, m.p. 180-183°, has been described
(L. Fitzpatrick, T. Sala, and M.V. Sargent, J. chem. Soc.,
Perkin I, 1980, 85).

The following xanthones, some of them new, have been
isolated, but have only been recorded by their chemical
names. Four B-ring *O*-free trioxygenated xanthones, 2,3-
-dimethoxy-1-hydroxy-, 1,2,3-trimethoxy-, 1-hydroxy-2,3-
-methylenedioxy-, and 1-methoxy-2,3-methylenedioxy-xanthone.
Two B-ring *O*-free glucosyloxyxanthones, 1-glucosyloxy-2-
-hydroxy-3-methoxy- and 1-glucosyloxy-2,3-methylenedioxy-
xanthones, and a pentaoxygenated xanthone, 2,3,6,7-
-di(methylenedioxy)-1-methoxyxanthone (6) have been isolated
from the flowering top of *Polygala triphylla* (S. Ghosal,
P.C. Basumatari, and S. Banerjee, Phytochem., 1981, 20,
489); 3,5-dihydroxy-1-methoxyxanthone and its 3-*O*-
-rutinosyl derivative from the aerial parts of *Canscora
decussata* (Ghosal *et al.*, *ibid.*, 1976, 15, 1041);

(6)

1-hydroxy-3,7,8-trimethoxy-, 1,7-dihydroxy-3,8-dimethoxy-,
1,8-dihydroxy-3,7-dimethoxy-, 3-methoxy-1,7,8-trihydroxy-,
3,8-dimethoxy-7-hydroxy-1-primeveroside-, 1-primeveroside-
-3,7,8-trimethoxy-, 3,7-dimethoxy-1-hydroxy-, 1,7-dihydroxy-
-3-methoxy-, 1,3-dihydroxy-7-methoxy-, and 1,3,7-trihydroxy-
-xanthones from the flowers of *Gentiana ciliata* L., a
chromatographic method has been devised, which distinguishes
1,7-dihydroxy-3-methoxyxanthone from the isomeric 1,3-
-dihydroxy-7-methoxyxanthone (M. Massias, J. Carbonnier, and
D. Molbo, Bull. Mus. Natl. Hist. Nat., Sci. Phys.-Chim.,
1976, 10, 45); 1,2,3,4,6,7-hexamethoxyxanthone and the
benzyldihydrochalcone, uvaretin, from the roots of *Uvaria*

Kirkii (B. Tammami *et al*., Phytochem., 1977, <u>16</u>, 2040);
3,7-dimethoxy-1-hydroxy-, 1-hydroxy-3,7,8-trimethoxy-, 1-
-hydroxy-4,6,8-trimethoxy-, 1,8-dihydroxy-3,7-dimethoxy-,
and 1,8-dihydroxy-3,5-diurethoxy-xanthone from
Lomathogonium carinthiacum (T. Sorig, L. Toth, and
G. Bujtas, Pharmazie, 1977, <u>32</u>, 803); 3,6-dihydroxy-1,5,7-
-trimethoxy- and 1,6-dihydroxy-3,7-dimethoxy-5-glucosyloxy-
-xanthone from the flowering tops of *Canscora decussata*,
also confirmed the structure of 6,7-dimethoxy-1,3,5-
-trihydroxyxanthone (Ghosal and K. Biswas, Phytochem.,
1979, <u>18</u>, 1029); 2-hydroxy-, 2-methoxy-, 4-hydroxy-, 1-
-hydroxy-7-methoxy-, 3-hydroxy-2-methoxy-, 1,5-dihydroxy-,
and 1,7-dihydroxy-xanthones, and a new naturally occurring
xanthone, 2,6-dihydroxyxanthone from the timber of *Mammea
acuminata* (W.M. Bandaranayake *et al*., Indian J. Chem.,
1980, <u>19B</u>, 463); 1-hydroxy-3,7,8-trimethoxyxanthone,
gentiacaulein and other compounds from *Gentianopsis paludosa*
(B.-C. Zang *et al*., Chung Ts'ao Yao, 1980, <u>11</u>, 149); 3,7,8-
-trihydroxyxanthone-1-*O*-β-laminaribioside from the fern
Asplenium adiantum-nigrum (F. Imperato, Phytochem., 1980,
<u>19</u>, 2030); the unusual xanthone, 2-(3-methylbut-2-enyl)-
-1,3,5-trihydroxyxanthone (7) along with other xanthones from
the heartwood of *Calophyllum tomentosum* (S. Karunanayake,
S. Sotheeswaran, and M.U.S. Sultanbawa, Phytochem., 1981,
<u>20</u>, 1303); 1-hydroxy-3,5-dimethoxy-, 1-hydroxy-3,7-
-dimethoxy-, 1-hydroxy-2,3,5-trimethoxy-, 1-hydroxy-3,4,7-
-trimethoxy-, 1-hydroxy-2,3,4,5-tetramethoxy-, 1-hydroxy-
-2,3,4,7-tetramethoxy-, and 1,8-dihydroxy-3,5-dimethoxy-
-xanthones from the roots of *Frasera albomarginata*
(Gentianaceae); and 1-hydroxy-2,3,5-trimethoxy-, 1-hydroxy-
-2,3,4,5-tetramethoxy-, 1-hydroxy-2,3,4,7-tetramethoxy-,
1,3-dihydroxy-4,5-dimethoxy-, 1,7-dihydroxy-2,3-dimethoxy-,
and 1,7-dihydroxy-2,3,4-trimethoxy-xanthones from the roots
of *Frasera speciosa* (D.L. Dreyer and J.H. Bourell, *ibid*.,
p.493).

(7)

The new glucoxanthones, 7-β-D-glucopyranosyl-3-methoxy-
-1,5,6-trihydroxy-, 2-β-D-glucopyranosyl-1-hydroxy-3,5,6-
-trimethoxy-, and 2-β-D-glucopyranosyl-1-hydroxy-7-methyl-
-3,5,6-trimethoxy-xanthones along with other xanthones,
flavones, and triterpenes have been isolated from *Hoppea*
dichotoma (Gentianaceae) (Ghosal, D.K. Jaiswal, and Biswas,
ibid., 1978, **17**, 2119). It has been found the whole plant
of *Swertia angustifolia* collected at different stages of
growth contains fourteen tetraoxygenated, five
pentaoxygenated xanthones, and xanthone 1-O-glucosides.
They are broadly based on 1,3,5,8- and 1,3,7,8-oxygenated
systems, with an added O function at C-4 in some compounds,
and represent a number of methylation patterns (Ghosal,
P.V. Sharma, and Jaiswal, J. pharm. Sci., 1978, **67**, 55).
The anti-Mycobacterium tuberculosis activity of some
naturally occurring xanthones from *Canscora decussata*
Schult and *Swertia purpurascens* Wall and some synthetic
analogues has been investigated (Ghosal, Biswas, and
R.K. Chaudhuri, *ibid.*, p.721; Massias, Carbonnier, and
Molbo, Bull. Mus. Natl. Hist. Nat., Sci. Phys.-Chim., 1977,
13, 55).
The following xanthones have been found in the flowers of
five species belonging to the genus *Gentianella*; 1,8-
-dihydroxy-3,7-dimethoxyxanthone in *G. tenella*; 1-hydroxy-
-3,7,8-trimethoxyxanthone in *G. tenella, G. campestirs*, and
G. ramosa; 4,7-dimethoxy-1,3,8-trihydroxyxanthone in *G.
bellidifolia*; 1,3,5,8-tetrahydroxyxanthone and the
corresponding 8-glucoside in *G. campestris, G. ramosa, G.
bellidifolia*, and *G. germanica*; 5-methoxy-1,3,8-trihydroxy-
xanthone and its 8-glucoside in *G. tenella, G. campestris*,

G. ramosa, *G. Bellidifolia*, and *G. germanica*; 5-methoxy-
-1,3,8-trihydroxy- and 1,8-dihydroxy-3,5-dimethoxyxanthones
in *G. campestris*, *G. ramosa*, and *G. bellidifolia*; 4,5-
-dimethoxy-1,3,8-trihydroxyxanthone in *G. campestris*, *G.*
ramosa, *G. bellidifolia*, and *G. germanica*; and its 1-
-glucoside in *G. campestris*, *G. ramosa*, and *G. germanica*
(*idem, ibid.*, p.23).

2-Hydroxy-5-methoxy- and 2,5-dimethoxy-3-hydroxy-
-xanthones have been isolated from the roots of *Hypericum*
androsaemum and represent oxygenation patterns not
previously found in nature. Another novel compound, 8-
-prenyl-1,3,6,7-tetrahydroxyxanthone, along with 3-hydroxy-
-2-methoxy-, 3-methoxy-1,5,6-trihydroxy-, 1,3,5,6-tetra-
hydroxy-, and 1,3,6,7-tetrahydroxy-xanthones have also been
obtained (H. Nielsen and P. Arends, J. nat. Prods., 1979,
42, 303).

A number of known naturally occurring xanthones have been
isolated from new sources and the following are some
examples. Euxanthone and the biflavanoids, volkensiflavone
and morelloflavone from the heartwood of *Garcinia indica*
(P.J. Cotterill, F. Scheinmann, and G.S. Puranik,
Phytochem., 1977, **16**, 148); gentioside, isogentisin, and
mangiferin from leaves of *Gentiana X marcailhouana* Ry.
(Luong Minh Duc, P. Fombasso, and A. Jacot-Guillarmod,
Helv., 1980, **63**, 244); decussatin from the bark of
Anthocleista vogelli (D.A. Okorie, Phytochem., 1976, **15**,
1799); 1-glucosyldecussatin, 1-glucosylgentiacaulein, 8-
-primeverosylisogentiacaulein, 3-glucosylisogentiacaulein,
1-glucosylswertianin, and 7-primeverosylswertianin from
leaves of *Gentiana ciliata* L. (M. Goetz, F. Maniliho, and
Jacot-Guillarmod, Helv., 1978, **61**, 1549); decussatin,
gentiakochianin, swertiaperennin, and gentiacaulein from
the aerial parts of *Gentiana barbata* (G.G. Nikolaeva *et al.*,
Khim. Prir. Soedin., 1980, 255); bellidifolin from the
aerial parts of *Swertia angustifolia* and *S. paniculate*
(M.I. Khan, N. Ahmed, and M.H. Haqqani, Planta Med., 1977,
32, 280); mangiferin from *Gentiana ariasnensis* (C -H. Chang
and H -C. Yen, T'ai-wan Yao Hsueh Tsa Chih, 1975, **27**, 38),
Gentiana asclepiadea (Goetz, K. Hostettmann, and
Jacot-Guillarmod, Phytochem., 1976, **15**, 2014), and the
aerial parts of *Gentiana schistocalux* (G.G. Nikolaeva *et al.*,
Khim. Prir. Soedin., 1980, 833); and maniferin *O*-glucoside
from the leaves of *Rigella inusitate* and *R. immaculata*
(R.E. Ballard and R.W. Cruden, Biochem. syst. Ecol., 1978,
6, 139).

Jacareubin, 6-deoxyjacareubin, 1,5-dihydroxy-6-(3-
-methylbut-2-enyl)xanthone, and 1,7-dihydroxy-3,6-
-dimethoxyxanthone along with other products have been
isolated from the timber of *Calophyllum inophyllum*
(V. Kumar, S. Ramachandran, and Sultanbawa, Phytochem.,
1976, <u>15</u>, 2016). The xanthones obtained from *Calophyllum
inophyllum* and *Mesua Ferrea* have been screened for various
pharmacological effects (C. Gopalakrishnan *et al.*, Indian
J. Pharmacol., 1980, <u>12</u>, 181). Maclurin, 1,5-dihydroxy-
xanthone, 1,7-dihydroxyxanthone, and other compounds have
been isolated from the fruit of *Garcinia xanthochymus*
(R.K. Baslas and P. Kumar, Curr. Sci., 1979, <u>48</u>, 814).

Athyriol, *3-methoxy-1,6,7-trihydroxyxanthone* has been
prepared by the selective demethylation of 6,7-dihydroxy-
-1,3-dimethoxyxanthone, obtained by the condensation of
2,4,5-trihydroxybenzoic acid with phloroglucinol dimethyl
ether (D.K. Bhardwaj, S.C. Jain, and R. Singh, Indian J.
Chem., 1978, <u>16B</u>, 150). The synthesis of hydroxyxanthone
3-O-β-glycosides by the condensation of the appropriate
aglycone with acetyl-α-D-glycosyl bromide in pyridine in
the presence of silver carbonate, including that of
gentioside, *1-hydroxy-7-methoxy-3-0-β-primeverosylxanthone*,
has been reported (V.M. Chari, R. Klapfenberger, and
H. Wagner, Z. Naturforsch., Anorg. Chem., Org. Chem., 1978,
<u>33B</u>, 946). For the synthesis of xanthone 1-O-β-glycosides
see Chari *et al.* (Helv., 1979, <u>62</u>, 678).

A convenient one-step synthesis of dihydropyranoxanthones
involves condensation of hydroxyxanthones with 2-methylbuta-
-1,3-diene (isoprene) in the presence of orthophosphoric
acid, for example, 1-hydroxyxanthone and isoprene affords
3,4-dihydro-2,2-dimethyl-2H,12H-pyrano[2,3-a]xanthen-12-one
(8), m.p. 119-120°, which on dehydrogenation with
dichlorodicyanobenzoquinone yields 2,2-dimethyl-2H,12H-
-pyrano[2,3-a]xanthen-12-one (9), m.p. 116-117°. Some
condensations furnish mixtures of products.

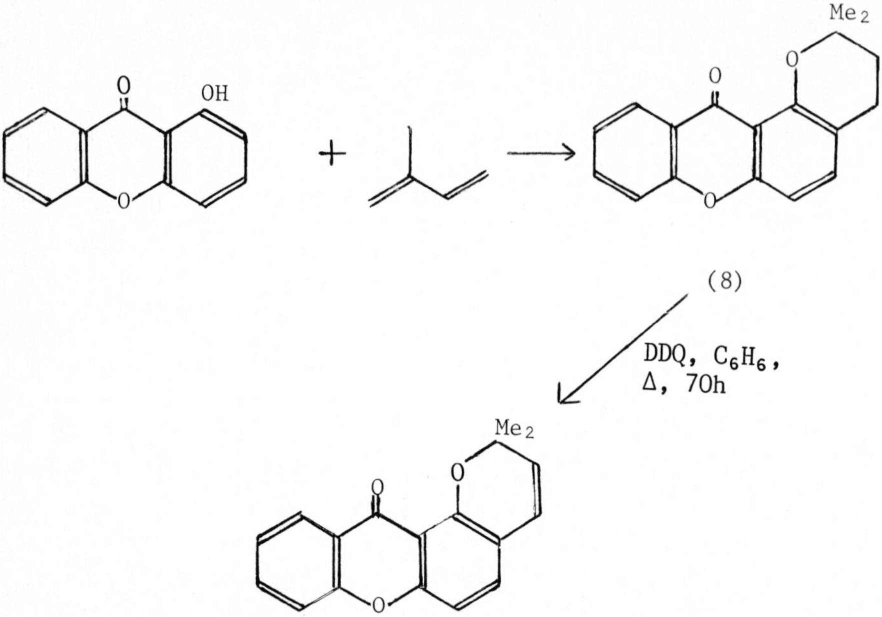

(8)

DDQ, C_6H_6,
Δ, 70h

(9)

2,7-Desoxyosajaxanthone, 5-O-methyl-6-dexoyjacareubin, and
7-O-methylosajaxanthone are obtained by dehydrogenation of
the corresponding dihydro derivatives with DDQ. Some
dihydropyranoxanthones fail to dehydrogenate, but their
methyl ethers undergo facile dehydrogenation with DDQ to
give the corresponding pyranoxanthones. 3,4,7,8-Tetrahydro-
-2,2,6,6-tetramethyl-2H,6H,14H-dipyrano[2,3-a:2',3'-c]-
xanthen-14-one (10), m.p. 158-159°, 3,4,7,8-tetrahydro-10-
-methoxy-2,2,6,6-tetramethyl-2H,6H,14H-dipyrano[2,3-a:2',3'-
-c]xanthen-14-one (11), m.p. 180-181°, and 3,4,7,8-tetra-
hydro-12-methoxy-2,2,6,6-tetramethyl-2H,6H,14H-dipyrano-
[2,3-a:2',3'-c]xanthen-14-one (12), m.p. 228-229°, have
been obtained by this method and the preparation of a number
of dihydropyranoxanthones and pyranoxanthone has been
reported (V.K. Ahlawalia, R.S. Jolly, and A.K. Tehim, J.
chem. Soc., Perkin I, 1983, 1229).

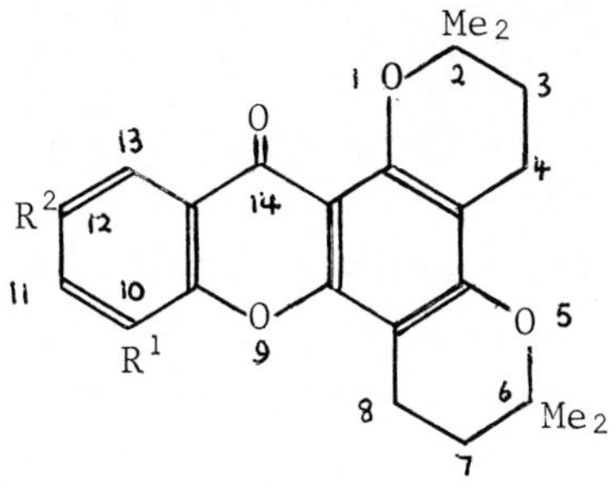

(10) $R^1 = R^2 = H$
(11) $R^1 = OMe, R^2 = H$
(12) $R^1 = H, R^2 = OMe$

Toxyloxanthone B isolated from the stem bark of *Toxylon pomiferum* Raffin, *Maclura pomifera* Schneid, and *M. avrantiaca* Nutt (Osage Orange) has now been shown to have structure (13). This follows from a reconsideration of the [1]H-nmr spectrum of its trimethyl ether, *3,3-dimethyl-5,9,11--trimethoxy-3H,12H-pyrano[3,2-a]xanthen-12-one* (15), m.p. 192-193°, and its unambiguous total synthesis. 5,9-Dimethoxy--3,3-dimethyl-11-hydroxy-3H,12H-pyrano[3,2-a]xanthen-12-one (14), m.p. 227-228° (Cotterill and Scheinmann, *ibid.*, 1980, 2353).

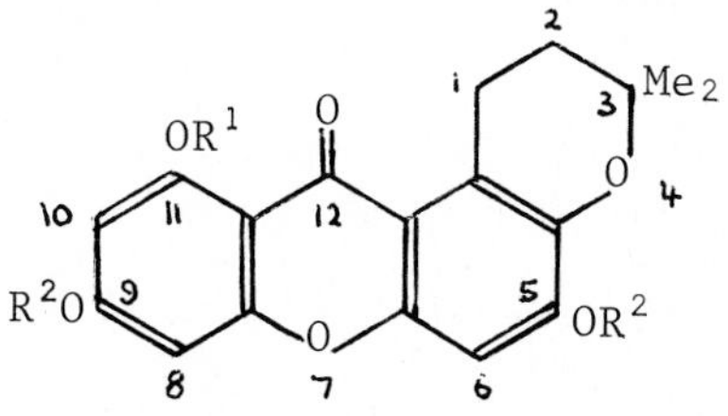

(13) $R^1 = R^2 = H$
(14) $R^1 = H, R^2 = Me$
(15) $R^1 = R^2 = Me$

Furo[3,2-b]xanthone (16) has been prepared from 3-
-hydroxy-4-methylxanthone and furo[2,3-c]xanthone (17) from
3-hydroxyxanthone (K.P. Sanghvi, R.J. Patolia, and
K.N. Trivedi, J. Indian chem. Soc., 1979, 56, 52).

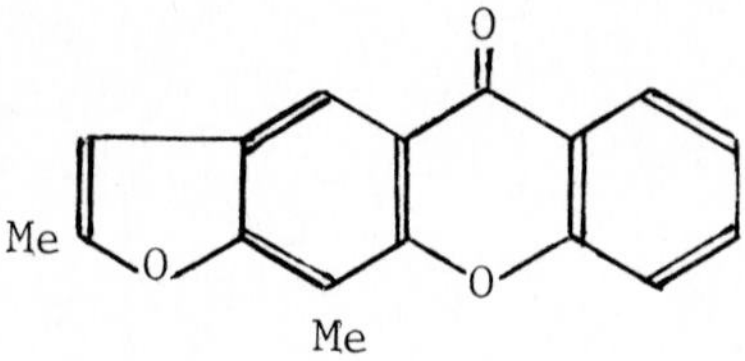

(16) 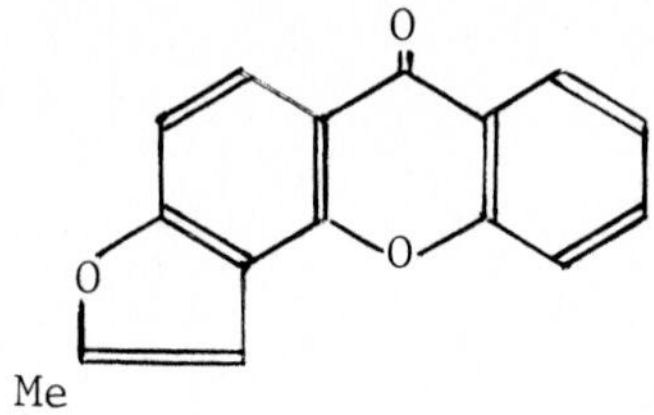   (17)

Two approaches to the synthesis of ustocystin A (18) have
been investigated (C.P. Gorst-Allman, M.J. Nolte, and
P.S. Steyn, S. Afr. J. Chem., 1978, 31, 143).

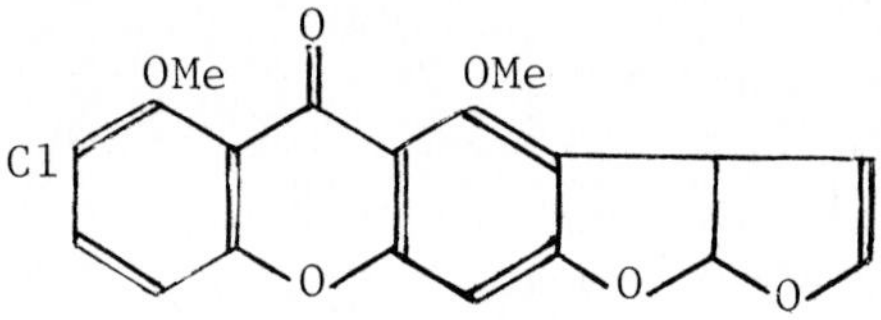

(18)

The dialdehyde, 2,8-bis(2-oxoethyl)-1-hydroxy-3,6,7-
-trimethoxyxanthone (19) obtained by the ozonolysis of 3,6-
-dimethyl-O-mangostin has been prepared from 1-hydroxy-
-3,6,7-trimethoxyxanthone via 2,8-diallyl-1,7-dihydroxy-
-3,6-dimethoxyxanthone. Also reported are the ^{13}C-nmr
spectral data of some 1,3,6,7-tetraoxygenated xanthones
(M.S.B.H. Idris, A. Jefferson, and Scheinmann, J. chem.
Soc., Perkin I, 1977, 2158).

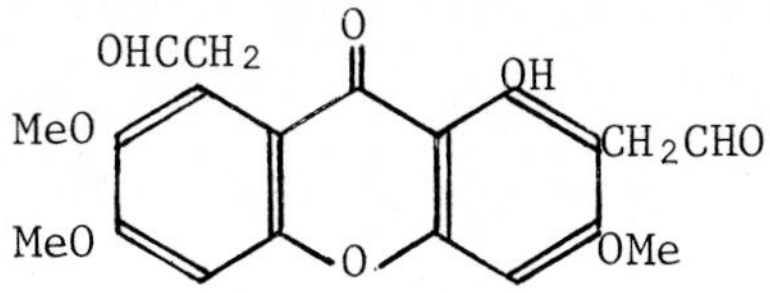

(19)

Secalonic acids D (20) and F (21), $C_{32}H_{30}O_{14}$, m.p. 218-221°
(hot stage), 253-256° (evacuated capillary), $[\alpha]_D^{20}$ + 202°
(pyridine) have been isolated from *Aspergillus aculeatus*
Iizuka grown on white corn (R. Anderson *et al.*, J. org.
Chem., 1977, **42**, 352).

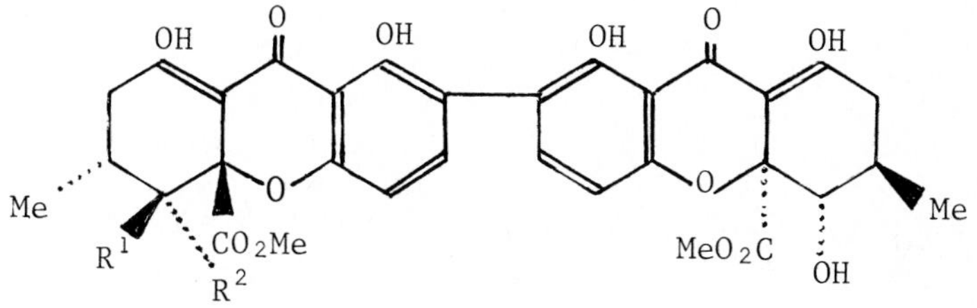

(20) R^1=OH, R^2=H
(21) R^1=H, R^2=OH

Secalonic acids A and E,   secalonic acid G (22), and
other products, have been obtained from five strains of
Pyrenochaeta terrestris. *Aspergillus aculeatus* produces
secalonic acids B, D, and F. From the combinations of
secalonic acids produced in organisms so far examined it

was concluded that precursor tetrahydroxanthone units were formed in pairs differing in stereochemistry only at C-5 or at the *trans*-invariant C-6:C-10a positions. A possible biosynthetic pathway has been discussed (I. Kurobane, L.C. Vining, and A.G. McInnes, J. Antibiot., 1979, <u>32</u>, 1256).

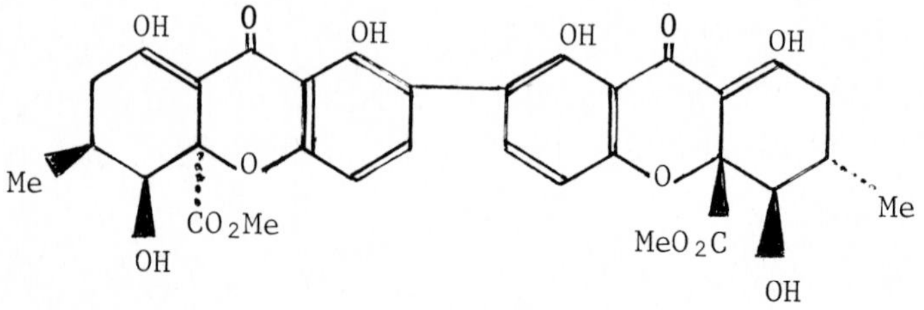

(22)

The biosynthesis of ravenelin has been studied because of a structural relationship to secolonic acid A and sterigmatocystin (23), an aflatoxin precursor (J.G. Hill, T.T. Nakashima, and J.C. Vederas, J. Amer. chem. Soc., 1982, <u>104</u>, 1745).

(23)

The use of high-performance liquid chromatography in the separation of natural xanthone glycosides and of the aglycones gentisin and isogentisin has been developed (M.J. Pettei and K. Hostettmann, J. Chromatogr., 1978, 154, 106). The interaction of lanthanide salts with phenols in dimethyl sulphoxide has been investigated as a tool, for the interpretation of [1]H-nmr spectral data, in the determination of the structure of naturally occurring polyphenolic compounds, including xanthones (D. Davoust *et al.*, Org. mag. Res., 1978, 11, 547). The ^{13}C-nmr spectral data have been determined for thirty six naturally occurring xanthones and all the chemical shifts assigned (P.W. Westerman *et al.*, *ibid.*, 1977, 9, 631), also the data for fourteen naturally occurring polysubstituted xanthones, including aglycones as well as *O*- and C-glycosides have been reported, along with a study in detail of that for synthetic 1,3,5,6-tetramethoxyxanthone (I. Miura, Hostettmann, and K. Nakanishi, Nouv. J. Chim., 1978, 2, 653). For a review of naturally occurring xanthones see Y -B. Yang (Yun-nan Chih Wu Yen Chiu, 1980, 2, 345) and for one on naturally occurring oxygen-ring compounds, including xanthones, R.D.H. Murray (Aromat. heteroaromat, Chem., 1977, 5, 472).

388

8. 6H-Dibenzo[b,d]pyran, 3,4-benzochromene and its derivatives

8,9-Disubstituted 7-hydroxy-6H-dibenzo[b,d]pyran-6-ones (oxaphenanthrenes) (1) have been prepared from appropriate salicylideneacetones and acetoacetates (F. Eiden and P. Gmeiner, Arch. Pharm., 1987, 320, 213).

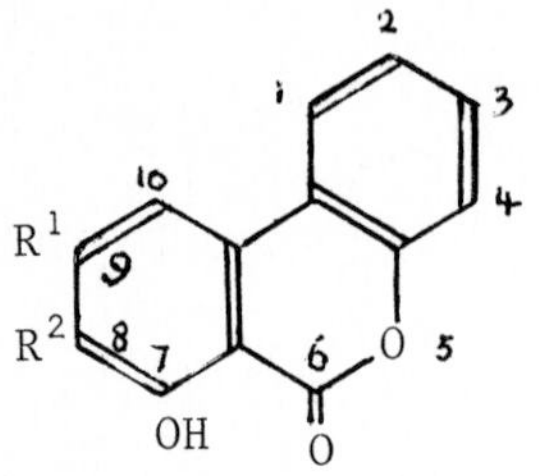

R^1=Me,CH_2CHMe$_2$
R^2=H,Me,Ph,CO_2Et

(1)

Oxidation of 1,3,7-trinitrobenzo[b]indan-5-one (2) with 30% hydrogen peroxide in sulphuric acid affords 3,8,10-trinitro-6H-dibenzo[b,d]pyran-6-one (3), which on heating in dimethyl sulphoxide yields 3,7-dinitrodibenzo[b,d]furan-1-carboxylic acid (4), and in dimethylformamide, 1-dimethylamido-3,7-dinitrodibenzo[b,d]furan (5) (A.M. Andrievskii, A.N. Poplavskii, and K.M. Dyumaev, Khim. Geterotsikl. Soedin., 1982, 703).

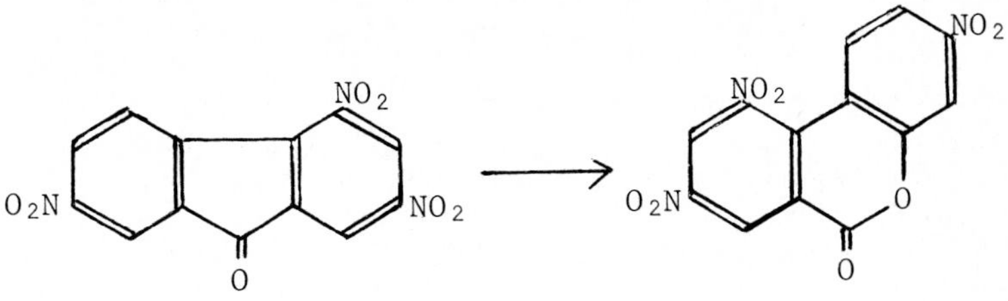

(2)   (3)

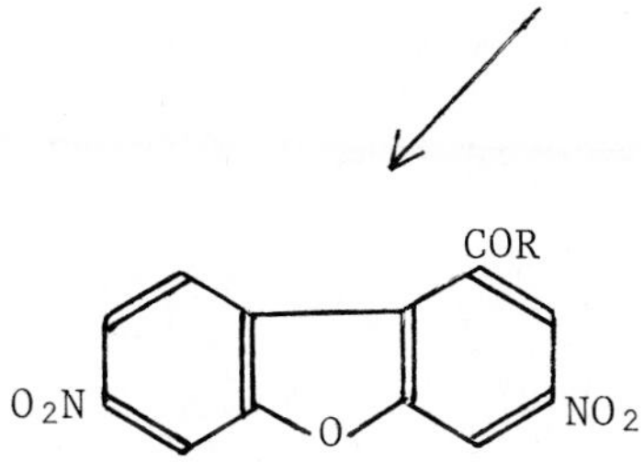

(4) R=OH
(5) R=NMe$_2$

9. Naphthopyrans, benzochromenes and some of their
   derivatives

A number of dihalogenodihydronaphthopyrans have been
obtained by either, the addition of halogen to the related
naphthopyrans, or by replacing the hydroxyl group in the
corresponding halogenohydrins by halogen. The halogenohydrins are obtained from the related halogenoketones.
Some trihalogenodihydronaphtho[1,2-b]pyrans have also been
prepared. Several dihydropyran derivatives have been
obtained directly and indirectly from dihalogenodihydronaphthopyrans. These include halogenomethoxy and halogenohydrins [R. Livingstone, Chemistry of Heterocyclic
Compounds, 1981, 36 (Chromans and Tocopherols), 168, 181].

The naphthoquinone (2) readily available by the Michael reaction between 2-hydroxy-1,4-naphthoquinone (1) and methyl vinyl ketone, is reduced by excess sodium tetrahydridoborate in ethanol, followed by air oxidation to quinone (3), which cyclizes in aqueous sulphuric acid to give 3,4-dihydro-2--methyl-2H-naphtho[2,3-b]pyran-5,10-dione (4), m.p. 121-122°, and in benzene containing boron trifluoride etherate to yield 3,4-dihydro-2-methyl-2H-naphtho[1,2-b]pyran-5,6--dione (5), m.p. 163-164°. 3,4-Dihydro-8-methoxy-2-methyl--2H-naphtho[1,2-b]pyran-5,6-dione, m.p. 171-172°; 3,4--dihydro-8-methoxy-4-methyl-2H-naphtho[1,2-b]pyran-5,6-dione, m.p. 127-128°; 3,4-dihydro-8-methoxy-2-methyl-2H-naphtho-[2,3-b]pyran-5,10-dione, m.p. 163-164°; 3,4-dihydro-8--methoxy-4-methyl-2H-naphtho[2,3-b]pyran-5,10-dione, m.p. 140-141° (R. Cassis, R. Tapia, and J. Valderrama, J. heterocyclic Chem., 1982, 19, 381).

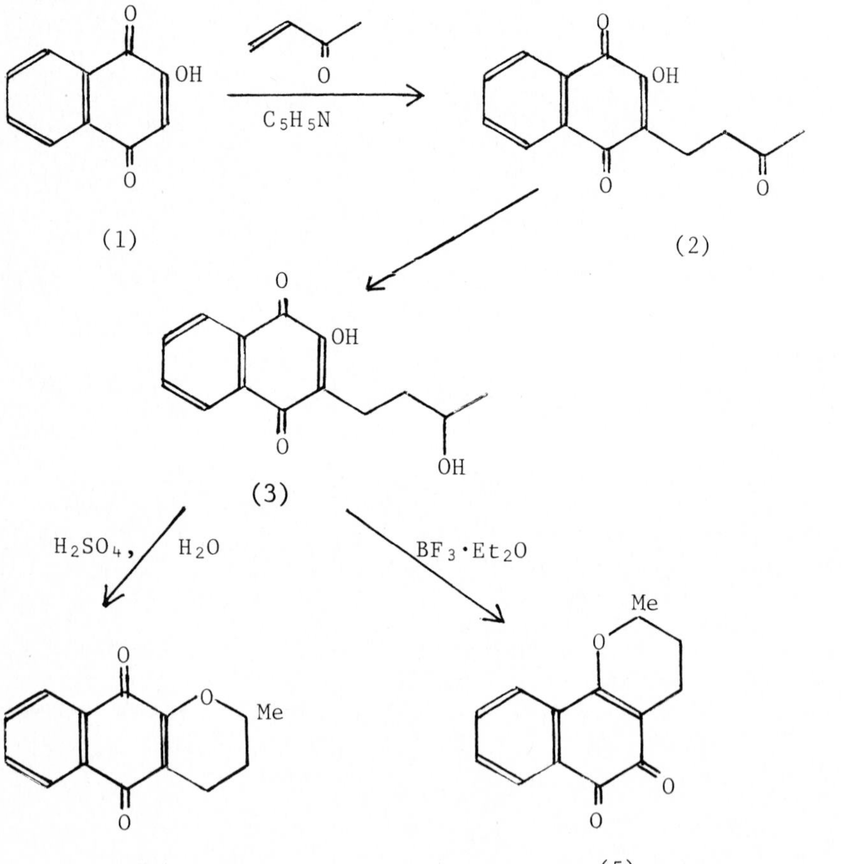

The Grignard reaction between ethyl 3-oxo-3H-naphtho[2,1-b]-
pyran-2-carboxylate (6) and phenylmagnesium bromide
furnishes 2-benzoyl-3-hydroxy-3-phenyl-3H-naphtho[2,1-b]-
pyran (7), whereas with excess phenylmagnesium bromide it
gives 2-(diphenylhydroxymethyl)-3-hydroxy-3-phenyl-3H-
-naphtho[2,1-b]pyran (8) and 1,3-diphenyl-2-(diphenylhydroxymethyl)-1H-naphtho[2,1-b]pyran (9). The reaction of
compound (6) with the appropriate alkylmagnesium bromide
affords derivatives (10) (A.M. Islam *et al*., Indian J.
Chem., 1981, <u>20B</u>, 924).

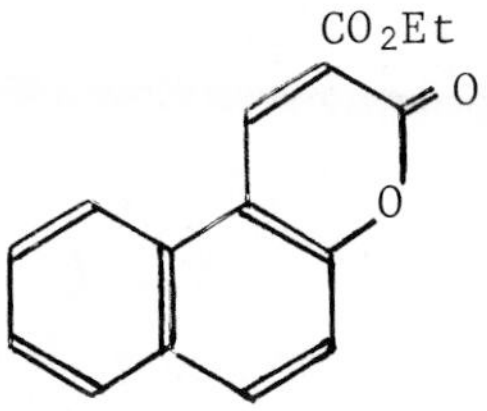

(6)

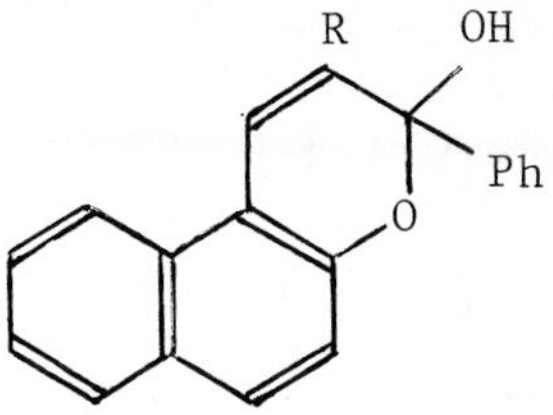

(7) R=COPh
(8) R=Ph$_2$COH

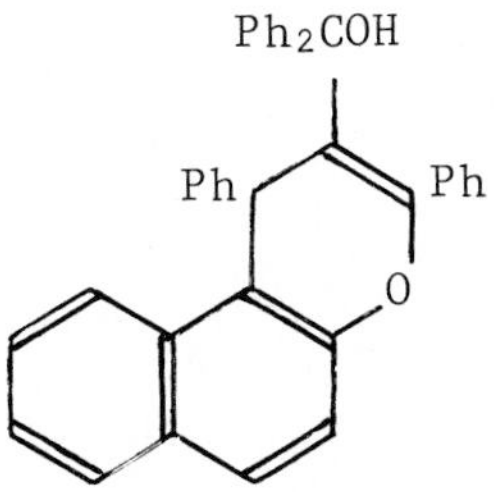

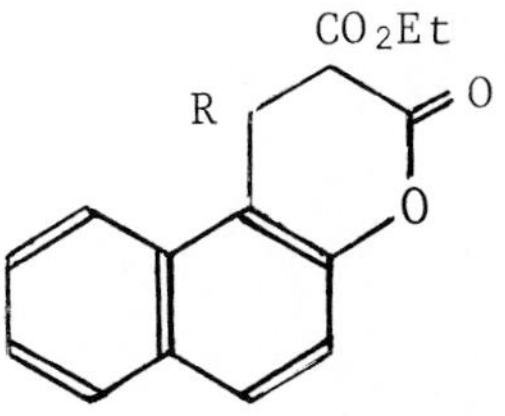

R=Pr,EtCHMe

(9)   (10)

Condensation of ethyl 8-bromo-3-oxo-3H-naptho[2,1-b]-pyran-2-carboxylate (11) with ketones, Me_2CO, MeCOEt, or Et_2CO at 170° in the presence of ammonium acetate or methylamine affords bromotrialkyldihydronaphthopyrano-pyridinediones (12; R^1, R^2, R^3 = H, Me). At room temperature the products are tetrahydronaphthooxazocine-carboxylates (13; R^1, R^2, R^3 = H, Me) (A.H. Bedair *et al.*, Acta Pharm. Jugosl., 1986, **36**, 363).

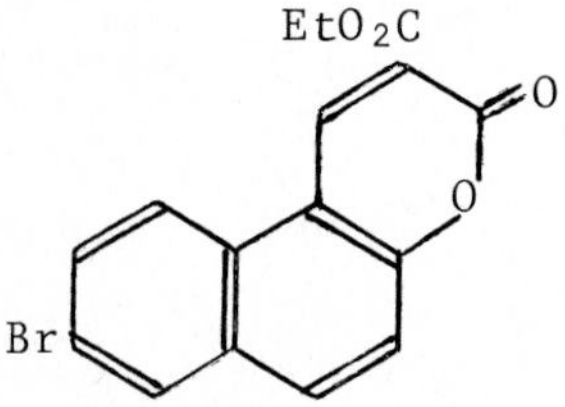

(11)

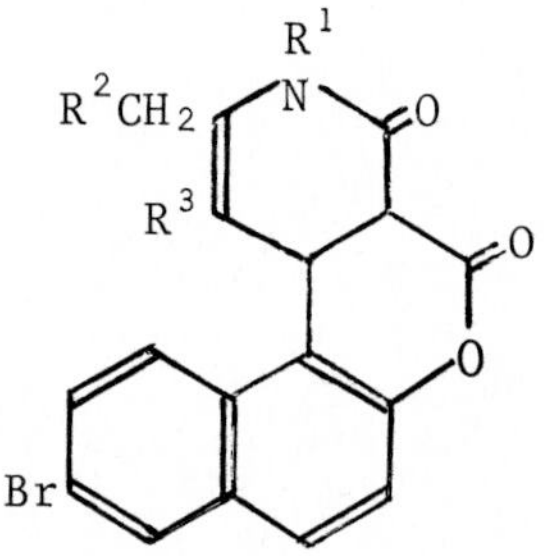

(12)

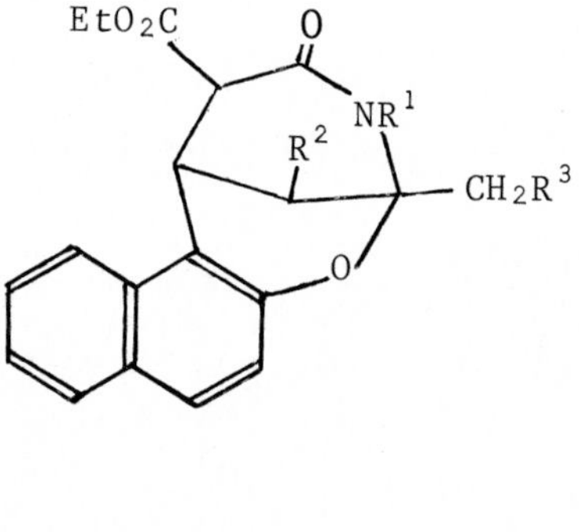

(13)

For the amidation of ethyl 3-oxo-3H-naphtho[2,1-b]pyran-
-2-carboxylate (6) and the preparation of a number of
related derivatives see A.M.S. El-Sharief, *et al*., [Egypt.
J. Chem., 1982 (Pub. 1983), 25, 41] and for some reactions
of 1-(4-bromobenzoyl)-2,3-dihydro-3-1H-naphtho[2,1-b]pyran-
3-one, M. El-Kady, *et al*., (Pol. J. Chem., 1982, 56, 1387).
Cyclocondensation of arylidene-cyanoacetates or
-malononitriles with 2-naphthols in ethanol in the presence
of morpholine affords 3-amino-1-aryl-1H-naphtho[2,1-b]pyrans
(Yu.A. Sharanin and G.V. Klokol, Zh. org. Khim., 1982, 18,
2005).
The reaction of 6-bromo-1H,3H-naphtho[1,8-cd]pyran-1,3-
-dione(4-bromonaphthalic anhydride) (14) with dimethylformamide or dimethylacetamide in the presence of base yields 6-
-dimethylamino-1H,3H-naphtho[1,8-cd]pyran-1,3-dione (15).
Partial demethylation occurs when the reaction is carried
out in pyridine (H. Kazoka and I. Meirovics, Latv. PSR
Zinat. Akad. Vestis, Kim. Ser., 1982, 620).

(14) R=Br
(15) R=NMe$_2$

10. Benzo- and dibenzo-xanthene derivatives

Salicylaldehydato copper and 2-bromo-6-methoxynaphthalene
on heating at 205-208° yields a mixture of 3-methoxybenzo-
[a]xanthen-12-one (1), m.p. 188-190°, and 2-(6-methoxy-2-
-naphthyloxy)benzaldehyde, separable by chromatography. The
latter compound undergoes concomitant nuclear bromination
to afford 4-bromo-3-methoxybenzo[a]xanthen-12-one (2),
m.p. 225-227°, with copper (II) bromide in boiling
nitrobenzene under nitrogen (J.I. Okogun, J. chem. Soc.,
Perkin I, 1976, 2241). The reaction between salicylic acid
and 2-naphthol furnishes benzo[a]xanthen-12-one (3)
(Ng.P. Buu-Hoi and Ng.D. Xuong, J. org. Chem., 1951, 16,
1633).

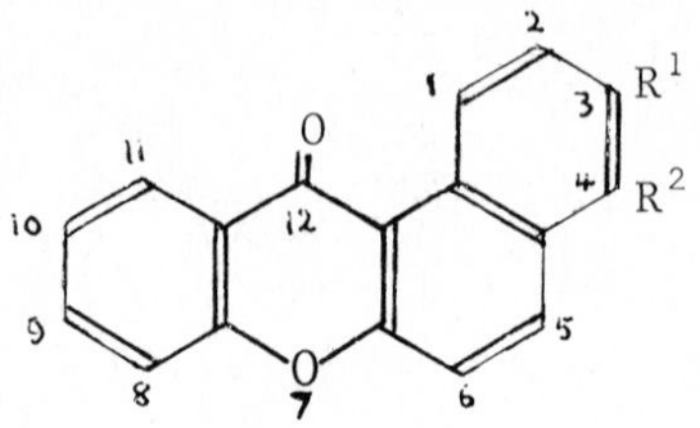

394

(1) R^1=OMe, R^2=H
(2) R^1=OMe, R^2=Br
(3) R^1=R^2=H

Reaction of benzoxanthone (3) with 4-MeC$_6$H$_4$SO$_2$NCO gives derivative (4), which on treatment with butylamine followed by hydrogen sulphide affords benzo[a]xanthen-12-thione (5). Dibenzo[a,i]xanthen-14-one (6) on reaction with Cl$_3$CCONCO gives derivative (7), which on similar treatment with butylamine and hydrogen sulphide yields dibenzo[a,i]xanthen--14-thione (8) (J.A. Van Allan, G.A. Reynolds, and J.F. Stenberg, J heterocyclic Chem., 1979, 16, 1661).

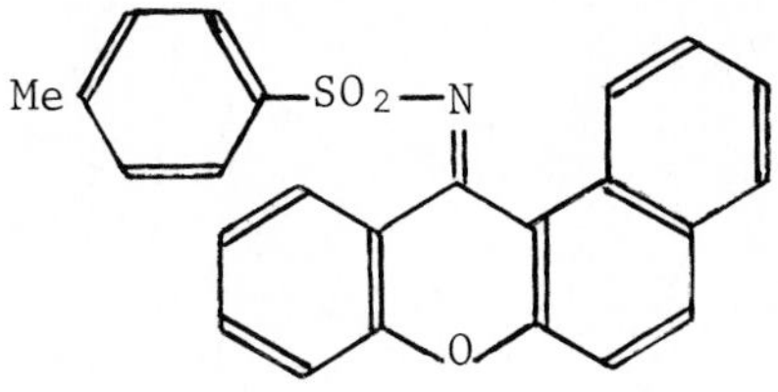

(4)

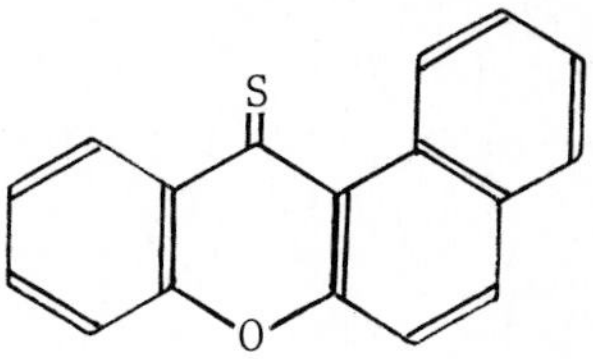

(5)

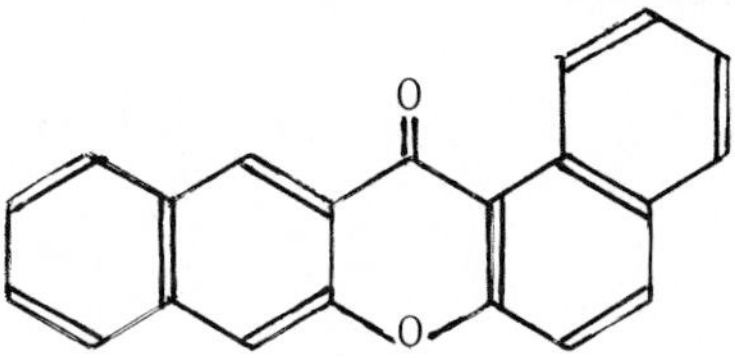

(6)

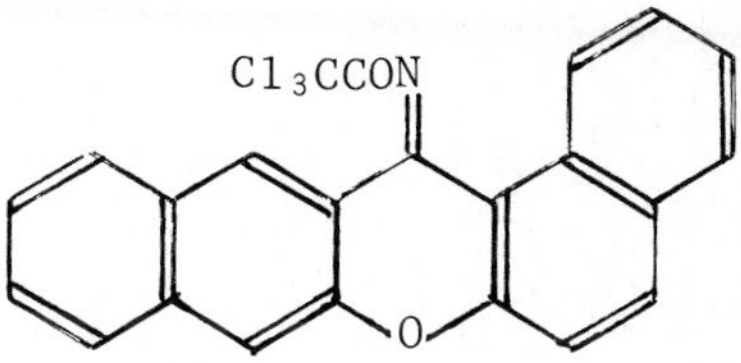

(7)

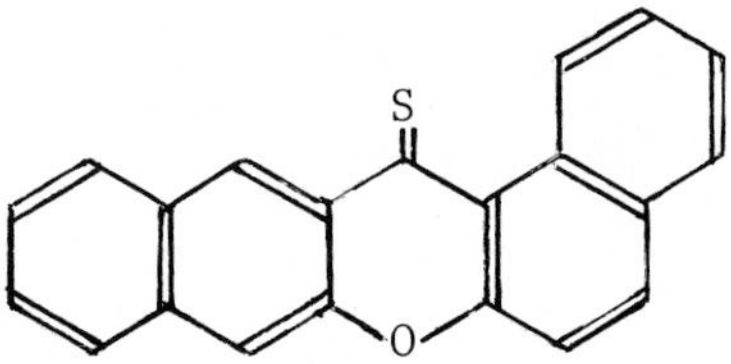

(8)

Bikaverin (9) on methylation with diazomethane gives monomethylbikaverin (10). Reductive acetylation of bikaverin (9), monomethylbikaverin (10) and 7,10-dihydro--3-methoxy-1-methyl-6,8,11-trihydroxybenzo[b]xanthen--7,10,12-trione (11) with acetic anhydride-sodium acetate in

the presence of zinc powder affords the corresponding
polyacetoxybenzoxanthen-12-ones (12, 13, and 14).
Bikaverin (9) on hydrolysis with aqueous sodium hydroxide
yields everninic acid and orcinol, whereas hydrolysis of
monomethylbikaverin (10) with alcoholic potassium hydroxide
furnishes the cyclopentenoxanthone (15) and its ozonolysis
gives xanthone (16) (T. Kato *et al*., Chem. pharm. Bull.,
1978, **26**, 209).

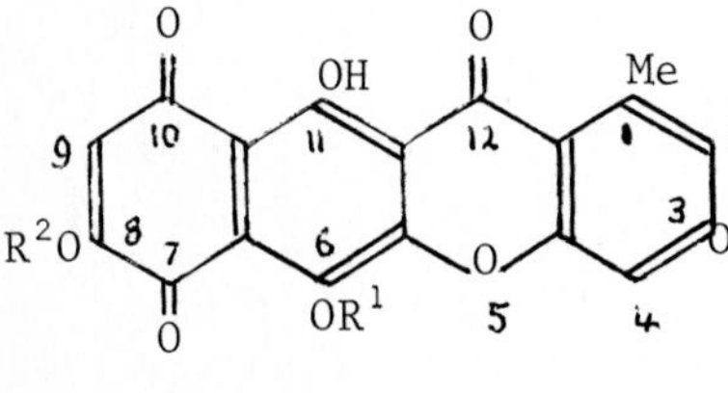

(9) R^1=H, R^2=Me
(10) R^1=R^2=Me
(11) R^1=R^2=H

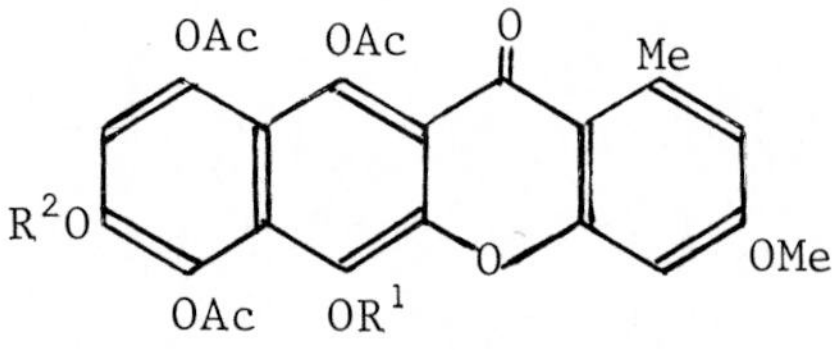

(12) R^1=Ac, R^2=Me
(13) R^1=R^2=Me
(14) R^1=R^2=Ac

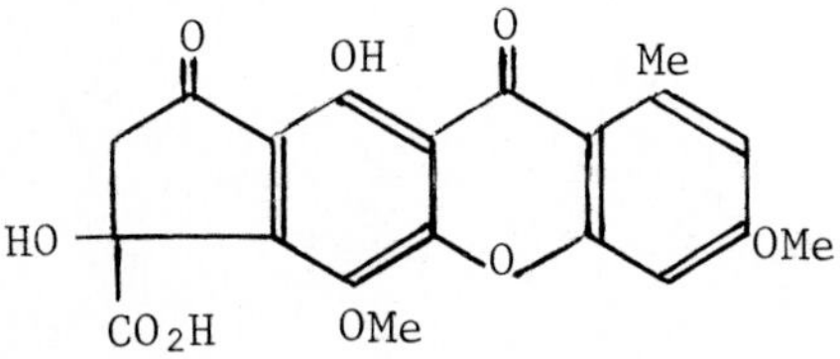

(15)

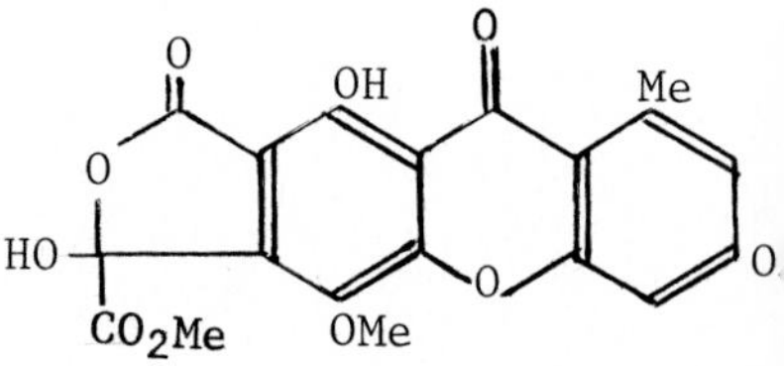

(16)

Xanthocycline, methyl 11-methoxy-7,8,9,10-tetrahydrobenzo-
[b]xanthen-12-one-9-carboxylate (18), a potential anti-tumor
agent with reduced cardiotoxicity has been prepared in six
steps from 2-methylchromone-3-carboxylic acid. Spiro
compound (19) rearranges on treatment with phosphorus
pentoxide in toluene to give the hydroxy ester (17), which
on methylation affords (18) (J.L. Charlton, V.A. Sayeed,
and G.N. Lypka, Canad. J. Chem., 1982, 60, 1996).

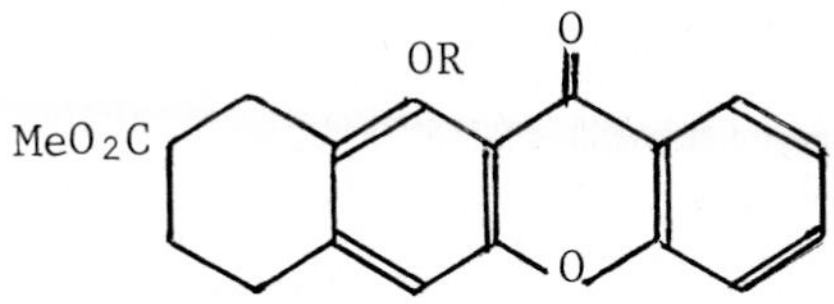

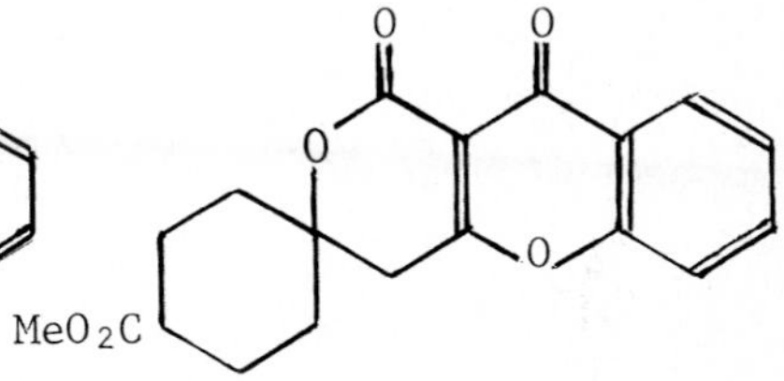

(17) R=H   (19)
(18) R=Me

Chapter 21

SIX-MEMBERED RING COMPOUNDS WITH ONE HETERO ATOM: SULPHUR,
SELENIUM, TELLURIUM, SILICON, GERMANIUM, AND TIN

R. LIVINGSTONE

1. Thiopyran derivatives

(a) Thiopyrans, thiopyrylium salts, and thiopyrones

Recent advances in the chemistry of thiopyrans, thiins, are
mainly concerned with the synthesis, structure, and
properties of their derivatives, with reference, in a
number of instances to their use as pharmaceutical and dye
intermediates.

(i) Thiopyrans
   1*H-Thiopyrans.*   1-Aryl-1*H*-thiopyrans, *1-arylthiabenzenes,
1-aryl-1H-thiins* (1) have been synthesized by condensing
PhC≡CCOPh or 2-acetylcyclohexanone with 4-$RC_6H_4S(O)Me=CH_2$
(R = H, Me, Cl).   An ylide-like bonding structure has been
proposed for the 1-arylthiopyran 1-oxide (1) (H. Matsuyama,
H. Takeuchi, and Y. Okutsu, Heterocycles, 1984, **22**, 1523).

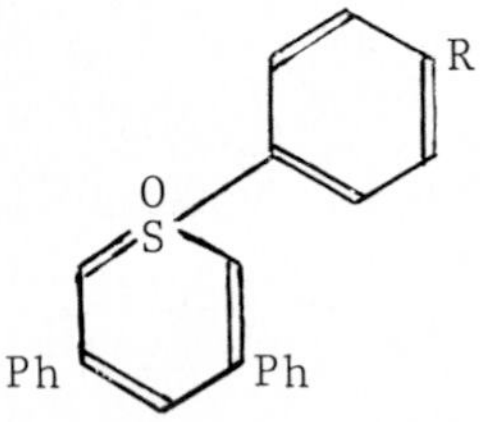

R=H,Me,Cl

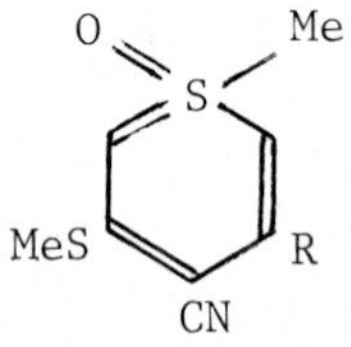

R=Ph, 4-BrC_6H_4, 4-ClC_6H_4,
3,4-$Cl_2C_6H_3$, 2-furyl,
2-thienyl, c-C_6H_{11} , PhCH=CH

(1)   (2)

Cyclization of $(MeS)_2C=C(CN)COR$ with $H_2C=S(O)Me_2$ affords
3-substituted 4-cyano-1-methyl-5-methylthio-1H-thiopyran 1-
-oxides (2) [W.D. Rudorf, Synth., 1984, 852; Rudorf and
U. Peger, Ger. (East) Pat. DD 216,238, 1984]. Some 1-
-(dimethylamino)-1H-thiopyrans 1-oxides, *1-(dimethylamino)-
thiabenzene 1-oxides* (3) have been synthesized and their
bonding properties discussed by comparison of their nmr
(^{1}H- and ^{13}C-) and ir-spectral data with those of the
known 1-methyl-1H-thopyran 1-oxides (4) (L. Weber, Z.
Naturforsch., Anorg. Chem., Org. Chem., 1985, 40B, 578).

	R^1	R^2	
(3)	Ph	Ph	(4)
	Ph	Me	
	Ph	t-Bu	
	t-Bu	t-Bu	

A comparative theoretical analysis has been made of 1H-
-thiopyran and thiacyclobutadiene and their properties have
been rationalized by perturbation MO theory (F. Bernardi *et
al.*, Tetrahedron, 1977, 33, 3061). For further application
of perturbation theory to 1H-thiopyran see W.C. Herndon and
C. Párkányi (*ibid.*, 1978, 34, 3419); for the synthesis and
reactions of 1H-thiopyrans, J Stackhouse (Diss. Abs. Int.
B, 1976, 37, 2258); and for reviews on the synthesis and
reactions of 1H-thiopyrans and related compounds, U. Eisner
(Org. Compd. Sulphur, Sellenium, Tellurium, 1977, 4, 319)
and M. Hori and H. Shimizu (Kagaku no Ryoiki, Zokan, 1980,
147). It has been confirmed that some materials once
identified as stable 1H-thiopyrans are oligmers of unknown
structure (C.C. Price and J. Follweiler, Heterocycles,
1978, 11, 219).

2H-Thiopyrans. 6,6-Dichloro-3-thiabicyclo[3.1.0]hexane 3,3-dioxides (5) on reduction with lithium tetrahydrido-aluminate afford the dechloro (6; R^3 = H) and the monochloro (6; R^3 = Cl) derivatives, several of which, along with some of the dioxides (5) undergo simultaneous elimination of hydrogen chloride and sulphur dioxide. Selective elimination of hydrogen chloride, under basic conditions using lithium diisopropylamide results in the formation of 2*H*-thiopyran 1,1-dioxides, α-*thiopyran 1,1--dioxides, 2H-thiin 1,1-dioxides* (7) (Y. Gaoni, J. org. Chem., 1981, **46**, 4502).

$$R^1 = H, Me; \quad R^1R^2 = (CH_2)_4, (CH_2)_5;$$

$$R^2, R^3 = H, Me, Ph; \quad R^3 = H, Cl$$

Some thiopyrans have been obtained from dihydrothiopyran 1--oxides by the vinylogous Pummer reaction (K. Praefcke and C. Weichsel, Ann., 1979, 784).

The reaction between 2,4,6-triarylthiopyrylium salts and sodium methoxide in methanol furnishes the corresponding 2--methoxy-2,4,6-triaryl-2*H*-thiopyran (T. Zimmermann and G.W. Fischer, J. prakt. Chem., 1986, **328**, 373).

5,6-Diphenyl-3-formyl-2*H*-thiopyran (9) is obtained by the cycloaddition reaction between the dicyclohexylammonium salt (8) of the appropriate monothio-β-dicarbonyl compound and acrolein. The mechanism is considered to involve a Michael addition followed by an aldol reaction (D. Greif, M. Pulst, and M. Weissenfels, Synth., 1987, 456).

$(c-C_6H_{11})_2N^+H_2\overline{S}CCPh=CPhCHO$

(8)

$+$ $\longrightarrow$

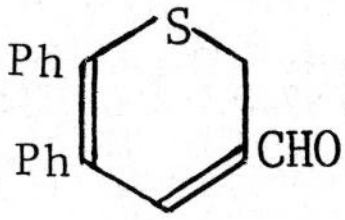

Ph, Ph substituted thiopyran with CHO

86%

$CH_2=CHCHO$

(9)

2-Acetonyl-2*H*-thiopyrans are obtained by treating the necessary thiopyrylium salt with acetone in the presence of the amine salt of a weak acid (e.g. piperidine acetate) (Zimmermann and Fischer, J. prakt. Chem., 1986, __328__, 573). 2-(1-Acetylalkylidene)-2*H*-thiopyrans, for example, 2-(1--acetylethylidene)-5,6-dimethyl-2H-thiopyran (10) and some related compounds have been synthesized (Pulst *et al.*, Z. Chem., 1987, __27__, 35).

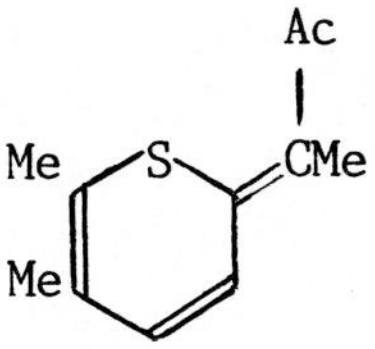

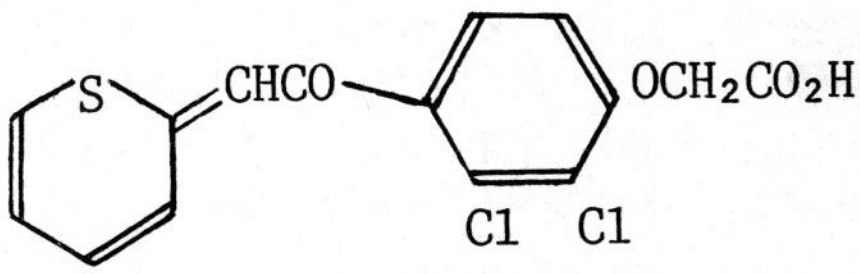

(10)

(11)

Also the thiopyranylidene analogue (11) of tienilic acid
has been prepared (A. Nuhrich *et al.*, Eur. J. med. Chem. -
Chim. Ther., 1986, **21**, 49) and the reaction between 2-
-formylmethylene-2*H*-thiopyrans and amines has been discussed
(Pulst, Weissenfels, and B. Hollborn, Z. Chem., 1984, **24**,
183).

A number of 2-amino-2*H*-thiopyrans (13), useful as dye
intermediates have been obtained from thiopyrylium salts
(12) and appropriate amines [Fischer and Zimmerman, Ger.
(East) Pat. DD 212,964, 1984].

ClO_4^-

$R^6, R^7 = alkyl$

$R^1, R^3, R^5 = aryl$
$R^2, R^4 = H, alkyl, aryl$

(12)   (13)

Treatment of $PhCMe=C(CN)_2$ with PhNCS yields 3-cyano-
-2-imino-4-phenyl-6-phenylamino-2*H*-thiopyran (14;
$R^1 = R^3 = Ph$, $R^2 = H$). Also reported is the preparation of
a number of related compounds (14) (K. Peseke, M. Aguila,
and I. Bohn, *ibid.*, 154,539, 1982) and of some 6-amino-4-
-aryl(2-furyl)-5-cyano-2H-thiopyran-2-ylidenecyanoacetic
esters (15) (Peseke, J.Q. Suarez, and C. Steffens, Z.
Chem., 1983, **23**, 406).

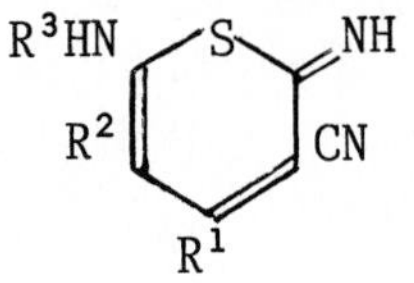

R³HN—S—NH, R², CN, R¹ (structure)

CN, H₂N—S—CCO₂R¹, NC, R² (structure)

R^1=aryl, heteroaryl
R^2=H, alkyl
R^3=aryl

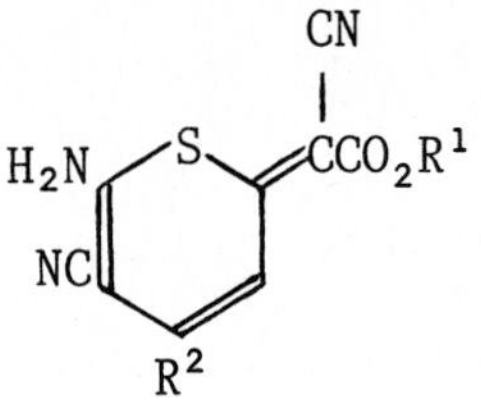

R^1=Me, Et
R^2=Ph, $4-MeOC_6H_4$, $4-ClC_6H_4$,
2-furyl

(14)   (15)

2H-Thiopyran reacts as a diene in the Diels-Alder
reaction with N-phenylmaleimide, maleic anhydride, and
methyl acrylate on heating at 150° in tubes sealed under
vacuum to yield the corresponding adducts (R.H. Fleming and
B.M. Murray, J. org. Chem., 1979, **44**, 2280).

Treatment of 2H-thiopyran with butyllithium and
tetramethylethylenediamine in THF yields 6-lithio-2H-
-thiopyran (16) which readily undergoes the standard
reactions of such compounds (R.H. Everhardus, R. Grafing,
and L. Brandsma, Rec. Trav. chim., 1978, **97**, 69). 6-
-Lithio-2H-thiopyran (16) on treatment with powdered sulphur
followed by bromomethylacetylene yields (17), which on
cyclization affords 2H-thieno[2,3-b]thiopyran (18) (Grafing
and Brandsma, Synth., 1978, 578).

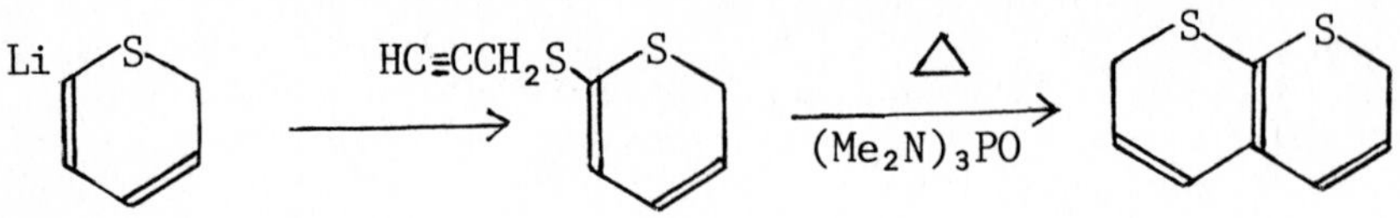

Li
HC≡CCH2S
(Me2N)3PO

(16) (17) (18)

Deprotonation of C-2 or C-6 of 2*H*-thiopyran is specific and depends on the base-solvent system. 6-Lithio-2*H*-thiopyran (16) is converted into the thermodynamically more stable charge-delocalized anion (19) *via* treatment with $(Me_2N)_3PO$ (*idem*, Rec. Trav. chim., 1978, 97, 208). The anion of 2*H*-thiopyran (19) reacts with *tert*-butyl bromide in liquid ammonia to give 2-*tert*-butyl-2*H*-thiopyran (20) and with cyclohexyl bromide to give a mixture of 2- and 4-cyclohexyl-2*H*-thiopyran (21) and (22) (Grafing, H.D. Verkruijsse, and Brandsma, Chem. Comm., 1978, 596).

Reductive cleavage of 2*H*-thiopyran with naphthylsodium, sodium in THF, or THF/HMPT gives the dianion $\bar{S}CH{=}CH{-}CH{-}CH\bar{C}H_2$ (Grafing and Brandsma, Rec. Trav. chim., 1979, 98, 520).

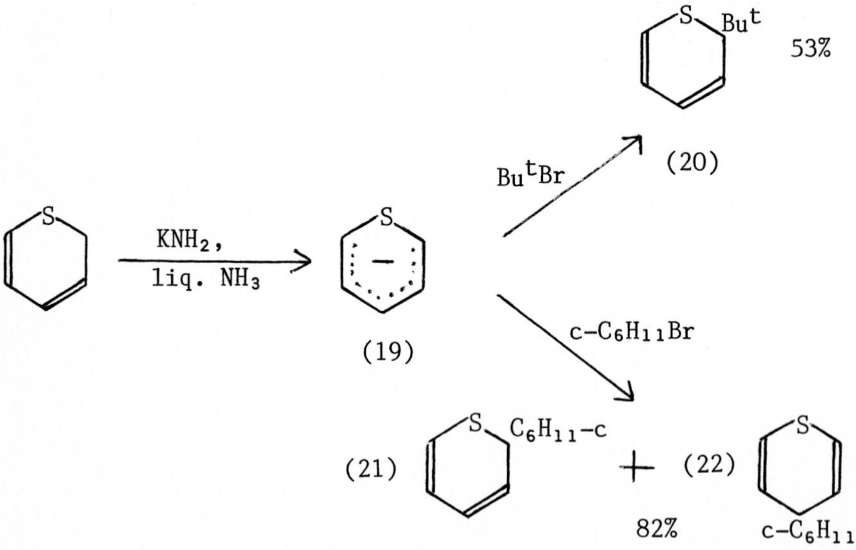

Bu^t
53%
(20)
Bu^tBr
KNH2,
liq. NH3
(19)
c-C6H11Br
(21)
C6H11-c
(22)
82%
c-C6H11

A quantum-chemical study has been made of the hydride
lability of hydrogen in thiopyrans and related compounds
(A.F. Pronin and V.G. Karchenko, Khim. Geterotsikl.
Soedin., 1977, 1206) and an investigation of the stability
of thiopyrans by the Huckel method has been reported. The
delocalizations energy of thiopyrans is increased by Me and
Ph substituents and the discrepancy between the calculated
instability and observed stability of thiopyrans results
from their nonplanar structure (Pronin and P.G. Zhbanov,
Issled. Obl. Sint. Katal. Org. Soedin., 1975, 18).

2-(α-Formylbenzylidene)-2H-thiopyrans (23) condense with
compounds containing an activated methylene group, (e.g.
$R^1CH_2R^2$ with R^1 = CN, R^2 = CN, CO_2Et, $4-O_2NC_6H_4$) to give
the corresponding unsaturated compounds (24). Similarly
with N nucleophilic reagents RNH_2 (R = Ph, $4-O_2NC_6H_4$, OH,
$NHCONH_2$) compounds (25) are formed (Pulst et al., Z. Chem.,
1983, $\underline{23}$, 147). For the preparation of several substituted
2-formylmethylene- and 2-(α-formylbenzylidene)-2H-thiopyrans
see $idem$ ($ibid$., p.146) and for some 2H-thiopyran-2-ylidene
thioketones, W. Hoederath and K. Hartke (Arch. Pharm., 1984,
$\underline{317}$, 938).

(23)

(24) Z = :CR^1R^2
(25) Z = :NR

The behaviour of 4-(trimethylsiloxy)-2*H*-thiopyran (26) as
an enophile is prone to yield Michael rather than Diels-
-Alder adducts as the dienophile becomes more reactive
(T. Harayama and M. Otsuyama, Yakugaku Zasshi, 1983, <u>103</u>,
1096). New approaches to the synthesis of 2*H*-thiopyrans
and related heterocycles have been reviewed (H. Khalil,
Diss. Abs. Int. B, 1978, <u>38</u>, 4813). For a review of 2*H*-
-thiopyrans and related compounds see U. Eisner (Org. Compd.
Sulphur, Sellenium, Tellurium, 1979, <u>5</u>, 324).

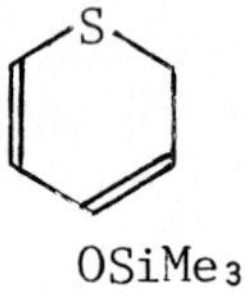

OSiMe₃

(26)

4H-Thiopyrans. Treatment of 2,4,6-tri-*tert*-butyl-4*H*-
-pyran with a mixture of hydrogen sulphide and hydrogen
chloride furnishes 2,4,6-tri-*tert*-butyl -4*H*-thiopyran, -γ-
-*thiopyran, -4H-thiin* (27) (S. Bohm *et al.*, Coll. Czech.
chem. Comm., 1987, <u>52</u>, 1305).

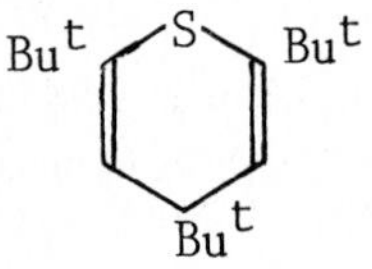

(27)

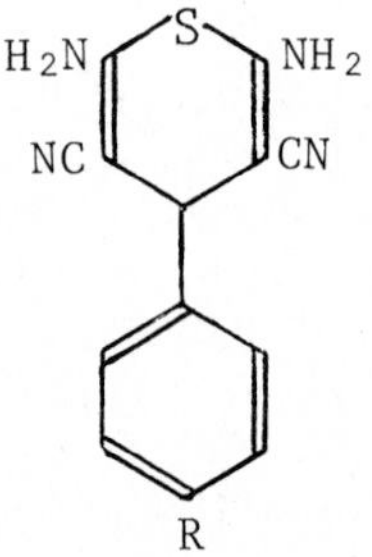

R = H, OMe, Cl

(28)

4-Aryl-2,6-diamino-3,5-dicyano-4H-thiopyrans (28) are obtained when 4-RC$_6$H$_4$CH=C(CN)$_2$ reacts with NCCH$_2$CSNH$_2$ in the presence of triethylamine in ethanol (G.E.H. Elgemeie *et al.*, Z. Naturforsch., Anorg. Chem., Org. Chem., 1986, **41B**, 781). Vinamidinium iodides (29) on treatment with methyl thioacetate in the presence of diisopropylethylamine in dichloromethane afford methyl cyclopenta[c]thiopyrancarboxylates (30) (T. Kaempchen, G. Moddelmog, and G. Seitz, Synth., 1984, 262).

R^1, R^3 = H, Cl, Bu^t
R^2 = H, Cl, CHO, Bu^t

(29)   (30)

A number of 4-[(un)substituted amino]-6-(4-nitrophenyl)--4H-thiopyran-3-carboxylic and -2,3-dicarboxylic acids and some of their esters and ethyl 2-[(un)substituted amino]-6--(4-nitrophenyl)-2H-thiopyran-3-carboxylates have been synthesized (J.B. Rasmussen, R. Shabana and S.O. Lawesson, Tetrahedron, 1982, **38**, 1705). The preparations of some 4H--thiopyran-3,5-dicarboxylic acids and their esters have been reported (S. Goldmann, G. Thomas and M. Schramm, Ger. Offen. DE 3,212,737, 1983).

The cyclization of 1,5-diketone (31) with hydrogen sulphide and acid (e.g. HClO$_4$ or F$_3$CCO$_2$H) yields mixtures of the thiopyran (32) and the thiopyrylium salt (33) (N.I. Kozhevnikova, Nukleofil'nye Reakts. Karbonil'nykh Soedin., 1982, 83). The preparation of other thiopyrans and thiopyrylium salts by the treatment of 1,5-diphenylpentane--1,5-diones with hydrogen sulphide and acid have been reported (A.L. Mikhailova *et al.*, *ibid.*, p.64).

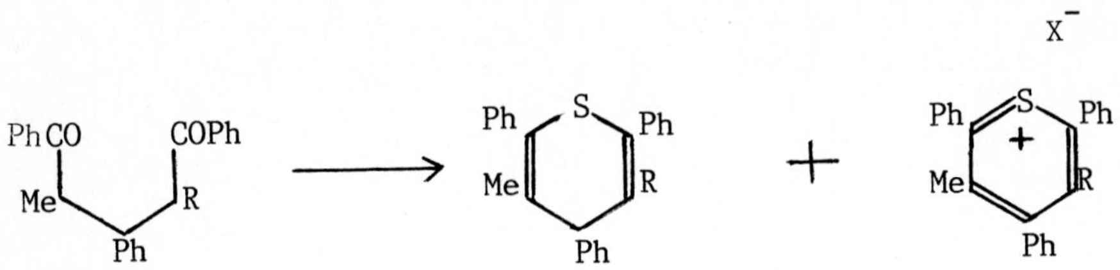

PhCO   COPh

Me   R

   Ph

R = H, Ph

(31)

(32)

X^-

X = ClO$_4$, CF$_3$CO$_2$

(33)

The oxidation of 4H-thiopyrans by hydrogen peroxide in acetic acid affords either 1,1-dioxides or thiopyrylium salts. Thus, 3,5-dimethyl-2,4,6-triphenyl-4H-thiopyran gives the 1,1-dioxide, whereas 2,4,6-triphenyl-4H-thiopyran yields the corresponding thiopyrylium salt (Kozhevnikova and V.G. Kharchenko, Khim. Geterotsikl. Soedin., 1985, 1042). Oxidation of 3,5-dimethyl-2,4,6- -triphenyl- and 2,6-diphenyl-3,4,5-trimethyl-4H-thiopyrans with HClO$_4$-AcOH in the presence of oxygen gives the corresponding thiopyryllium perchlorate (*idem, ibid.*, 1983, 1689). Oxidation of 4H-thio-pyrans (34) with potassium permanganate in acetone gives the corresponding 4H-thio-pyran-4-ones (35), which on oxidation with hydrogen peroxide in acetic acid afford the related 1,1-dioxides, readily converted to their 2,4-dinitrophenylhydrazones (A.M. Plotnikov, A.D. Shebaldova and Kharchenko, *ibid.*, 1985, 1489).

$R^1 = R^4 = Ph, \ R^2 = R^3 = H$
$R^1R^2 = (CH_2)_4, \ R^3 = Me, \ R^4 = Ph$
$R^1R^2 = R^3R^4 = (CH_2)_4$

(34)   (35)

The catalytic hydrogenation of alkyl and aryl substituted 4H-thiopyrans and dihydrothiopyrans using palladium at 100° and 60 atmospheres affords tetrahydrothiopyrans, *thiacyclohexanes* (N.S. Smirnova *et al.*, Issled. Obl. Sint. Katal. Org. Soedin, 1975, 17). The ionic hydrogenation of 4H-thiopyrans (36) with Et_3SiH/F_3CCO_2H produces the tetrahydrothiopyrans (37) and treatment of the thiopyrans (36) with F_3CCO_2H/O_2 yields the thiopyrylium salts (38) in addition to the tetrahydro derivatives (37) (Kharchenko and Kozhevnikova, Khim. Geterotsikl. Soedin., 1983, 200).

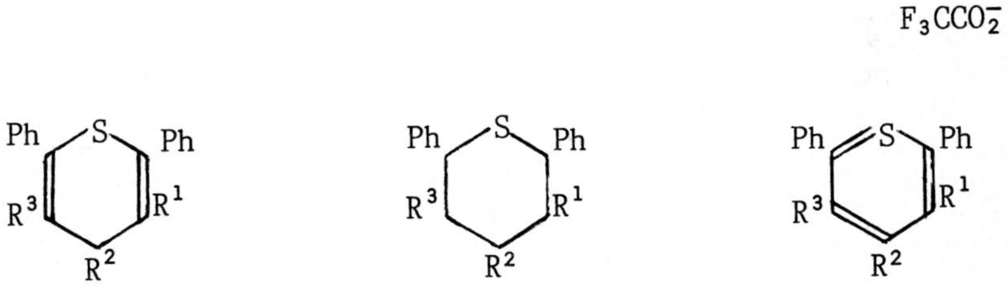

$F_3CCO_2^-$

$R^1 = R^3 = Me, \ R^2 = Me, \ Et$
$R^1 = H, Me, \ R^2 = Ph, \ R^3 = H$

(36)   (37)   (38)

Quenching of 2,6-diphenyl-4-lithio-4H-thiopyran with
trimethylsilyl chloride gives 2,6-diphenyl-4-trimethylsilyl-
-4H-thiopyran (39), which on treatment with a combination
of butyllithium and potassium $tert$-butoxide oxide in THF
slightly below $-20°$ furnishes the metallated derivative (40).
The reaction of compound (40) with a variety of aldehydes
and ketones provides a route to Δ^4-2,6-diphenyl-4H-thiopyrans
(e.g. 41) (C.H. Chen, J.J. Doney and G.A. Reynolds, J. org.
Chem., 1982, 47, 680).

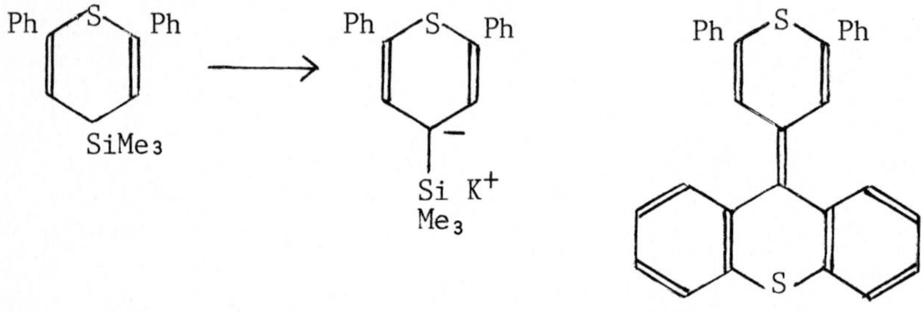

(39)   (40)   (41)

Treatment of 2,6-diphenyl-4H-thiopyran 1,1-dioxide with
toluene-4-sulphonohydrazide in boiling ethanol gives 4-
-diazo-2,6-diphenyl-4H-thiopyran 1,1-dioxide (42), m.p.
131° (decomp.).   It is a remarkably stable compound and
can be stored indefinitely in the solid state at ambient
conditions. Single-crystal X-ray analysis of (43) has been
reported (J.L. Fox, Chen, and H.R. Luss, $ibid.$, 1986, 51,
3551).

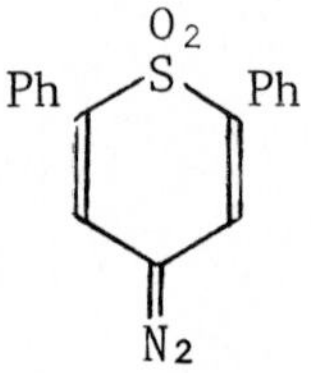

(42)

The uv-spectral data of 4*H*-thiopyran and related heterocycles have been assigned by comparison with recorded spectra of similar compounds and with the aid of *ab initio* and semiempirical calculations (F.P. Colonna, G. Distefano and V. Galasso, J. electron Spec. Relat. Phenom., 1980, 18, 75).

Bis(thiopyranylidenes). Bis(thiopyranylidenes) are also referred to as bis(chalcogenathiopyranylidenes). Bis-2,6--diaryl-4-bithiopyranylidenes, *2,2',6,6'-tetraaryl-4--bithiopyranylidenes,* are prepared from the 2,6-diarylthiopyrylium salts and a catalytic amount of triphenylphosphine in pyridine, for instance, 2,2',6,6'-tetraphenyl-4*H*-bithiopyranylidene (43). Some limitations of the method have been recorded (Reynolds, Chen, and J.A. Van Allan, J. org. Chem., 1979, 44, 4456). 4-Bibenzo[b]thiopyranylidene (44) and the 2-bibenzo[b]thiopyranylidenes (45) have also been prepared.

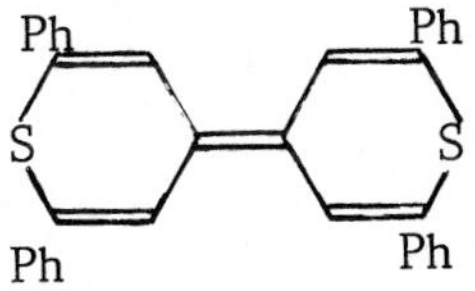

(43) 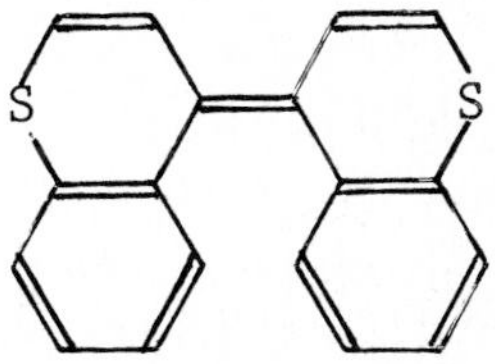   (44)

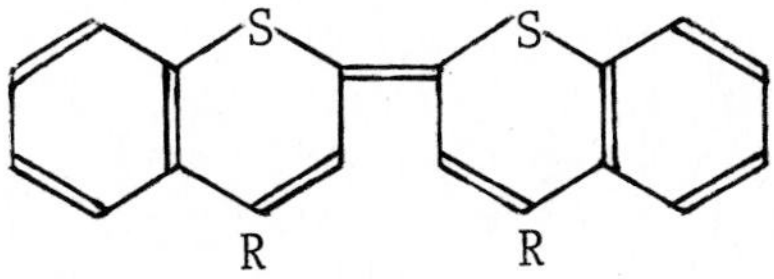

R = H, Ph

(45)

Diethyl 2,4,6-trioxoheptanedioate (acetonedioxalic ester) reacts with hydrogen sulphide and hydrogen chloride in an autoclave to afford 2,2',6,6'-tetracarbethoxy-4H-bithiopyranylidene (46). If the reaction is carried out in conventional glassware a mixture is obtained of compound (46) and 2,2',6,6'-tetracarbethoxy-4-(4'-pyranylidene)-4H-thiopyran (47) (D.J. Sandman, T.J. Holmes, and D.E. Warner, *ibid.*, p.880).

(46)   (47)

2,2',6,6'-tetra-*tert*-butyl-4H-bithiopyranylidene, $\Delta^{4,4'}$-*2,2',6,6'-tetra-tert-butyl-4-(thiopyranyl)-4H-thiopyran*, $\Delta^{4,4'}$-*2,2',6,6'-tetra-tert-butyl-4-(chalcogenathiopyranyl)-4H-chalcogenathiopyran* (49), m.p. 257-258°, is obtained from thione (48) and copper powder in boiling xylene. Along with other related compounds containing O, Se, and Te heteroatoms and Ph, Bu^t, and Me substituents they are characterized electrochemically by two reversible one-electron oxidations to give the radical-cations and dication states (M.R. Detty *et al.*, Tetrahedron, 1985, **41**, 4853). 2,2',6,6'-Tetraaryl-4H-bithiopyranylidenes, *2,2',6,6'-tetraaryl-$\Delta^{4,4'}$-bis(4H-thiopyrans)* and related pyrans are of interest as electron-donating materials that form conducting charge-transfer salts with suitable acceptors, such as, tetracyanoquinodimethane. 2,2',6,6'-Tetraphenyl-4-bithiopyranylidene has been prepared, by coupling the related thiopyrylium perchlorate with zinc in acetonitrile (E.M. Gluzman, L.V. Gavrilko and V.A. Starodub, Elektron. org. Mater., 1985, 40).

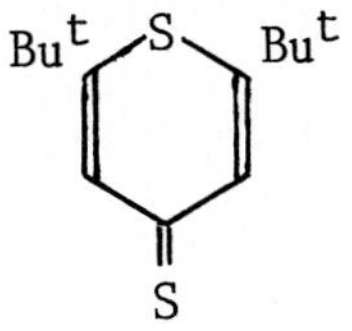

(48)

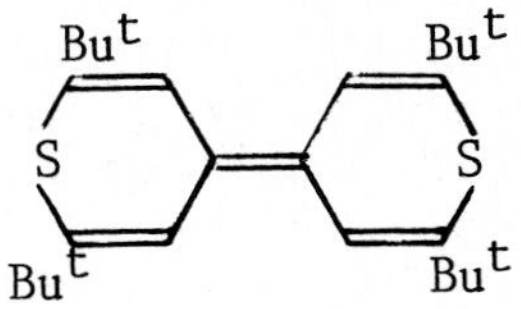

(49)

The redox potentials of some 2*H*- and 4*H*-bithiopyranylidenes have been determined (S. Es-Seddiki *et al*., Bull. Soc. chim. Fr., 1984, 241).

4,4'-(Ethanediylidene)bis(4*H*-thiopyrans) (52) and related compounds have been synthesized by a general method based on the reaction between phosphonate anions (50) and aldehydes (51), for example, derivatives (53) (Reynolds and Chen, J. org. Chem., 1981, **46**, 184).

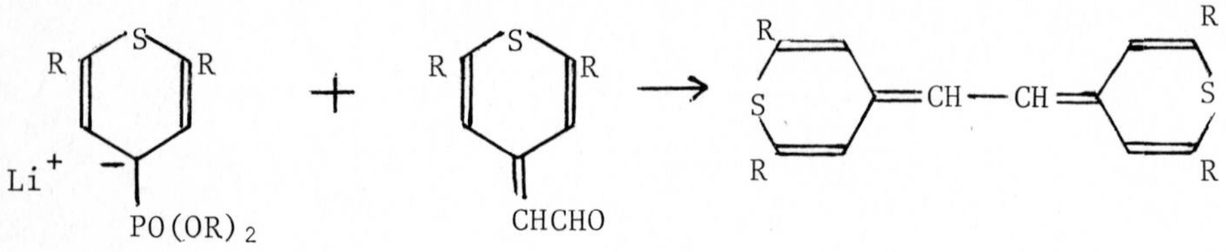

R=e.g., Ph

(50)　　　　　　　　(51)　　　　　　　　(52)

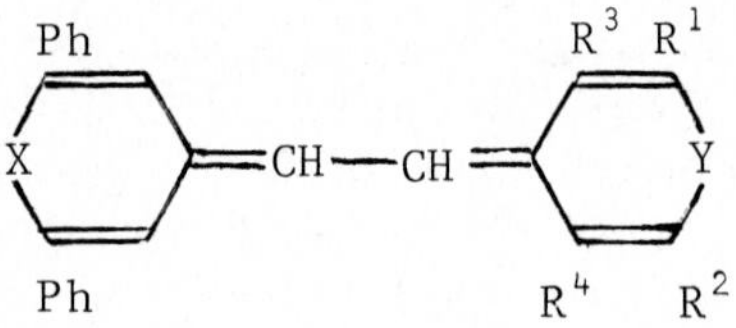

(53)

R^1	R^2	R^3	R^4	X	Y	m.p $^\circ$C	Yield %
Ph	Ph	H	H	S	S	301-302	29
Ph	Ph	H	H	O	S	257-258	6.2

For the synthesis of 1,2-bis(2,6-diphenyl-4*H*-thiopyran-4-
-ylidene)ethene (54) see K. Nakasuji, M. Nakatsuka, and
I. Murata (Chem. Comm., 1981, 1143).

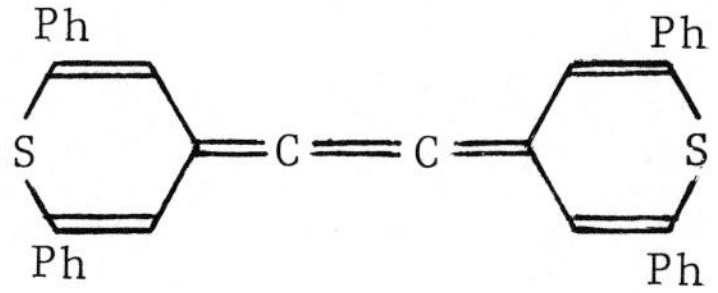

Ph   Ph

S   =C═C═   S

Ph   Ph

(54)

Electron donor molecules of the cumulene type (55), (56)
and (57) (K. Nakasuji, K. Takatoh and I. Murata, Chem.
Letters, 1982, 1727) and (58) and the peri-condensed type
(60) have been prepared, and their uv-spectra recorded and
their oxidation potentials determined. Compound (60;
R = Ph) is obtained on cyclization of naphthalene
derivative (59), furnished by the reaction between 1,5-
-naphthalenedithiol and ω-bromoacetophenone [Murata and
Nakasuji, Stud. org. Chem. (Amsterdam), 1986, 25, (New
Synth. Methodol. Funct. Interesting Compd.), 335].

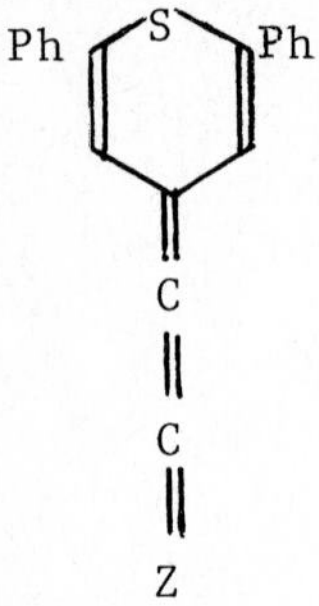

(55)

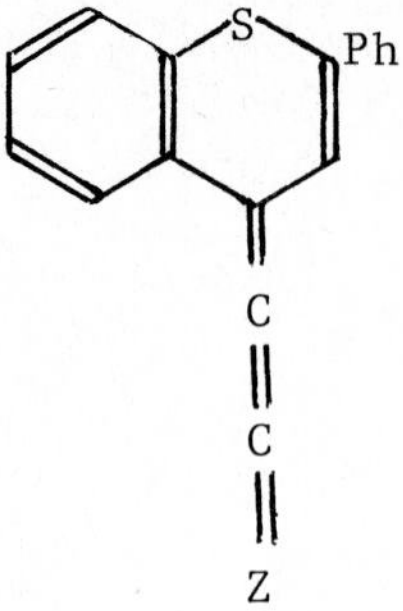

(56)

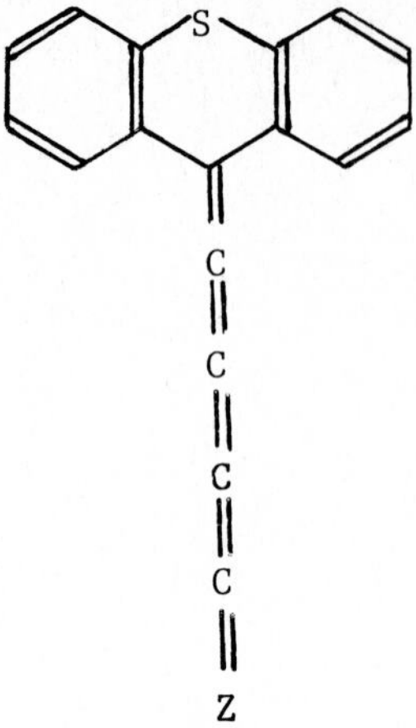

(57)

$$Z =$$

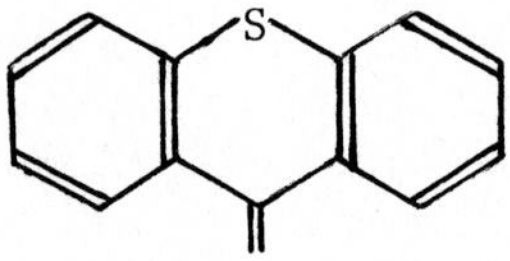

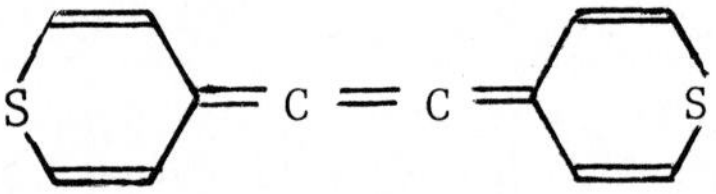

(58)

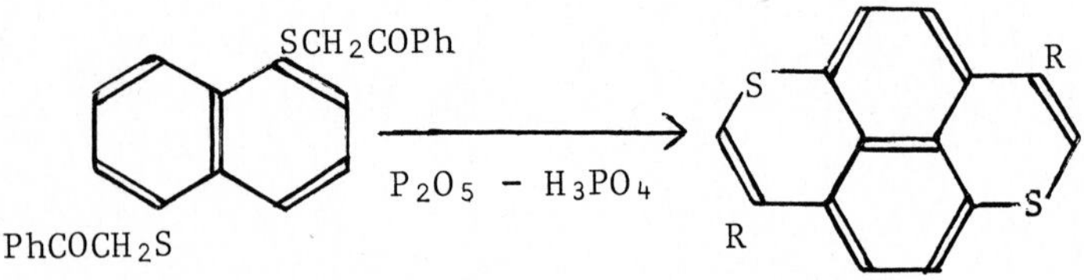

$$\xrightarrow{P_2O_5 - H_3PO_4}$$

R=H,Ph

(59)

(60)

(ii) Thiopyrylium salts

Treatment of the appropriate pentane-1,5-dione (1) with H_2S in acidic medium in the presence of Ph_3C^+ generated in *situ* affords 2,4,6-tri-*tert*-butylthiopyrylium perchlorate (2). 4-Aryl-2,6-di-*tert*-butylthiopyrylium perchlorates may also be prepared by this method [V.C. Cordischi, G. Doddi, G. Ercolani, J. chem. Res. (S), 1985, 62].

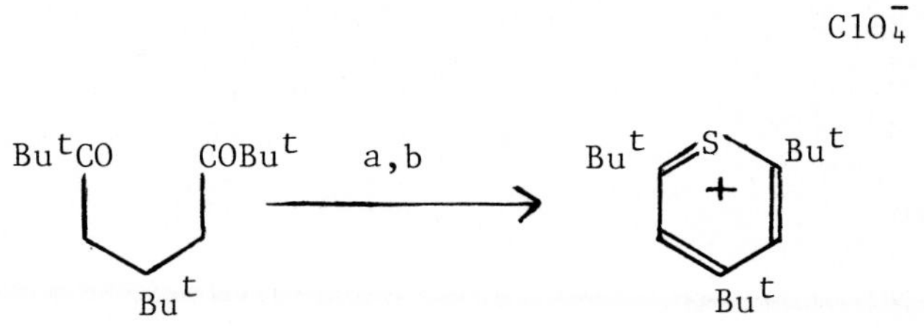

(1)   (2)

Reagents:- (a) Ph_3COH, Ac_2O, H_2S; (b) $HClO_4$, Ac_2O

2,6-Diarylthiopyrylium salts (4) are obtained on treating 1,5-diketones (3) with phosphorus pentasulphide and lithium perchlorate in boiling acetic acid (V. Gionis and H. Strzelecka, Synth. Comm., 1984, **14**, 775).

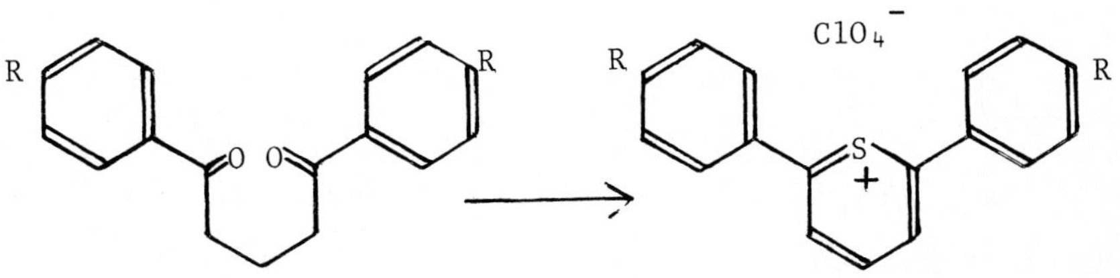

R=H, alkyl, alkoxy

(3)   (4)

They are also formed on treating aryl methyl ketones with ethyl orthoformate and hydrogen sulphide in acetic anhydride containing perchloric acid. Some of the corresponding pyrylium salt are formed along with the thiopyrylium salt (4). In a similar manner 2,6-di-*tert*-butylthiopyrylium is prepared from *tert*-butyl methyl ketone (pinacolone) (Doddi and Ercolani, Synth., 1985, 789).

The unsaturated ketone (5) on treatment with phosphoryl chloride and sodium perchlorate gives 5,6-diphenyl-2-methyl-thiopyrylium perchlorate (6; $R^1 = R^2 = Ph$, $R^3 = H$) (90%). Several 2-alkyl-5,6-diarylthiopyrylium salts (6) have been similarly prepared [B. Hollborn *et al.*, Ger. (East) Pat. DD 240,745, 1986].

$$ClO_4^-$$

PhCSH   COMe

Ph

$\longrightarrow$

R^1 S CH_2R^3
        +
R^2

$R^1 = R^2 = aryl$
$R^3 = H, alkyl$

(5)

(6)

Polysubstituted 4*H*-thiopyrans (7) on reaction with perchloric acid and oxygen yield the corresponding thiopyrylium salts (8) (V.G. Kharchenko, N.I. Kozhevnikova and N.V. Voronina, Khim. Geterotsikl. Soedin., 1979, 562).

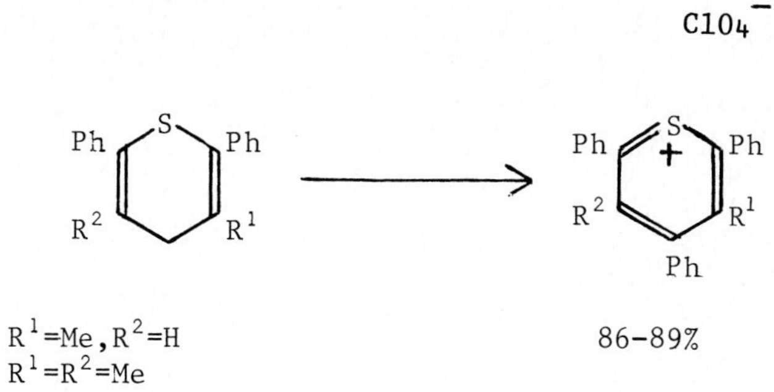

ClO_4^-

R^1=Me,R^2=H
R^1=R^2=Me

86-89%

(7)   (8)

For the preparation of 2,6-diaminothiopyrylium salts see
J. Liebscher, H. Hartmann, and B. Abegaz [Ger. (East) Pat.
DD 159,639, 1983] and Liebscher, Abegaz and A. Areda (Z.
Chem., 1983, 23, 403).

2,4-Diaryl-5,6,7,8,tetrahydrobenzo[b]thiopyrylium
tetrafluoroborates, *2,4-diarylcyclohexa[b]thiopyrylium
tetrafluoroborates*(9) are oxidized by potassium
ferricyanide in alcoholic sodium hydroxide to yield dimers
(10), which on treatment with 70% perchloric acid afford
the diperchlorates (11) (S.K. Klimenko, N.N. Ivanova and
N.N. Sorokin, Zh. org. Khim., 1987, 23, 2019).

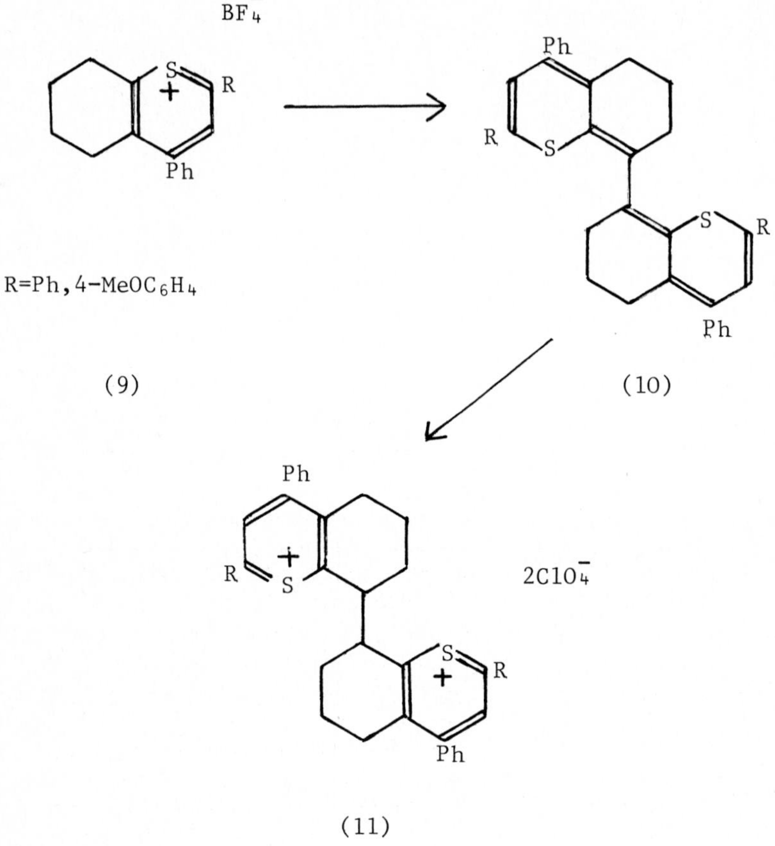

BF$_4^-$

R=Ph,4-MeOC$_6$H$_4$

(9)

(10)

2ClO$_4^-$

(11)

Catalytic reduction of thiopyrylium salts (12) over a platinum-group catalyst, especially 10% palladium/carbon gives the corresponding tetrahydrothiopyran (13) (Kharchenko, O.A. Bozhenova and A.D. Shebaldova, *ibid.*, 1982, 18, 2435).

R=H,Me,X=BF$_4$, CF$_3$CO$_2$
R=Ph, X=Cl

(12)   (13)

   The reaction between 2,4,6-triarylthiopyrylium salts and
methanolic sodium methoxide gives 2-methoxy-2,4,6-triaryl-
-2H-thiopyrans (14). 2,4,6-Triphenylthiopyrylium salts
react with nitromethane and with ethyl cyanoacetate to yield
1,3,5-triphenylbenzene (15; R = H) and 1-cyano-2,3,4-
-triphenylbenzene (15; R = CN), respectively (G.W. Fischer
and T. Zimmermann, Z. Chem., 1983, $\underline{23}$, 333). 2,4,6-
-Triarylthiopyrylium salts (16) with dimethylamine in ether
afford the corresponding 2-(dimethylamino)-2,4,6-triaryl-
-2H-thiopyrans (17). Ring transformation occurs with 2-
-(dimethylamino)-2,4,6-triphenyl-2H-thiopyran (17; R^1 = H,
R^2 = Ph) on reaction with nitromethane and ethyl cyanoacetate
to produce benzene derivatives (15; R = H) and (15;
R = CN), respectively, whereas reactions with methyl iodide
and benzoyl chloride results in cleavage of the Me$_2$N group
and formation of a thiopyrylium salt and (*idem*, J. prakt.
Chem., 1986, $\underline{328}$, 567).

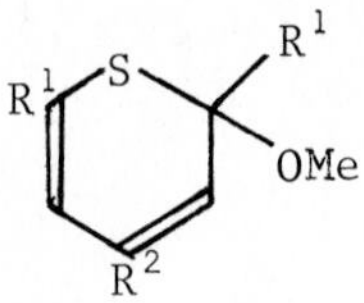

R¹,R²=Ph, Ar

(14)

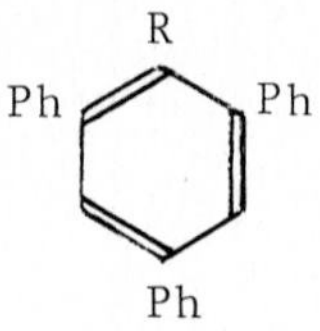

R=H,CN

(15)

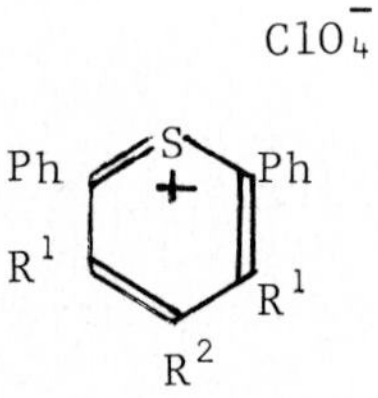

ClO₄⁻

R¹=H,Me
R²=Ph, 4-MeC₆H₄, 4-MeOC₆H₄
    4-ClC₆H₄, 4-BrC₆H₄

(16)

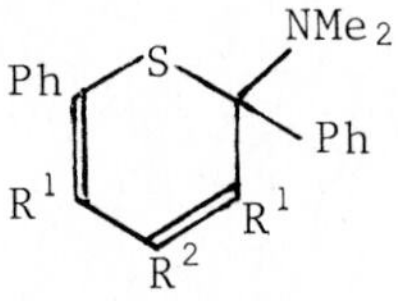

(17)

For the preparation of 2-(dialkylamino)-2,4,6-triaryl-2*H*--thiopyrans (19) from the related thiopyrylium salts (18) see Fischer and Zimmermann (Z. Chem., 1983, **23**, 144).

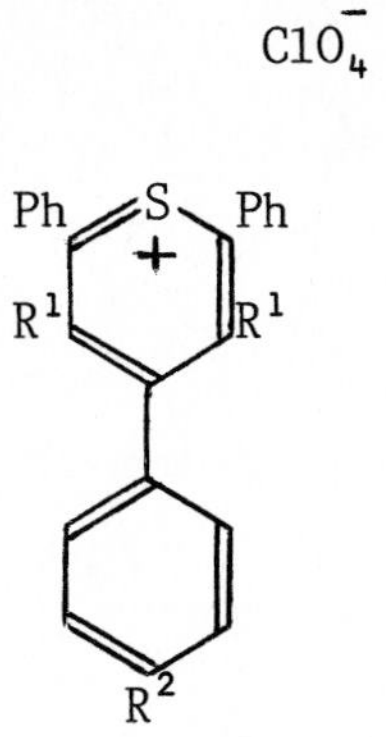

ClO_4^-

Ph S Ph
+
R^1 R^1

R^2

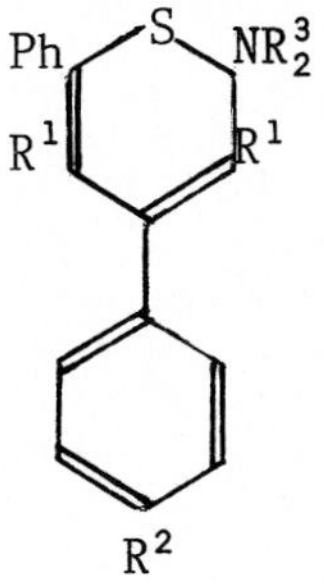

Ph S NR_2^3
R^1 R^1

R^2

R^1 = H, R^2 = H, Me, MeO, Cl, Br R^3 = Me, NR_2^3 = morpholino
R^1 = Me, R^2 = H

(18)   (19)

Nitrosation of the cyclohexa[b]thiopyrylium salts (20) with aqueous sodium nitrite in acetic acid containing ethanol affords the nitroso derivative (21), which rearranges immediately to furnish the oximino salt (22) (S.K. Klimenko *et al*., Zh. org. Khim., 1985, **21**, 2617).

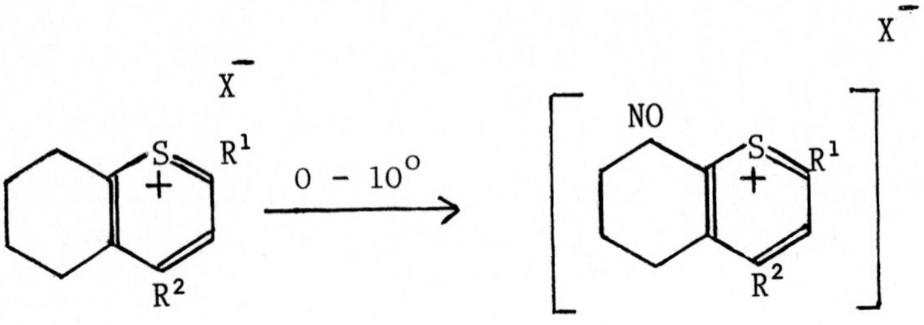

R^1 = Ph, 4-MeOC$_6$H$_4$, R^2 = H, X = ClO$_4$
R^1 = R^2 = Ph, X = ClO$_4$, BF$_4$

(20)

(21)

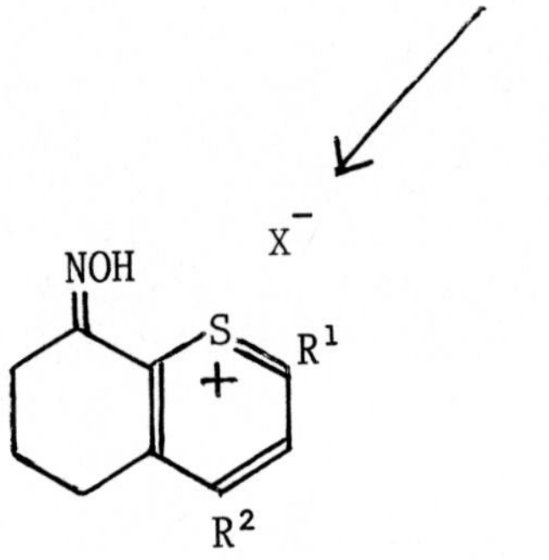

(22)

(iii) Thiopyranones

2H-Thiopyran-2-ones, 2H-thiin-2-ones. Treatment of 4,6-
-diphenylpyridine-2-thione with hydroxylamine in pyridine
yields the oxime (1) of 4,6-diphenyl-2*H*-thiopyran-2-one.
The reaction in methanolic sodium methoxide affords the
oxime of the corresponding 2*H*-pyran-2-one [P.M. Fresneda,
P. Molina and A. Soler, An. Univ. Murcia, Cienc., 1975
(Pub. 1978), **32**, 5].

3-Butyryl-4-hydroxy-6-methyl-2*H*-thiopyran-2-one and *N*-
-bromo-succinimide when irradiated in carbon tetrachloride
containing benzoyl peroxide affords 6-bromomethyl-3-butyryl-
-4-hydroxy-2*H*-thiopyran-2-one (2; R^1 = Pr). 3-Acyl-4-
-hydroxy-6-methyl-2*H*-thiopyran-2-ones have been converted
into bromo derivatives (2) and derivatives (3) and (4),
which show bactericidal and fungicidal activity (O. Caputo
et al., Farmaco, Ed. Sci., 1982, **37**, 213).

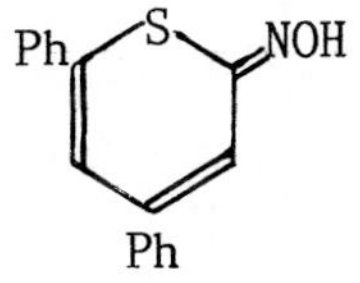

(1)

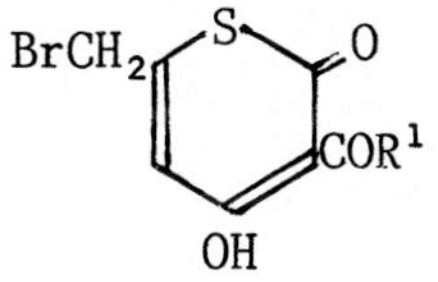

R^1 = C$_{3-7}$alkyl

(2)

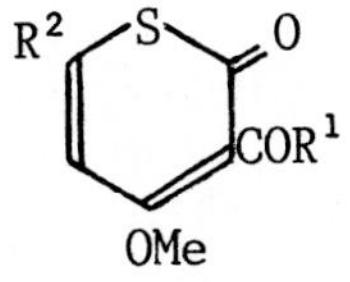

R^2 = CH$_2$Br, CHBr$_2$

(3)

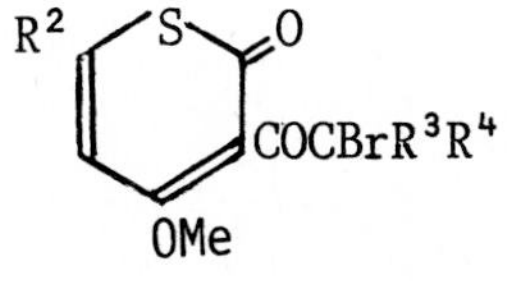

R^3 = H, Et
R^4 = Et, Bu, hexyl

(4)

The radical bromination using *N*-bromosuccinimide and the
ionic bromination with bromine in acetic acid of 4-methoxy-
-6-methyl-2*H*-thiopyran-2-one has been investigated (Caputo,
L. Cattel and F. Viola, *ibid*., 1979, **34**, 869).

4H-Thiopyran-4-ones, 4H-thiin-4-ones.
   2,6-Diphenyl-4*H*-thiopyran-4-one (2) is prepared, by
adding to a solution of 1,5-diphenylpenta-1,4-diyn-3-one
(1) in THF and sodium ethoxide in ethanol a solution
obtained, by adding sodium ethoxide in ethanol to a
solution prepared by adding lithium triethylborohydride in
THF to sulphur under nitrogen (M.R. Detty, B.J. Murray and
M.D. Seidler, J. org. Chem., 1982, **47**, 1968).

(1)   (2)

   The unusual selenium dioxide oxidation of 4-chloro-2,6-
-diphenyl-3-formyl-2*H*-thiopyran (3) gives 2,6-diphenyl-3-
-formyl-4*H*-thiopyran-4-one (4).   Thiopyran (3) is obtained
either by treating 3,4-dihydro-2,6-diphenyl-2*H*-thiopyran-
-4-one with phosphoryl chloride in DMF or preferably by
allowing phosphoryl chloride to react with 3,4-dihydro-2,6-
-diphenyl-3(dimethylaminomethylene)-2*H*-thiopyran-4-one
(C.H. Chen and G.A. Reynolds *ibid*., 1979, **44**, 3144).

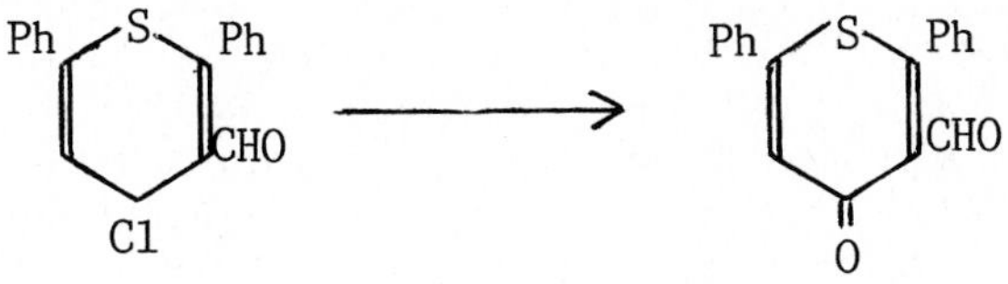

(3)   (4)

2-Amino-2-methylacrylonitrile with hydrogen sulphide in benzene on treatment with an aqueous solution containing sodium sulphide, potassium hydroxide, and tetra-*tert*-butyl ammonium bromide or cyanide gives 3-cyano-2,6-dimethyl-4H-thiopyran-4-one (5; X = o) (65%). In absence of the aqueous layer 3-cyano-2,6-dimethyl-4H-thiopyran-4-thione (5; X = s) (30%) is formed and without hydrogen sulphide 3-cyano-2,6-dimethyl-4H-pyran-4-one (24%) is obtained (J.P. Roggero *et al.*, Heterocycles, 1985, <u>23</u>, 153).

Sulphur dichloride with 1,5-diphenylpenta-1,4-diyn-3-one (1) gives 3,5-dichloro-2,6-diphenyl-4H-thiopyran-4-one (6) and with [Me$_2$C(OMe)C≡C]$_2$CO yields 3,5-dichloro-2,6-di-(2-methoxyisopropyl)-4H-thiopyran-4-one (7) (N.M. Morlyan, E.L. Abagyan and Sh.O. Badanyan, Arm. Khim. Zh., 1979, <u>32</u>, 318).

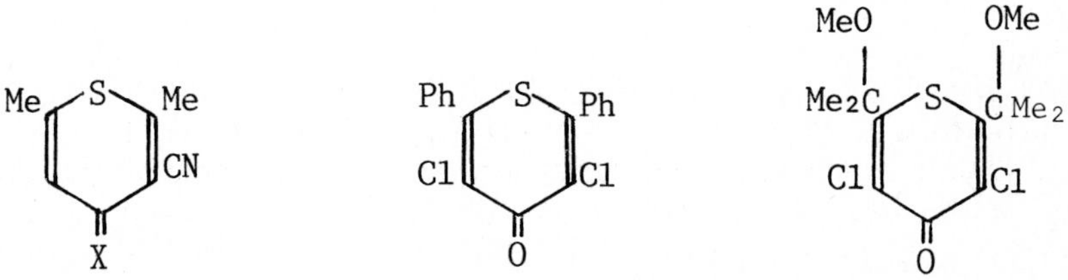

(5)   (6)   (7)

Depending on the amount of bromine used, the bromination of 2,6-diphenyl-4H-thiopyran-4-one 1,1-dioxide with bromine in acetic acid gives either 3-bromo- (8) or 3,5- -dibromo-2,6-diphenyl-4H-thiopyran-4-one 1,1-dioxide (9). Chlorination with chlorine in tetrachloromethane affords a mixture of the 3-chloro and the 3,5-dichloro derivative. Also bromination of 4H-thiopyran 1,1-dioxide gives a mixture the 3-bromo and the 3,5-dibromo derivative, and that of 3,5-diphenyl-4H-thiopyran-4-one 1,1-dioxide a mixture of 2-bromo- and 2,6-dibromo-3,5-diphenyl-4H- -thiopyran-4-one 1,1-dioxide. 3,5-Dibromo-2,6-diphenyl-4H- -thiopyran-4-one 1,1-dioxide has been converted into the 3- -amino-5-bromo-2,6-diphenyl-4H-thiopyran-4-one 1,1-dioxide (10) (W. Reid and H. Bopp, Ann., 1978, 1280).

R = morpholino, piperidino

(8)   (9)   (10)

Some 6-amino-2-methylthio-4H-thiopyran-4-ones (11) are prepared by treating $R^2CH_2COCR^1=C(SMe)_2$ with sodium hydride and then with 4-$R^3C_6H_4NCS$. If the latter compound is replaced by EtNCS (when R^2=H) under similar conditions the 6-ethylamino derivative (12) is obtained (L.W. Singh, H. Ila and H. Junjappa, Synth., 1985, 531).

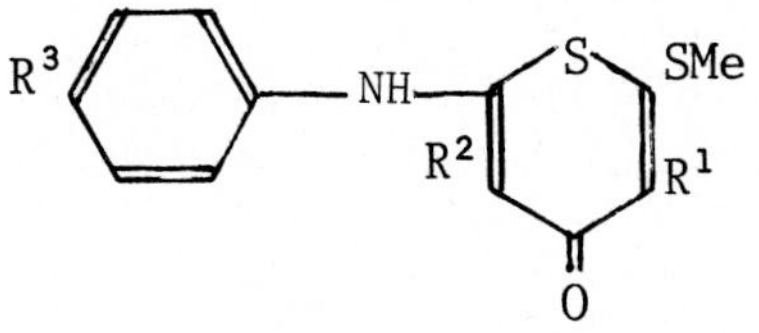

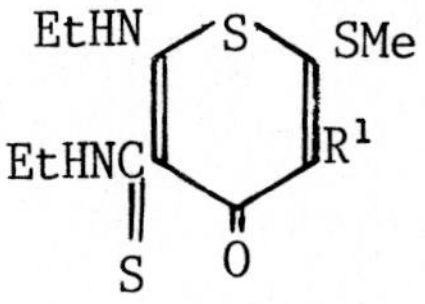

$R^1 = R^2 = H$, Me
$R^3 = H$, Me, Cl

(11)   (12)

2H-Thiopyran-2-thiones and 4H-thiopyran-4-thiones.
2*H*-Thiopyran-2-thiones (15) are prepared by reacting
dithiolethiones (13) with enamines (14). Uv-spectral data
have been recorded (F. Ishii, M. Stavaux and N. Lozac'h,
Bull. Soc. chim. Fr., 1977, 1142).

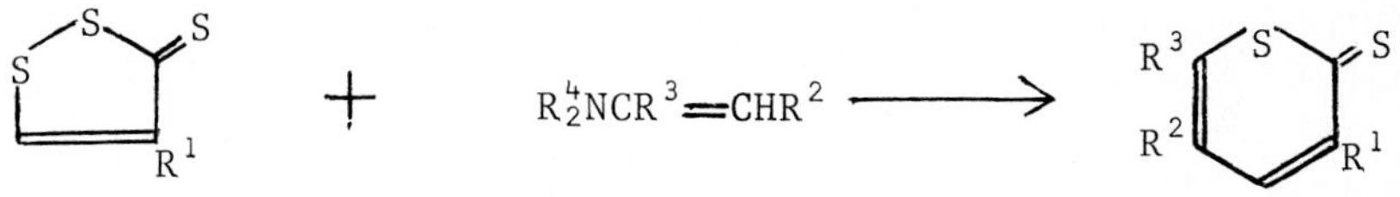

R^1=H,Et,Ph, R^2=H
   4-MeC$_6$H$_4$ R^3=H,Me,Et
                         R^2R^3=C$_3$-C$_5$alkylene
                         R^4=Et,R$_2^4$N=1-pyrolidinyl

   (13)   (14)   (15)

Reduction of the 5-[2-(dimethylamino)vinyl]-3-methylthio-
-1,2-dithiole ylium (16) with potassium tetrahydridoborate
in methanol yields 2-methylthio-4H-thiopyran-4-thione (17),
which on boiling with dimethylamine in ethanol gives 4-
-dimethylamino-2H-thiopyran-2-thione (18). In contrast
reduction of 5-[2-(dimethylamino)vinyl]-3-benzylthio-1,2-
-dithiole ylium affords 4-benzyl-2H-thiopyran-2-thione *via*
the kinetic product 2-benzylthio-4H-thiopyran-4-thione.
The mechanisms of these rearrangements have been discussed
(M. Barreau and C. Cotrel, Tetrahedron Letters, 1981, **22**,
4507).

(16)   (17)   (18)

(b) Di- and tetra-hydro-thiopyrans and -thiopyranones

*(i) Dihydrothiopyrans, dihydrothiins, dihydrothiapyrans
and related 3- and 4-ones*
3,4-Dihydro-2H-thiopyrans. In some reports these
compounds are referred to as 5,6-dihydro-4H-thiopyrans. 1-
-Oxides and 1,1-dioxides of 3-methyl- and 2,3-dimethyl-3,4-
-dihydro-2H-thiopyrans have been synthesized *via* an
intramolecular Horner-Wittig reaction (M. Mikolajczyk,
M. Popielarczyk and S. Grzejszczak, Z. Chem., 1984, **24**,
377). 2-Acetyl-3,4-dihydro-6-methyl-2H-thiopyran (2) may
be obtained by a Cope rearrangement of dihydropyran (1),
(K.B. Lipkowitz and B.P. Mundy, Tetrahedron Letters, 1977,
3417).

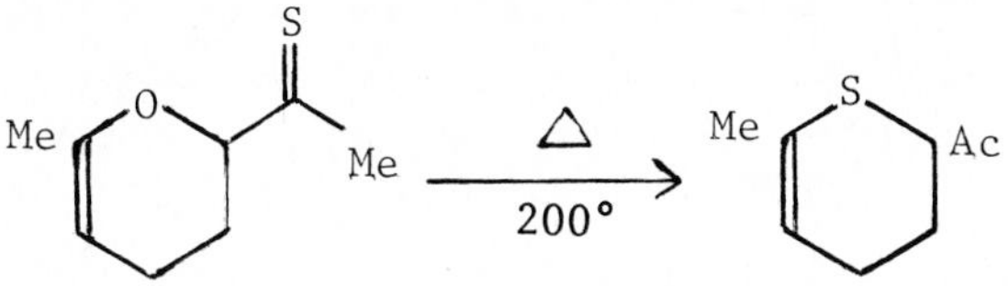

(1)   (2)

The α,β-unsaturated thiones (3) react with the parent
ketones (4) to yield 3-arylcarbonyl-3,4-dihydro-2,4,6-
-triaryl-2H-thiopyrans (5) and 2-arylcarbonyl-3,4-dihydro-
-3,4,6-triaryl-2H-thiopyrans (6). In these cycloaddition
reactions the α,β-unsaturated thione and the parent ketone
may serve as a diene or dienophile (T. Karakasa and
S. Motoki, J. org. Chem., 1979, **44**, 4151). 3-Benzoyl-3,4-
-dihydro-2,4,6-triphenyl-2H-thiopyran, m.p. 187.5-188.5°;
3-benzoyl-3,4-dihydro-3,4,6-triphenyl-2H-thiopyran, m.p.
143.5-144.5°.

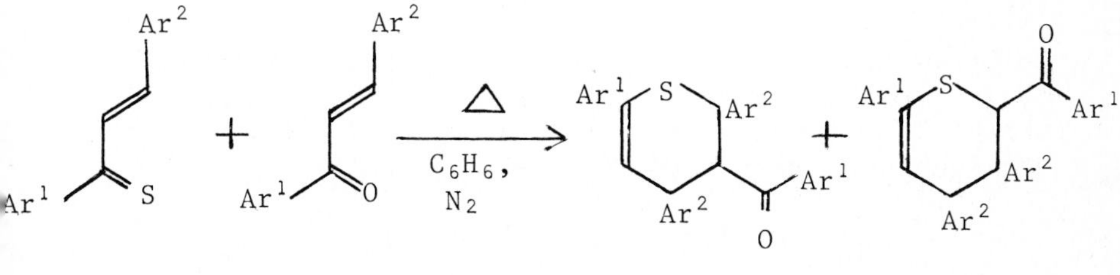

(3)   (4)   (5)   (6)

432

The addition of arylmagnesium bromides (8) to thioamide
vinylogues (7) leads to E-thiochalcones which dimerize
regio- and stereospecifically to the 3,4-dihydro-2,4,6-
-triaryl-2H-thionpyran-3-yl aryl thioketones (9)
(J.P. Guemas and H. Quiniou, Sulphur Letters, 1984, <u>2</u>, 121).

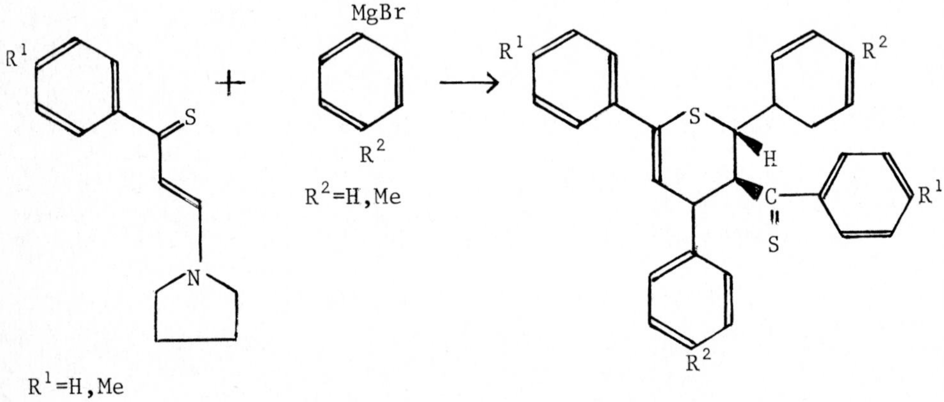

R^2=H,Me

R^1=H,Me

(7)   (8)   (9)

Diels-Alder reaction between the unsaturated thioamides
(1) and acrylonitrile, ethyl acrylate, and 1-nitro-2-
-phenylethene affords the 3,4-dihydro-2H-thiopyrans (11),
(12) and (13), respectively (B.Y. Ried et al., ibid., 1987,
<u>6</u>, 105).

$$R^1CH=C(CN)CNH_2$$
$$\overset{S}{\underset{\parallel}{}}$$

$$\underset{NC}{\overset{H_2N}{}}\;\text{(ring)}\;\underset{R^2}{\overset{R^3}{}}$$
$$R^1$$

R^1=2-furyl,2-thienyl

(10)

(11) R^2=CN, R^3=H
(12) R^2=CO$_2$Et, R^3=H
(13) R^2=NO$_2$,R^3=Ph

3,4-Dihydro-2H-thiopyran-3- and -4-ones. 3,4-Dihydro-2*H*-
-thiopyran-3-ones (15) and 5,6-dihydro-2*H*-thiopyran-5-ones
(16) have been obtained by the aldol condensation of the
diketones (14) catalysed by toluene-4-sulphonic acid
(J. Nakayama, K. Hidaka and M. Hoshino, Heterocycles, 1987,
26, 1785).

$$R^1CH_2COCHR^2SCHR^3COR^4 \longrightarrow \underset{R^4}{\overset{R^3}{}}\;\text{(ring)}\;\underset{O}{\overset{R^2}{}} \;+\; \underset{O}{\overset{R^2}{}}\;\text{(ring)}\;\underset{R^4}{\overset{R^3}{}}$$
$$R^1 \qquad\qquad R^1$$

R^1,R^2,R^3=H,Me
R^4=Me,Et,Ph

(14)   (15)   (16)

The treatment of 2-butyltetrahydrothiopyran-4-one (17) in
dichloromethane containing pyridine with *N*-chlorosuccinimide
gives a mixture of 2-butyl-3,4-dihydro-2*H*-thiopyran-4-one
(18) and 6-butyl-3,4-dihydro-2*H*-thiopyran-4-one (19). A
number of approaches to the synthesis of 6-alkyl-3,4-
-dihydro-2*H*-thiopyran-4-ones have been described and 5-
-chloro-3,4-dihydro-2*H*-thiopyran-4-one (20), m.p. 80-81°,
has been converted into 2-butyl-3-chloro-tetrahydrothio-
pyran-4-one (21) (V.K. Kansal and R.J.K. Taylor, J. chem.
Soc., Perkin I, 1984, 703).

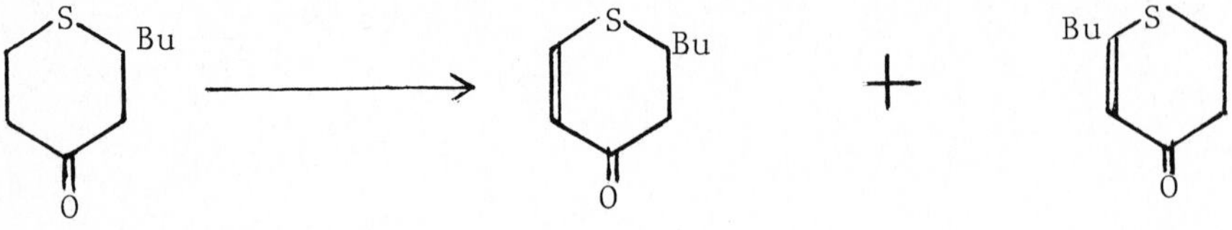

(17)   (18)   (19)

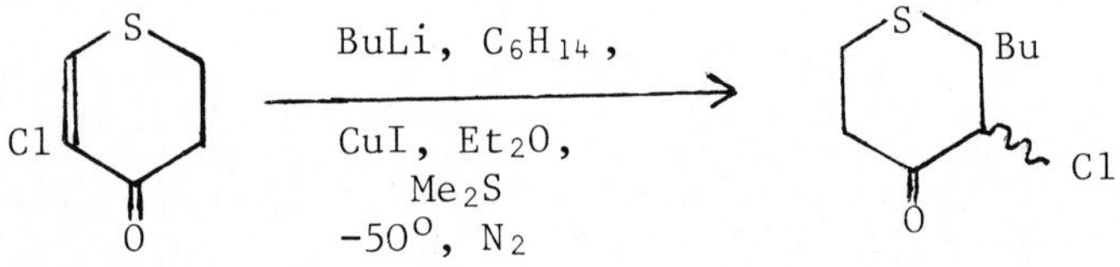

(20)   (21)

The Pummerer rearrangement of 3-carbomethoxytetrahydrothio-
pyran-4-one 1-oxide (22) to give 3-carbomethoxy-3,4-dihydro-
-2H-thiopyran-4-one (23) and 5-carbomethoxy-3,4-dihydro-2H-
-thiopyran-4-one (24), is effected by chlorotrimethylsilane
in boiling tetrachloromethane. Under standard conditions
in boiling acetic anydride, the acetates (25) and (26) are
obtained. Related chlorotrimethylsilane-induced Pummerer
rearrangements are also reported (S. Lane, S.J. Quick and
Taylor, Tetrahedron Letters, 1984, 25, 1039; J. chem.
Soc., Perkin I, 1984, 2549).

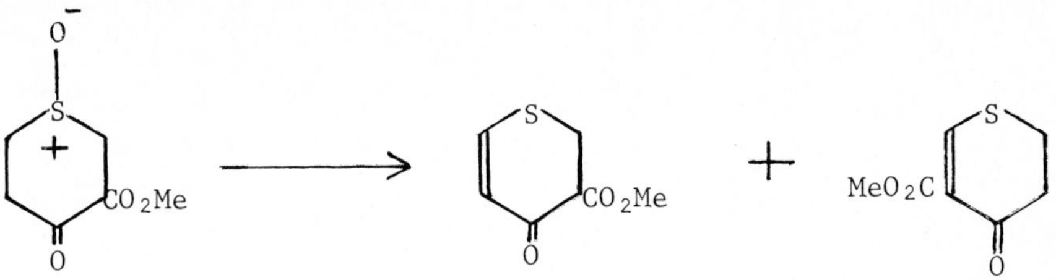

(22)   (23)   (24)

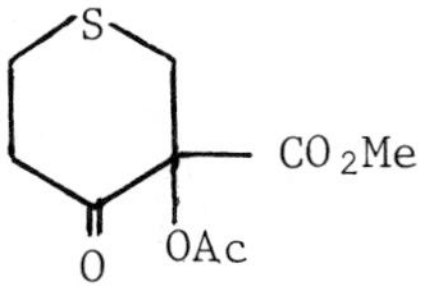

(25) 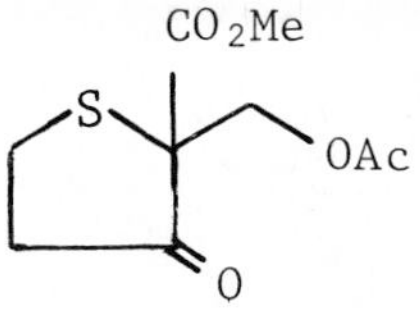   (26)

The ring cleavage of 2,6-disubstituted 5-cyano-3,4-
-dihydro-2*H*-thiopyrans in the presence of sodium hydride
has been described (M. Augustin and G. Jahreis, Z. Chem.,
1978, 18, 91) and some 2,4-substituted 6-amino-3,5-dicyano-
-3,4-dihydro-3-thioamido-2H-thiopyrans have been synthesized
(J.S.A. Brunskill, A. De and D.F. Ewing, J. chem. Soc.,
Perkin I, 1978, 629).

5,6-Dihydro-2H-thiopyrans. Tetrahydrothiopyran-4-mesylate
(27) on mild dehydromesylation using Woelm W-200 (Super 1)
alumina in dichloromethane affords 5,6-dihydro-2*H*-thiopyran
(28), b.p. 76-78°/62mm. *trans*-2,6-Diphenyl-5,6-dihydro-2*H*-
-thiopyran, m.p. 34.1°, b.p. 104-115°/ 1μ; *cis*-2,6-
-diphenyl-5,6-dihydro-2*H*-thiopyran, m.p. 74°, and 2,6-bis-
(4-methoxyphenyl)-5,6-dihydro-2*H*-thiopyran have also been
prepared (C.H. Chen *et al.*, J. heterocyclic Chem., 1978,
15, 289).

436

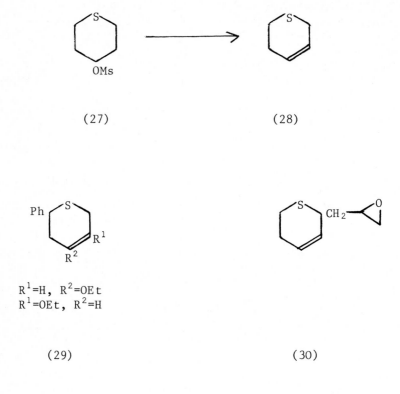

(27)   (28)

R^1=H, R^2=OEt
R^1=OEt, R^2=H

(29)   (30)

Thiobenzaldehyde generated by the thermolysis of
PhCH$_2$S$^+$(O$^-$)SCH$_2$Ph is trapped efficiently by 2,3-dimethyl-
butadiene to give 5,6-dihydro-3,4-dimethyl-6-phenyl-2H-
-thiopyran. By using appropriate trapping reagents
dihydrothiopyrans (29) have been obtained (J.E. Baldwin and
R.C.G. Lopez, Tetrahedron, 1983, 39, 1487).

The reaction of 5,6-dihydro-2H-thiopyran (28) with 2-
-butyllithium and epichlorohydrin in THF at -78° gives 2-
-(2,3-epoxypropyl)-5,6-dihydro-2H-thiopyran (30) (S. Torii,
H. Tanaka and Y. Tomotaki, Japan Tokkyo Koho 79 38,110,
1979).

Anion (31) obtained from 5,6-dihydro-4-phenyl-2H-thiopyran
1-oxide on treatment with primary alkyl halides at -78°
furnishes *trans*-2-alkyl-4-phenyl-5,6-dihydro-2H-thiopyran
1-oxide exclusively. However, with 2-propyl iodide at -78°
to 20°, anion (31) gives a mixture of the 2-*cis* and the
2-*trans* derivative. It has been indicated that the lack of
stereoselectivity is due to anion exchange reactions at the
higher temperature required to effect alkylation. Adducts
obtained from reaction between anion (31) and benzaldehyde
have been shown to be 2-*cis* and 2-*trans* (R.L. Crumbie,
D.D. Ridley and P.J. Steel, Austral. J. Chem., 1985, 38,
119).

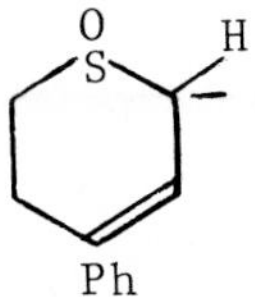

(31)

The spiro compound (34) is formed when 2,3-dimethylbuta-
-1,3-diene reacts with α-oxo sulphine (33), obtained by
treating trimethylsilyloxyindene (32) with thionyl chloride
(B.G. Lenz, H. Regeling and B. Zwanenburg, Tetrahedron
Letters, 1984, 25, 5947).

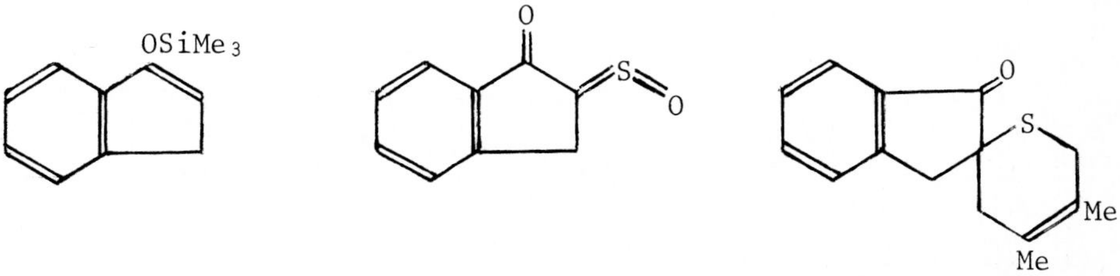

(32)   (33)   (34)

1(Z),3-Butadienyl sulphoxides (35) on treatment with lithium diisopropylamide are converted stereospecifically into 2-lithio-5,6-dihydro-2H-thiopyran 1-oxides (36) (M. Reglier and S.A. Julia, *ibid*., 1985, 26, 2655).

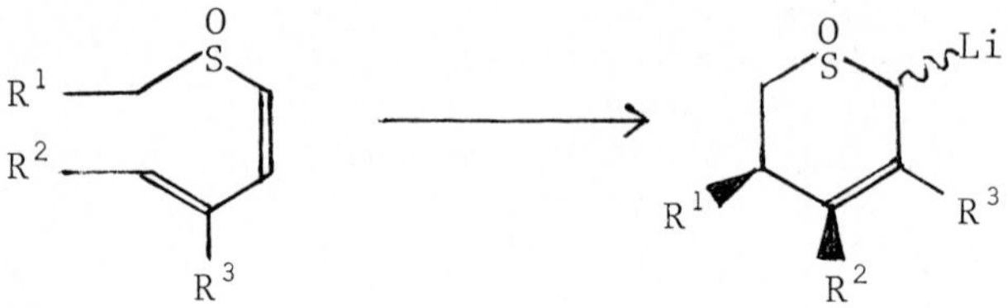

R^1=Ph, Me$_2$C=CH, R^2=H, R^3=Me
R^1=Ph, R^2R^3=(CH$_2$)$_4$

(35)   (36)

Thioaldehydes generated by the photolysis of phenacyl sulphides (37) and bearing electron withdrawing groups are excellent dienophiles and can be trapped by 2-ethoxy- or 2- -*tert*-butyldimethylsilyoxy- butadiene to give 5,6-dihydro- -2H-thiopyrans (38) and (39) (E. Vedejs, T.H. Eberlein and D.L. Varie, J. Amer. chem. Soc., 1982, 104, 1445).

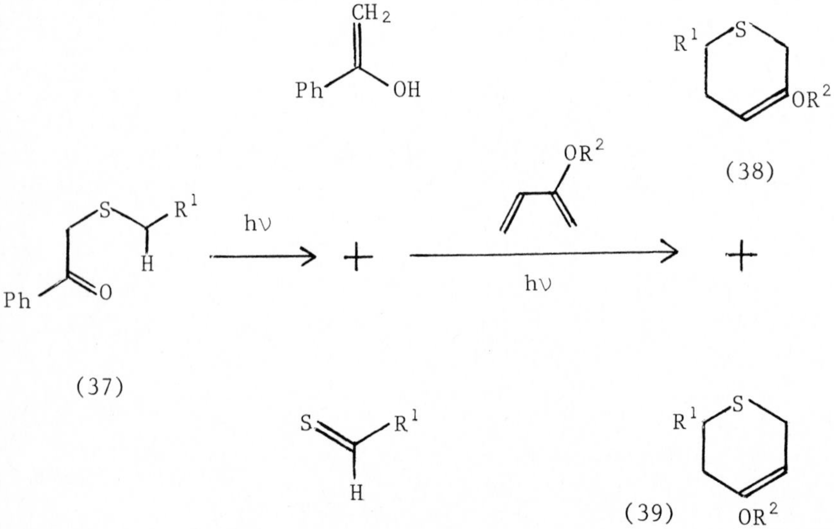

(37)

(38)

(39)

R^1	R^2	Yield (isolated) %	Product Ratio (38):(39)
CN	Et	76	13:1
CN	Bu^tMe_2Si	74	18:1
COPh	Et	54	*
COPh	Bu^tMe_3Si	64	12:1
CO_2Pr^i	Bu^tMe_3Si	62	6:1
CO_2Me	Bu^tMe_3Si	58	*
COMe	Bu^tMe_3Si	55	3.5:1
COMe	Et	33	7:1

* Not determined.

Cyclization of acrolein with hydrogen sulphide in
dichloromethane containing copper turnings and
triethylamine at -10° furnishes 5,6-dihydro-2*H*-thiopyran-
-3-carboxaldehyde (40; R = H). The preparation of 5,6-
-dihydro-2,6-dimethyl-2*H*-thiopyran-3-carboxaldehyde (40;
R = Me) has also been reported (B. Unterhalt and M. Ghori,
Z. Lebensm. -Unters. Forsch., 1980, <u>170</u>, 34). Aldehydes
(40; R = H,Me) have also been obtained by treating
acrolein or crotonaldehyde with hydrogen sulphide in a
mineral oil (b.p. > b.p. of starting material or end
product). They are useful as intermediates for
pharmaceuticals and plant protective agents (M. Sauerwald
et al., Ger. Offen. DE 3,427,404, 1986).

(40)

Methylation of 2-(ethoxycarbonyl)-1,3-dithiolane (41) with methyl fluorosulphonate yields sulphonium salt (42), which on treatment with solid potassium carbonate in the presence of 2,3-dimethylbuta-1,3-diene affords the cycloadduct, ethyl 5,6-dihydro-3,4-dimethyl-6-methylthio-2*H*-thiopyran-6-carboxylate (45). A labile ylide (43) is apparently formed as the intermediate which undergoes cyclo reversion to (44) under nearly neutral conditions. The preparation of a number of related 6-acyl-5,6-dihydro-6-
-methylthio-2*H*-thiopyrans (2-acyl-3,6-dihydro-2-methylthio-
-2H-thiopyrans) and their conversion to 6-acyl-5,6-dihydro-
-2H-thiopyrans has been reported (Vedejs *et al.*, J. org. Chem., 1980, **45**, 2601).

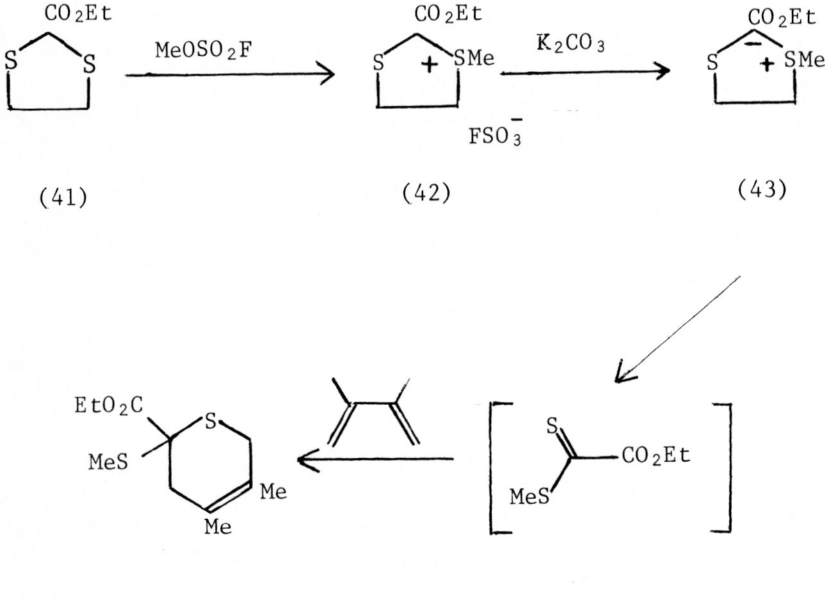

(41)   (42)   (43)

(45)   (44)

Cyclization of 2-aminoethyl phenyl ketone by hydrogen sulphide in alcohol at room temperature yields 3-benzoyl--5,6-dihydro-4-phenyl-2H-thiopyran (K.Kh. Tokmurzin *et al*., Khim. Geterotsikl. Soedin., 1984, 701).

The cycloadduct of anthracene and ethyl thioxacetate is oxidised by 3-chloroperbenzoic acid to a mixture of the *trans*- and the *cis*-sulphoxide (46). Heating the *trans*--sulphoxide with 2,3-dimethylbuta-1,3-diene in benzene at 60° gives anthracene and the *trans*-sulphoxide, *ethyl 5,6--dihydro-3,4-dimethyl-2H-thiopyran-6-carboxylate* 1-oxide (48), *via* the generation of a thioaldehyde S-oxide (47) by a retro Diels-Alder reaction (A.A. Freer, G.W. Kirby and R.A. Lewis, Chem. Comm., 1987, 718).

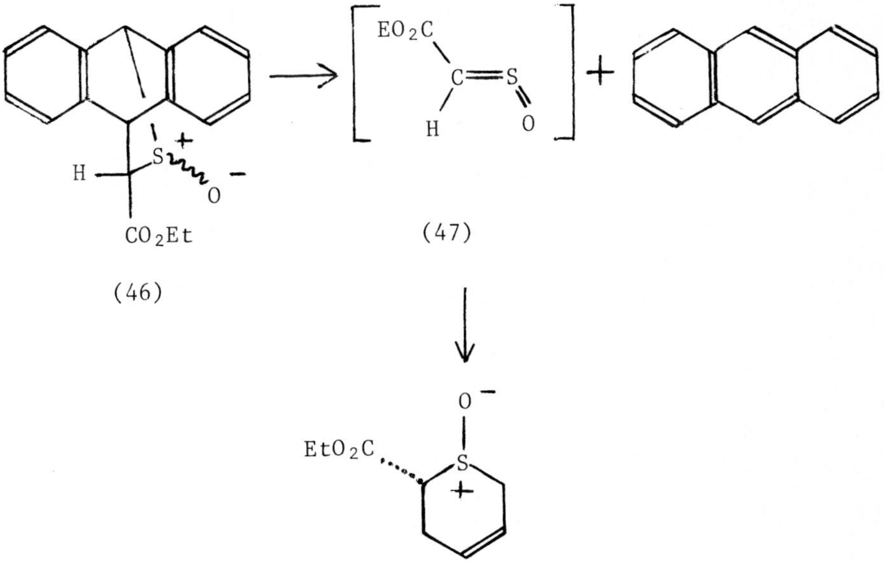

(46)   (47)

(48)

Betaines (49; R = Me) and (49; R = Ph) undergo an intramolecular Wittig reaction in boiling toluene to give ethyl 5,6-dihydro-3-methyl- (50; R = Me) and ethyl 5,6--dihydro-3-phenyl- (50; R = Ph) 2*H*-thiopyran-4-carboxylate, respectively (M. Barreau and G. Ponsinet, Tetrahedron Letters, 1985, **26**, 5451).

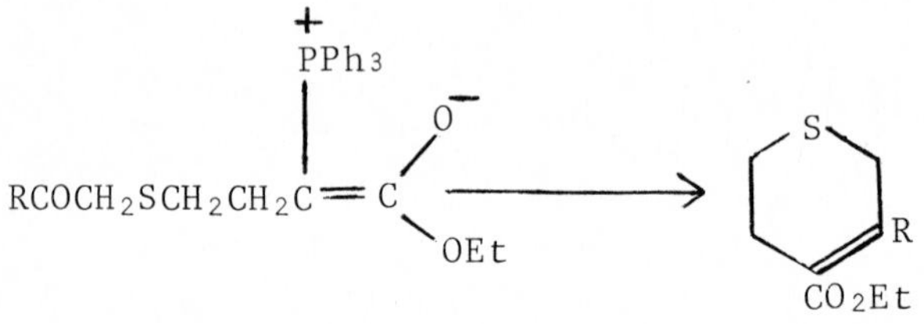

R=Me,Ph

(49)   (50)

Some properties and derivatives of 5,6-dihydro-2H--thiopyran-3- and -4-carboxylic esters have been discussed (U. Moser, G. Lambrecht and E. Mutschler, Arch. Pharm., 1983, **316**, 670).

The Diels-Alder reaction between cyanothioformamides (51) and 2,3-disubstituted butadienes (52) affords 3,4--disubstituted 6-amino-6-cyano-5,6-dihydro-2*H*-thiopyrans (53) (K. Friedrich and M. Zamkanei, Ber., 1979, **112**, 1867).

(51)

$+$

$\xrightarrow{\hspace{3cm}}$

R^1R^2N—[structure with CN, S, R^4, R^3]

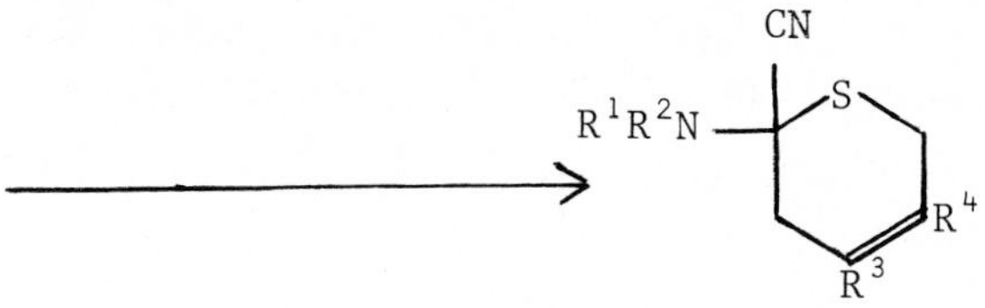

$H_2C=CR^3R^4C=CH_2$

$R^1 = Me, Ph, 4-O_2NC_6H_4$
$R^2 = H, Ac, Bz, COCF_3$
$R^3, R^4 = H,H;\ Me,H;\ Me,Me;\ Ph,Ph$

(52)   (53)

Some derivatives of 5,6-dihydro-6,6-dimethyl-4-hydroxy-
-2H-thiopyran-2-thione and -one and their corresponding
tautomers have been described (K. Schweiger, Monatsh.,
1982, **113**, 1283).
Methyl but-2-enedithioate (54) acts as a heterodiene in
undergoing cycloaddition to cyclopentadiene as dienophile
to give 4-methyl-2-methylthio-4a,7a-dihydro-4H-cyclopenta-
[b]thiopyran-6-ene (55) (K.R. Lawson, A. Singleton, and
G.H. Whitham, J. chem. Soc., Perkin I, 1984, 865).

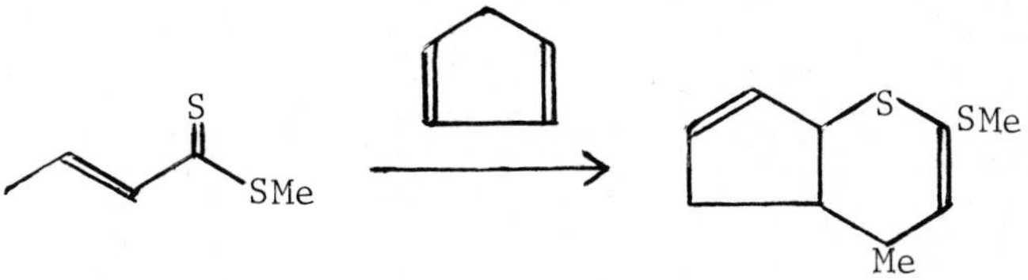

[reaction scheme with structures]

(54)   (55)

Cyclopenta[b]thiopyrans (56) (Y. Aso *et al.*, Tetrahedron Letters, 1981, <u>22</u>, 3061) and (57) (Y. Hasegawa *et al.*, *ibid.*, p.3065) have been prepared and some of their chemical and physical properties have been discussed.

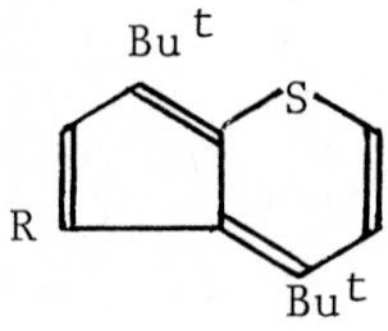

R=OMe,SBu^t

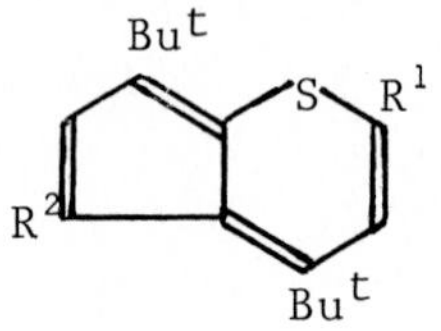

R^1=H,R^2=OMe,SBu^t
R^1=OMe,R^2=H

(56)   (57)

(ii) Tetrahydrothiopyrans, thianes, tetrahydrothiapyrans
and related 4-ones

Tetrahydropyrans. 4-Alkyl- and 4-phenyl-tetrahydrothio-pyrans (2) are obtained by the cyclization of 1,5-dibromo-alkanes (1) with sodium sulphide (N.P. Volynskii and L.P. Shcherbakova, Izv. Akad. Nauk SSSR, Ser. Khim., 1979, 1077).

BrCH$_2$ CH$_2$Br   Na$_2$S $\longrightarrow$

R=Me, Bu, pentyl, hexyl, c-hexyl, Ph

(1)   (2)

R^1=Me,R^2=Me,CH$_2$D,Bu,C$_5$H$_{11}$,Ph
R^1=CH$_2$D,R^2=Bu

(3)

Similarly, appropriate 1,5-dibromoalkanes have been converted into 2-methyl-4-alkyl(phenyl)tetrahydrothiopyrans (3) and their mono- and di-oxides have been prepared. It is found that these tetrahydrothiopyrans and their dioxides do not follow the von Auwers-Skita rule; *ie*. the *cis* isomers have a lower b.p., refractive index, and density, and a higher m.p. than do the *trans* isomers (L.I. Perepelitchenko, Chem. Abs., 1984, **100**, 209586y; Volynskii and Perepelitchenko, Neftekhimiya, 1982, **22**, 400). The method has been applied to the synthesis of 4,4- -dimethyl- and 2,4,4-trimethyl-tetrahydrothiopyrans (K. Nagasawa and A. Yoneta, Chem. pharm. Bull., 1985, **33**, 5048). 2-Propyltetrahydrothiopyran has been prepared and an effective method for purifying dialkyl sulphides has been developed (Volynskii, Neftekhimiya, 1979, **19**, 259).

2,6-Diphenyl- and 2,4,6-triphenyl-tetrahydrothiopyrans and related derivatives (5) and compounds (6) are obtained by the hydrogenation of related thiopyrylium salts (4) over 10% palladium/carbon (O.B. Bozhenova, Nukleofil'nye Reakts. Karbonil'nykh Soedin., 1982, 125).

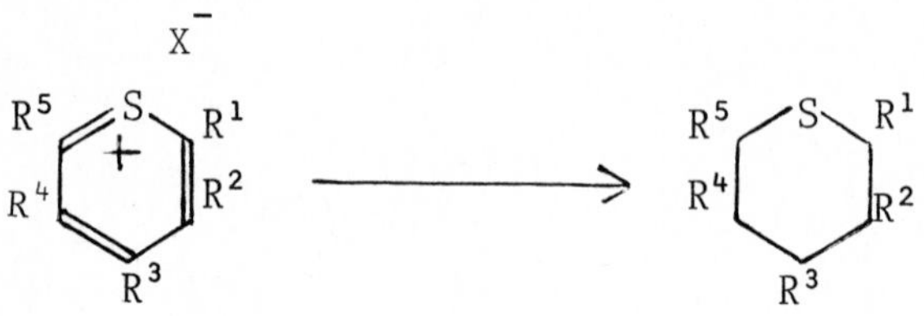

$R^1 = R^3 = R^5 = Ph$, $R^2 = R^4 = H$, $X = CF_3CO_2$, Cl, BF_4, I
$R^1 = R^5 = Ph$, $R^2 = R^4 = H$, $R^3 = H$, Me, $X = BF_4$
$R^1 = Ph$, $R^2 = H$, Ph, $R^3 = H$, $R^4R^5 = (CH_2)_4$, $X = BF_4$

(4)   (5)

(4) $\longrightarrow$

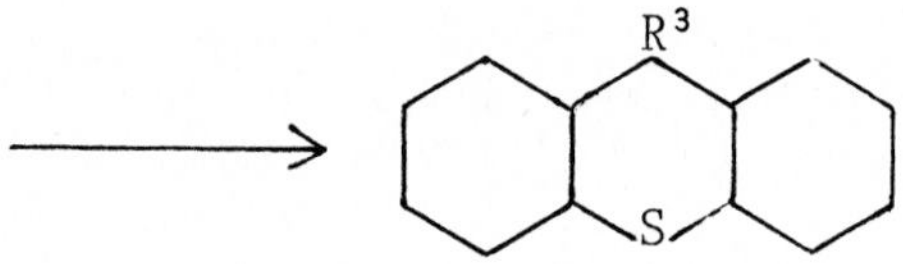

$R^1R^2 = R^4R^5 = (CH_2)_4$
$R^3 = H$, Me, $X = BF_4$
$R^3 = Ph$, $X = SbCl_6$

(6)

The catalytic reduction of 4H-thiopyran 1,1-dioxides (7) over palladium/carbon at 20-120° and 5.07-10.13 MPa pressure affords tetrahydrothiopyran 1,1-dioxides (8) (V.G. Kharchenko, A.M. Plotnikov, and A.D. Shebaldova, Khim. Geterotsikl. Soedin., 1983, 1058).

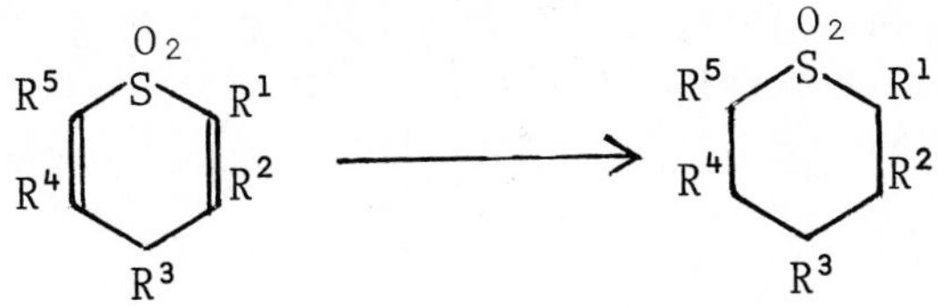

$$R^5 \underset{R^4}{\overset{O_2}{\underset{R^3}{\bigcirc}}} \overset{R^1}{R^2} \longrightarrow R^5 \underset{R^4}{\overset{O_2}{\underset{R^3}{\bigcirc}}} \overset{R^1}{R^2}$$

$R^1 = R^5 = Ph, \ R^2 = R^4 = Me, \ R^3 = H$
$R^1 = R^3 = Ph, \ R^2 = H, \ R^4R^5 = (CH_2)_4$
$R^1R^2 = R^4R^5 = (CH_2)_4, \ R^3 = Me, Ph$

(7)   (8)

The photochemical rearrangement of the cyclopropyl ethanethiol (9) in pentane at $-70°$ yields a mixture (65%) containing 4,4-dimethyltetrahydrocyclopropane[c]thiopyran (10) (98%) and 3,3-dimethyl-2-methylenetetrahydrothiopyran (11) (2%) after 19h. If the reaction is carried out at $36°$ for 11h a mixture (82%) containing (10) (69%), (11) (12%) and 3-isopropyl-2-methylenetetrahydrothiophene (19%) is obtained (J.L. Stein and M.P. Crozet, Compt. rend., 1985, 300, 59).

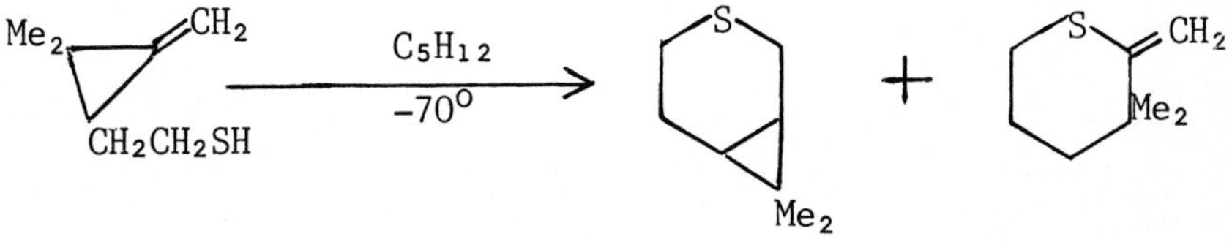

$$Me_2 \overset{CH_2}{\underset{CH_2CH_2SH}{\triangle}} \xrightarrow[-70°]{C_5H_{12}} \underset{Me_2}{\bigcirc} + \underset{Me_2}{\overset{S}{\bigcirc}} CH_2$$

(9)   (10)   (11)

A synthesis of 2-vinyltetrahydrothiopyran has been reported (A.O. Tosunyan *et al.*, Khim-Farm. Zh., 1978, <u>12</u>, 56).

The 3-cyclohexenone derivatives (12) (B. Zeeh *et al.*, Ger. Offen. DE 3,601,066, 1987) and (13) (R. Becker *et al.*, *ibid.*, 3,412,794, 1985; Eur. Pat. Appl. EP 136,647, 1985) of tetrahydrothiopyran have been prepared and their use as herbicides and plant growth regulators has been investigated.

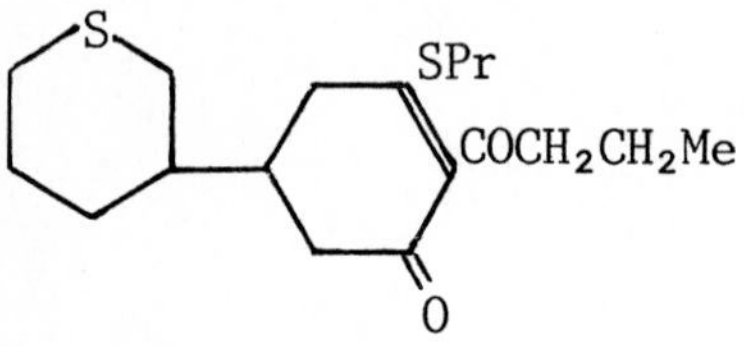

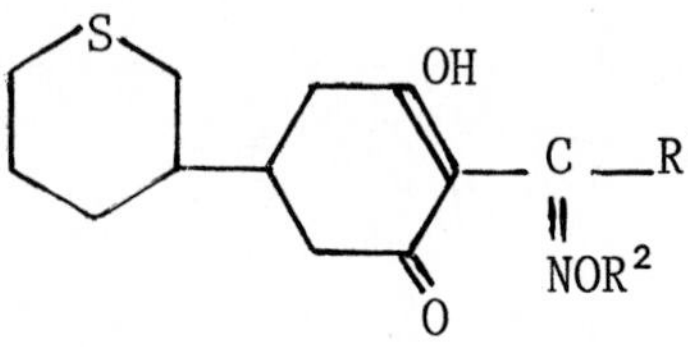

R^1 = alkyl, R^2 = chloroalkenyl

(12) (13)

The synthesis of the first optically active cyclic dithiohemiacetal, 4,4-dimethoxy-2-mercaptotetrahydrothiopyran (14) has been described. Optical resolution is achieved *via* diastereoisomeric thio esters, which are separated by fractional crystallization. Hydrolysis yields the enantiomerically pure dithiohemiacetals, which are configurationally stable (G. Dittrich *et al.*, Sulphur Letters, 1984, <u>2</u>, 261).

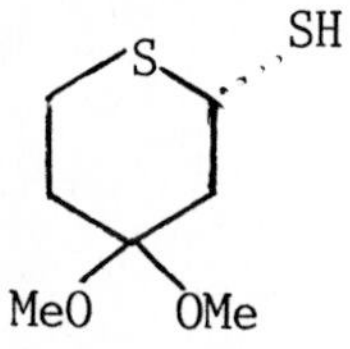

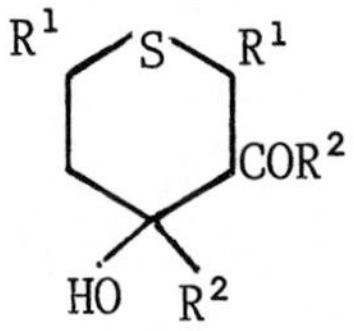

R^1 = Ph, 2-furyl
R^2 = H, Ph, 2-methyl-5-pyridyl

(14)

(15)

Cyclization of 2-aminoethyl methyl ketone by hydrogen sulphide in alcohol at room temperature gives 3-acetyl-4-
-hydroxy-4-methyltetrahydrothiopyran (K.Kh. Tokmurzin *et al.*, Khim. Geterotsikl. Soedin., 1984, 701). 3-Acyl-2,6-
-diphenyl(2-furyl)tetrahydrothiopyrans (15) are obtained by treating R^2COMe with R^1CHO in aqueous alcohol containing sodium sulphide (*idem*, U.S.S.R. Pat. SU 1,121,263, 1984).

The 1,3-dianion formed from the sulphone (16) reacts selectively in a controlled manner with 1-bromo-3-
-chloropropane to give compound (17), which on cyclization affords 2-benzoyl-tetrahydrothiopyran 1,1-dioxide (18), m.p. 139-141°. The benzoyl derivative (18) may be converted to 2-chlorotetrahydrothiopyran 1,1-dioxide (19), m.p. 72-73°; 2-benzoyl-2-chlorotetrahydrothiopyran, m.p. 94.5-96° (J.S. Grossert, J. Hoyle, and D.L. Hooper, Tetrahedron, 1984, **40**, 1135).

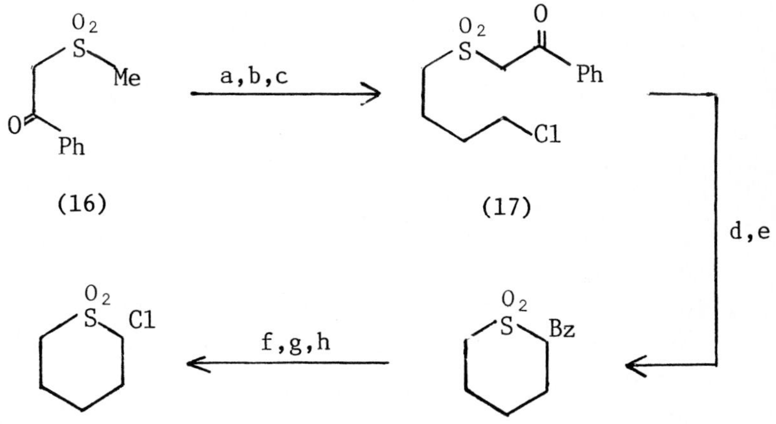

Reagents:- (a) Na, THF; (b) n-BuLi; (c) Br(CH$_2$)$_3$Cl;
(d) NaI,Me$_2$CO; (e) NaH, THF; (f) NaH;
(g) NCS; (h) KOH, EtOH.

The reaction between disulphur dichloride and hexa-1,5-
-diene affords a mixture containing 5-chloro-2-chloromethyl-
tetrahydrothiopyran (20) (40%) and 2,5-di(chloromethyl)-
tetrahydrothiophene (21) (60%). The former compound (20)
irreversibly isomerizes at 20° to yield a thermodynamically
stable mixture of (20) (65%) and (21) (35%). The 1,1-dioxide
of (20) on treatment with triethylamine gives 3,4-dihydro-2-
-methylene-2H-thiopyran 1,1-dioxide (22) (G.A. Tolstikov et
$al.$, Zh. org. Khim., 1983, $\underline{19}$, 48).

R = CH$_2$Cl

(20)   (21)

(22)   (23)

Di-(formylmethyl) sulphide on treatment with ethyl
cyanoacetate in the presence of piperidine, followed by
acetylation furnishes ethyl 4-cyano-3,5-diacetoxytetra-
hydrothiopyran-4-carboxylate (23) (F.J. Lopez Aparicio,
F. Zorrilla Benitez, and F. Santoyo Gonzalez, An. Quim.,
1985, $\underline{81C}$, 178).

Although distyryl sulphides and sulphoxides do not react with activated methylene compounds as do the corresponding sulphones, they do so if activated by two additional ester groups to give tetrahydrothiopyrans and 1-oxides, respectively, for instance, $[4-MeC_6H_4CH=C(CO_2Me)]S(O)_n$ reacts with $NCCH_2R$ in benzene under basic conditions to give 4-subsituted dimethyl 4-cyano-3,5-di(4-methylphenyl)-tetrahydrothiopyran-2,6-dicarboxylate (24) (H. Yamamura and H.H. Otto, Arch. Pharm., 1978, 311, 762).

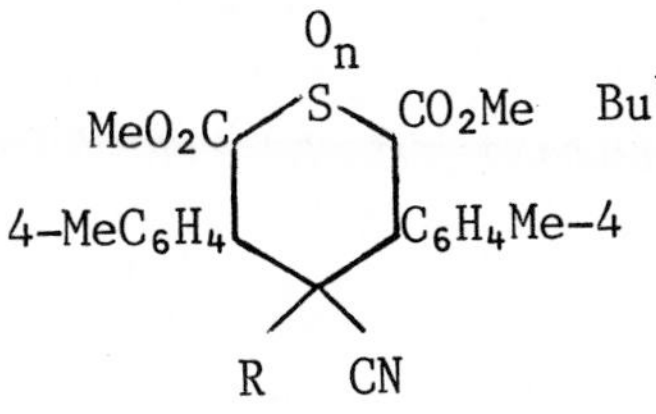

$R = CN, CO_2Me$
$n = 0,1,2$

(24)   (25) 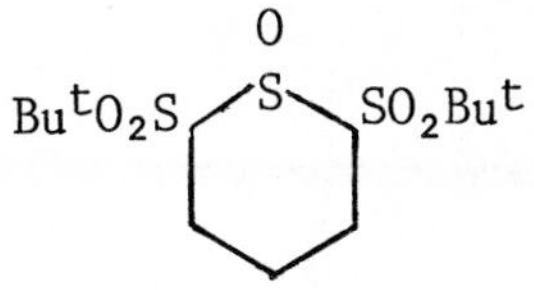   (26)

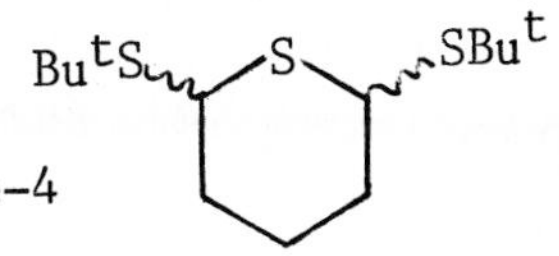

For the preparation of the diastereomers of 2,6-di(*tert*-butylthio)tetrahydrothiopyran (25) and their oxidation product, di-*tert*-butyl tetrahydrothiopyran-2,6-sulphinate 1-oxide (26) see Lopez Aparicio, Zorrilla Benitez, and Santoyo Gonzalez (An. Quim., 1983, 79C, 158) and for some 2-thioamidotetrahydrothiopyran derivatives, J.C. Aloup, D. Farge, and C. James, (Eur. Pat. Appl. EP 97,584, 1984).

Reduction of the enamines (27) and (28) by anhydrous
formic acid affords the 4-amino derivatives (29)
(S.A. Vartanyan and E.A. Abgaryan, Arm. Khim. Zh., 1984,
37, 316). Similar reduction have been carried out with the
related pyrans.

R^1, R^2, R^4 = H, Me
R^3 = pyrrolidino, morpholino, piperidino

(27)   (28)   (29)

2-Alkyltetrahydrothiopyran (30) have been converted into
the *cis* and the *trans* 1-oxides and 1-tosylimides
(sulphilimines) by various stereoselective methods. *cis*-
-1-Tosylimides (31) have been formed by using *tert*-BuOCl
and TsNH⁻ in a two-stage process, while the 2-alkyltetra-
hydrothiopyrans have been converted by chloramine-T
predominantly into *trans*-1-tosylimides (32). 1-Oxides
enriched in *cis*-1-oxide (33) and *trans*-1-oxide (34) have
been obtained by different methods of oxidation. The
preferred conformations of 1-tosylimides have been
determined by analysis of ^{13}C-nmr spectroscopic and
crystallographic X-ray data. The ^{13}C-nmr spectra show the
conformation of the 1-tosylimides and 1-oxides to be
analogous (I. Jalsovszky *et al.*, Tetrahedron, 1986, 42,
5649).

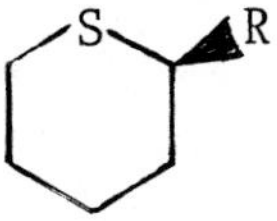

Ts
N

S R

(30)

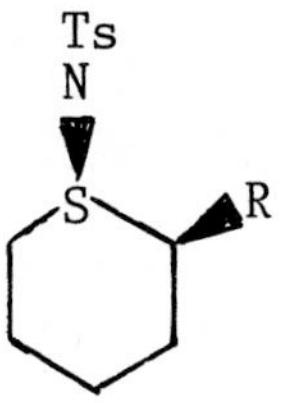

Ts
N

S R

(31)

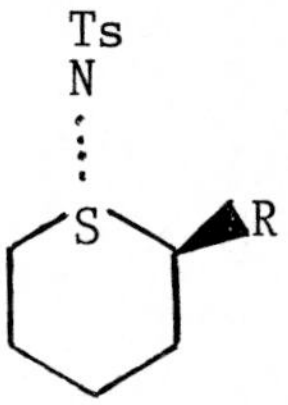

Ts
N

S R

(32)

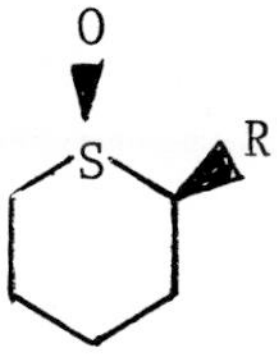

O
S R

(33)

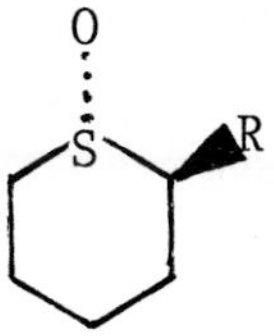

O
S R

R = Me, Et, Pr^i, Bu^t

(34)

The oxidation of tetrahydrothiopyran by atmospheric oxygen in isopropanol containing copper (II) chloride at 130° and 50 atmospheres yields mainly the 1-oxide (L.V. Vlasova, L.B. Avdeevam and R.I. Dudina, Neftekhimiya, 1980, 20, 122). Electrochemical oxidation of 4-(4-chlorophenyl)tetrahydrothiopyran in aqueous organic solvents in the presence of halide salts as electrolytes, producing halonium ions, gives selectively the *trans*-1-oxide. A preferential formation of the *cis*-1-oxide is obtained using acidic electrolytes (M. Kimura *et al.*, Tetrahedron Letters, 1984, 25, 4665). Treatment of 2-aryltetrahydrothiopyran (35) and *cis*-2,4-diaryltetrahydrothiopyran (36) with hydrogen peroxide in acetic acid for 7 min at 60° gives the corresponding 1,1-dioxides (37) (S.K. Klimenko *et al.*, Khim. Geterotsikl. Soedin., 1987, 463).

(37)

(35) R^1 = Ph, 4-MeOC$_6$H$_4$, R^2 = H
(36) R^1 = Ph, R^2 = Ph, 4-MeOC$_6$H$_4$,
    3,4-(MeO)$_2$C$_6$H$_3$
    R^1 = 4-MeOC$_6$H$_4$, R^2 = Ph,
    4-MeOC$_6$H$_4$

R^1 = R^2 = H
R^1 = H, R^2 = OH
R^1R^2 = OCH$_2$CH$_2$O

(38)

Tetrahydrothiopyrans 1-oxides (38) undergo facile deoxygenation on treatment with a mixture of zinc and dichlorodimethylsilane to give the corresponding tetrahydrothiopyran. 3,4-Dihydro-2H-benzo[b]thiopyran 1--oxides are deoxygenated in a similar manner (K. Nagasawa et al., Heterocycles, 1987, 26, 2607).

Catalytic hydrogenation of 2,6-diphenyl-4-ethynyl-4--hydroxytetrahydrothiopyran 1,1-dioxide over Pd/CaCO$_3$ affords 2,6-diphenyl-4-ethyl-4-hydroxytetrahydrothiopyran 1,1-dioxide. The reduction of related sulphides, 1-oxides and 1,4-di(tetrahydrothiopyranyl)buta-1,3-diynes over Pd/CaCO$_3$ and Raney nickel catalysts have been discussed (D.K. Kiyashev et al., Izv. Akad. Nauk Kaz. SSR, Ser. Khim., 1979, 73).

The Grignard reaction has been used to convert 4-cyano-
-2,2-dimethyl- and 4-cyano-2-ethyl-2-methyl- tetrahydrothiopyran into the related 4-acyl derivatives (39)
(R.A. Kuroyan *et al.*, Arm. Khim. Zh., 1983, **36**, 190) and
the Reformatskii, Darzens, and Wittig reactions have been
applied to alkyl substituted 4-formyltetrahydrothiopyrans
to synthesize substituted propionic, glycidic, and acrylic
acid derivatives, for example, ethyl 3-(2,2-dimethyltetra-
hydrothiopyran-3-yl)-3-hydroxypropanoate (40) (Kuroyan,
N.S. Arutyunyan and S.A. Vartanyan, *ibid.*, 1979, **32**, 546).

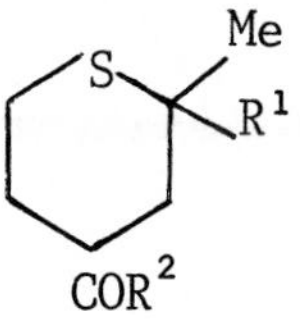

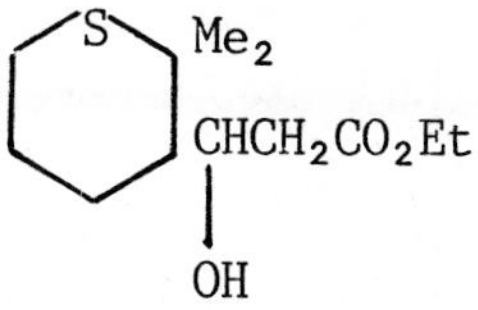

R^1 = Me, Et
R^2 = Me, Et, Pr, Bu, C_5H_{11}, Ph
        2-, 3-, 4-MeC_6H_4, 4-$MeOC_6H_4$

(39)   (40)

Aminomethylation of 2,6-diphenyl-4-ethynyl-4-hydroxytetra-
hydrothiopyran and its 1-oxide and 1,1-dioxide with
formaldehyde and a secondary amine yields 1-(aminomethyl)-2-
-(2,6-diphenyl-4-hydroxytetrahydrothiopyran-4-yl) acetylene
(41) (N.N. Godovikov *et al.*, Izv. Nauk Kaz. SSR, Ser.
Khim., 1982, 78).

R_2N = Et_2N, morpholino, n=0
R_2N = Et_2N, morpholino,
    Bu_2N, n=1
R_2N = Et_2N, morpholino,
    Bu_2N, n=2

(41)

(42) R^1 = Me, R^2 = lone pair
(43) R^1 = lone pair, R^2 = Me

The crystal and molecular structure of 1-(2,5-
-dihydroxyphenyl)tetrahydrothiopyranium bromide have been
determined from X-ray crystallographic and ir-spectral data
(D.S. Yufit, Yu.T. Struchkov, and L.R. Barykina, Zh. org.
Khim., 1987, 23, 1717).

Sulphonic ylide hexafluorophosphates (42) and (43)
undergo highly stereoselective but nonstereospecific [2,3]
sigmatropic ring enlargement. The six-membered methylide
from (42), where the carbanionic and vinyl moieties are on
the same side of the ring, reacts faster than the ylide
from (43); however both give (E)-5-methylthiacyclonon-4-
-ene as the exclusive product (V. Ceré et al., J. org.
Chem., 1979, 44, 4128).

Tetrahydrothiopyran-4-ones. 3,5-Dialkyl- and 3-methyl-5-
-phenyl- tetrahydrothiopyran-4-ones (44) are obtained by the
reaction of the appropriate dialkyl ketone with
formaldehyde and potassium hydrogen sulphide (V.I. Dronov
et al., Zh. org. Khim., 1985, 21, 1102).

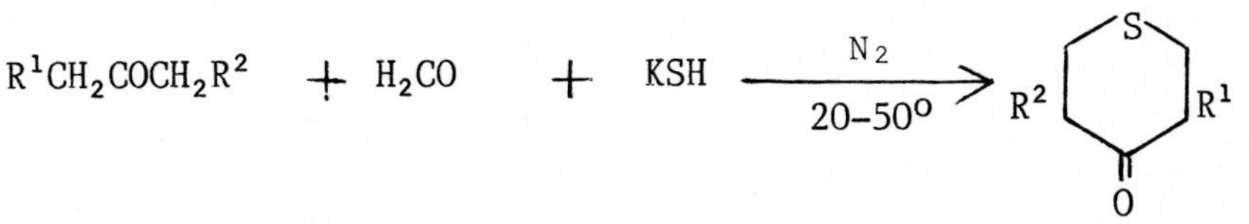

$$R^1CH_2COCH_2R^2 \quad + \quad H_2CO \quad + \quad KSH \quad \xrightarrow[20-50^o]{N_2} \quad$$

$R^1 = R^2 = $ Et, Pr, Bu
$R^1 = $ Me, $R^2 = $ Ph

(44)

The treatment of the unsaturated ketones $Me_2C=CHCOCH_2CH_2NEt_2$, $Me_2C=CHCOC(Me)_2NC_5H_{10}$, and $PhCH=CHCOCH_2CHPhNC_5H_{10}$ with hydrogen sulphide in ethanol affords 2,2-dimethyl-, 2,2,6,6--tetramethyl-, and 2,6-diphenyl- tetrahydrothiopyran-4-one, respectively (K.Kh. Tokmurzin *et al.*, Izv. Akad. Nauk Kaz. SSR, Ser. Khim., 1986, 92). Methyl 2-methyl-3-(2,2--dimethyltetrahydrothiopyran-4-one-5-yl)propanoate (47) is prepared by the addition of the enamine (45) to the unsaturated ester (46). Hydrolysis of (47) yields the corresponding acid. (E.A. Abgaryan and S.A. Vartanyan, Arm. Khim. Zh., 1984, **37**, 567).

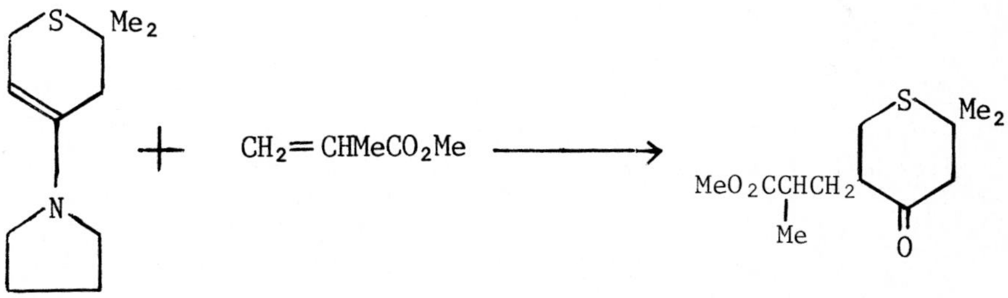

$$+ \quad CH_2=CHMeCO_2Me \quad \longrightarrow$$

(45)   (46)   (47)

The organocuprate conjugate addition reactions of 3,4-
-dihydro-4H-thiopyran with LiCuR$_2$ (R = Me, Bu),
BrMgCu(Bu)C≡CPr, MeMgBr-Cu(OAc)$_2$.H$_2$O, and RCu-Bu$_3$P [R = Me,
Bu, CH=CH(CH$_2$)$_6$OSiMe$_2$Bu^t] afford 2-substituted
tetrahydrothiopyran-4-ones (48). New methods have been
developed for the synthesis of a variety of 2-substituted
tetrahydrothiopyran-4-ones and the corresponding 1-oxides
and 1,1-dioxides and approaches to the synthesis of 2,3-
-disubstituted tetrahydrothiopyran-4-ones have been
discussed (R.J. Batten et $al.$, J. chem. Soc., Perkin I,
1982, 1177).

R = Me, Bu,
  CH=CH(CH$_2$)$_6$OSiMe$_2$Bu^t

R = Ph, 2-, 3-ClC$_6$H$_4$

(48)   (49)

The condensation of 2,6-diphenyltetrahydrothiopyran-4-
-ones with aldehydes affords 3,5-dimethylene derivatives
(49), which have been oxidised to their 1-oxides. Their
reactions with bromine, Grignard reagents, and hydroxylamine
have been reported [A.A. El-Barbary and F.M.E. Abdel-Megeid,
J. Chin. chem. Soc. (Taipei), 1984, 31, 377]. Some
derivatives of 3,5-bis(arylmethylene)-2,6-diphenyltetra-
hydrothiopyran-4-thiones have been synthesized (El-Barbary,
S.M. Hussain, and M.M.S. El-Morsy, Croat. Chem. Acta, 1985,
58, 79). Also for the preparation of some derivatives of
3,5-dibenzylidene-2,6-diphenyltetrahydrothiopyran-4-one and
-4-thione see El-Barbary and A.M. El-Reedy (Proc. Pak.
Acad. Sci., 1984, 21, 37).
   Treatment of 2-benzoyl-3-phenyl-3-(2,2-dimethyltetrahydro-
thiopyran-4-one-5-yl)prop-1-ene (50) with active methylene
compounds gives the octahydrobenzo[c]thiopyrans (51)
(G.V. Pavel and M.N. Tilichenko, Zh. org. Khim., 1978, 14,
2369). Similar studies have been made with the related
tetrahydropyran derivatives.

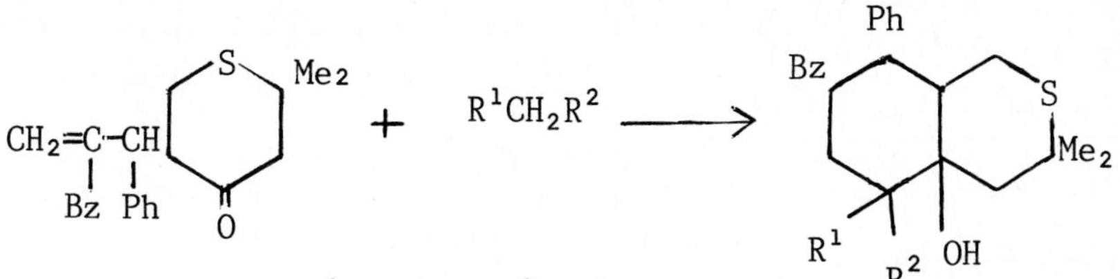

$$R^1 = NO_2, \quad R^2 = H$$
$$R^1 = R^2 = CN$$
$$R^1 = CN, \quad R^2 = CONH_2$$

(50)   (51)

2-Alkyl- and 2-alkyl-2-methyl-4-acetyltetrahydrothiopyran (54) are obtained when the related tetrahydrothiopyran-4-one (52) reacts with ethyl α-chloropropionate in the presence of sodium, followed by treatment of the product (53) with sodium hydroxide and then hydrochloric acid (R.H. Kuroyan, A.I. Markosyan, and S.A. Vartanyan, Arm. Khim. Zh., 1979, 32, 801).

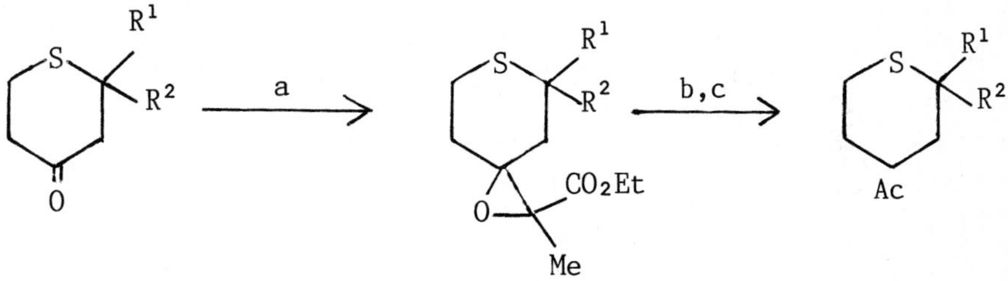

R^1 = H, Me
R^2 = Me, Et, Pr, Pr^i

(52)   (53)   (54)

Reagents:- (a) MeCHClCO$_2$Et, Na; (b) NaOH; (c) HCl

The structure of (R)-2,*trans*-6-diphenyl-*cis*-3-methyltetra-
hydrothiopyran-4-one and (R)-2,*trans*-6-diphenyl-*cis*-3-ethyl-
tetrahydrothiopyran-4-one have been determined by single-
-crystal X-ray diffraction studies. The nmr ([1]H- and [13]C-)
and ir-spectral data of a number of tetrahydrothiopyran-4-ones
and -4-ols indicate that the heterocyclic ring is predominantly
in the chair formation (K. Ramalingam *et al*., J. org.
Chem., 1979, **44**, 477).

The reduction of the *cis*-2,6-diaryltetrahydrothiopyran-4-
-one oximes (55) with lithium tetrahydridoaluminate affords
the epimeric 4-aminotetrahydrothiopyrans (56) and (57),
whereas similar reduction of the *trans*-2,6-diaryltetrahydro-
thiopyran-4-one oximes (58) yields only the aminotetra-
hydrothiopyrans (59). The stereochemistry of each of the
amino compounds (56), (57), and (59) has been determined
from their [1]H-nmr spectral data. Compounds (57) and (59),
each possessing an equatorial amino group are stronger bases
than compounds (56) which each have an axial amino group
(V. Baliah and N. Bhavani, Indian J. Chem., 1978, **16B**, 776).

R = Ph, $4\text{-MeOC}_6\text{H}_4$, $4\text{-ClC}_6\text{H}_4$, $2\text{-ClC}_6\text{H}_4$, $4\text{-MeC}_6\text{H}_4$

The thermal and photolytic decomposition of tetrahydrothio-
pyran-3- and -4-one tosylhydrazones have been investigated
(T. Ogata et $al.$, Nippon Kagaku Kaishi, 1987, 1370).

Irradiation of 3,4-dihydro-2H-thiopyran-4-one (60) in
benzene saturated with 1,1-dimethylethene yields a mixture
of 3-(2-methylallyl)tetrahydrothiopyran-4-one (61) and the
cyclo adduct (62) in the ratio of 45:55, respectively
(E. Anklam, S. Lau, and P. Margaretha, Helv., 1985, 68,
1129).

| (60) | (61) | (62) |

A characteristic reaction is observed in the solid state,
on heating a ground mixture of cis-2,6-diphenyl-1-methyl-4-
-oxotetrahydrothiopyranium tetrafluoroborate (63) and an
alkali halide. These reactions occur via ion exchange. In
potassium chloride the reaction products are cis-2,6-
-diphenyltetrahydrothiopyran and 1,5-diphenyl-5-
-(methylthio)-(E)-1-penten-s-$trans$-3-one, in potassium
bromide mainly dibenzalacetone, in potassium iodide 1,5-
-bis(methylthio)-1,5-diphenylpentan-3-one, 1,5-diphenyl-5-
-(methylthio)-(E)-1-penten-s-$trans$-3-one, dibenzalacetone,
and cis-2,6-diphenyltetrahydrothiopyran, and in sodium
chloride the latter three compounds. Other observations
concerning these reactions have been reported and the
reactions of $trans$-2,6-diphenyl-1-methyl-4-oxotetrahydrothio-
pyranium tetrafluoroborate (64) and some alkali halides
have been reported (K. Tokuno et $al.$, Yakugaku Zasshi,
1985, 105, 828; 1986, 106, 187, 193).

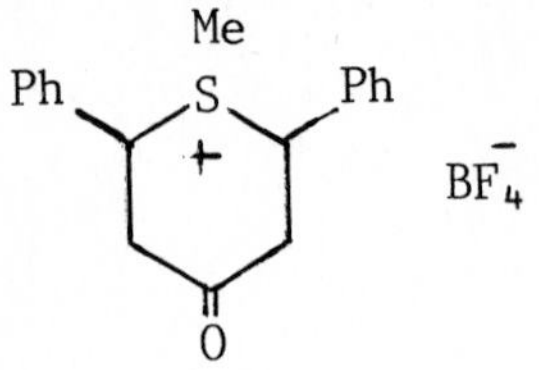

(63)

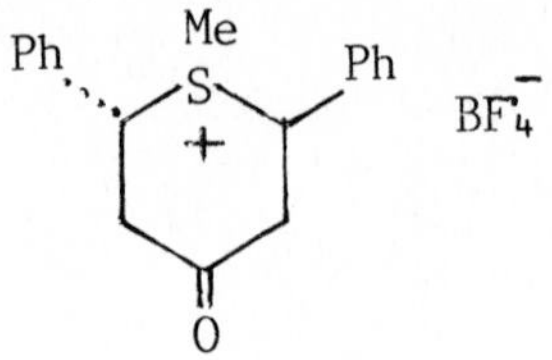

(64)

Some substituted tetrahydrothiopyran-2,4-diones and their metal salts, useful as herbicides, have been synthesized, for instance, 6,6-dimethyltetrahydrothiopyran-2,4-dione in pyridine on treatment with zinc chloride and then butyryl chloride yields 3-butyryl-6,6-dimethyltetrahydrothiopyran--2,4-dione (65) which on oximation gives the dione (66) (H.J. Wroblowsky *et al.*, Ger. Offen. DE 3,421,351, 1985).

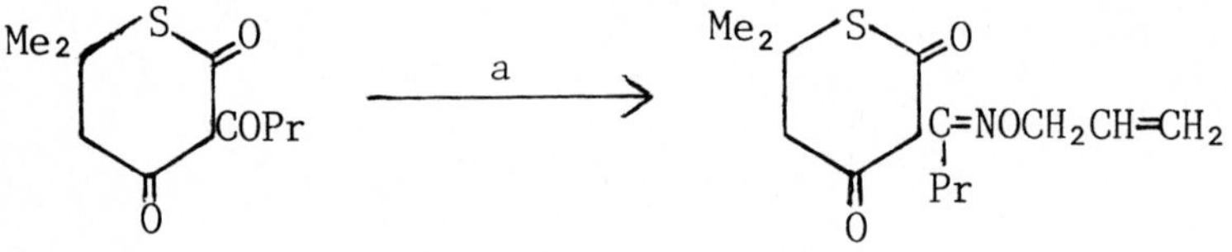

(65)   (66)

Reagents:- (a) $H_2C=CHCH_2ONH_2 \cdot HCl$, MeOH, MeONa

4-Alkyltetrahydrothiopyran-3,5-diones (68) and related tetrahydropyrandiones, useful in the preparation of D--homosteroid analogues with potent antiimplantation activity, are prepared from anhydrides (67) by sequential methanolysis to diacid mono esters, acid chloride formation, alkylation by dialkylcadmiums, and cyclization of the resulting keto esters (T. Terasawa and T. Okada, Brit. Pat., 1,552,599, 1979).

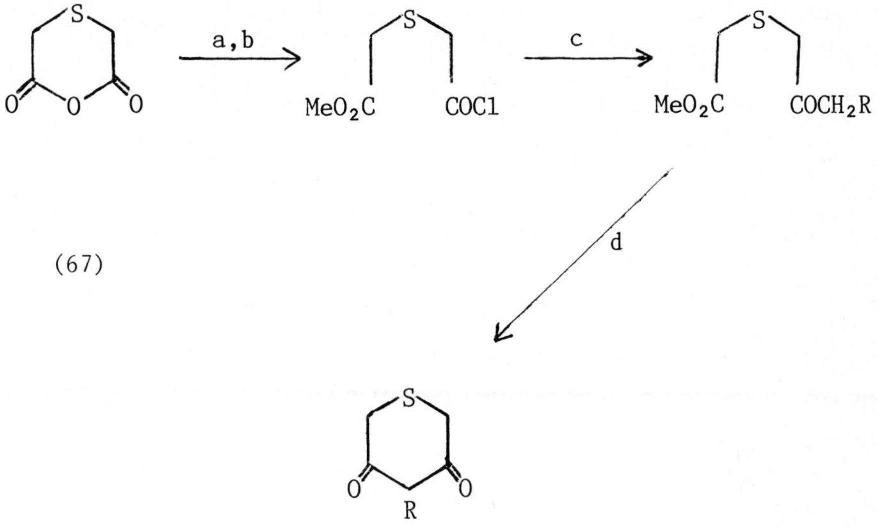

(67)

R = C$_{1-5}$ alkyl

(68)

Reagents:- (a) MeOH; (b) SOCl$_2$; (c) RCH$_2$MgBr, CdCl$_2$;
(d) NaH, THF

2. Benzo[b]thiopyrans, 1-benzothiopyrans, benzo[b]thiins, thiochromenes, 5,6-benzothiapyrans and derivatives

(a) Benzo[b]thiopyrans and benzo[b]thiopyranones and related naphtho derivatives

(i) Benzo[b]thiopyrans, 1-benzothiopyrans, benzo[b]thiins, thiochromenes

2,2-Dimethyl-2*H*-benzo[b]thiopyrans (4), the sulphur analogues of precocenes are synthesized by the following route. Michael addition of thiophenol to 3,3-dimethyl-acrylic acid in piperdine affords sulphide (1), which undergoes Friedel-Crafts cyclization in polysphosphoric acid at 50° to yield thiochroman-4-one (2). Reduction of ketone (2) to the 4-hydroxy compound (3) using sodium borohydride, followed by dehydration on heating with toluene-4-sulphonic acid in benzene furnishes the 2,2--dimethyl-2H-benzo[b]thiopyran (4) (J. Tercio *et al.*, Synth., 1987, 149).

(1)

R = H, 6-, 7-, 8-MeO,
6-Me, 6,7-(MeO)$_2$,
6,7-(MeO)EtO

(3)

(2)

(4)

The preparation of 6-methoxy-2,2-dimethyl-2H-benzo[b]thio-pyran, b.p. 125-130°/0.4 torr, and 6,7-dimethoxy-2,2--dimethyl-2H-benzo[b]thiopyran, b.p. 140-145°/0.4 torr, sulphur analogues of the natural insect antijuvenile hormones precocene I and II, by a related method has been described. The ^{13}C-nmr spectral data of the above and related compounds have also been reported (F. Camps *et al.*, J. heterocyclic Chem., 1983, **20**, 1115).

4-Trimethylsilyloxy-2H-benzo[b]thiopyran 1,1-dioxide (5) on treatment with thionyl chloride in dichloromethane/ether at 0° for 10 min yields the expected sulphine (6), which is trapped by 1,1-dimethylethene or norbornene to give the novel [4+2] cycloadducts (7) and (8), respectively (I.W.J. Still and F.J. Ablenas, Chem. Comm., 1985, 524).

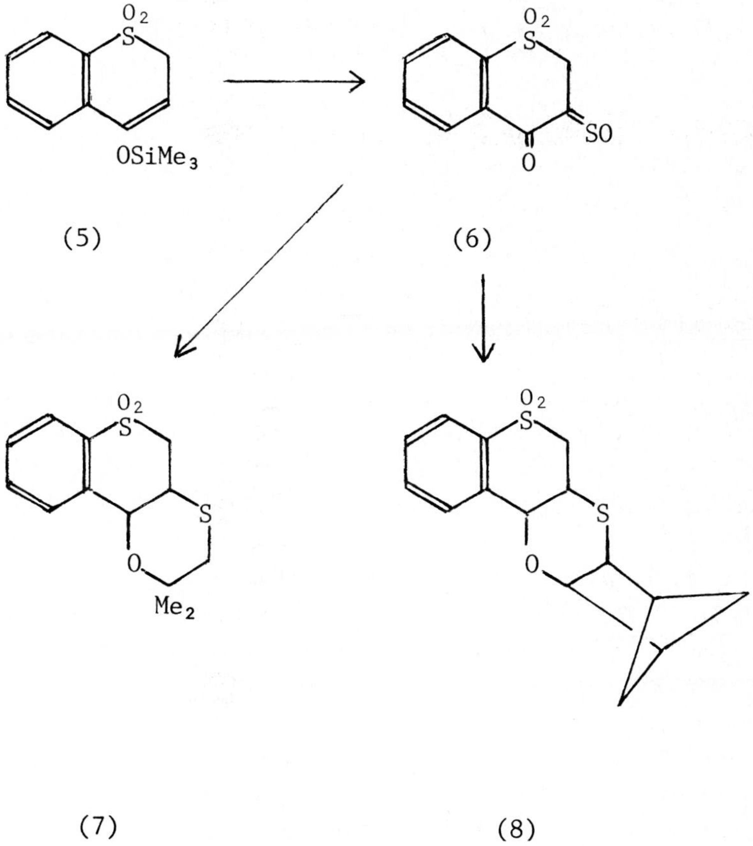

(5)

(6)

(7)

(8)

Cyclization of 2-(3-arylpropynoyl)cyclohexane-1,3-diones with perchloric acid or boron trifluoride etherate in the presence of hydrogen sulphide gives benzothiopyrylium salts, *thiochromylium salts*. The cyclization of 1,5--diketones to thiochromones and their conversion into the corresponding benzothiopyrylium salts has been discussed (L.I. Markova and K.M. Korshunova, Nukleofil'nye Reakts. Karbonil'nykh Soedin., 1982, 127).

(ii) 2H-Benzothiopyran-2-ones, 2H-benzothiin-2-ones, thio-coumarins

The general name in use for this class of compounds is thiocoumarin. A new, versatile thiocoumarin synthesis has been developed that is based on the Claisen rearrangement of allylic (2) or propargylic aryl ethers (6), which are at the same time monothio ortho esters. The routes diagrammatically presented below, proceed first by reaction between a benzenethiol (1) and an orthoacrylate and an orthopropiolate. The resulting arylthio ortho esters (2) and (6) are then converted to the 2-mercaptodi-hydrocinnanmate (3) and 2-mercaptocinnamate (7) respectively. Ring closure of the dihydrocinnamate (3) affords the 3,4-dihydrothiocoumarin (4), which following dehydrogenation yields thiocoumarin (5), obtained directly by ring closure of intermediate (7) (J.A. Panetta and H. Rapoport, J. org. Chem., 1982, 47, 2626).

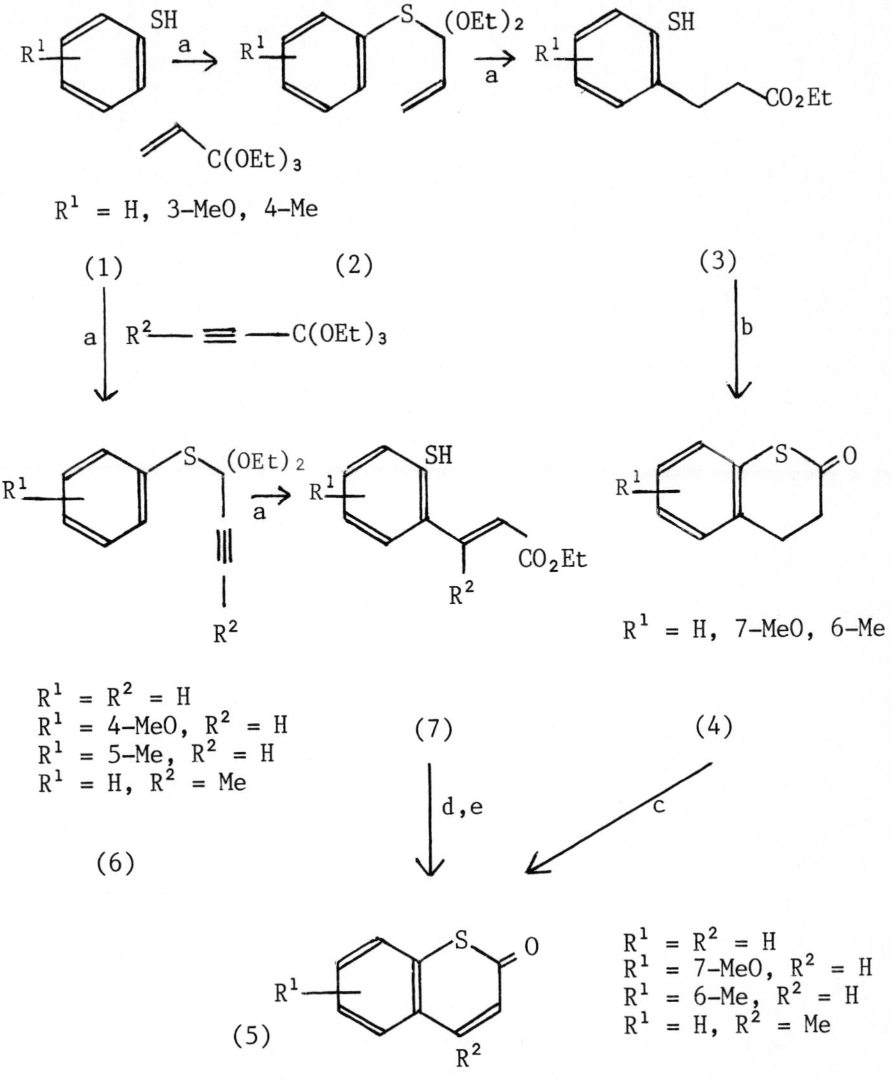
SH
a
(OEt)2
a
SH
CO2Et
C(OEt)3
R1 = H, 3-MeO, 4-Me
(1)
(2)
(3)
a
R2—≡—C(OEt)3
b
(OEt)2
a
SH
S
O
CO2Et
R2
R2
R1 = H, 7-MeO, 6-Me
R1 = R2 = H
R1 = 4-MeO, R2 = H
R1 = 5-Me, R2 = H
R1 = H, R2 = Me
(7)
(4)
d,e
c
(6)
S
O
(5)
R2
R1 = R2 = H
R1 = 7-MeO, R2 = H
R1 = 6-Me, R2 = H
R1 = H, R2 = Me

Reagents:- (a) pivalic acid, p-cymene, Δ;
(b) Ph_2O, 4-$MeC_6H_4SO_3H$; (c) Ph_2O, 10%Pd/C, Δ;
(d) 95% EtOH, KOH, Δ, HCl;
(e) polyphosphoric acid 100-103°

4-Methyl-, m.p. 121-122°; 6-methyl-, m.p. 84-85°; 7-methoxy-thiocoumarin, m.p. 102-103°.

4-Methylthiocoumarins (9) are prepared by the cyclocondensation of aryl acetothioacetate (8) in the presence
of aluminium chloride at 80-90°. 4-Methyl-2H-naphtho-
[1,2-b]thiopyran-2-one (10) is obtained in a similar
manner. The ir-, nmr-, and mass-spectral data of these 4-
-methylthiocoumarins have been reported (H. Nakazumi,
A. Asada, and T. Kitaom, Chem. Letters, 1979, 387). For
the synthesis of 3H-naphtho[2,1-b]thiopyran-3-ones(11) see
T. Manimaran and V.T. Ramakrishnan (Indian J. Chem., 1979,
18B, 78).

(8)

R = 7-Me, 6-Me, 7-Cl

(9)

R = H, Me

(10)   (11)

3-Phenyl-, 6-methyl-3-phenyl, 4-aryl-, and 4-aryl-6-
-methyl-thiocoumarins have been synthesized (M. Natarajan
and V.T. Ramakrishnan, *ibid.*, 1984, 23B, 720).

The substitution reaction between 3-bromo-4-hydroxythio-
coumarin (12) and an alkyl thiol or thiophenol gives the
corresponding 4-hydroxy-3-mercaptothiocoumarin (13)
(G.M. Vishnyakova, T.V. Smirnova, and O.N. Fedorova, Izv.
Vyssh. Uchebn. Zaved., Khim. Khim. Tekhnol., 1983, 26,
554).

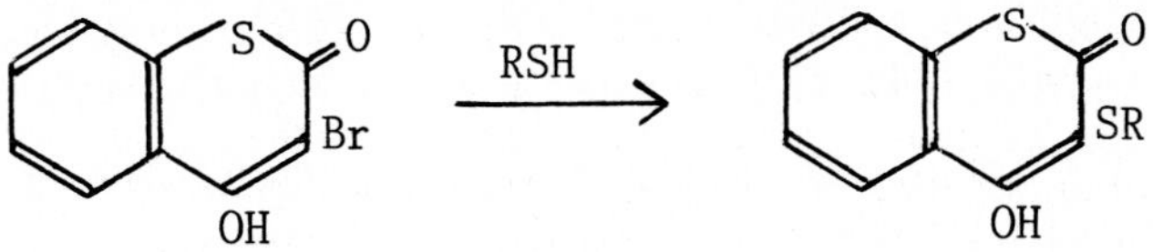

RSH →

R = Et, Pr, Pr^i, Bu, Me$_2$CHCH$_2$, Ph

(12)   (13)

Substituted 2-chloroacetophenones (14) react with carbon disulphide in the presence of sodium hydride to form 4--hydroxydithiocoumarin anions. Kinetic protonation furnishes 4-hydroxydithiocoumarins (15), which tautomerize to 2--mercaptothiochromones (16) on treatment with strong acid. Both tautomers are quite stable and do not interconvert readily.   Alkylation of compounds (15) and (16) affords S--alkyl derivatives only, while acylation with acid anhydrides gives mixtures of O- and S- acyl products (J.E. Anderson-McKay and A.J. Liepa, Austral. J. Chem., 1987, 40, 1179).

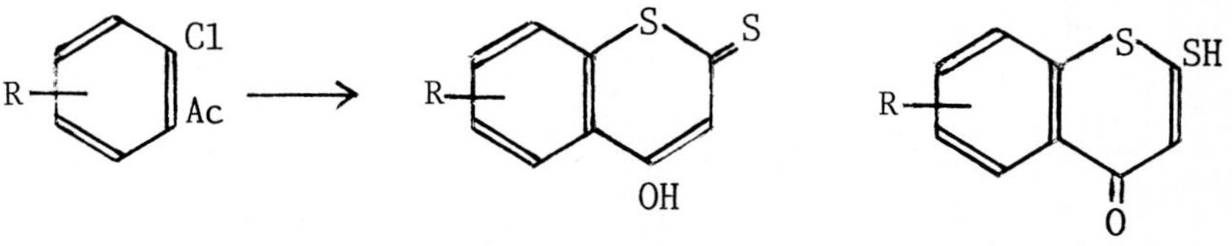

R = H, 4-Cl, 5-Cl   R = H, 6-Cl, 7-Cl

(14)   (15)   (16)

Condensation of dihydrobenzo[b]thiophene-2,3-dione (17) with malononitrile yields the 3-substituted 2-imino-2*H*- -benzo[b]thiopyran-4-carboxylic acid (18), but the reaction of dione (17) with ethyl cyanoacetate also affords the imide derivative (19) of thiocoumarin. Similar reactions of dione (17) with $RCH_2C(NH_2)=CRCN$ (R = CN, CO_2Et) give 3-(2-amino-1,1-dicyanoethen—2-yl)-2-imino-2*H*-benzo[b]- thiopyran-4-carboxylic acid (20) and 3-(2-amino-1-cyano-1- -ethoxycarbonylethen-2-yl)thiocoumarin-4-carboxylic acid (21), respectively (M.M.M. Sallam *et al*., J. prakt. Chem., 1985, 327, 333).

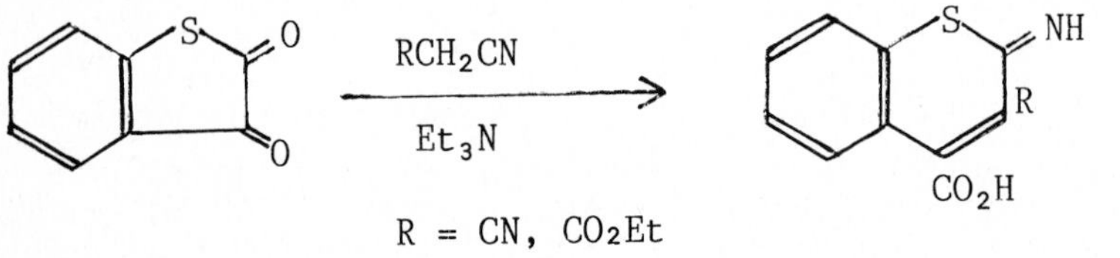

(17)   (18)

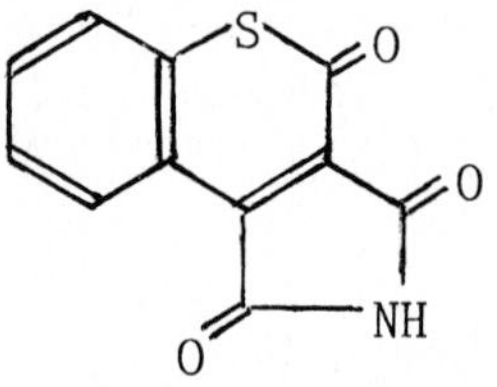

(19)

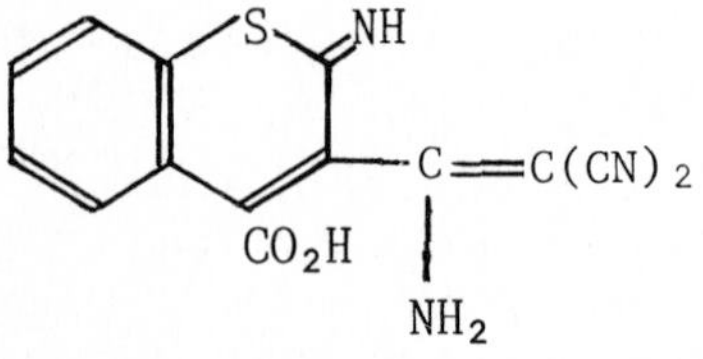

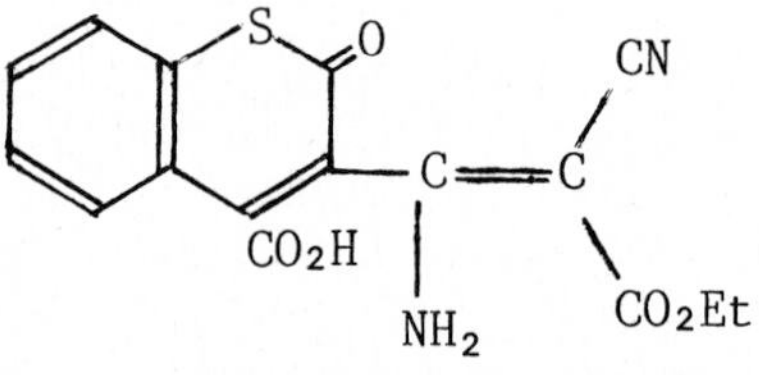

(20)   (21)

Thiocoumarin-3-carboxamides (22) have been prepared. The
amide (22; R = 4-MeOC$_6$H$_4$) reacts with 2,3-dimethylbuta-
-1,3-diene and with 1,2-dimethylenecyclohexane to give the
[4 + 2] cycloadducts (23) and (24), respectively
(H. Gotthardt and N. Hoffman, Ann., 1985, 529).

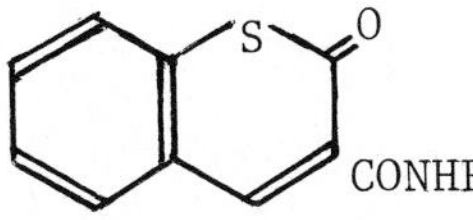

R=PhCH$_2$,Ph, 4-MeC$_6$H$_4$,
4-MeOC$_6$H$_4$, 4-ClC$_6$H$_4$

(22)

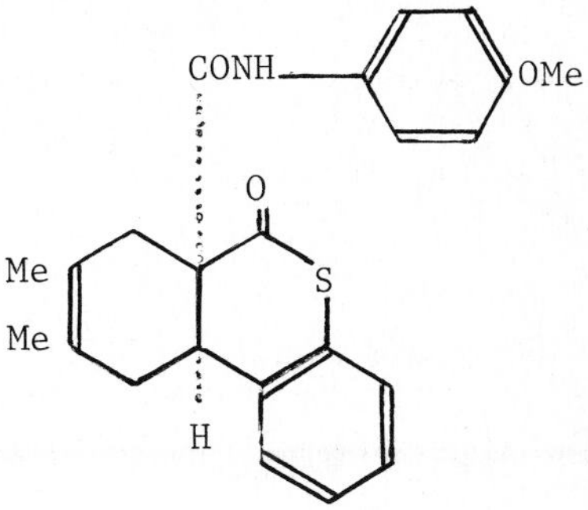

(23)

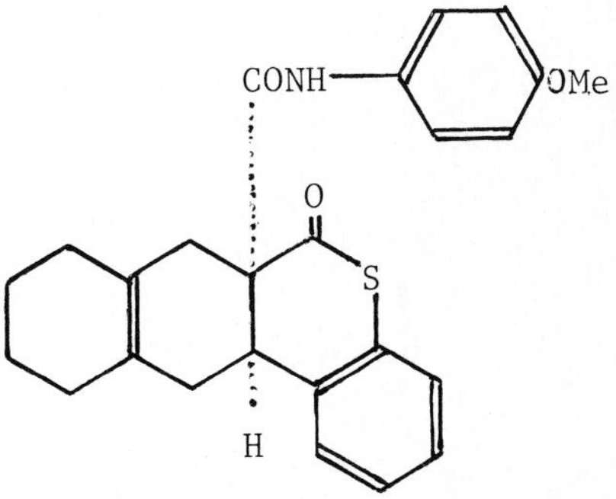

(24)

Cycloaddition occurs on irradiating (≥ 300 nm) 4-methoxy-
thiocoumarin and isobutene in acetonitrile to yield the
tetrahydrocyclobutabenzothiopyranone (25), which on
treatment with boron trifluoride etherate yields the 1,2-
-dihydrocyclobuta[c]benzo[b]thiopyran-3(3H)-one (26).
(C. Kaneko, T. Naito, and T. Ohashi, Heterocycles, 1983,
20, 1275).

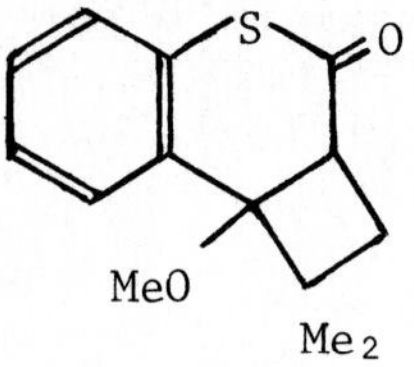

472

(25) 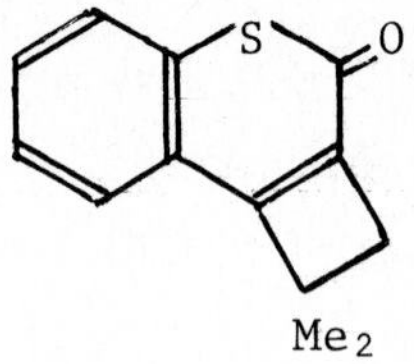   (26)

Michael addition of 4-hydroxythiocoumarin with
arylideneindandiones (27) in methanol containing sodium
methoxide affords [3-[α-(indan-1,3-dione-2-yl)-
-(subsituted)benzyl]-4-hydroxythiocoumarins (28). Related
derivatives of coumarin have also been prepared.
(V.N. Marshalkin, T.V. Smirnova, and K.V. Yushchenko, Zh.
org. Khim., 1985, 21, 2427).

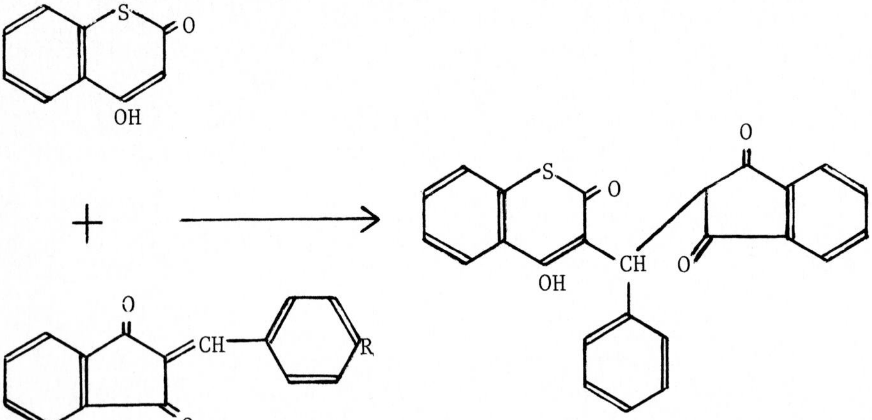

R=e.g., H, Me, OMe, Cl, NO$_2$

(27)   (28)

4-Hydroxythiocoumarin condenses with 3-(4'-bromo-4-
-biphenylyl)-1,2,3,4-tetrahydro-1-naphthol in acetic acid
containing sulphuric acid to afford the 4-hydroxy-3-
naphthylthiocoumarin (29), which along with other related
derivatives, act as rodenticides (J.J. Berthelon, Fr.
Demande FR 2,562,893, 1985).

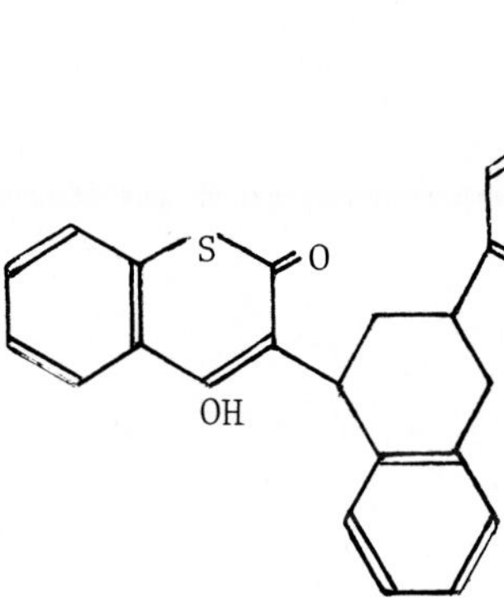

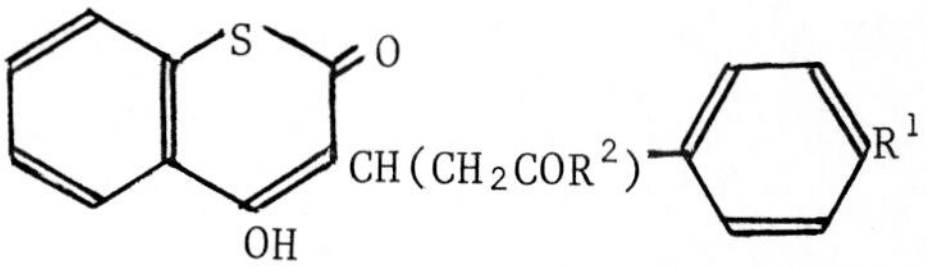

R^1=H, Br, Cl, OMe, OEt, OPr^i
R^2=Me, Et

(29)   (30)

Treatment of 4-hydroxythiocoumarins with α,β-unsaturated
ketones yields 3-substituted 4-hydroxythiocoumarins (30)
(G.M. Vishnyakova *et al.*, Izv. Vyssh. Uchebn. Zaved., Khim.
Khim. Tekhnol., 1979, **22**, 283).

4-Hydroxythiocoumarin with phosphorus pentasulphide in boiling dioxane or toluene gives di(dithiocoumarin-4-yl)-sulphide (31; X = S). The same reaction with 4-hydroxy-coumarin yields sulphide (31; X = O) (T.V. Smirnova, Vishnyakova, and B.F. Andronov, Chem. Abs., 1979, $\underline{90}$, 54771d).

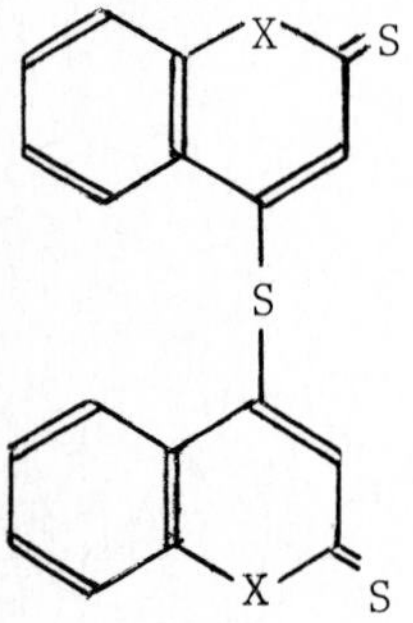

R^1, R^2 = H,Me

(32)

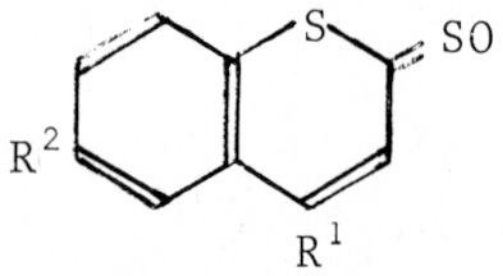

(33)

X=S,O

(31)

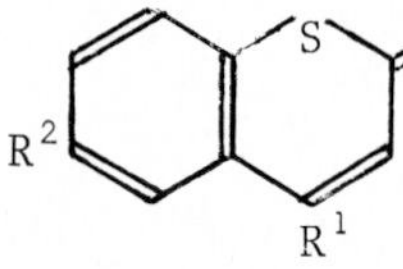

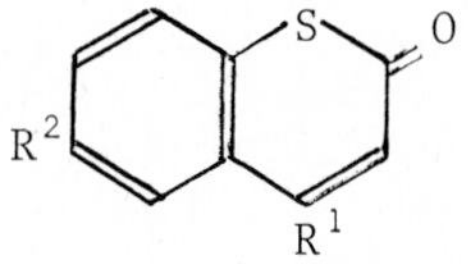

(34)

Dithiocoumarins, $2H$-benzo[b]thiopyran-2-thiones (32) undergo peracid oxidation to give sulphoxides (33), which on further oxidation afford the related thiocoumarins (34) or polymeric products (K. Buggle and B. Fallon, Monatsh., 1987, $\underline{118}$, 1197).

(iii) 4H-Benzothiopyran-4-ones, 4H-benzothiin-4-ones, thiochromones and 2-phenyl-4H-benzothiopyran-4-ones, thioflavones

S-Aryl oxobutanethioates (1) undergo cyclization in the presence of polyphosphoric acid to give 2-methylthio-chromones (2) or 7-methoxy-4-methylthiocoumarin (3) from (1; R = 3-MeOC$_6$H$_4$) (H. Nakazumi and T. Kitao, Chem. Letters, 1978, 929).

MeCOCH$_2$COSR

R=Ph, 3- or 4-MeC$_6$H$_4$,
 3- or 4-MeOC$_6$H$_4$,
 2-naphthyl

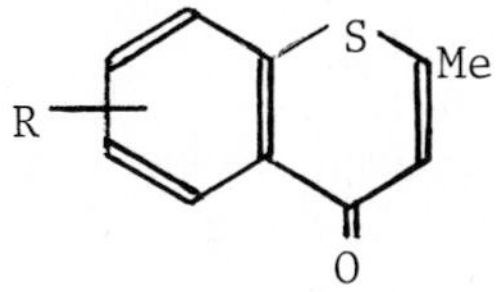

R=H, 6- or 7-MeO
 6-MeO,5,6-benzo

(1)   (2)   (3)

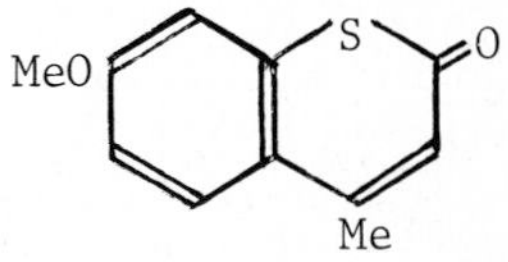

The *S*-esters (4) are very versatile intermediates and they can be transformed into different heterocyclic systems, for instance, on treatment with sodium hydride in the presence of air they give benzothiophene (5) and 2--hydroxythiochromone (6) derivatives. In the absence of air only the 2-hydroxythiochromones (6) are produced (C.K. Law *et al.*, J. org. Chem., 1987, 52, 1670).

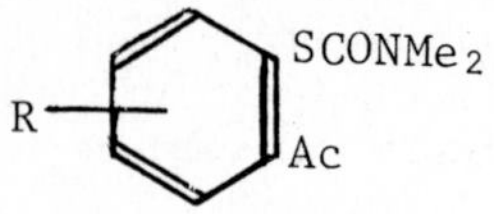

R=H,5-MeO,3-HO,
   3-MeO,3-AcO,3-CF$_3$SO$_3$

(4)

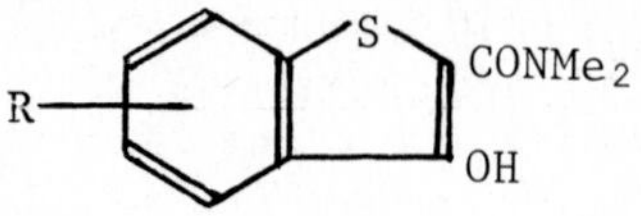

(5)

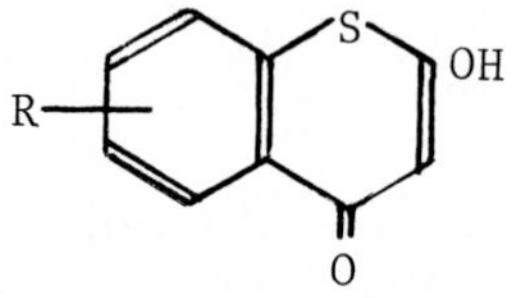

(6)

Thiochromone-3-carboxylic acids (7) substituted in the benzene ring are prepared from the appropriately substituted thiophenol and diethyl ethoxymethylenemalonate in polyphosphoric acid, followed by heating with water (Mitsui Toatsu Chemicals, Inc. Japan Kokai Tokkyo koho JP 57, 185,279 [82,185,279], 1982).

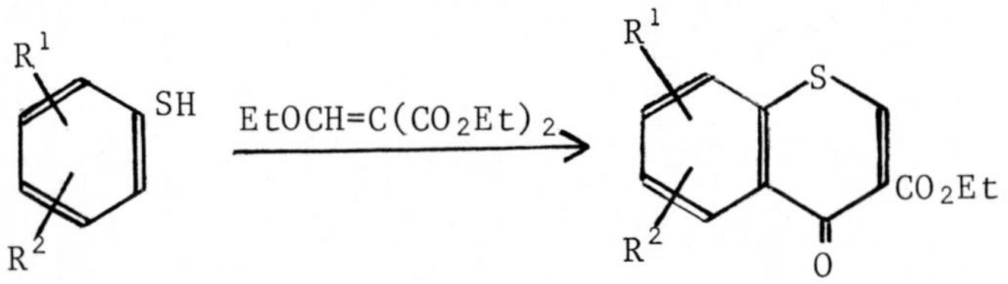

R^1,R^2=H,Me,Cl,MeO

(7)

A novel irreversible Wessely-Moser rearrangement occurs on heating ethyl 5-mercapto-2-carboxylate (8) with 2M sodium hydroxide on a steam bath, followed by acid work-up to afford 5-hydroxythiochromone-2-carboxylic acid (9) (J.L. Suschitzky, Chem. Comm., 1984, 275).

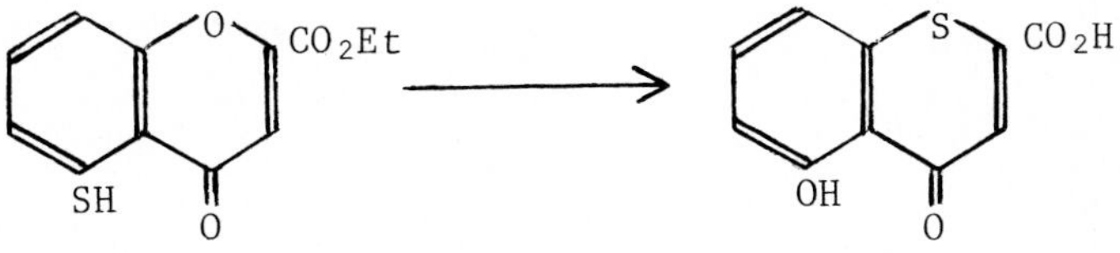

(8)   (9)

Methylation of 2-methylthiochromones (10) with dimethyl sulphate yields blue coloured dimers, benzothiopyrylium perchlorates (11), while methylation of 7-methoxy-4--methylthiocoumarin gives 2,7-dimethoxy-4-methylbenzo[b]-thiopyrylium perchlorate, a yellow salt (Nakazumi and Kitao, Bull. chem. Soc., Japan, 1979, 52, 160).

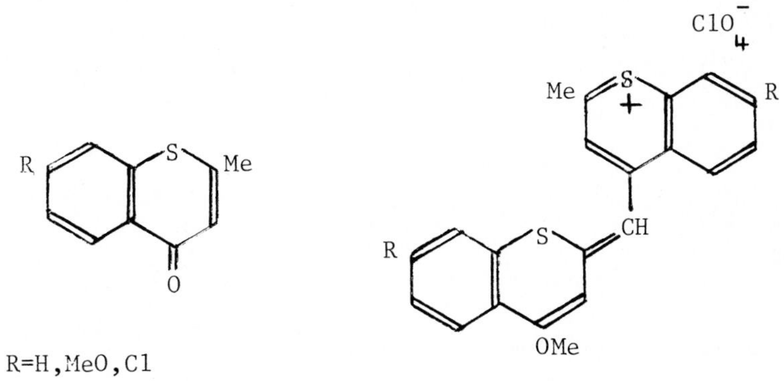

R=H,MeO,Cl

(10)   (11)

Treatment of 5- and 7-methoxy-2-methylthiochromones with lithium diisopropylamide followed by electrophiles gives 2-substituted thiochromones (J. Hirao *et al.*, *ibid.*, 1985, 58, 2203).

Cyclocondensation of 3-formylthiochromone (12) with *o*-phenylenediamines (13) and *o*-aminobenzenethiol (15) affords benzodizaepines (14) and benzothiazepine (16), respectively (Nakazumi *et al.*, Chem. Express, 1986, 1, 21).

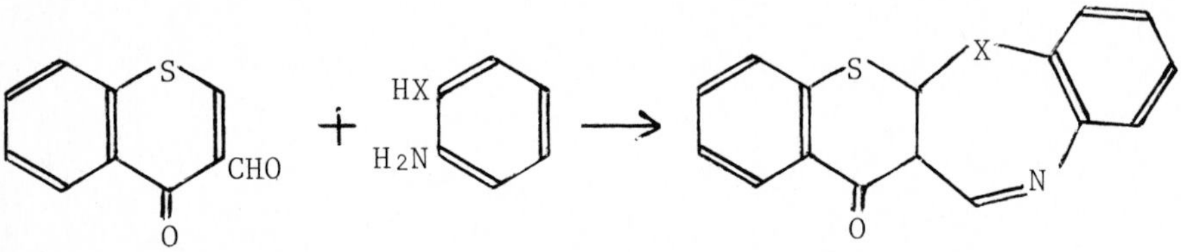

(12)	(13) X=NMe,NPh	(14) X=NMe,NPh
	(15) X=S	(16) X=S

2-Amino-3-hydroxy(or methoxy)thiochromones (17) are obtained by reduction of the 2-nitro or 2-phenylazo derivatives or by the reaction of the 2-bromo derivatives with amines. 3-Methoxy-2-(4-methylpiperazinyl)thiochromone affects the central nervous system (F. Eiden and G. Felbermeir, Arch. Pharm., 1983, 316, 921).

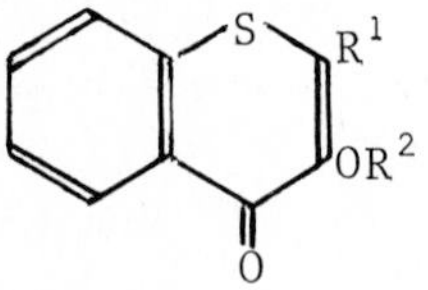

$R^1=NH_2, R^2=H$
$R^1=NMe_2$, piperidino, morpholino, 4-methylpiperazino, $R^2=Me$

(17)

Aminomethylation of 3-hydroxythiochromone with formaldehyde and an amine affords the 2-aminomethyl-3--hydroxythiochromone (18). 2-(Dimethylamino)-3-hydroxythiochromone (18; $R^1 = R^2$ = Me) on heating yields the spiro compound (19) (*idem, ibid.*, p.1034).

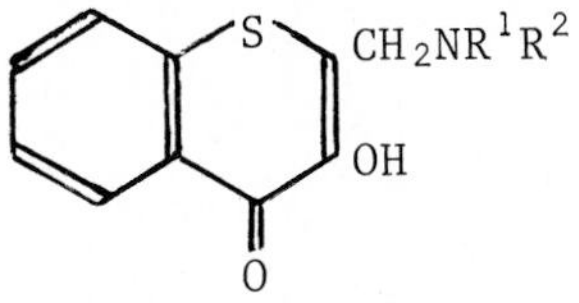

$R^1=R^2=Me,Et$
$R^1=H,Me,R^2=Ph$
NR^1R^2=piperidino, morpholino,
4-methyl-1-piperazino,
4-[3-(trifluoromethyl)phenyl]--1-piperazino

(18)

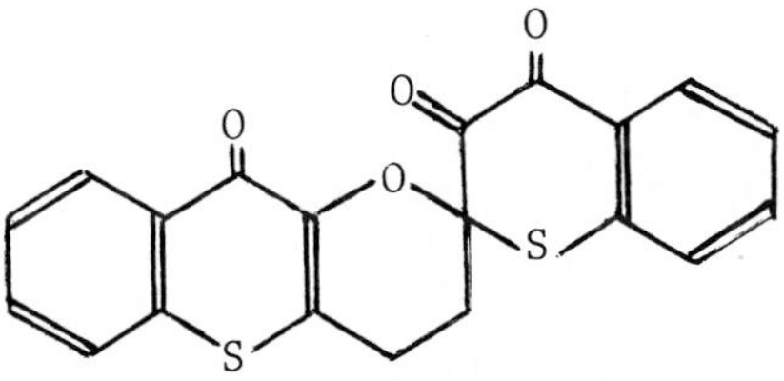

(19)

2-Acetylaminothiochromone (20) on boiling with phosphoryl chloride gives 2-acetylimino-4-chloro-2H-benzo[b]thiopyran (21), which on amination with aniline yields 2-acetylamino--4-phenylimino-4H-benzo[b]thiopyran (22), but on treatment with piperidine affords the tautomeric derivative, 2--acetylimino-4-piperidino-2H-benzo[b]thiopyran (23) (O.M. Glozman, L.A. Zhmurenko, and V.A. Zagorevskii, Khim. Geterotsikl. Soedin., 1978, 622).

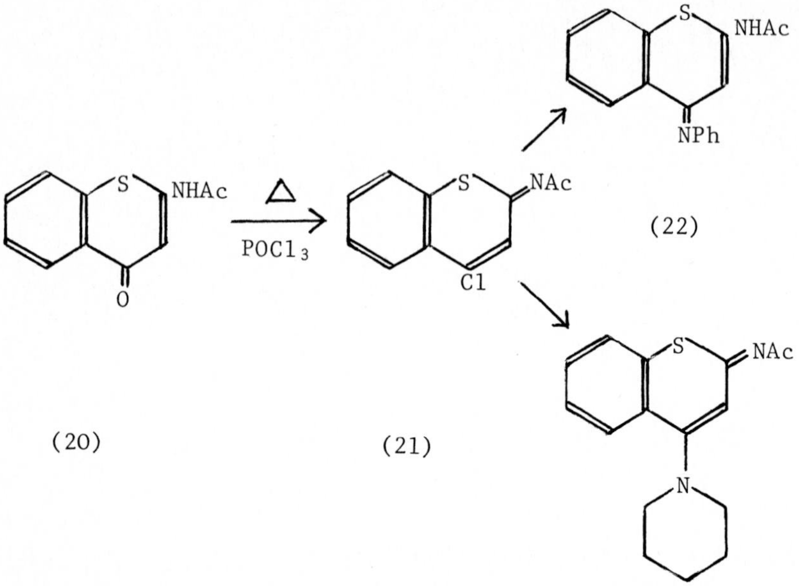

(20)   (21)

(22)

(23)

A number of derivatives of thiochromone including, 7-
-aminoalkoxythiochromones and related chromones (J.C. Jaen
and L.D. Wise, Eur. Pat. Appl. EP 190,015, 1986); 5-
-aminothichromones and chromones and their derivatives
obtained by acylation with ethyl oxalyl chloride, followed
by hydrolysis of the resulting ester (T.S. Abram,
P. Norman, and B.T. Warren, *ibid.*, 163,227, 1985); 7-
-amino-2-methylthiochromones and related 4-methylthio-
coumarins (P.G. Ferrini, *ibid.*, 5,141, 1979); and 3-
-ureidothiochromones, for example, 3-(3,3-dimethylureido)-
thiochromone, and related chromones (R. Giraudon, Brit. UK
Pat. Appl. 2,009,740, 1979) have been reported.

2-Phenyl-4H-benzo[b]thiopyran-4-ones, 2-phenyl-4H-benzo[b]-thiin, thioflavones. Thioflavones (1) are obtained by treating 2-MeS(ClCO)C$_6$H$_3$R with phenylethyne, followed by cyclization of the resulting product with hydrogen bromide. Related compounds containing an O or Se heteroatom have also been prepared (A.J. Luxen, L.E.E. Christiaens, and M.J. Renson, J. organometal. Chem., 1985, **287**, 81).

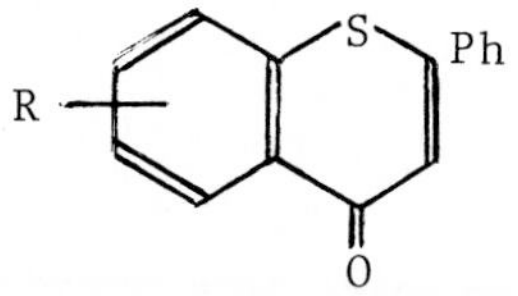

R=H, 7-F, 6-NO$_2$, 6-Me, 6-MeO

(1)

Some 3-(substituted methyl)thioflavones (2) have been prepared and tested for antimicrobal activity. 3--Acetoxymethylthioflavone shows the most antimicrobial potency *in vitro* against *Trichophyton mentagrophytes*, also most of the thioflavone 1,1-dioxides (3) show antimicrobial activity against fungi (H. Nakazumi, T. Ueyama, and T. Kitao, J. heterocyclic Chem., 1985, **22**, 1593).

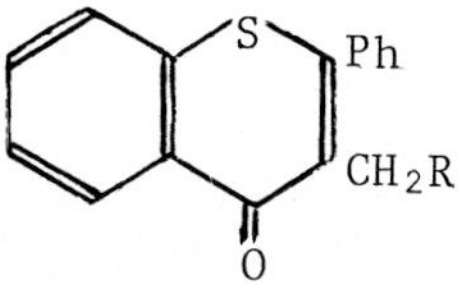

R=Cl, OH, OAc, OMe
OEt, OBu, O(CH$_2$)$_5$Me
O(CH$_2$)$_7$Me

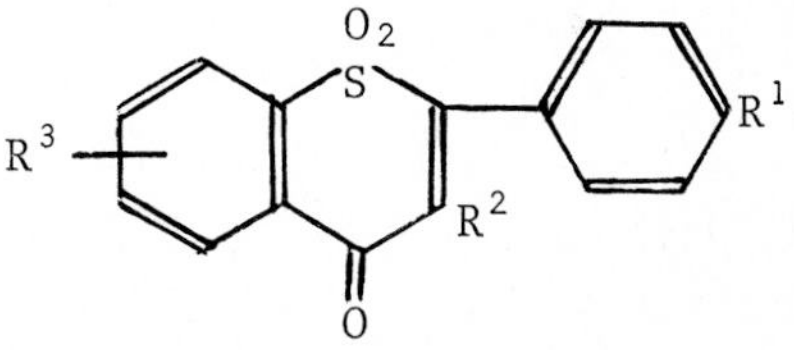

R^1=H,OMe
R^2=H,CH$_2$Cl, OAc
R^3=H,Me,OH,OMe

(2)   (3)

Some 3-substituted thiochromones and thioflavones (4) have been prepared and 3-formylthioflavone exhibits weak antimicrobial activity against *Trichophytons Candida* (Nakazumi *et al.*, *ibid.*, p.821.

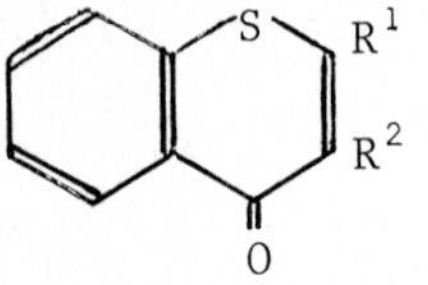

R^1=H, Ph
R^2=CHO, CN, CH=CHCO$_2$H, CO$_2$H

(4)

8-Formyl-thiochromones, -chromones, and -thioflavones (5) have been prepared (S. Goldmann, Ger. Offen. DE 3,311,004, 1984), and some appropriate 8-formylthioflavones have been converted to Schiff bases (6) which react with β-keto esters to give 2-thiochromenylideneacetoacetates (7), useful as intermediates in drug synthesis (G. Franckowiak and Goldman, *ibid.*, 3,517,950, 1986).

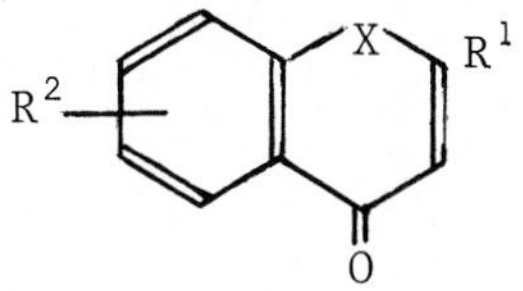

e.g. R^1=H, alkyl, alkoxycarbonyl, Ph,
      Ar, heteroarom
      R^2=H, halogeno
      X =S,O

(5)

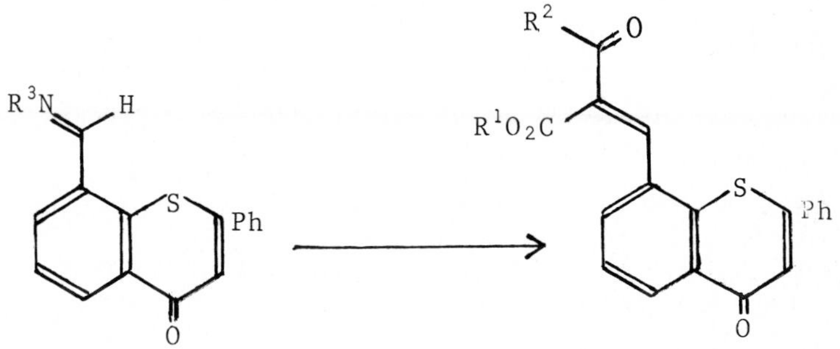

R^3=C $\leqslant$ 6 alkyl
   e.g., Bu

(6)

R^1=C $\leqslant$ 4 alkyl
R^2=C $\leqslant$ 4 alkyl (un)substituted
   with halogeno or C $\leqslant$ 7
   acyloxy

(7)

Thioflavone reacts with thionyl chloride in benzene to
afford 3-chlorothioflavone, which on oxidation with
hydrogen peroxide in acetic acid yields 3-chlorothioflavone
1,1-dioxide. Thioflavone on boiling with sulphuryl
chloride in tetrachloromethane, followed by treatment with
methanol or ethanol gives the 3-alkoxy-2,3-dichlorothio-
flavanone (8), which on oxidation with hydrogen peroxide in
acetic acid gives the corresponding 1,1-dioxides
(J.R. Merchant, G. Martyres, and S. Dike, Chem. and Ind.,

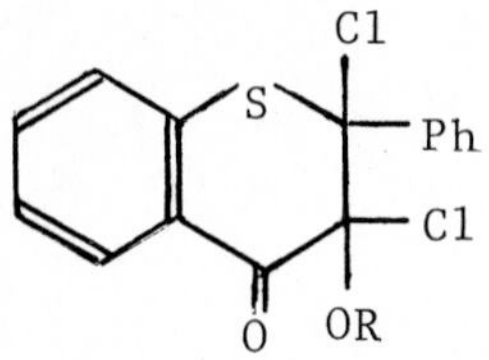

R=Me, Et

(8)

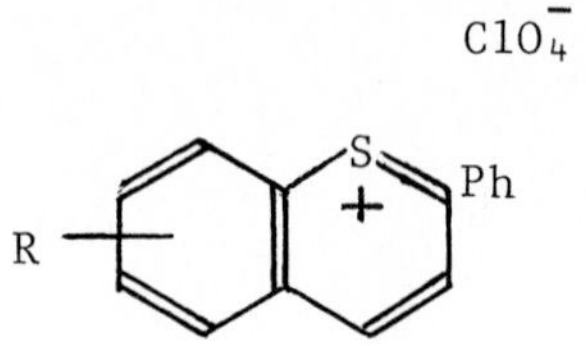

R=H, 5-, 6-, 7-MeO

(9)

A convenient synthesis of thioflavylium perchlorates, *2--phenylbenzo[b]thiopyrlium perchlorates* (9) is by reduction of the appropriate thioflavone with $LiAlH_4$ - $AlCl_3$ to the corresponding 4H-thioflavene, which on treatment with perchloric acid in acetic acid in the presence of dichloro-dicyanobenzoquinone gives the thioflavylium salt (9) (Nakazumi *et al.*, Bull. chem. Soc., Japan, 1983, <u>56</u>, 1251).

   3-Phenyl-4H-benzo[b]thiopyran-4-ones, 3-phenyl-4H-benzo-[b]thiin-4-ones, isothioflavins. Isothioflavones (11) are obtained by the Meerwein reaction of thiochromones (10) with 4-nitrobenzenediazonium ion in acetone in the presence of the catalyst, copper (II) chloride hydrate, and sodium acetate. Reduction of the 4'-nitro derivatives (11) using reductive iron and concentrated hydrochloric acid and boiling in ethanol affords the 4'-amino derivative (12) (*idem*, J. heterocyclic Chem., 1985, <u>22</u>, 821).

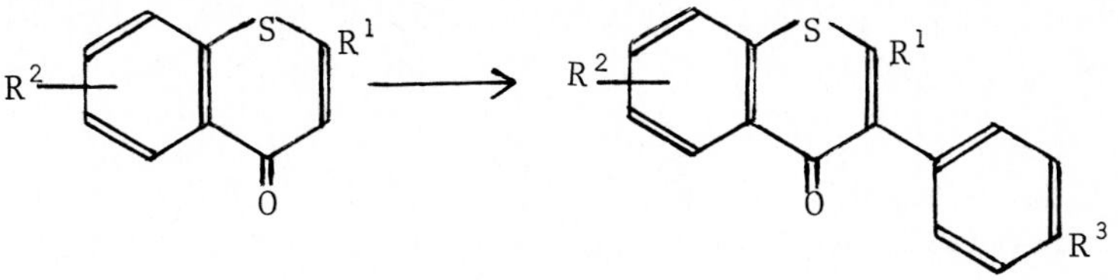

(10)

(11) $R^3 = NO_2$
(12) $R^3 = NH_2$

R^1	R^2	(11) m.p. °C	(12) m.p. °C
Me	H	173-176	145-147
Me	5-Me	167-170	132-135
Me	6-Me	169-171	140-142
Me	7-Me	192-194	179-181
Me	8-Me	182-187	168-172
H	H	221-224	

2-Aminoisothioflavones (13) are obtained on boiling RCH_2CN, $2\text{-}HSC_6H_4CO_2Me$ and Bu^tONa in pyridine (Yu.M. Volovenko *et al.*, Khim. Geterotsikl. Soedin., 1982, 1047).

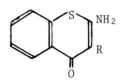

R=Ph, $2\text{-}ClC_6H_4$, $3\text{-}BrC_6H_4$, $4\text{-}EtOC_6H_4$, $3,4\text{-}(MeO)_2C_6H_3$, $2,3,4\text{-}(MeO)_3C_6H_2$

(13)

(b) 2,3-Dihydro-4H-benzo[b]thiopyrans, 2,3-dihydro-4H-1-
   -benzothiopyrans, 2,3-dihydro-4H-benzo[b]thiins,
   thiochromans and 2,3-dihydro-4H-benzo[b]thiopyran-4-
   -ones, 2,3-dihydro-4H-1-benzothiopyran-4-ones, 2,3-
   -dihydro-4H-benzo[b]thiin-4-ones, thiochroman-4-ones

All the above names are in use, but the ones generally
found in the literature are thiochroman and thiochroman-4-
-one, and these will, therefore, form the basis of the
nomenclature in this section.

(i) Thiochromans, 2,3-dihydro-4H-benzo[b]thiopyrans
6-Methylthiochroman has been obtained in 85% yield by the
Wolff-Kishner reduction of 6-methylthiochroman-4-one
(N. Mitsumori, H. Suya, and T. Tsumura, Japan Kokai Tokkyo
Koho 79, 100,393, 1979). Some thiochromans with one or two
alkyl or alkoxy substituents on the benzene ring are
prepared by condensing the related thiochroman-4-one with
hydrazine hydrate in low-boiling alcohol and heating the
resulting hydrazide in a high-boiling alcohol in the
presence of alkali at 120-150°, for example, boiling a
mixture of 6-methylthiochroman-4-one and hydrazine hydrate
in methanol, followed by heating the reaction mixture with
diethylene glycol and sodium hydroxide at 130° yields 6-
-methylthiochroman (1) (Hokko Chemical Industry Co., Ltd.,
Japan Kokai Tokkyo Koho JP 58, 177,987 [83,177,987],
1983).

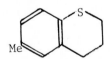

(1)

Benzothiete (2), readily obtained by ring contraction of a number of benzothiophene derivatives, undergoes thermal or photochemical cycloreversion to a thioquinone-methide (3) which can be trapped by a dienophile to give thiochromans and 4*H*-benzo[b]thiopyrans, (e.g., 4, 5 and 6). With relatively unreactive dienophiles such as 2-cyclohexenone considerable amounts of dimer (7) contaminate the product (K. Kanakarajan and H. Meier, J. org. Chem., 1983, **48**, 881).

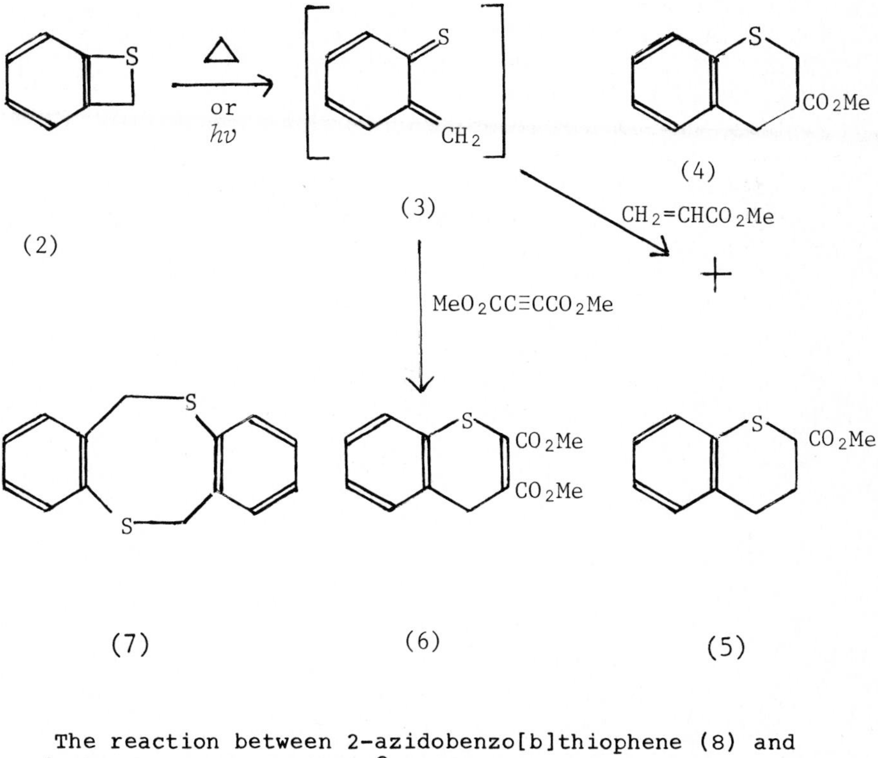

The reaction between 2-azidobenzo[b]thiophene (8) and *cis*-butene for ~50h at 45° yields 4-cyano-2,3-dimethylthio-chroman (9) and the aziridine (10) (P. Spagnolo and P. Zanirato, Chem. Comm., 1985, 1441).

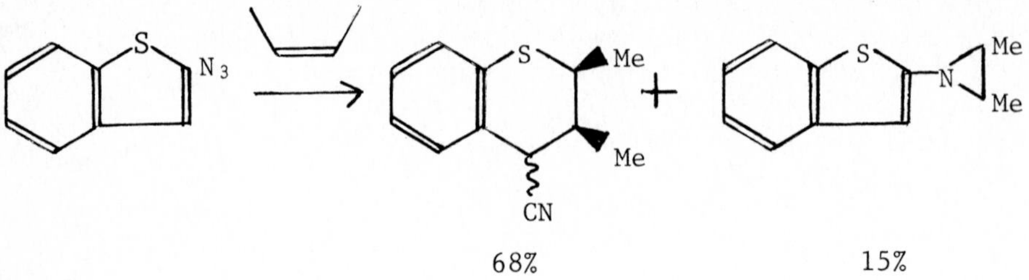

68%   15%

(8)   (9)   (10)

7-(Alkylhydroxymethyl)-6-methylthiochromans (11) have
been prepared by the reductive alkylation of 7-acetyl-6-
-methylthiochroman using the appropriate alkyllithium
(M.D. Vorozhtsova, I.U. Numanov, and V.S. Askenov, Doklady
Akad. Nauk Tadzh. SSR, 1986, 29, 219).

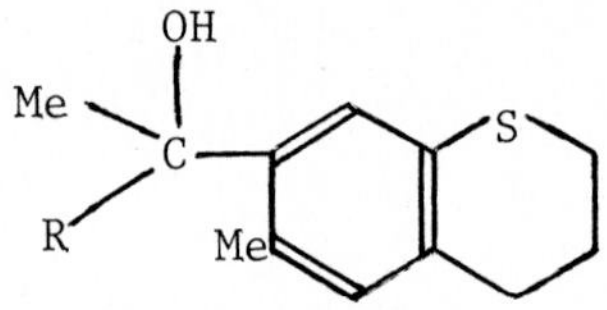

R=Et, Bu

(11)

Ethyl 2-chloro-S-phenylmercaptoacetate (12) undergoes a
polar cycloaddition reaction with styrene and tin (IV)
chloride to yield ethyl 4-phenylthiochroman-2-carboxylate
(13), with phenylacetylene, ethyl 4-phenyl-2H-benzo[b]thio-
pyran-2-carboxylate (14) is obtained (Y. Tamura et al.,
Tetrahederon Letters, 1981, 22, 3773).

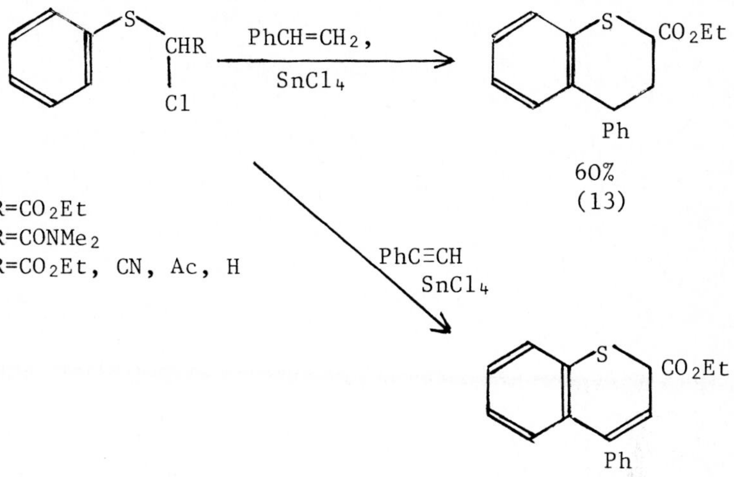

(12) R=CO$_2$Et
(15) R=CONMe$_2$
(16) R=CO$_2$Et, CN, Ac, H

The reaction of N,N-dimethylamide (15) with styrene in the presence of tin (IV) chloride affords 2-dimethylamido--4-phenylthiochroman (17; R = CONMe$_2$) also obtained from N,N-dimethyl-2-(phenylsulphinyl)acetamide and styrene under Pummerer reaction conditions, but in lower yield. Other thiochroman derivatives (17) have been obtained in a similar manner from appropriate chlorophenylmercaptomethanes (16) and the reaction of amide (15) with (E)-1,2-diphenylethene affords 2-dimethylamido-3,4-diphenylthiochroman (18) (H. Ishibashi et al., Chem. pharm. Bull., 1985, 33, 90).

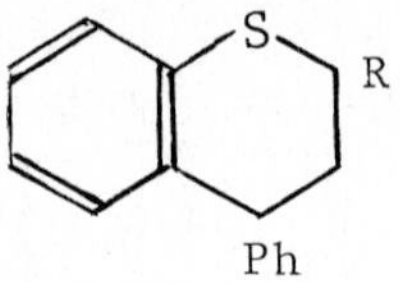

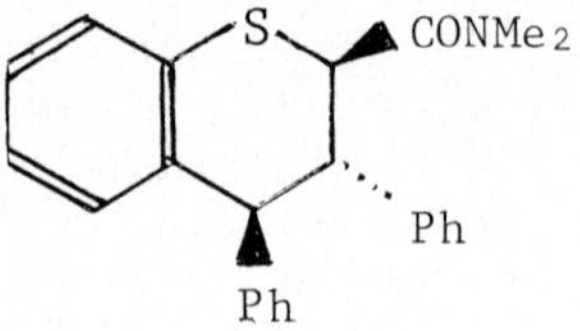

R=CONMe$_2$, CN, Ac, H

(17)   (18)

The cyclocondensation of α,β-unsaturated ketones (19)
with thiophenol in the presence of tin (IV) chloride (1.5
equivalents) furnishes thiochromans (20). Acrylaldehyde
condenses with thiophenol in the presence of 0.2 equivalents
tin (IV) chloride to yield (PhS)$_2$CH(CH$_2$)$_2$SPh, which on
treatment with 1.5 equivalents tin (IV) chloride affords 4-
-phenylmercaptothiochroman (20; R^1 = R^2 = H) (J. Cossy,
F. Henin, and C. Leblanc, Tetrahedron Letters, 1987, 28,
1417).

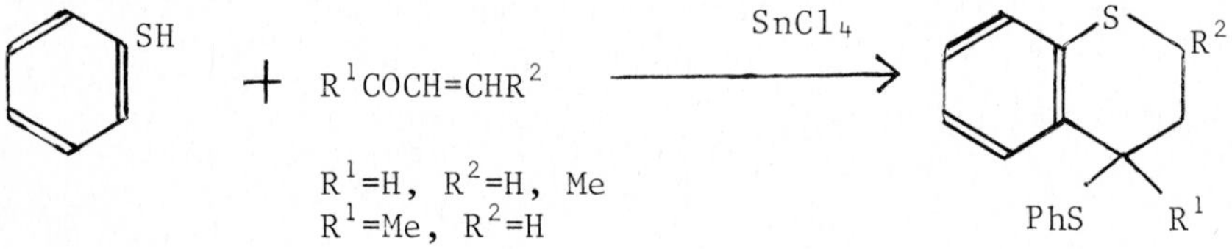

R^1=H, R^2=H, Me
R^1=Me, R^2=H

(19)   (20)

Some 3-aminothiochroman and 3-aminochroman derivatives
have been prepared and studied for use as serotoninergic
agonists (Ciba-Geigy A.-G. Japan Kokai Tokkyo Koho JP 62
59,273 [87   59,273], 1987).

The yields of 2,3-dihydro-1H-naphtho[2,1-b]thiopyran (21)
obtained by the acid-catalyzed reaction between 2-
-naphthalenethiol and acrylaldehyde depends on reaction
conditions. The yield of (21) can vary between 43 and 96%.
Other products obtained are dihydronaphthothiopyran (22)
and the intermediate 3-(2-naphthylthio)propionaldehyde,
which cyclizes with thiols RSH(R = 2-naphthyl, Ph, PhCH$_2$, Bu)
to yield dihydronaphthothiopyrans (23) (T. Nakazawa *et al*.,
Nippon Kagaku Kaishi, 1985, 725; Nakazawa, M. Shibazaki,
and K. Itabashi, *ibid*., 1987, 45).

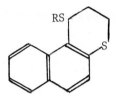

(21) R=2-naphthyl
(22) RS=H
(23) R=2-naphthyl, Ph,
     PhCH$_2$, Bu

The reaction of 2-(1,3-diphenyl-1-propanon-3-yl)-1-
-tetralone with hydrogen sulphide in acetic acid containing
trifluoroacetic acid, hydrogen bromide or chloride gives
5,6-dihydro-2,4-diphenyl-4*H*-naphtho[1,2-b]thiopyran (24),
also obtained by lithium aluminium hydride reduction of
5,6-dihydro-2,4-diphenylnaphtho[1,2-b]thiopyrylium
trifluoroacetate (25; X = F$_3$CCO$_2$). The original reaction
in the presence of perchloric acid affords 2,4-diphenyl-
-2,3,5,6-tetrahydro-4*H*-naphtho[1,2-b]thiopyran (26) and
salt (25; X = ClO$_4$). The tetrahydro derivative (26) and
salt (25; X = F$_3$CCO$_2$) are obtained when the reaction is
carried out in trifluoroacetic acid, also treatment of
dihydro derivative (24) with trifluoroacetic acid yields
the same products (25; X = F$_3$CCO$_2$ and 26). Dihydronaphthothiopyran (24) on reduction with hydrogen over Pd/C
or with triethylsilane in trifluoroacetic acid gives the
tetrahydronaphthothiopyran (26), which may be converted
into the 1,1-dioxide (O.V. Fedotova, A.P. Kriven'ko, and
G. Kharchenko, Zh. org. Khim., 1978, **14**, 1782).

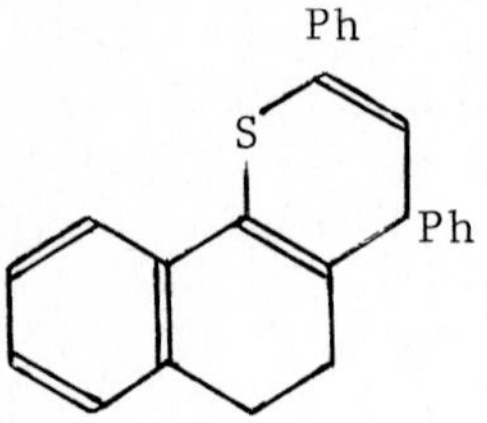

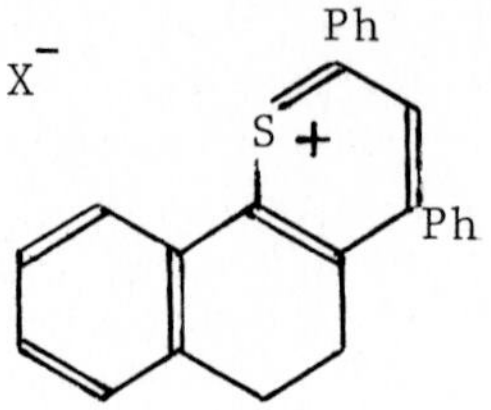

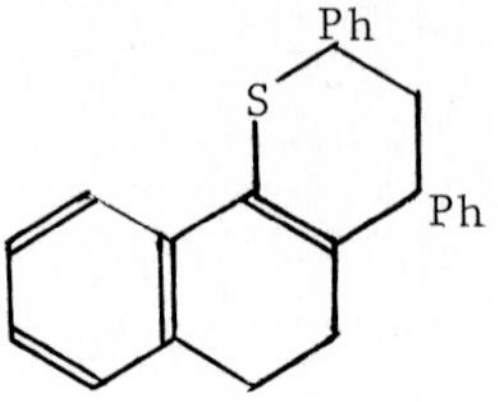

X=F$_3$CCO$_2$,ClO$_4$

(24)   (25)   (26)

4-(4-Hydroxyphenyl)-2,2,4-trimethylthiochroman and some
of its derivatives and 2,3-dihydro-4-(4-hydroxyphenyl)-
-2,2,4-trimethyl-4H-naphtho[1,2-b]thiopyran have been
prepared and subjected to structural studies (A.D.U. Hardy,
J.J. McKendrick, and D.D. MacNicol, J. chem. Soc., Perkin
II, 1979, 1072).

Cycloaddition of 2-arylmethylenetetralin-1-thione S-
-oxides (27) with olefins (28) gives mixtures of 4-aryl-
-2,3,5,6-tetrahydro-4H-naphtho[1,2-b]thiopyran 1-oxides
(29) and (30) (T. Karakasa, S. Satsumabayashi, and
S. Motoki, Nippon Shika Daigaku Kiyo, Ippan Kyoiku-kei,
1983, 12, 117). Also see J.N. Chatterjea, K.R.R.P. Singh,
and C. Bhakta [Natl. Acad. Sci. Letters (India), 1981, 4,
83].

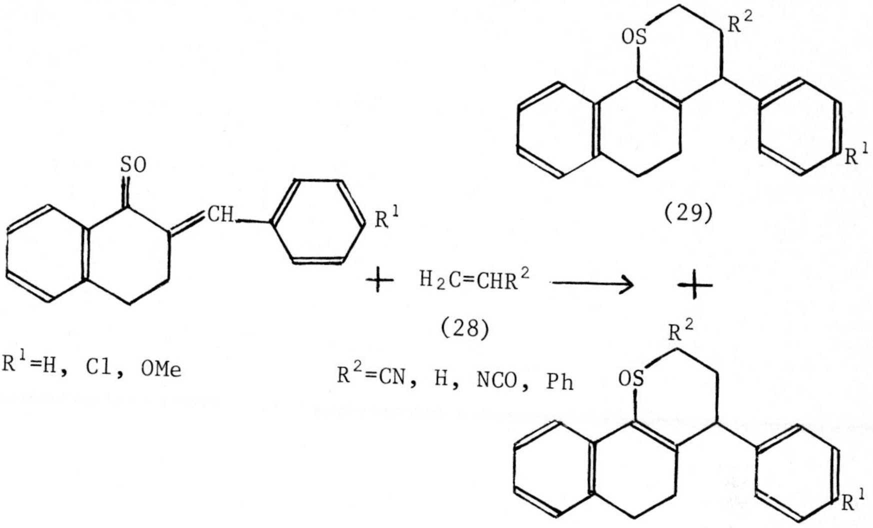

(29)

(27)

(28)

R^1=H, Cl, OMe

R^2=CN, H, NCO, Ph

(30)

Thiochroman 1-oxide on treatment with zinc and dichloro-dimethylsilane is deoxygenated to give thiochroman (K. Nagasawa *et al.*, Heterocycles, 1987, 26, 2607). Reduction of acylthiochromans (31) by sodium tetrahydrido-borate yields the corresponding secondary alcohols (32) (M.D. Vorozhtsova *et al.*, Doklady Akad. Nauk Tadzh. SSR, 1984, 27, 26).

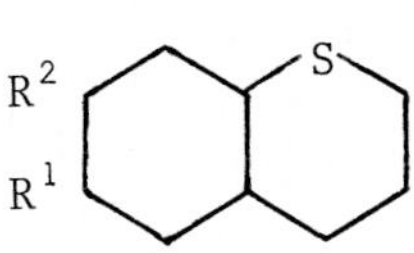

(31) R^1=Ac, EtCO, R^2=H
R^1=Me, R^2=Ac
(32) R^1=MeCHOH, EtCHOH, R^2=H
R^1=Me, R^2=MeCHOH

Hydroxythiochroman is etherified by N,N-diethyl-2-chloro-propionamide and sodium hydride to yield N,N-diethyl-2-(4--thiochromanyloxy)propionamide (33) an insect repellant (N.S. Bunker and R.Y. Wong, U.S. Pt. US 4,452,810, 1984). The reaction of 8-hydroxythiochroman with epichlorohydrin and sodium hydroxide affords 8-(2,3-dihydroxypropoxy)thiochroman, which on treatment with *tert*-butylamine, Ph_3P^+NMePh I^- and sodium hydride in DMF gives 8-(3-*tert*--butylamino-2-hydroxypropoxy)thiochroman (34). A number of related derivatives (35) have been prepared (C. Malen and P. Roger, Swiss Pat. CH 623,818, 1981). Related preparation of these derivatives, and others have been reported (Science Union at Cie.-Societe Francaise de Recherche Medicale, Japan Kokai 78 25,571, 1978).

$$R^3NHCH_2CH(OH)CH_2O$$

OCHMeCONEt$_2$

(33)

(34) $R^1=R^2=H$, $R^3=Bu^t$
(35) $R^1=H$, alkyl, $R^2=H$, halogeno, $R^3=$alkyl c-alkyl

Acylation of thiochroman and 6-methylthiochroman with an acid chloride in the presence of aluminium chloride yields a 6-acylthiochroman and an 8-acyl-6-methylthiochroman, respectively. Using acids as the acylating agent give lower yields and acid anhydrides require twice as much catalyst (R. Usmanov, I.U. Numanov, and N. Radzhabov, Doklady Akad. Nauk Tadzh. SSR, 1979, 22, 117).

Substituted thiochroman 1,1-dioxides (36) react with chlorosulphonic acid in dichloromethane to yield the sulphonyl chlorides (37), which on treatment with anhydrous ammonia in dichloromethane afford 7-sulphonamidothiochroman 1,1-dioxides (38) (Hokko Chemical Industry Co., Ltd., Japan Kokai Tokkyo Koho JP 58,194,881 [83,194,881], 1983).

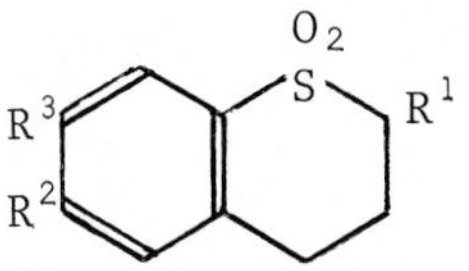

R^1, R^2=H,H; H,Me; Me,Cl;
Et, Me; H,Cl

(36) R^3=H
(37) R^3=SO$_2$Cl
(38) R^3=SO$_2$NH$_2$

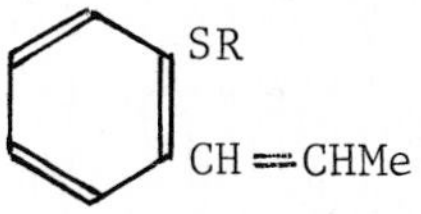

(39)

Treatment of thiochroman with lithamide in hexametapol in presence of electrophilic reagents, for example, bromoethane yields the benzene derivative (39; R = Et). Analogous treatment with iodomethane using potassium *tert*-butoxide in DMSO gives derivative (39; R = Me) (E.A. Karakahnov, K.N. Kreindel, and E.A. Runova, Vestn. Mosk. Univ., Ser. 2, Khim., 1983, 2, 286).

The decarboxylation of the thiochroman-3-carboxylic acids (40) results in an unexpected ring contraction to yield the benzothiophene derivatives (41) and (42) (B. Gopalan, K. Rajagopalan, and K.K. Balasubramanian, Sulphur Letters, 1984, 2, 167).

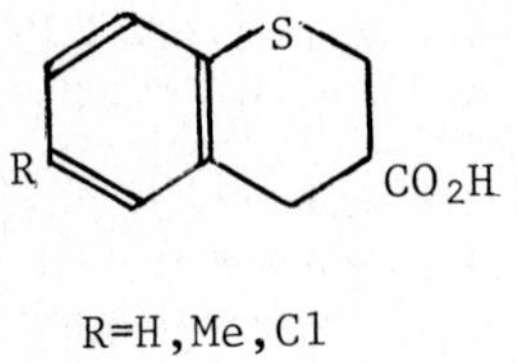

R=H,Me,Cl

(40) 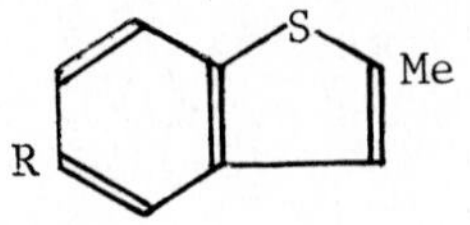   (41) 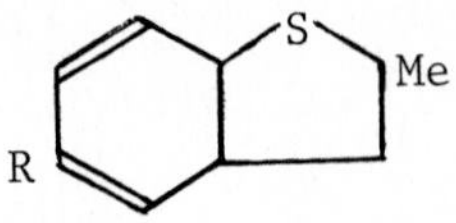   (42)

Thiochroman and 2-methylthiochroman undergo ring contraction on heating over a alumina-silica mixture at 350-500° to give methyl- and ethyl- benzothiophene, respectively, and their dihydro derivatives (Karakhanov *et al*., Vestn. Mosk. Univ., Ser. 2, Khim., 1978, 19, 362).

Treatment of thiochroman with diethyl diazomalonate in the presence of a copper-bronze catalyst affords the ylide (43), which on acetolysis gives thiochroman, 2-acetoxythio-chroman, diethyl malonate, $AcOCH(CO_2Et)_2$, and $2\text{-}[AcO(CH_2)_3]$-$C_6H_4SCH(CO_2Et)_2$ (R. Pellicciari *et al*., Gazz., 1978, 108, 671).

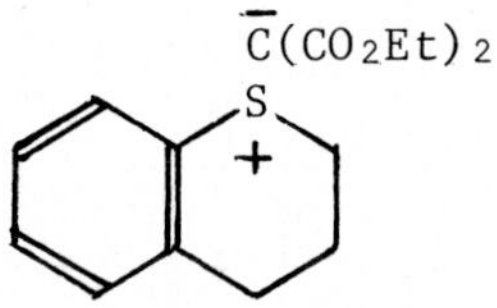

(43)

*(ii) Tetra- and hexa-hydrobenzothiopyrans and hexa-
-hydrothiochromans*

trans,trans-4a,5,6,7,8,8a-Hexahydro-4*H*-benzopyrans (1) on
treatment with hydrogen sulphide in an acetic acid acetic
anhydride mixture containing hydrogen chloride are
converted into *trans,trans*-8a-mercapto-2-phenyl-4-aryl-
-4a,5,6,7,8,8a-hexahydro-4*H*-benzothiopyrans, *trans,trans*-1-
-mercapto-3-phenyl-5-aryl-2-thiabicyclo[4.4.0]dec-3-enes
(2). With excess of the acetal (1) the 5,6,7,8-tetrahydro-
-4H-benzothiopyrans (3) are obtained and in the absence of
acetic anhydride the bicyclononenone (4) (∼9.5%) is formed
(S.K. Klimenko *et al.*, Khim. Geterotsikl. Soedin., 1986,
28).

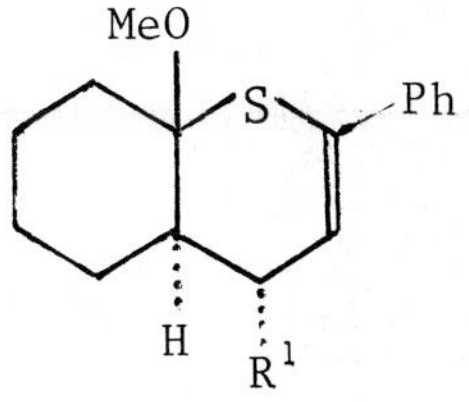

R¹=4-MeOC₆H₄,Ph

(1)   (2)

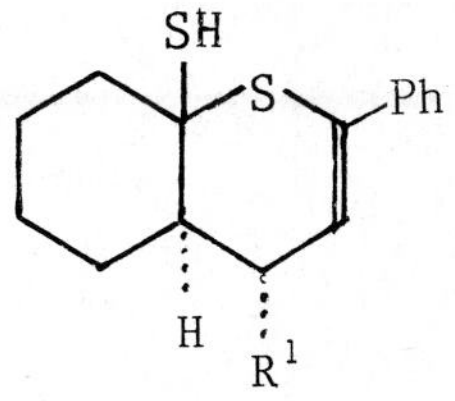

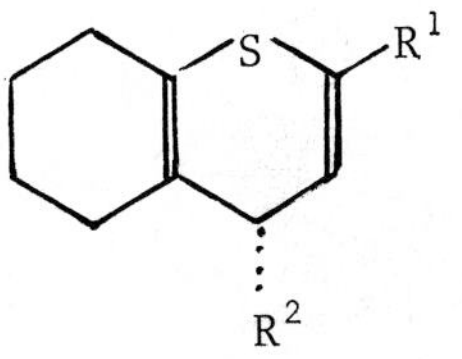

R¹=4-MeC₆H₄, R²=Ph
R¹=Ph,R²=4-MeC₆H₄

(3)   (4)

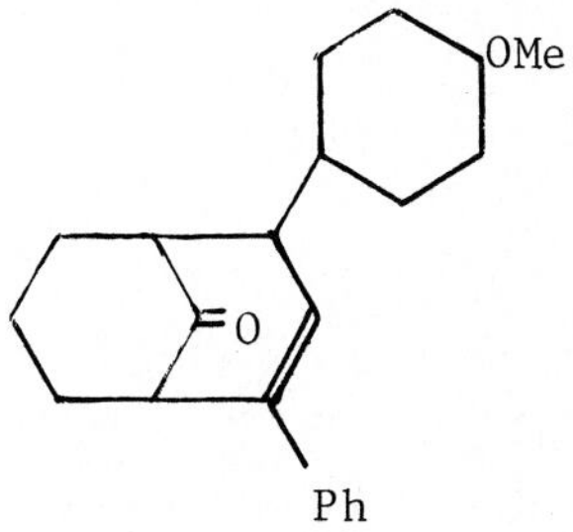

Disproportionation and simultaneous ionic reduction of tetrahydrobenzothiopyrans (5) and (6) and hexahydrobenzothiopyran (7) with trifluoroacetic acid yields the corresponding 2,4-diphenyl-5,6,7,8-tetrahydrobenzothiopyrylium salt and 2-thiabicyclo[4.4.0]decane derivative (Klimenko, T.V. Stolbova, and V.G. Kharchenko, *ibid.*, 1981, 1338).

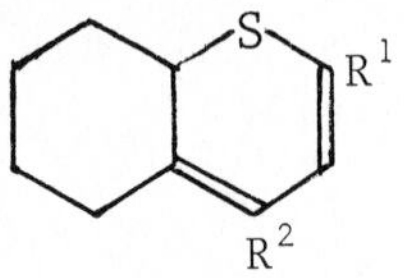

(5) $R^1=R^2=Ph$
(8) $R^1,R^2=H,Ar$

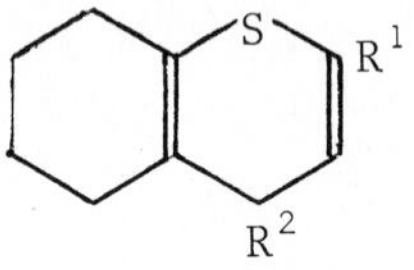

(6) $R^1=R^2=Ph$
(9) $R^1,R^2=H,Ar$

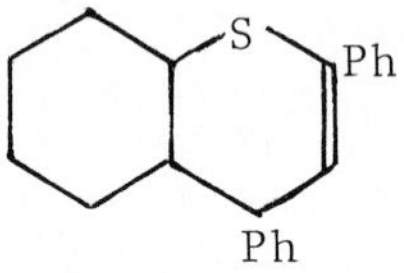

(7)

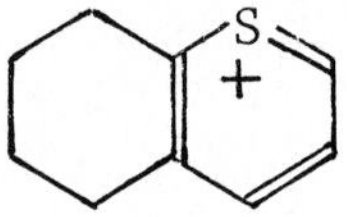

X^-

$R^1,R^2=H,Ar$
$X=BF_4,CF_3CO_2$

(10)

Configurational and conformational investigations have been carried out with references to these transformations by the application of ^{13}C-nmr spectral data (Klimenko *et al.*, *ibid.*, 1984, 898). The hydrogenation of compounds (8) and (9) and tetrahydrobenzothiopyrylium salts (10) over Pd/C at 50 or 100 atmospheres yields *cis*-1-thiadecalins, octahydrobenzothiopyrans, with substituents R^1 and R^2 in the equatorial orientation. They have been oxidized to sulphoxides and sulphones (Klimenko, T.I. Tyrina, and N.N. Sorokin, *ibid.*, 1987, 614).

1,5-Diketones (11) on treatment with hydrogen sulphide and hydrogen chloride in methanol gives *trans*-hexahydro-4*H*--benzothiopyrans (13). Diketone (12) besides giving the *trans* compound (13; R^1, R^2 = Ph, H) also affords some of the *cis*-isomer. 2-Phenyl-5,6,7,8- tetrahydrobenzothiopyrylium perchlorate has been prepared (Klimenko *et al.*, *ibid.*, 1985, 1194).

(11) R^1,R^2=4-MeOC$_6$H$_4$,Ph;
      2-naphthyl,Ph;
      2-naphthyl,4-MeOC$_6$H$_4$

(12) R^1,R^2=Ph,H

(13) R^1,R^2=4-MeOC$_6$H$_4$,Ph;
      2-naphthyl,Ph;
      2-naphthyl,4-MeOC$_6$H$_4$;
      Ph,H

The reaction of "semicyclic" 1,5-diketones similar to (11) with hydrogen sulphide and trifluoroacetic acid yields a mixture of the corresponding 5,6,7,8-tetrahydrobenzothiopyrylium salt and hexahydrothiochroman. The products are the result of the disproportionation of the intermediate 5,6,7,8-tetrahydro-4*H*-benzothiopyrans. The trifluoroacetates have been converted to perchlorates (T.V. Stolbova, Klimenko, and V.G. Kharchenko, Zh. org. Khim., 1980, **16**, 178).

Alkylation of 8a-mercapto-2-phenyl-5-aryl-4a,5,6,7,8,8a--hexahydro-4*H*-benzothiopyrans (2; R^1 = Ph, 4-MeOC$_6$H$_4$, 3,4-(MeO)$_2$C$_6$H$_3$) with acrylic acid in hydrogen chloride gives the benzothiopyranium chlorides (14), which on treatment with chloroplatinic acid in acetic acid/acetic anhydride yields the corresponding hexachloroplatinates. These compounds have been tested for antimicrobial activity (*idem*, Khim.-Farm. Zh., 1983, **17**, 167).

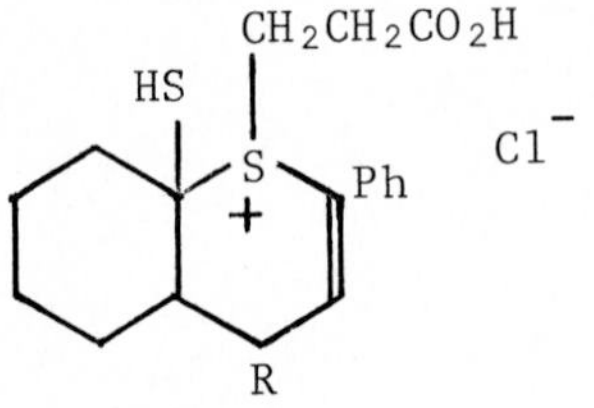

HS / CH$_2$CH$_2$CO$_2$H

Cl$^-$

R=Ph, 4-MeOC$_6$H$_4$,
    3,4-(MeO)$_2$C$_6$H$_3$

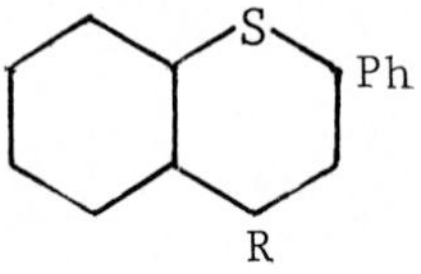

R=H,Ph

(14)   (15)

Hexahydrothiochromans (15) have been obtained by the hydrogenation of the corresponding 5,6,7,8-tetrahydrobenzo-thiopyrylium salts in the presence of a platinum-group metal, especially 10% Pd/C (Kharchenko, O.A. Bozhenova, and A.D. Shebaldova, *ibid*., 1982, <u>18</u>, 2435).

The cyclic 10-membered (*E*)-homoallylic sulphoxide (16) undergoes regiospecific butyllithium-promoted transannular cyclization furnishing hexahydrothiochroman 1-oxide (17) (V. Cere *et al*., J. org. Chem., 1986, <u>51</u>, 4850).

(16)   (17)

The synthesis and stereochemistry of *trans*-2-methyl-4-
-ethynyl-4-hydroxyhexahydrothiochroman 1-sulphoxides and
trans-2-methyl-4-ethynyl(acetate)-4-hydroxyhexahydrothio-
chroman 1,1-dioxides (V.V. Kokhomskaya *et al*., Vestsi Akad.
Navuk BSSR, Ser. Khim. Navuk, 1984, 76), and the synthesis,
stereochemistry and ^{1}H-nmr spectroscopy of *trans*-2-methyl-
-4-hydroxyhexahydrothiochroman and its oxidation by 27%
hydrogen peroxide to give a mixture of two epimeric 1-
-oxides (*idem, ibid*., 1986, 59) have been discussed.

Lithiation of 2-phenylhexahydrothiochroman 1,1-dioxide,
2α-phenyl-*cis*-1-thiadecalin 1,1-dioxide (18), with
butyllithium in THF, followed by hydrolysis of the
lithiated product yields a 9:1 mixture of the 2-phenyl-
-*trans*-1-thiadecalin 1,1-dioxides (19) and (20). The
hydrolysis with deuterium oxide and the catalytic effect of
sodium deuteroxide have been investigated (Klimenko *et al*.,
Zh. org. Khim., 1985, 21, 1918).

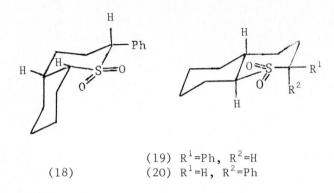

(18)

(19) R^1=Ph, R^2=H
(20) R^1=H, R^2=Ph

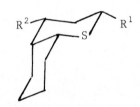

R^1=Ph, R^2=H, Ph
    4-MeOC$_6$H$_4$,
    3,4-(MeO)$_2$C$_6$H$_3$
R^1=4-MeOC$_6$H$_4$, R^2=H,
    Ph,4-MeOC$_6$H$_4$

(21)

Oxidation of hexahydrothiochromans (21) by hydrogen peroxide in acetic acid for a day at 20° yields the corresponding 1-oxide (*idem*, Khim. Geterotsikl. Soedin., 1987, 463).

2α-Phenyl-2,4-*ortho*-(4,5-dimethoxybenzo)-*cis*-1--thiadecalin (22), a product of intramolecular rearrangement is obtained together with the usual products of disproportionation, 5,6-tetramethylenethiopyrylium salts and 2α-phenyl-4α-(3,4-dimethoxyphenyl)-*cis*-1-thiadecalin, from the intermediate 2-phenyl-4-(3,4-di-methoxyphenyl)--5,6-tetramethylene-4*H*-thiopyran, by the reaction of 1-

-phenyl-3-(3,4-dimethoxyphenyl)-3-(2-oxocyclohexyl)-1-
-propanone with hydrogen sulphide and an acid such as
trifluoroacetic, perchloric or boron trifluoride etherate
(*idem*, *ibid*., 1986, 199).

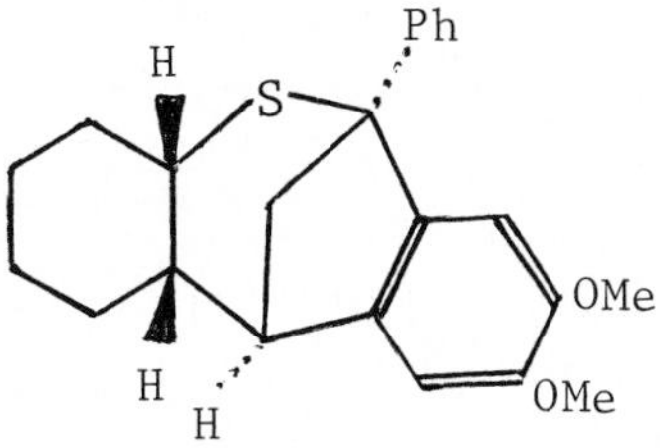

(22)

(*iii*) *2,3-Dihydro-4H-benzo[b]thiopyran-3- and -4-ones,*
2,3-dihydro-4H-1-benzothiopyran-3- and 4-ones,
thiochroman-3- and -4-ones

Thiochroman-3-ones. Thiochroman-3-ones (2) are obtained
by the Dieckmann cyclization of 2-[(carboxymethyl)thio]-
phenylacetic acids to yield 3-acetoxy-2H-benzothiopyrans
(1), which are subsequently hydrolyzed. Thiochroman-3-ones
(2; R^1 = R^2 = H, R^3 = Me, Bz) are obtained *via* enamine (3)
(P.D. Clark and D.M. McKinnon, Canad. J. Chem., 1982, 60,
243).

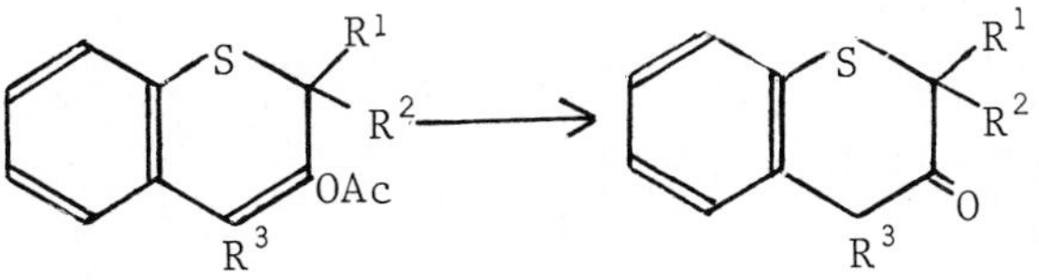

R^1=H,Me,Ph,R^2=R^3=H
R^1=R^2=Me,R^3=H

(1)   (2)

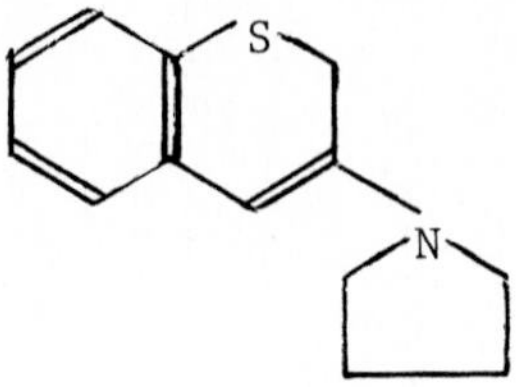

(3)

Thiochroman-4-ones. Condensation of the appropriate acetophenone (4; R = OH, OMe) with $Me_2N^+=CHCH=CHNMe_2.ClO_4^-$ affords 5-hydroxy- and 5-methoxy- 2-(1-oxoeth-2-yl)thio-chroman-4-one (5), respectively. Related chromanones have been obtained in a similar manner and these compounds are key products for the synthesis of 6-heterotetracyclines (R. Kirchlechner, Ber., 1985, 115, 2461).

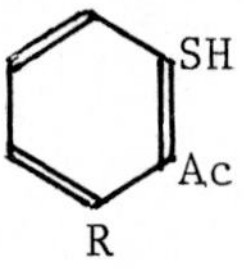

R=OH,OMe

(4)

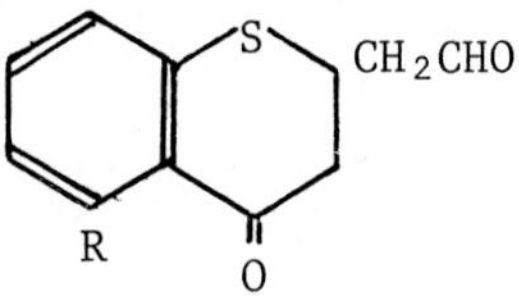

(5)

Thiochroman-4-ones (7) with substituents in the 6-, 8- and 6,8-positions are obtained by treating the related arylthiopropionic acid (6) with concentrated or fuming sulphuric acid in dichloromethane (Hokko Chemical Industry Co., Ltd., Japan Kokai Tokkyo Koho JP 58,198,483 [83,198,483], 1983).

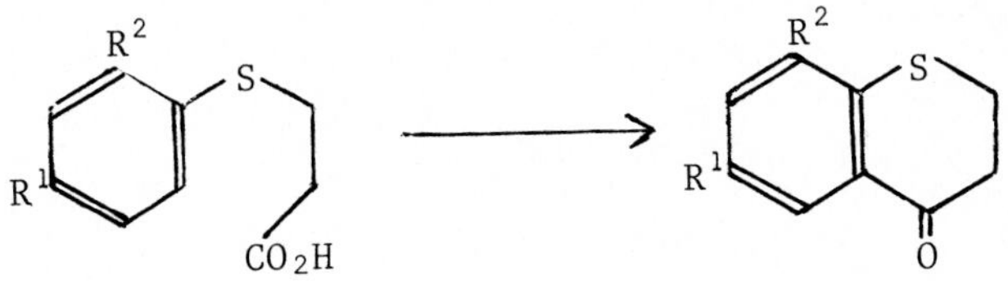

$$R^1, R^2 = Me,H; \ Cl,Me;$$
$$H,F; \ EtO,H; \ H,H;$$
$$NO_2,H; \ H,Cl; \ MeO,H$$

(6)   (7)

The trimethylsilyl enol ethers of thiochroman-4-one and its 1,1-dioxide react with carbonyl compounds and titanium tetrachloride under mild conditions, to give 3-alkylidene derivatives (e.g., 8). Other Lewis acids, except boron trifluoride, are ineffective (I.W.J. Still and F.J. Ablenas, Canad. J. Chem., 1984, **62**, 2535).

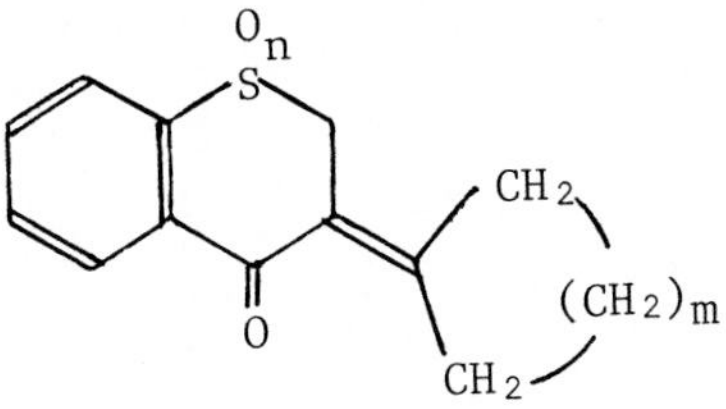

n=0,2; m=2,3

(8)

A compound (9) corresponding to a dimer of 3-methylene-
thiochroman-4-one 1,1-dioxide (10) is obtained on boiling
thiochroman-4-one 1,1-dioxide with paraformaldehyde and
dimethylamine hydrochloride in 2-propanol. Under these
conditions neither (10) nor the expected Mannich base, 3-
-(dimethylaminomethyl)thiochroman-4-one 1,1-dioxide are
isolated from the reaction mixture. 3-Methylenethiochroman-
-4-one 1,1-dioxide (10) can be prepared by sublimation of
dimer (9) at 230-250°, but redimerization slowly occurs at
room temperature (M.H. Holshouser and L.J. Loeffler, J.
pharm. Sci., 1982, <u>71</u>, 715).

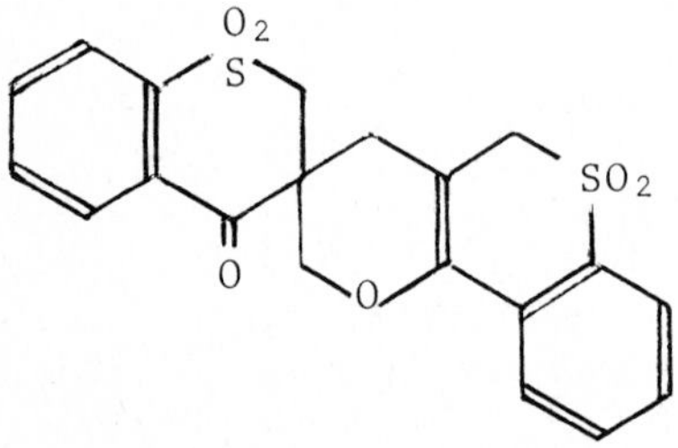

(9) 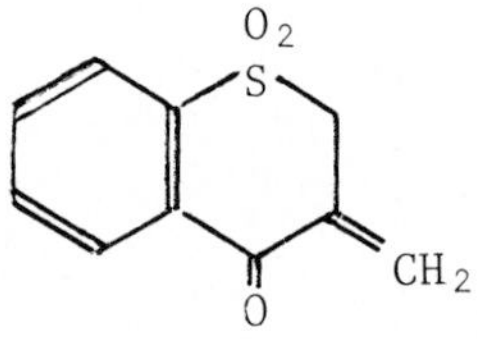   (10)

3-Benzylidenethiochroman-4-ones (11) are prepared by
treating thiochroman-4-one with 4-substituted benzaldehydes
in the presence of pyridine. Similarly 3-benzylidenechroman-
-4-ones are obtained from chroman-4-one (A. Levai and
J.B. Schag, Pharmazie, 1979, <u>34</u>, 749). Also see G. Wagner
et al. (*ibid.*, p.55).

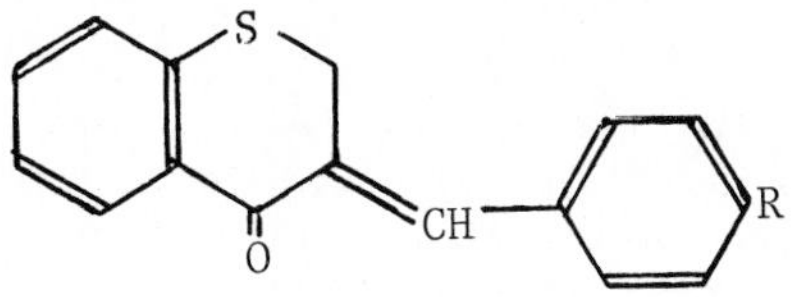

R=NMe$_2$,OH,OMe,H,
Me,Pr^i,Cl,Br

(11)

7-Chlorothiochroman-4-one 1,1-dioxide oxime is obtained by reacting the ketone and hydroxylamine. It possesses diuretic and anti-hypertensive activity comparable to that of furosemide (E. Bouchara, Fr.Demande 2,354,096, 1976).

6-Substituted thiochromans (12) on treatment with acrylonitrile undergo cyanoethylation to give derivatives (13), which on hydrolysis afford 6-substituted thiochroman-4-one-3,3-di-propionic acids (14) (T. Pomorski and B. Sedzimrska, Pol. J. Chem., 1984, 58, 941).

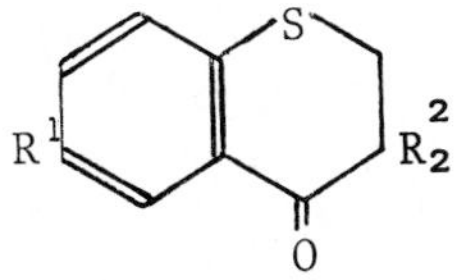

(12) R^1=Cl,Me,R^2=H
(13) R^1=Cl,Me,R^2=CH$_2$CH$_2$CN
(14) R^1=Cl,Me,R^2=CH$_2$CH$_2$CO$_2$H

3-Bromo- and 3,3-dibromo- thiochroman-4-one on treatment with xenon difluoride yield 3-bromo-4H-benzothiopyran-4-one and 3,3-dibromo-2-fluorothiochroman-4-one, respectively (M. Zupan and B. Zajc, J. chem. Soc., Perkin I, 1978, 965).

The optical resolution of $(\pm)$-2-phenylthiochroman-4-one has been achieved using the optically active reagent (+)-5--(α-phenylethyl)semioxamazide (A.L. Tokes, Ann., 1987, 1007).

Reduction of *trans*-2-methylhexahydrothiochroman-4-ones, epimeric at C-2 by lithium tetrahydridoaluminate, sodium tetrahydridoborate, or aluminium isopropoxide gives *trans*--4-hydroxy-2-methylhexahydrothiochromans (15), epimeric at C-4. Alcohols (15) have been converted into their acetates, 1,1-dioxides and 1,1-dioxide acetates (V.V. Kokhomskaya *et al.*, Vesti Akad. Navuk BSSR, Ser. Khim. Navuk, 1985, 47).

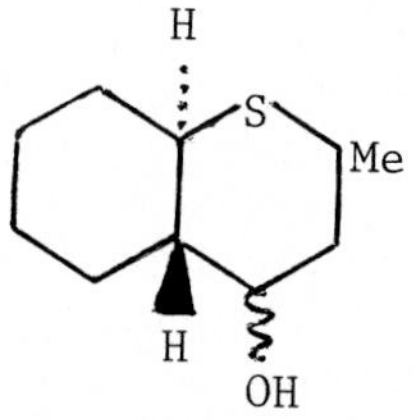

(15)

The addition of sodium acetylene, vinyl- or ethyl--magnesium bromide or phenyllithium to the carbonyl group of *trans*-2-methylhexahydrothiochroman-4-one (16) affords epimeric mixtures of products (17). Similar reactions with the axial methyl isomer (18) affords only alcohols having an axial hydroxyl group. Structures of the products have been confirmed by hydration, hydrogenation and acetylation studies (*idem, ibid.*, 1984, 63).

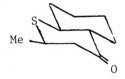 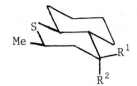

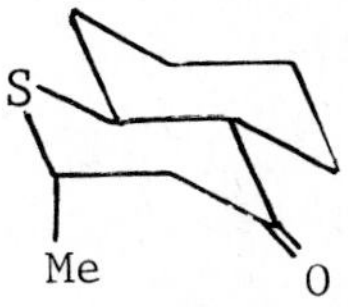

(16) R^1, R^2 = C≡CH, OH;
         OH, C≡CH;
         OH, CH=CH₂; CH=CH₂, OH;
         Et, OH; OH, Et;
         Ph, OH; OH, Ph

(18)

(17)

3. 1H-Benzo[c]thiopyran, 1H-2-benzothiopyran, isothio-chromene, 3,4-benzothiopyran and its derivatives

(a) 1H-Benzo[c]thiopyran, isothiochromene and derivatives

The reaction of 1-cyano-4-(2,4,6-trimethylphenyl)naphtho-[2,1-c]thiopyran (1) with 3-chloroperbenzoic acid and methanol affords the methoxylated derivative (2) (60%) and the ring-contracted naphtho[c]thiophene (3) (34%) (H. Ohmura and S. Motoki, Chem. Letters, 1984, 1973).

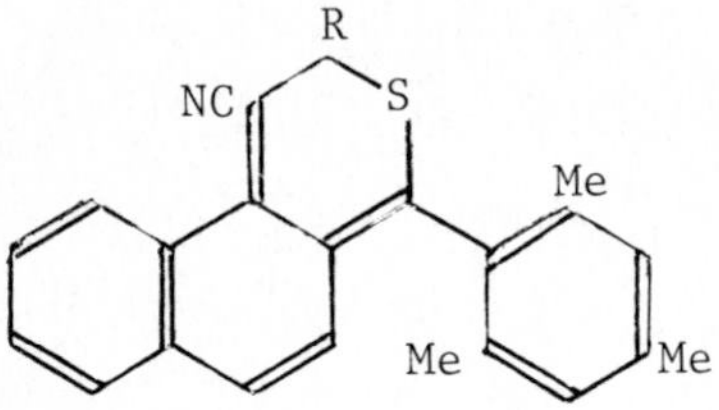

(1) R=H   (3)
(2) R=OMe

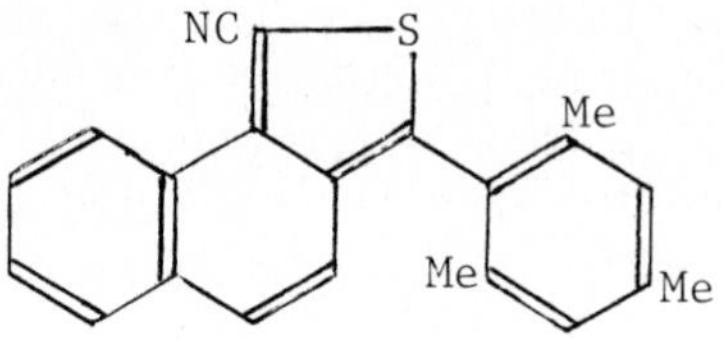

The X-ray structural analysis of 1-benzoyl-2-methyl-1H-
-benzo[c]thiopyran (4) shows that the heterocyclic ring has
a half-boat conformation with the apex at sulphur. The
methyl group is in an axial position, *trans* to the benzoyl
group and the negative charge is delocalized towards the
oxygen atom (M. Hori *et al.*, Tetrahedron Letters, 1979,
4315).

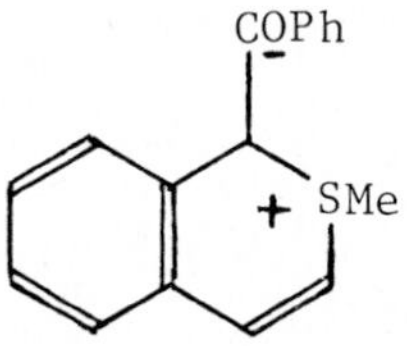

(4)

Dimethyl 5,6,7,8-tetrahydro-1H-benzo[c]thiopyran-1-thione-
-3,4-dicarboxylate has been prepared (S.R. Ramadas,
C.N. Pillai, and R. Jayalakshmi, Sulphur Letters, 1985, 4,
1).

(b) 3,4-Dihydro-1H-benzo[c]thiopyran, isothiochroman and its derivatives

Thermal and photochemical reactions of 2,4,6-tri-*tert*-
-butylthiobenzaldehyde yield 4,4-dimethyl-6,8-di-*tert*-
-butylisothiochroman(1) in high yield (R. Okazaki *et al.*,
Tetrahedron Letters, 1984, 25, 849).

(1)

R^1, R^2, R^3=alkyl
R^4=CN, CO_2H, hydroxyalkyl
R^5=H, AcO; R^4R^5=$CH_2CH_2NR^6$
R^6=alkyl, alkoxycarbonyl
n=0,1

(2)

A number of isothiochroman derivatives (2), which are
effective analgesics have been prepared (Hori *et al.*, Japan
Kokai Tokkyo Koho JP 61,227,580 [86,227,580], 1986).
(S)-(+)-4-Ethylisothiochroman has been prepared from (S)-
-(+)-phenylethylacetic acid and its formation of a complex
with mercury(II) chloride has been reported [M. Bavia,
P. Biscarini, J. chem. Res., (S), 1987, 44].
The preparation and reactions of isothiochroman-1-
-carboxaldehyde (3), for example, the formation of a
cyanhydrin and its condensation with secondary amines to
furnish enamines have been reported. Transformation into
acetals and Wittig reactions have also been described. The
3-benzothiepin derivative (4) results from the formylation
of isothiochroman 2-oxide (H. Boehme and P.N. Sutoyo,
Ann., 1982, 1643).

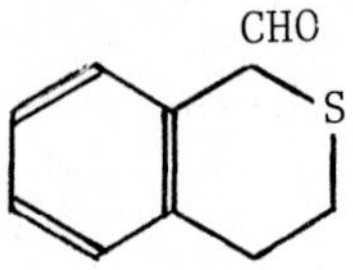

(3)

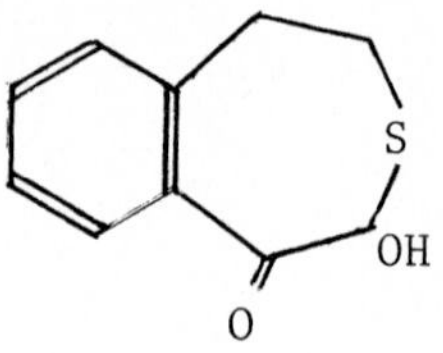

(4)

Pummerer cyclization of sulphoxide (5) by trifluoroacetic anhydride in methanol free dichloromethane gives 1-benzoyl-isothiochroman (6), which on alkylation or arylation yields the corresponding isothiochromanium salt, for instance, treatment with methyl iodide and silver perchlorate in dichloromethane affords 1-benzoyl-2-methylisothiochromanium perchlorate (7). Deprotonation of salt (7) with triethyl-amine in ethanol at 0° gives ylide (8), which on boiling in benzene or toluene under nitrogen undergoes a novel 1,4--rearrangement to give the enol ether (9). Some related conversions have been carried out on derivatives of 1H--benzo[c]thiopyran (T. Kataoka *et al.*, J. chem. Soc., Perkin I, 1983, 2913).

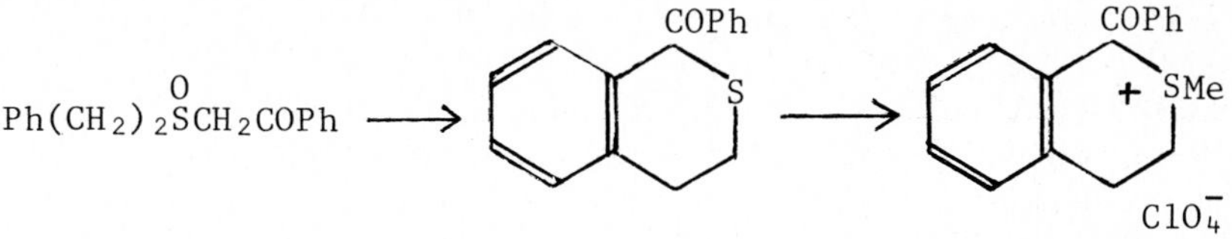

$$Ph(CH_2)_2SCH_2COPh \longrightarrow$$

(5)　　　　　　　　(6)　　　　　　　　(7)

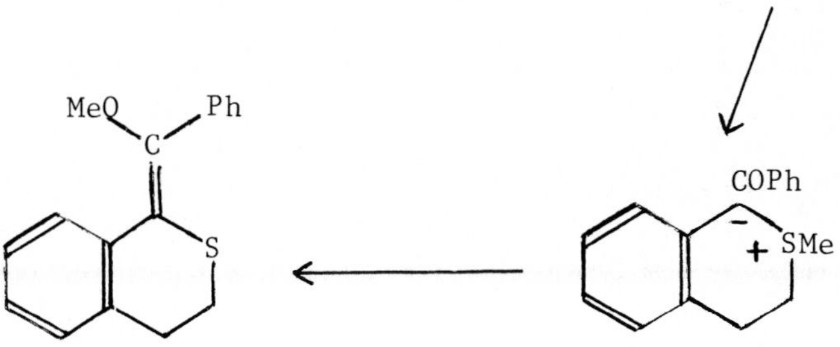

(9)　　　　　　　　(8)

Ylide (8) reacts with acids or thiols, for example,
acetic acid and thiophenol, to yield ring-opened sulphides
(10) and (11), respectively. However, with 2-methoxyphenol
or succinimide, it gives benzoxanthionin (12) (Hori *et al.*,
Tetrahedron Letters, 1982, **23**, 2597).

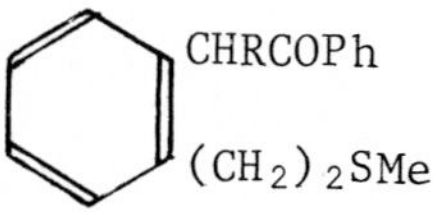

(10) R=AcO (89%)
(11) R=PhS (95%)

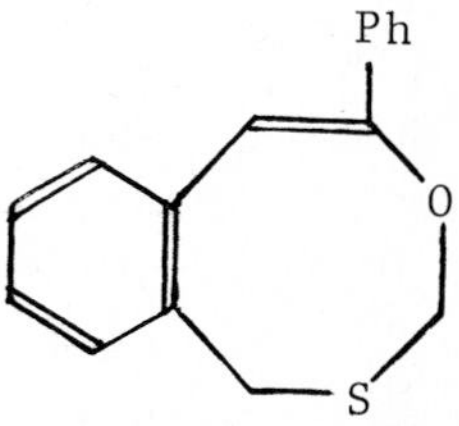

(12)

The carbanion (13), generated from isothiochroman and butyllithium reacts with alkyl or silyl halides to give the 1-alkyl or silyl derivatives (14). With carbon disulphide the carbanion (13) affords dithiolate (15), which is transformed with alkyl halides into the ketene mercaptals (16). Treatment of carbanion (13) with aldehydes and ketones leads to secondary and tertiary alcohols, respectively. It is oxidized by iodine to 1,1'-bisthiochroman (Boehme and Sutoyo, Phosphorus Sulphur, 1982, __13__, 235).

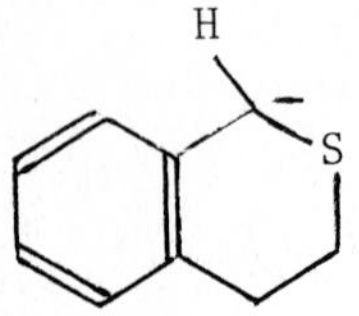

(13)

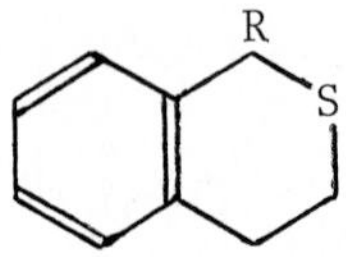

R=Me,Pr^i,
(EtO)$_2$CHCH$_2$,
PhCH$_2$,Me$_3$Si

(14)

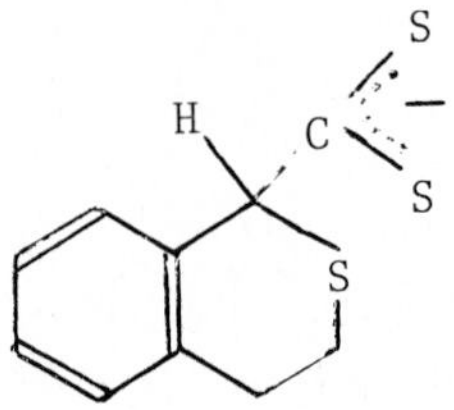

(15)

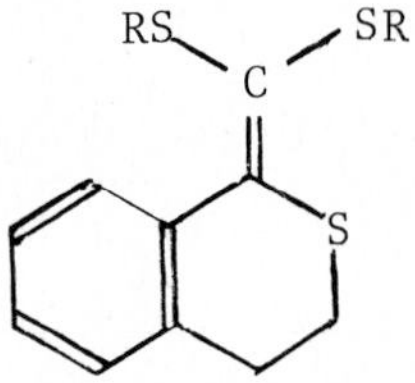

R=Me,Et
R$_2$=(CH$_2$)$_3$

(16)

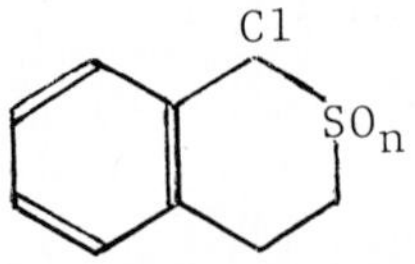

(17) n=1
(18) n=0

The dehydrohalogenation of 1-chloroisothiochroman 2-oxide
(17) produces isothiochroman-1-one, and hydrolysis of the
2-oxide (17) yields 2-(OHSCH$_2$CH$_2$)C$_6$H$_4$CHO, which
disproportionates to (2-OCHC$_6$H$_4$CH$_2$S)$_2$ and
2-OCHC$_6$H$_4$CH$_2$CH$_2$SSO$_2$CH$_2$CH$_2$C$_6$H$_4$CHO-2. The hydrolysis
of 1-chloroisothiochroman (18) has also been investigated
(Boehme and H.J. Wilke, Ann., 1978, 1123).

The thermolysis of the 1-cyano-2-methylisothiochroman
ylide (19) in toluene affords 1-cyano-1-methyl- (20; R=Me)
and 1-benzyl-1-cyanoisothiochromans (20; R=PhCH$_2$), whereas
the reaction in ethanol yields the spiro compound (21)
(Hori *et al*., Tetrahedron Letters, 1983, **24**, 3733).

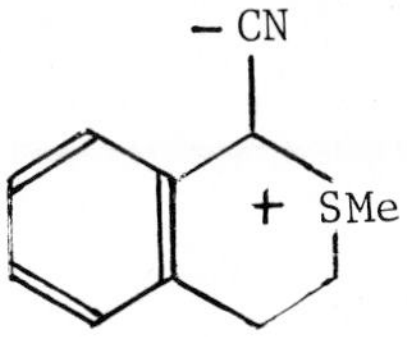

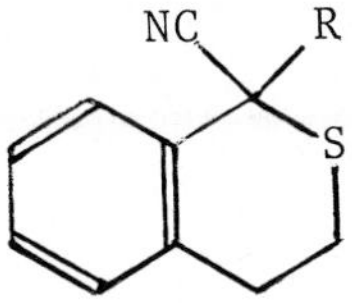

R=Me,PhCH$_2$

(19)   (20)   (21)

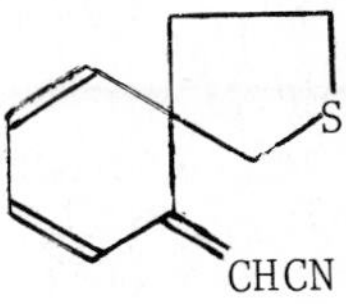

Acetolysis of the isothiochroman sulphonium ylide (22)
gives diethyl 1-isothiochromanylmalonate (R. Pellicari *et
al*., Gazz., 1978, **108**, 671).

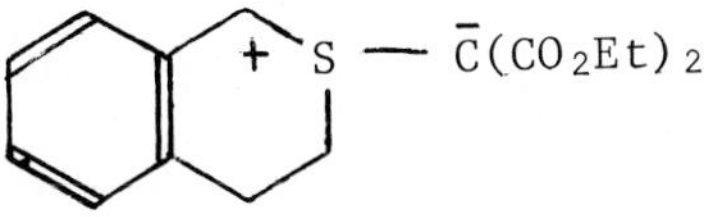

(22) 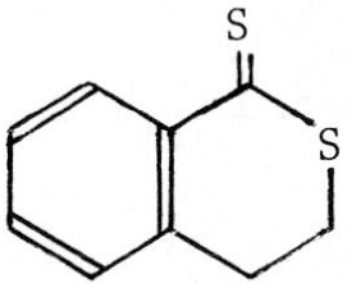   (23)

516

The reaction between 2-phenylethanol and carbon
bisulphide promoted by aluminium chloride and benzoyl
chloride gives dihydrodithioisocoumarin (23). A much lower
yield of (23) is obtained without the use of benzoyl
chloride, whose role is probably to modulate the Lewis
acidity of the aluminium chloride, and thus reduce the
degradation of the starting materials and products
(M. Czarniecki and R.Q. Kluttz, Tetrahedron Letters, 1979,
4893).
    Isothiochroman-3-one (24) is prepared by the
cyclocondensation of 2-(2-bromomethylphenyl)acetyl chloride
with sodium hydrogen sulphide, using tetrabutylammonium
hydrogen sulphate as a phase-transfer catalyst (C. Lemaire
et al., J. heterocyclic Chem., 1983, 20, 811).

(24)

    Isothiochroman-4-one (25) condenses with aryl aldehydes,
in the presence of piperidine to give the 3-methyleneiso-
thiochroman-4-ones (26). Derivatives (26; R=Ph, $MeOC_6H_4$)
on cyclo-condensation with phenylhydrazine afford the
pyrazoles (27), and addition of a phenyl Grignard reagents,
to them yields the 3-benzylisothiochroman-4-ones (28)
[M.I. Ali, M.A.F. El-Kaschef, and A.G. Hammam, Egypt. J.
Chem., 1977 (pub. 1979), 20, 323].

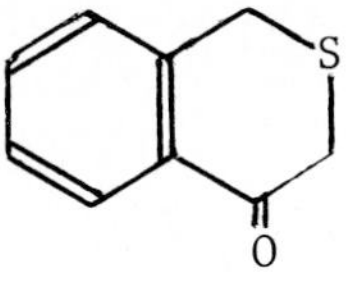

(25)

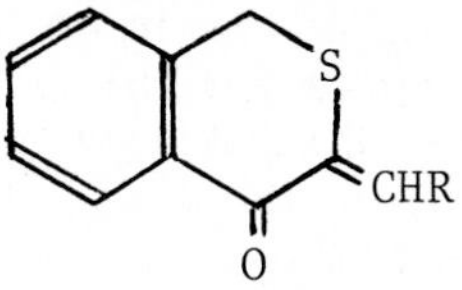

R=Ph,4-MeOC$_6$H$_4$,
   4-ClC$_6$H$_4$,4-O$_2$NC$_6$H$_4$,
   4-Me$_2$NC$_6$H$_4$,3,4-(MeO)$_2$C$_6$H$_3$,
   3,4-(OCH$_2$O)C$_6$H$_3$

(26)

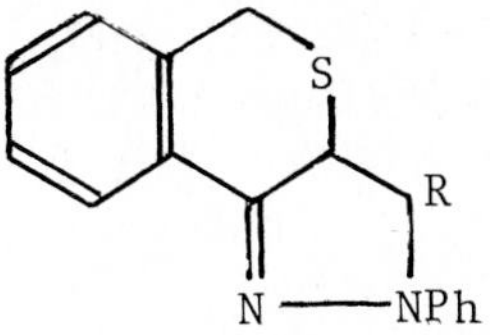

R=Ph,4-MeOC$_6$H$_4$

(27)

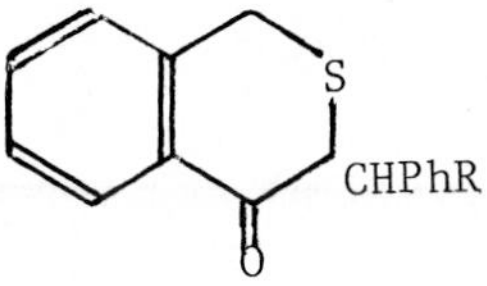

R=Ph,4-MeOC$_6$H$_4$

(28)

Lithiation of isothiochroman-4-one, followed by reaction with carbon disulphide and methyl iodide yields derivative (29), which on treatment with lithium dimethylcuprate gives 3-*tert*-butylisothiochroman-4-one (30) (O.H. Johansen and K. Undheim, Acta. Chem. Scand., 1979, <u>B33</u>, 460).

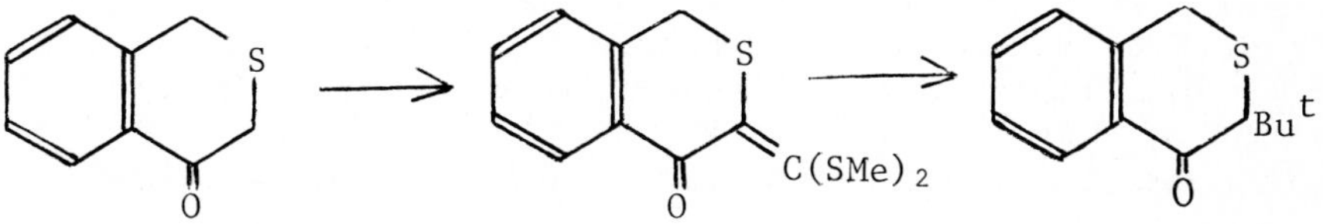

(29)   (30)

2-[α-(Methylsulphonylmethyl)benzyl]cyclohexanone with
potassium hydroxide in boiling methanol gives 3 isomeric
8a-hydroxy-4-phenylhexahydroisothiochroman 2,2-dioxides
(8a-hydroxy-4-phenyl-2-thiadecaline 2,2-dioxides)
(S. Fatutta and A. Risaliti, J. chem. Soc., Perkin I, 1978,
1321).

The crystal structure and conformation of 4β-acetoxy-4a-
-methylhexahydroisothiochroman-7-one (4β-acetoxy-4a-methyl-
-2-thiadecaline-7-one) (31) have been determined from X-ray
data. The two six-membered rings are normal chair
conformers and the ring junction is trans (H. Koyama and
Y. Yokouchi, Cryst. Struct. Comm., 1982, 11, 1959).

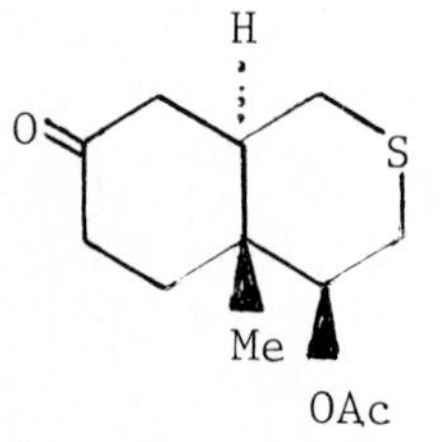

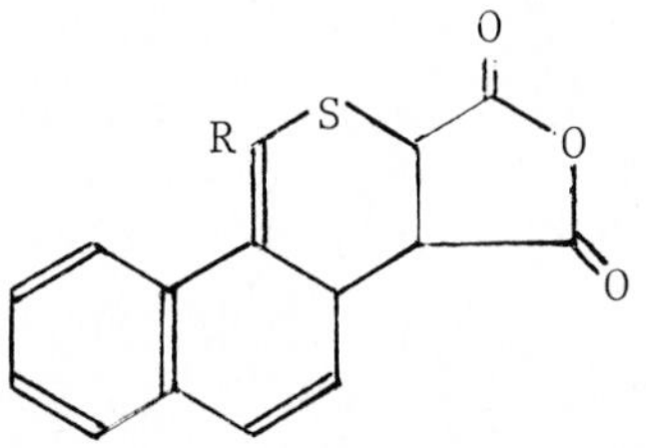

$R=Ph,2,4,6-Me_3C_6H_2$

(31)   (32)

Aryl 1- or 2-naphthyl thioketones react with maleic
anhdride, norbornene, and norbornadiene to give 1,4-cyclo-
adducts, for instance, the naphtho[1,2-c]thiopyran
derivative (32) (H. Ohmura and S. Motoki, Phosphorus
Sulphur, 1983, 16, 19).

4. Dibenzothiopyran, thioxanthene and derivatives

(a) Thioxanthenes

9-Mesityl-10-methyl-10-thiaanthracene generated by proton abstraction from 9-mesityl-10-methylthioxanthenium perchlorate (1; R^1=mesityl, R^2=Me) with sodium hydride in THF undergoes thermal rearrangement to yield 9-mesityl-3--methylthioxanthene (2; R^1=mesityl, R^2=Me). 3-Alkyl-9--arylthioxanthene (2) are obtained in a similar manner from appropriate 9-bulky aryl group substituted 10-alkylthiaanthracenes (S. Ohno *et al.*, J. org. Chem., 1984, **49**, 2472).

R^1=mesityl, R^2=Me,Et,Pr
R^1=duryl, R^2=Me,Et,Pr

(1)   (2)

9-Cyano- and 9-carboethoxy-10-methyl-10-thiaanthracenes (4) are prepared in high yield by proton abstraction from the corresponding thioxanthenium salts (3) with sodium hydride in THF under N_2. In THF at 50° the thiaanthracenes (4) undergo thermal 1,4-rearrangement to give the corresponding 9-cyano- and 9-carboethoxy- 9-methyl-thioxanthenes (5). Reactions of the thiaanthracenes with electrophiles, for example, dimethyl acetylenedicarboxylate or tetracyanoethene have been described (M. Hori *et al.*, *ibid.*, 1980, **45**, 2468).

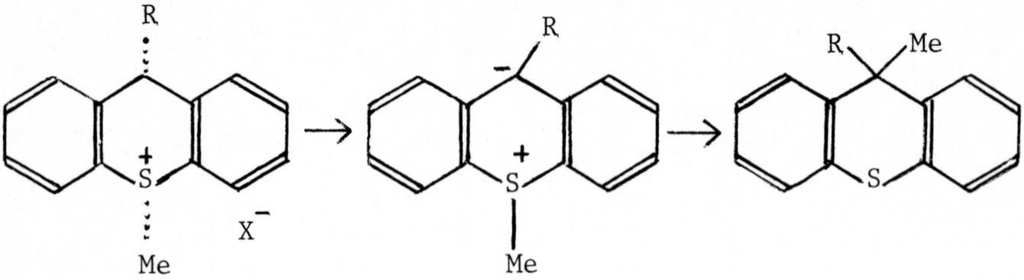

R=CN,X=ClO$_4$
R=CN,X=BF$_4$
R=CO$_2$Et,X=ClO$_4$

(3)   (4)   (5)

Treatment of (4-Me$_2$NC$_6$H$_4$)$_2$CH$_2$ with fuming sulphuric acid (23-30% concentration) at 60-80° in the presence of inorganic salts of mercury furnishes 3,6-bis-(dimethylamino)- -thioxanthene 10,10-dioxide (6) (B.I. Stephanov, A.B. Kostitsyn and N.N. Bychkov, U.S.S.R. Pat. SU 121,264, 1984). Its structure has been confirmed as the "tub" conformation by [1]H-nmr spectral data and HMO calculations (Kostitsyn et al., Zh. obschch. Khim.,1986, 56, 1282).

Several 10-alkyl-9-arylthioxanthenium salts (7) have been prepared by the reaction of the appropriate 9-arylthio-xanthene with an alkyl halides in the presence of silver salts. The stereochemistry of the sulphonium salts (7) has been determined from their [1]H-nmr spectral and X-ray analytical data of cis-10-methyl-9-phenylthioxanthenium tetrafluoroborate (Ohno et al., J. org. Chem., 1984, 49, 3151).

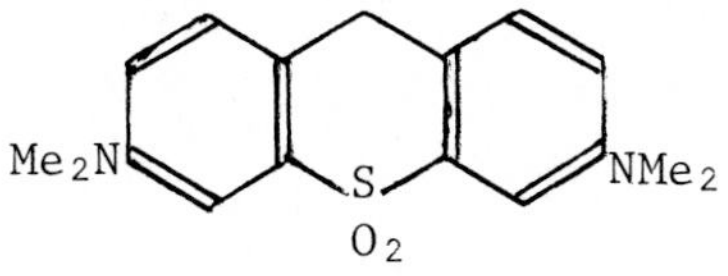

(6)

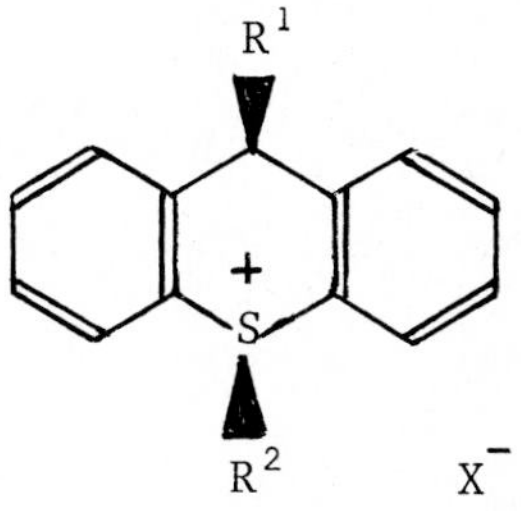

R^1=Ph, 2,4,6-Me$_3$C$_6$H$_2$
 2,3,5,6-Me$_4$C$_6$H
R^2=Me,Et,Pr
X=BF$_4$,ClO$_4$

(7)

9-Substituted and 9,9-disubstituted thioxanthenes (8) are obtained when thioxanthenes (8; R^2=H) react with dimethyl diazomalonate or when thioxanthenium biscarbomethoxymethylides (9; R^4=Me) are treated with base. The reaction is solvent-dependent. Thus, (9; R^1 = R^3 = H, R^4 = Me) with base in DMF yields thioxanthene [8; R^1 = R^3 = H, R^2 = CH(CO$_2$Me)$_2$]; (9; R^1 = R^3 = H, R^4 = Me; R^1 = H, R^3 = Cl, R^4 = Et) with base in toluene give only thioxanthenes [8; R^1 = R^3 = H, R^2 = CH(CO$_2$Me)$_2$; R^1= H, R^2 = CH(CO$_2$Et)$_2$, R^3 = Cl], respectively, whereas in DMF [8; R^1 = H, R^2 = CH(CO$_2$Me)$_2$, CH(CO$_2$Et)$_2$, R^3 = H, Cl] are formed (Y. Tamura, C. Mukai and M. Ikeda, Heterocycles, 1979, 12, 1179).

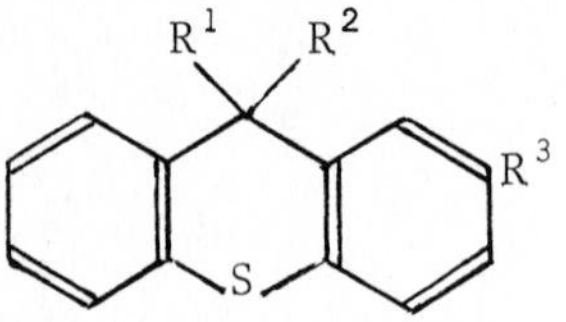

R^1=H,Me
R^2=CH(CO$_2$Me)$_2$
R^3=H,Cl

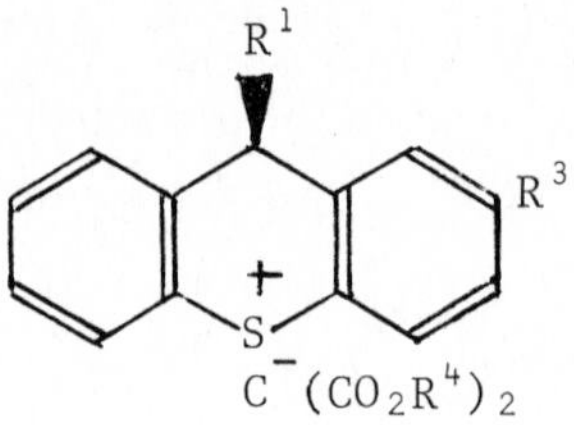

R^4=Me,Et

(8)

(9)

9-Alkylthioxanthene-*N*-(4-toluenesulphony)sulphilimines, *cis*- and *trans*-9-ethyl- (10), and *trans*-9-isopropyl-thio-xanthene-*N*-(4-toluenesulphonyl)sulphilimine (11) are prepared by tosylation of the corresponding 10-aminothio-xanthenium mesitylenesulphonates. In boiling dioxane · containing a small amount of hydrogen chloride they give the corresponding thioxanthenes and *cis*- and *trans*- (10) and *trans*- (11) on treatment with 1,8-diazabicyclo[5.4.0]-undec-7-ene in benzene rearrange to the corresponding 9--alkyl-9-(*N*-4-toluenesulphonamido)thioxanthenes. The rates of rearrangement have been discussed (Y. Tamura *et al.*, J. org. Chem., 1979, **44**, 1684).

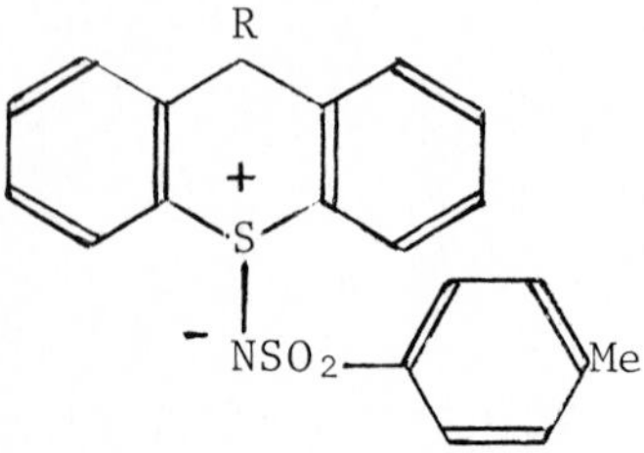

(10) R=Me,Et
(11) R=Pr^i

On treatment with 1,8-diazabicyclo[5.4.0]undec-7-ene in benzene 9-alkylidenethioxanthene-N-(4-toluenesulphonyl)-sulphilimines (12) rearrange to 9-(N-4-toluenesulphonamido)--9-vinylthioxanthenes (13). The rearrangement has been rationalized in terms of thioxanthylium ion intermediates (*idem, ibid.*, 1980, **45**, 2970).

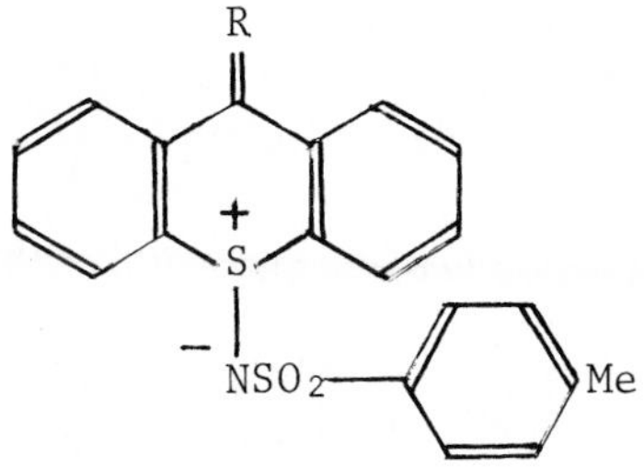

R=MeCH,EtCH,Me$_2$C

(12)

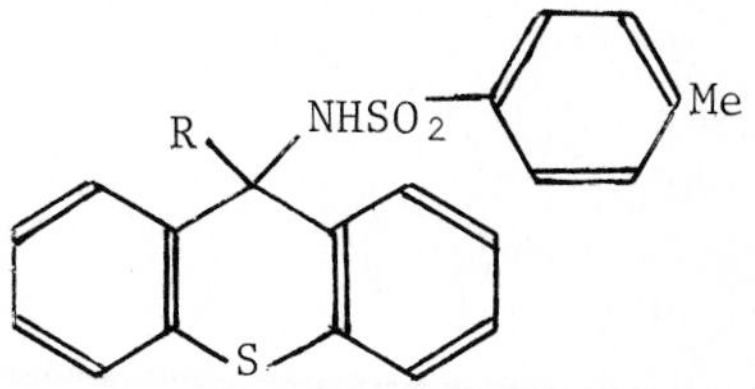

R=CH$_2$==CH,MeCH==CH,
CH$_2$==CMe

(13)

The N-tosylsulphilimines and bis(carbomethoxy)methylides of 1,4-dimethylthioxanthene and its 9-alkyl derivatives have been synthesized and their stereochemistry assigned by comparison of their nmr spectral data with those of 2,4--dimethylthioxanthene derivatives (Tamura, Mukai and Ikeda, Chem. pharm. Bull., 1982, **30**, 4069).

The rearrangement of optically active 10-aryl-10--thiaanthracenes (14) to 9-arylthioxanthenes (15) occurs with predominant racemization, but there is some retention of optical activity, in the product(15) (C.A. Maryanoff, K.S. Hayes and K. Mislow, J. Amer. chem. Soc., 1977, **99**, 4412).

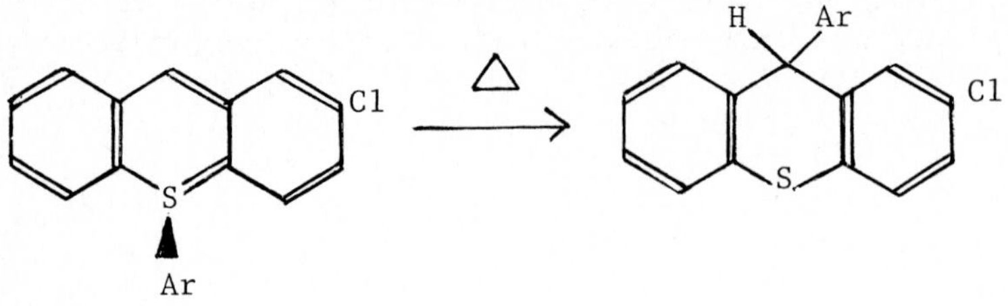

(14)   (15)

Peroxides (16) and (17) have been obtained by the photooxidation of 9-phenylthioxanthene in benzene. Similar products have been obtained by the photooxidation of the corresponding 1-oxide and 1,1-dioxide and related xanthenes (A. Goosen, S. African Pat. ZA 82 06,827, 1983).

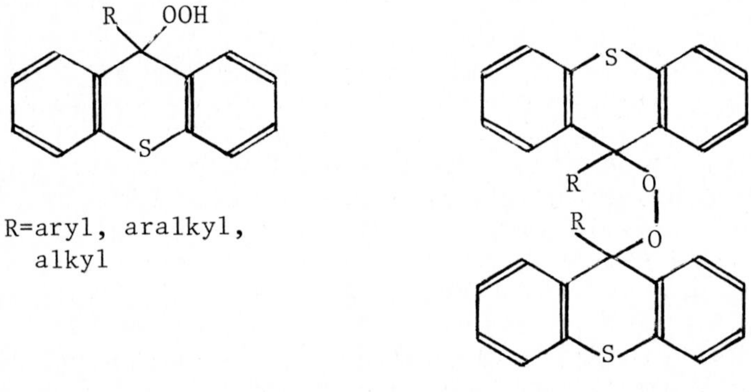

R=aryl, aralkyl, alkyl

(16)   (17)

Heating methyl thioxanthen-9-ylacetate (18) with hydrazine in ethanol affords hydrazide (19), which on boiling with benzaldehyde in methanol yields condensation product (20). Other related N-acyl derivatives (21) and condensation products (22) from hydrazide (19) have been prepared [N. Rasanu and O. Maior, Rev. Chim. (Bucharest), 1981, **32**, 1019].

CH$_2$CO$_2$Me   CH$_2$CONHNHR^1 CH$_2$CONHN=CHR^2

(18)

(19) R^1=H
(21) R^1=PhCO,
   PhNHCS

(20) R^2=Ph
(22) R^2=HOC$_6$H$_4$,
   O$_2$NC$_6$H$_4$,
   furyl

The structures of 2,4-dimethyl-,2,4,9-trimethyl-, and 2,4-dimethyl-9-isopropyl- thioxanthene 10,10-dioxides have been determined by X-ray diffraction. The central ring of the thioxanthenes is in a boat conformation and in the last two compounds the 9-substituents are in boat-axial conformations with respect to the central ring (S.S.C. Chu, V. Napoleone and T.L. Chu, J. heterocyclic Chem., 1987, **24**, 143).

The reaction of 9-methylenethioxanthene with n-propylthiol in the presence of di-*tert*-butyl peroxide gives 9-propyl-thiomethylenethioxanthene (23) rather than 9-thioxanthenyl-methyl sulphide regardless of the concentration of the thiol. With a low concentration of thiol in the presence of dibenzoyl peroxide the reaction yields compound (23), 1,2-bisthioxanthenylideneethane (24) and derivative (25), but only product (23) is formed at high concentrations of thiol (E.K. Kim, K. Kim and J.H. Shin, Bull. Korean chem. Soc., 1987, **8**, 380).

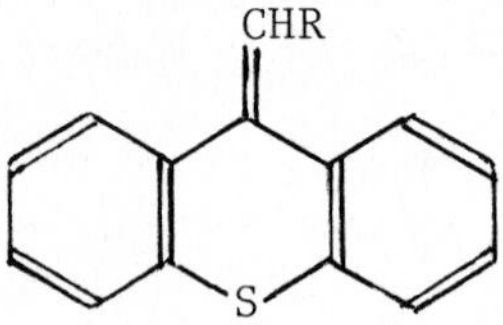

(23) R=SPr
(25) R=O$_2$CPh

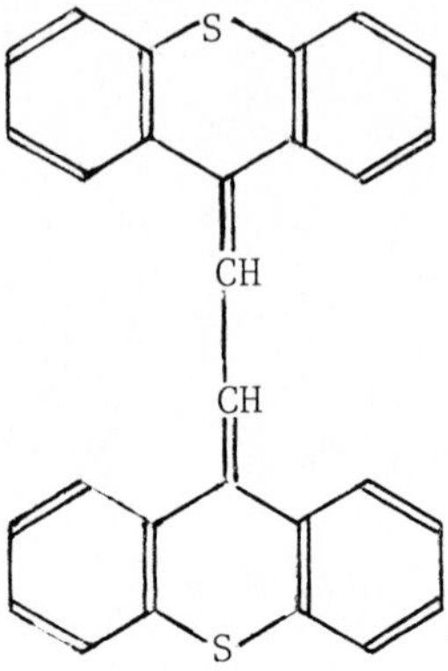

(24)

The 2-chlorothioxanthene derivatives (26), of which
(26; R=CH$_2$CH$_2$OH) and its hydrogen oxalate salt possess
tranquilizing activity without cataleptic or extrapyramidal
symptom side effects, have been prepared (M. Protiva and
V. Kmonicek, Czech. Pat. CS 235, 148, 1987). 1-Methyl-4-
-(2-bromothioxanthen-9-ylidene)piperidine on sequential
treatment with copper (I) cyanide in (Me$_2$N)PO, at 180° for
10h, and hydrolysis affords the thioxanthene-2-carboxylic
acid (27) (P.S. Anderson and D.C. Remy, Brit. Pat.
1,495,890, 1977).

R=Me,CO₂Et,H,
   CH₂CH₂OH

(26)   (27)

The preparations of 2-substituted 9-(3-amino-1-
-propylidene)thioxanthenes (28) and (29) (S. Tammilehto,
R. Pere and M.L. Haemaelaeinen, Acta. Pharm. Suec. 1986,
$\underline{23}$, 289) and of substituted 9-(3-dimethylamino-1-propyl-
idene)thioxanthenes (30) (Protiva et al., Coll. Czech,
chem. Comm., 1978, $\underline{43}$, 2656) have been described.
Dehydration of 2-chloro-9(3-dimethylaminopropyl)-9-
-hydroxythioxanthene with sulphuric acid at 100-200° gives
a mixture of the cis- and the $trans$- 2-chloro-9-(3-
-dimethylaminopropylidene)thioxanthene. Crystallization
from ethyl acetate petroleum ether with seeding yields the
$trans$-isomer, which contains 1-3% cis-isomer (J. Michalsky,
V. Sapara and J. Mayer, Czech. Pat. CS 198,338, 1982). For
the preparation of 9-(3-dimethylaminopropylidene)-6-fluoro-
-2-isopropylthioxanthene see Protiva, M. Rajsner and
J. Metysova (Czech. Pat. CS 202,229, 1982); for 3-alkyl-9-
-(3-dimethylaminopropylidene)thioxanthenes, Protiva,
Metysova and Z. Sedivy ($ibid.$, 224,820, 1984); for 9-(3-
-dimethylaminopropylidene)-1-ethylthioxanthene and its
hydrogen fumarate, Protiva and Sedivy ($ibid.$, 224,812, 1984);

528

for 2-(disubstituted amino)-9-(3-dimethylaminopropylidene)-
thioxanthenes, potential neuroleptics and tranquillizers,
V. Kmonicek *et al*., (Coll. Czech. chem. Comm., 1986, **51**,
937); for 2-chloro-9-(3-dimethylaminopropyl)-9-hydroxy-
thioxanthene, Sapara, Michalsky and Mayer (Czech. Pat.
175,004; 175,223, 1978); and for 9-(3-dimethylaminopropyl-
idene)-2-(methylthiomethyl)thioxanthene, Kmonieck, V. Bartl
and Protiva, (Coll. Czech. chem. Comm., 1984, **49**, 1722).

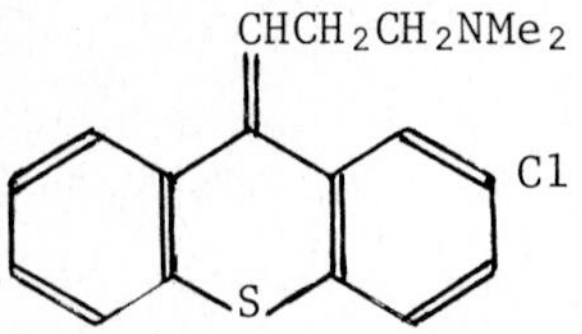

(28) 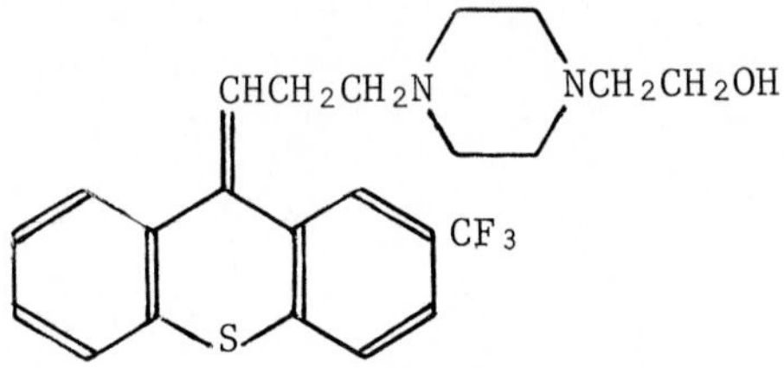   (29)

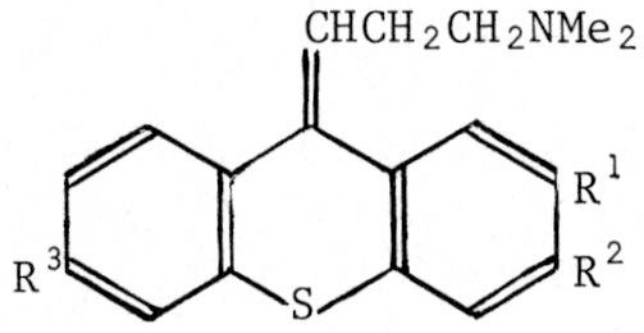

$R^1 = Cl, Pr^i, R^2, R^3 = H, F$

(30)

The synthesis of spiro(cyclohexane-1,9'-thioxanthene)
compounds (31) (Y. Tsuda, T. Kawata and T. Kenjo, Ger.
Offen. 2,818,329, 1979), (32) (*idem*, Japan Kokai Tokkyo
Koho 79, 144,373, 1979; Yoshitomi Pharmaceutical
Industries Ltd., Belg. Patt. 866,665, 1978) and related
compounds (*idem*, Neth. Appl. 78, 04,686, 1979) and the mass
spectra of thioxanthenium triiodide salts (33)
[M.M. El-Namaky and M.A. Salama, Egypt. J. Chem., 1975
(pub. 1978), **18**, 799] have been reported.

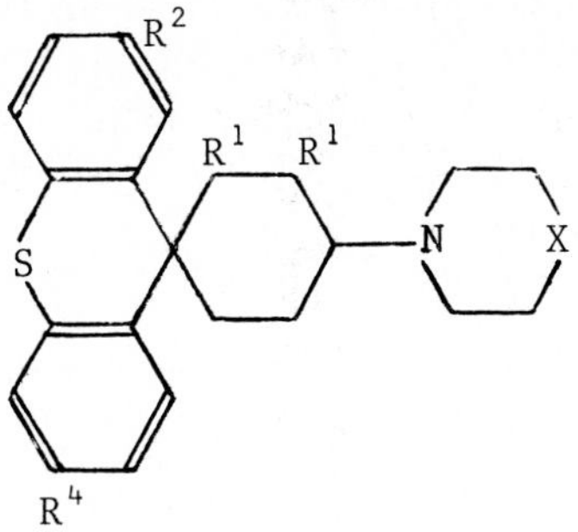

R^1=H; R^1R^1= bond
R^2=H,halogeno, CF_3,
    R^3,SR^3,SO_2R^3,$SO_2NR_2^3$
R^3=alkyl
R^4=H,F
X =NR^5,CR^6R^7
R^5=R^6=alkyl, hydroxyalkyl
R^7=H,OH

(31)

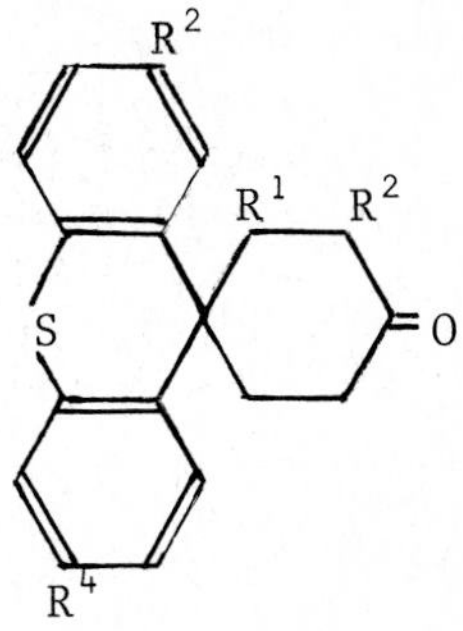

R^1=H; R R =bond
R^2=H, alkyl, halogeno,
    CF_3,OR^3,SO_2R^3
    $CO_2NR_2^3$,SO_2NH_2
R^3=alkyl,
R^4=H,F

(32)

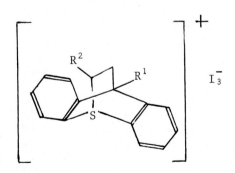

R^1=H,R^2=H,Me
R^1=Et,R^2=H

(33)

X-ray analytical data indicate that the acid-catalyzed isomerization of 9-benzyl-1,2,3,4,5,6,7,8-octahydrothio-xanthene (34) gives 3,4-benzo-5,9,7,8-bis(tetramethylene)--6-thiatricyclo[1.3.3]non-7-ene (35), (A.A. Shcherbakov *et al*., Khim. Geterotsikl, Soedin., 1979, 1470).

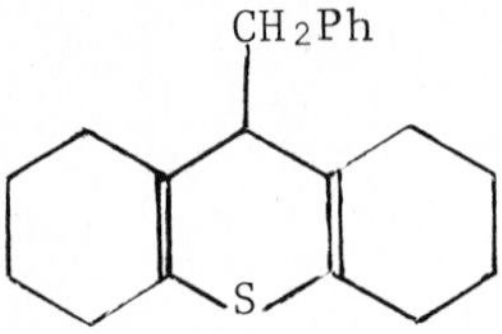

ÇH₂Ph

(34) 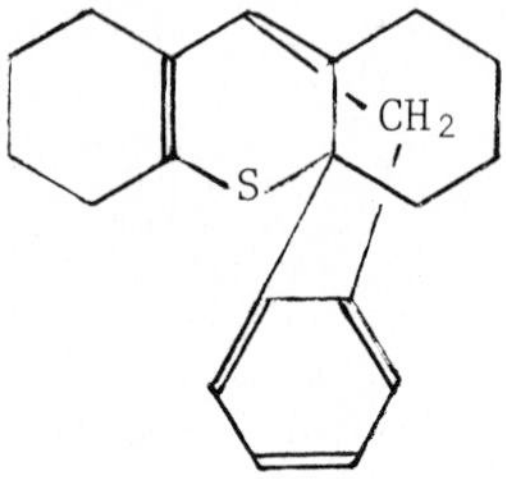   (35)

The reaction of 9-phenylthioxanthylium perchlorate (36) with phenyllithium yields eight products including thio-xanthenes (37) and their formation has been explained by a radical mechanism (Hori *et al*., J. chem. Soc., Perkin I, 1987, 187).

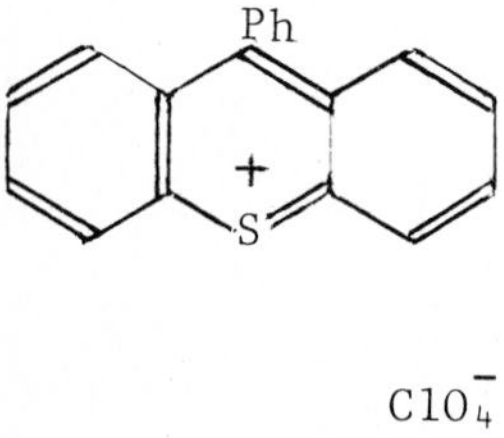

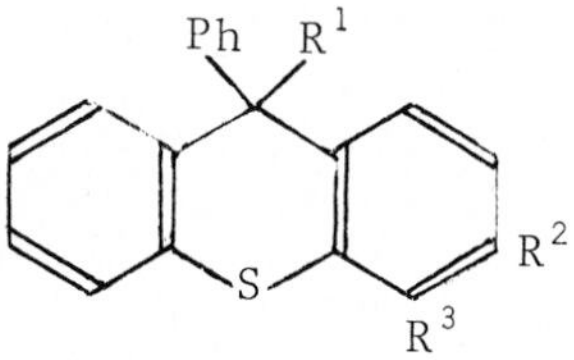

$R^1 = R^2 = R^3 = H$
$R^1 = Ph, R^2 = R^3 = H$
$R^1 = R^3 = H, R^2 = Ph$
$R^1 = R^2 = Ph, R^3 = H$
$R^1 = R^3 = Ph, R^2 = H$

(36)   (37)

The reactions of thioxanthylium ions with dimethyl-,
dibenzyl-, di-isopropyl-, and diphenyl- mercury in the air
affords 9,9'-methylenedithioxanthene, 9-benzyl-, 9-
-acetonyl-, and 9-phenyl- thioxanthenes, respectively
(D.M. Shin and K. Kim, Bull, Korean chem. Soc., 1981, 2,
114). Thioxanthylium perchlorate reacts with pyridine to
give 1-(9-thioxanthenyl)pyridinium perchlorate
(G.E. Ivanov, G.V. Pavlyuk and B.T. Kaminskii, Zh. org.
Khim., 1982, 18, 1996). Hydrogenation of thioxanthylium
and 9-methylthioxanthylium tetrafluoroborates in the
presence of 10% Pd/C gives the corresponding dodecahydro-
thioxanthenes (V.G. Kharchenko, O.A. Bozhenova and
A.D. Shebaldova, Zh. org. Khim., 1982, 18, 2435).

A novel cleavage of the excocylic double bond in bithio-
xanthylene (38) occurs on heating it at 270° with thiobenzo-
phenone to give 9-diphenylmethylenethioxanthene (39) and
thioxanthene-9-thione (40) (M.M. Sidky and L.S. Boulos,
Chem. and Ind., 1983, 118). A similar result is obtained
with bixanthylene.

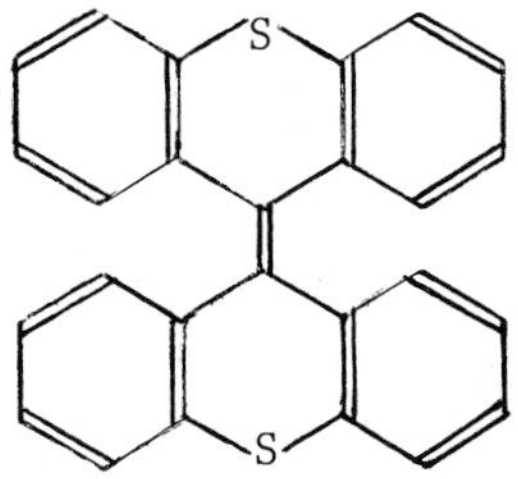

(38)

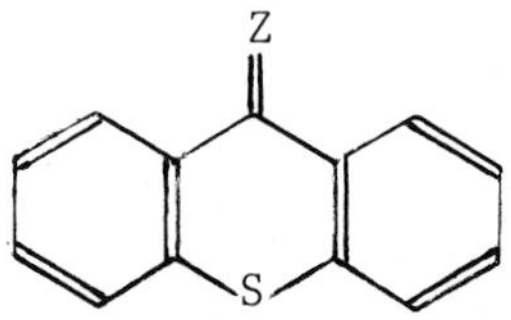

(39) Z=CPh$_2$
(40) Z=S

(b) Thioxanthones, thioxanthen-9-ones

Thioxanthone has been obtained by the cyclization of benzophenone with crystalline sulphur trioxide or its DMF or dioxane complexes at -30° (W. Ried and G. Oremek, Chem. -Ztg., 1980, **104**, 12) and 2-methylthioxanthone (2) has been formed in the photo-Friedel-Crafts reaction of the corresponding thioester (1) (J. Martens, K. Praefcke and U. Schulze, Synth., 1976, 532).

(1)   (2)

The 1-,2-,3-, and 4-methylthioxanthones and their dioxides have been prepared and a previous synthesis of 2- and 3-methylthioxanthones (H. Gilman *et al.*) has been shown to give the 2- and 4-isomers (I.D. Brindle and P.P. Doyle, Canad. J. Chem., 1983, **61**, 1869). Cyclocondensation of 2- -mercaptobenzoic acid with *o*-xylene in sulphuric acid gives a mixture of 1,2-, 2,3-, and 3,4- dimethylthioxanthones, with *m*-xylene the 1,3-isomer is obtained and with *p*-xylene the 1,4-isomer (I. Okabayashi *et al.*, Chem. pharm. Bull., 1987, **35**, 2545). Some 2,4-dialkylthioxanthones have been prepared (T. Shirosaki and S. Fukunago, Ger. Offen. DE 3,209,706, 1982).

Thiosalicyclic acid on cyclocondensation with phenol in concentrated sulphuric acid, followed by hydrolysis yields 2-hydroxythioxanthone. Some derivatives and related thio-xanthones have been reported (P. Heaton and J.R. Curtis,

Brit. UK Pat. Appl. GB 2,108,487, 1983). 1-[2-(diethyl-aminoethyl)amino]-7-hydroxy-4-methylthioxanthone has been obtained from 7-methoxy-4-methylthioxanthone (S. Archer, U.S. Pt. US 4,539,412, 1985). 2-Ethoxyxanthone is obtained on heating $2,5-(EtO)_2C_6H_3COC_6H_4OMe-2$ and sodium sulphide hydrate in DMF at $114-119^{\circ}$ for 6h (L. Avar, Ger. Offen. DE 3,523,680, 1986). Cyclocondensation of 2-chloro-sulphenylbenzoyl chloride with appropriate substituted benzenes in the presence of Lewis acid catalysts affords thioxanthenes (3), for example, 2-chlorosulphenylbenzoyl chloride and dimethoxybenzene in 1,2-dichloroethane in the presence of tin(IV) chloride and under nitrogen gives 1,4--dimethoxythioxanthone (3; $R^1 = R^4 = OMe$, $R^2 = R^3 = H$) (J.F. Honek, M.L. Mancini and B. Belleau, Synth. Comm., 1983, __13__, 977).

$R^1 = OMe, Me, H, OH$
$R^2 = H, Me$
$R^3 = H, Me$
$R^4 = OMe, Me, H, OH$

(3)

1-Fluoro-4-hydroxythioxanthone has been obtained by treatment of 2,2'-dithiosalicyclic acid with 4-fluorophenol and sulphuric acid and has been O-methoxymethylated and converted into 1-(substituted amino)-4-methoxymethoxythio-xanthones (Mancini and Honek, _ibid._, 1986, __16__, 1109). 2,5--$Cl_2C_6H_3COC_6H_4Cl$-2 in benzene on treatment with sodium sulphide hydrate and $Me(CH_2)_{15}P^+Bu_3.Br^-$ gives 2-chlorothio-xanthone. Other derivatives of thioxanthone have been prepared by this method from appropriately substituted benzophenones [L. Avar and E. Kalt, Patentschrift (Switz.) CH 642,652, 1984; Brit. UK Pat. Appl. 2,018,243, 1979]. For the preparation of a mixture of 2- and 4-chloro, 2,7--dichloro-, 4,7-dichloro-, and 7-chloro-2-phenyl- thioxanthone see L. Vacek and H.M. Foster (U.S. Pt. 4,101,558, 1978); for 2-chloro-6-(5-tetrazoly)thioxanthone, J.F. Batchlor

534

[Pat. Specif. (Aust.) AU 523,260, 1982]; and for 1,2,4-
-trichloro- and 2,4,5,6-tetrachloro- thioxanthone
derivatives, V. Bartl *et al.*, (Coll. Czech. chem. Comm.,
1983, **49**, 2295).

3-Bromothioxanthone 10,10-dioxide is obtained by the
cyclization of 5,2-Br(HO$_2$C)C$_6$H$_3$SO$_2$Ph in the presence of
sulphuric acid. It undergoes cyanation, followed by cyclocondensation with sodium azide to yield 3-(5-tetrazolyl)-
thioxanthone 10,10-dioxide. Other related derivatives have
been reported (P. Puigdellivol Llobet, Span. Pat. 464,944,
1978).

The reaction between chlorodinitrobenzoic acid (4) and
thiophenol in the presence of alkali yields sulphide (5),
which on cyclization affords 2,4-dinitrothioxanthone (6).
2,4,7-Trinitrothioxanthone (7) is prepared in a similar
manner using 4-nitrothiophenol (K. Sakai *et al.*, Japan
Kokai Tokkyo Koho 79 09,277, 1979).

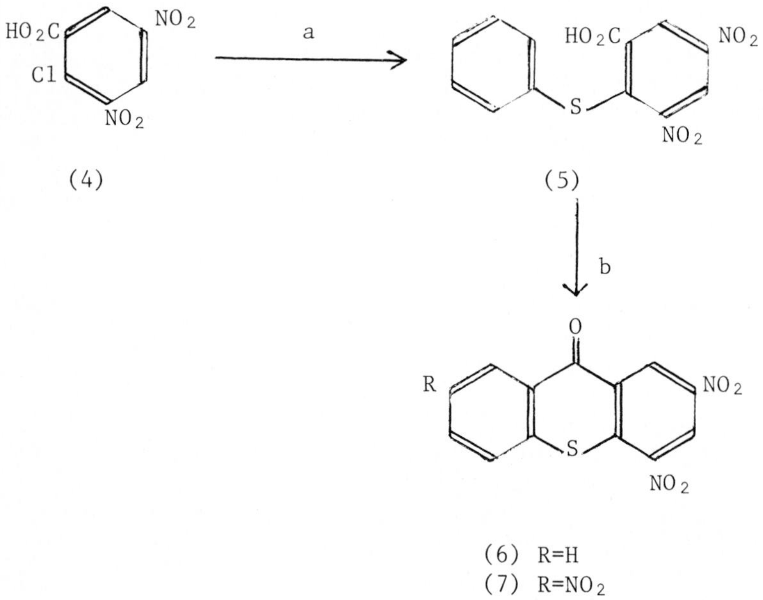

(4)   (5)

b

(6) R=H
(7) R=NO$_2$

Reagents:- (a) PhSH,MeOH,KOH; (b) PCl$_5$,PhNO$_2$,AlCl$_3$

Many 1-substituted 4-methyl-6-nitrothioxanthones, for example (8), have been synthesized for schistosomicidal evaluation. Derivatives (8; R = morpholino, NHCH$_2$CH$_2$NEt$_2$) show promising activity against *Schistosoma mansoni* in mice (M.M. El-Kerdawy, A.A. El-Emam and H.I. El-Subbagh, Arch. Pharmacol Res., 1986, **9**, 25).

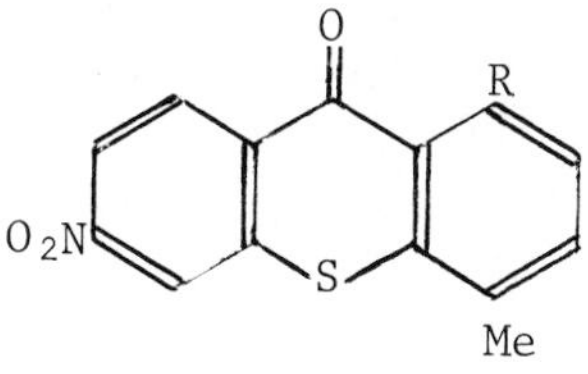

R=NMe$_2$,NEt$_2$,morpholino, piperidino

(8)

7-Substituted 3-nitrothioxanthone-1-carboxylic acids and their esters (9) and related xanthones have been prepared, and other derivatives have been obtained by their reactions with nucleophiles (W. Fischer and V. Kvita, Helv., 1985, **68**, 854).

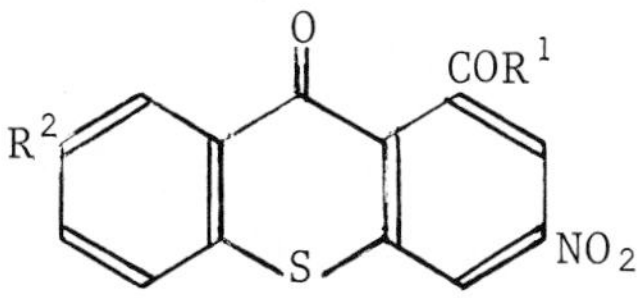

R^1=OH,OMe,OEt,OBu,
NBu$_2$,Cl
R^2=H,Me,OMe

(9)

A series of esters (10) of hycanthone, *1-(2-diethylamino-ethylamino)-4-(hydroxymethyl)thioxanthone*, and (11) of 7--hydroxyhycanthone have been synthesized and their anti--tumour activity and anti-schistosomal effects on HC-sensitive and HC-resistant schistosomes have been reported (S. Archer *et al.*, J. med. Chem., 1988, **31**, 254).

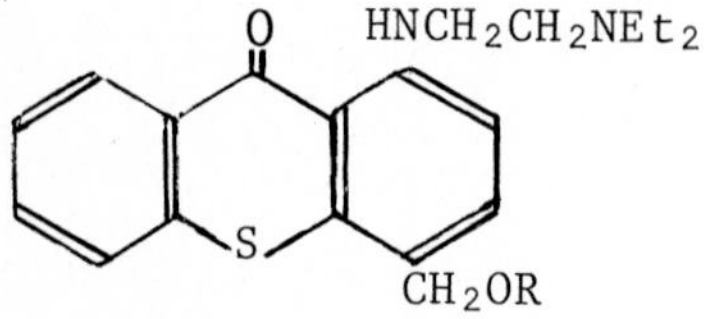

R=PhNHCO, PrNHCO,
  BuNHCO, MeCO, MeO$_2$C,
  (O$_2$N)$_2$C$_6$H$_3$CO

(10)   (11)

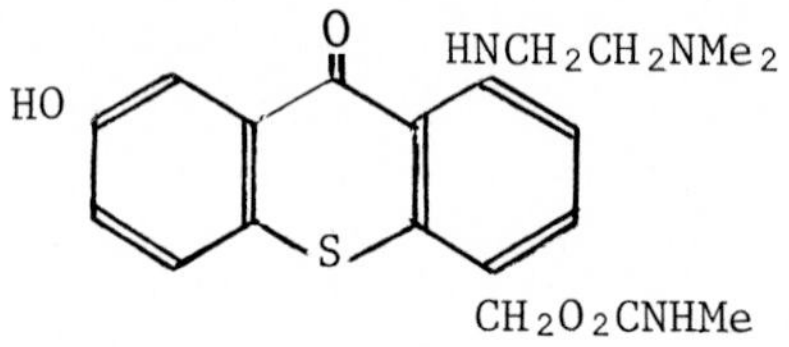

2,7-Dicarboxamidinothioxanthone (12) and the related
xanthone (12; S=O) and the parent thioxanthene and xanthene
derivatives have been obtained from the corresponding
dibromo derivatives *via* the dicarbonitriles. They show
anti-leishmanial and bactericidal activity (P.M.S. Chauhan
et al., Indian J. Chem., 1987, **26B**, 248).

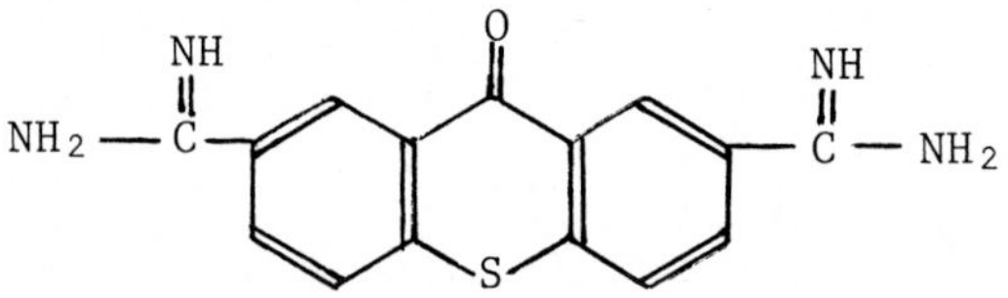

(12)

Thioxanthone-3-carboxylic acid on treatment with thionyl chloride is converted into the acid chloride, which reacts with aqueous methylamine to give 3-methylamidothioxanthone (13; $R^1 = R^3 = H$, R^2 = CONHMe). A number of related derivatives (13) have been prepared (M. Harfenist, C.T. Joyner and D.J. Heuser, Eur. Pat. Appl. EP 150,891, 1985).

R^1 or R^2=H the other
    = carbamoyl, amidino,
    (alkyl)-2-imidazolyl,
    (alkyl)tetrazol-5-yl
R^3= H, alkyl, alkenyl, OR^4
R^4= H, alkyl, alkenyl

n = 0,1,2

(13)

The cyclocondensation of 5-chloro-2-mercaptobenzoic acid with 5-(2,4-dichlorophenyl)-3-hydroxycyclohex-2-en-1-one in phosphoryl chloride produces 7-chloro-3-(2,4-dichlorophenyl)--1,2,3,4-tetrahydrothioxanthen-1,9-dione (14). Derivatives (15) and (16) have also been prepared along with (14) as sulphur isosteric analogues of the acridinedione antimalarial agent, floxacrine, but they were devoid of antimalarial activity against *P. berghei* in mice (J. Hung, D.J. McNamara and L.M. Werbel, J. heterocyclic Chem., 1983, 20, 1575).

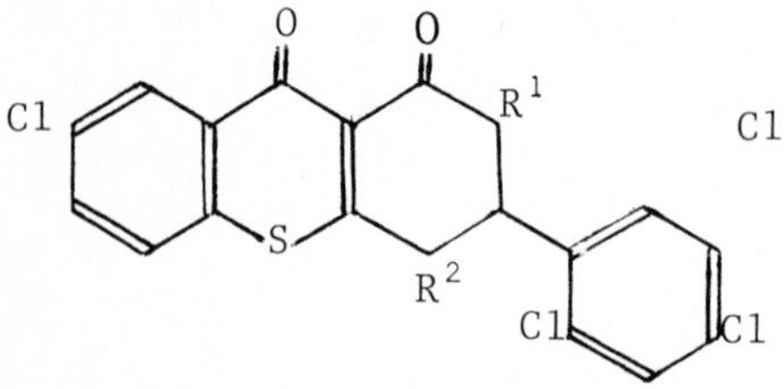

(14) $R^1=R^2=H$
(15) $R^1=Me, R^2=H$
     $R^1=H, R^2=Me$

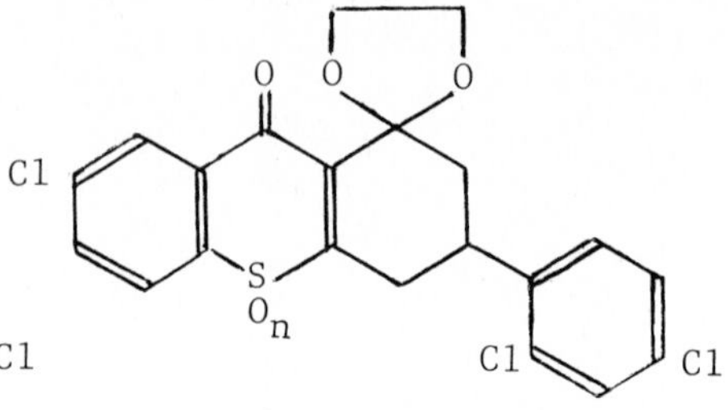

$n=C,1,2$

(16)

Thiothiene (Navane), N,N-*dimethyl-9-[3-(4-methyl-1-piper-azinyl)propylidene]thioxanthene-2-sulphonamide*, undergoes spontaneous photo-oxidation to give N,N-dimethylthio-xanthone-2-sulphonamide (L.E. Eiden and J.A. Ruth, Experientia., 1978, **34**, 1062).

(c) Dibenzo[bd]thiopyrans

6-Substituted 6-hydroxy-6H-dibenzo[b,d]thiopyrans(1) are obtained by the Grignard reaction between 6H-dibenzo[b,d]-thiopyran-6-one and an alkyl or phenylmagnesium halide. Dehydration of the 6-ethyl and 6-isopropyl alcohols yields 6-ethylidene- and 6-isopropylidene-6H-dibenzo[b,d]thiopyrans, respectively. In some cases dimerisation occurs (D.D. Ridley and M.A. Smal, Austral. J. Chem., 1983, **36**, 795).

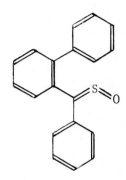

(1) R^1=OH,R^2=Me,Et,Pr^i,Ph
(3) R^1=OMe, R^2=Ph
(4) R^1=OH,R^2=H
(5) R^1=CO_2H, R^2=H

(2)

The sulphine (2) on stirring with aluminium chloride in dichloromethane, yields after treatment with methanol, 6-
-methoxy-6-phenyl-6H-dibenzo[b,d]thiopyran (3). (T. Saito et
$al.$, Nippon Kagaku Kaishi, 1987, 1385).

6H-Dibenzo[b,d]thiopyran 5-oxide on reaction with trifluoroacetic anhydride and lutidine gives 6-hydroxy-6H-
-dibenzo[b,d]thiopyran (4), which on treatment with trimethylcyanosilane affords 6-cyano-6H-dibenzo[b,d]-
thiopyran, converted on hydrolysis into 6H-dibenzo[b,d]-
thiopyran-6-carboxylic acid (5). The alcohol (4) with piperazine yields 6-(1-piperazinyl)-6H-dibenzo[b,d]thiopyran
(C. Banzatti, P. Melloni and P. Salvadori, Gazz., 1987,
117, 269). For the preparation of several 6,6-
-disubstituted 6H-dibenzo[b,d]thiopyrans with a number of substituents on the benzene rings see Melloni, Salvadori
and P.P. Lovisolo (Ger. Offen. DE 3,606,900, 1986), and for
6H-dibenzo[b,d]thiopyran-6-one 5,5-dioxide, K. Schank and
F. Werner (Ann., 1983, 1739).

The reaction of 6-phenyl-6H-dibenzo[b,d]thiopyran (6)
with diphenyldiazomethane in boiling benzene gives dibenzothiepin (7) (58%) and 6-diphenylmethyl-6-phenyl-6H-
-dibenzo[b,d]thiopyran (8) (29%). Similarly 6H-dibenzo[b,d]-
thiopyran affords dibenzothiepin (9) (41%) and 6-diphenyl-
methyl-6H-dibenzo[b,d]thiopyran (10) (L. Benati,
P.C. Montevecchi and P. Spagnolo, J. chem. Soc., Perkin I,
1982, 917).

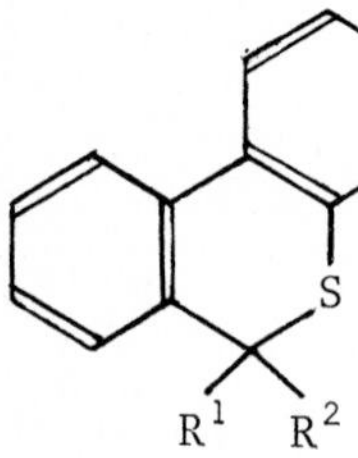

(6) R^1=H, R^2=Ph
(8) R^1=CHPh$_2$, R^2=Ph
(10) R^1=CHPh$_2$, R^2=H

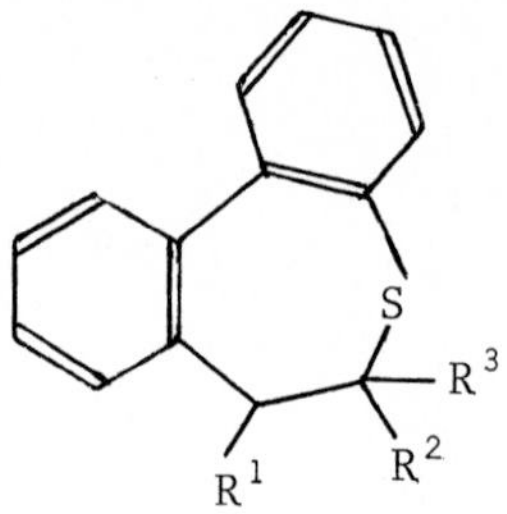

(7) R^1=R^2=R^3=Ph
(9) R^1=H, R^2=R^3=Ph

The ylides, 5-alkyl-6-cyanodibenzo[b,d]thiopyrans, 9-
-alkyl-10-cyano-9-thiaphenanthrenes (11), are obtained in
good yield by proton abstraction from the corresponding 6H-
-dibenzo[b,d]thiopyranium salts with triethylamine in
ethanol. Treatment of ylide (11; R^1=Me) with dimethyl or
diethyl acetylenedicarboxylate in benzene yields three 1:1
adducts, spiro compound (12), dibenzothiocin derivative
(13), and benzothiocinium ylide (14; R^1=Me). Similar
treatment of ylide (11; R^1=Et) affords dibenzothiepin (15),
as the major product and ylide (14, R^1=Et) (M. Hori *et al.*,
J. org. Chem., 1987, **52**, 3668; Chem. Comm., 1985, 883).

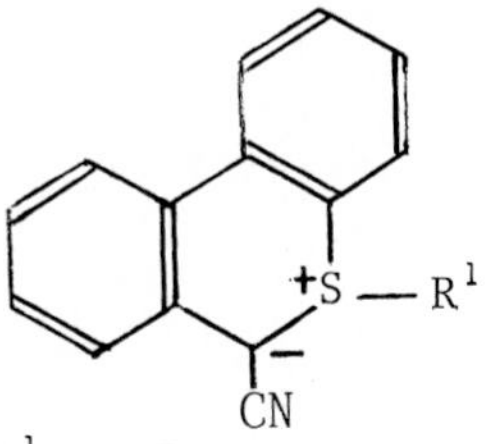
+S — R1
CN

R^1=Me,Et

(11)

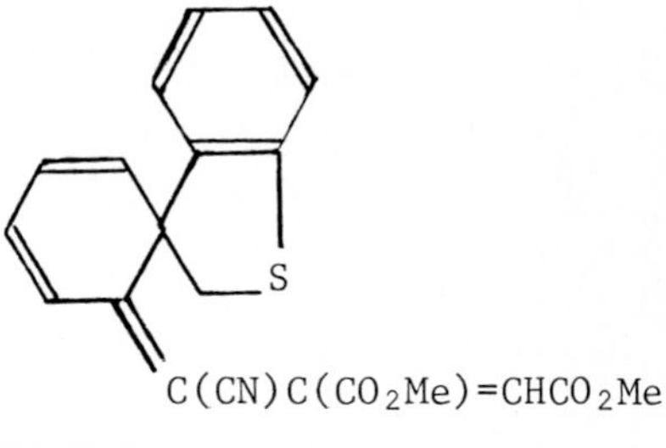
S
C(CN)C(CO2Me)=CHCO2Me

(12)

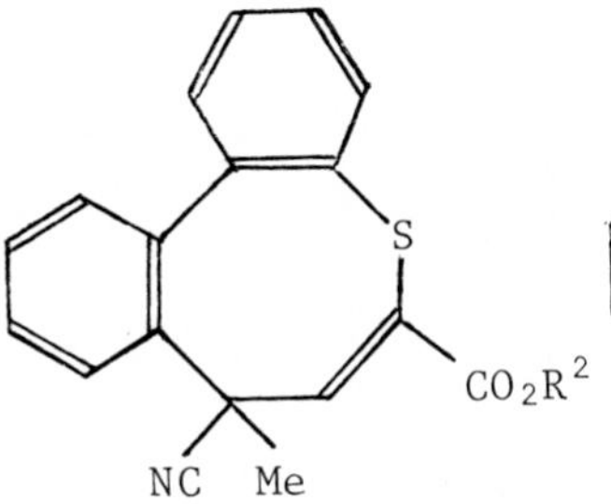
S
CO2R2
NC Me

R^2 = Me, Et

(13)

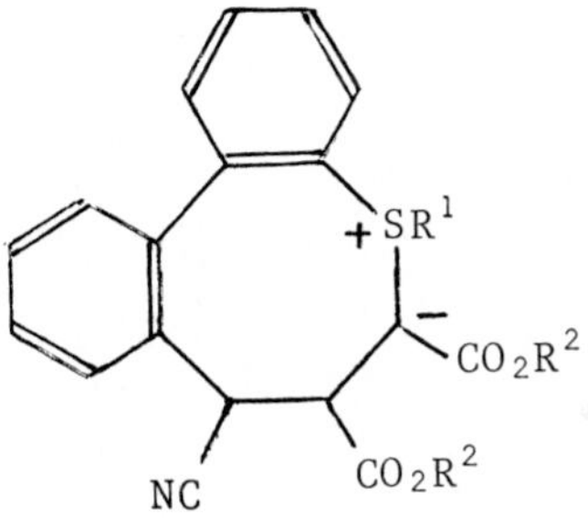
+SR1
CO2R2
CO2R2
NC

R^2 = Me,Et

(14)

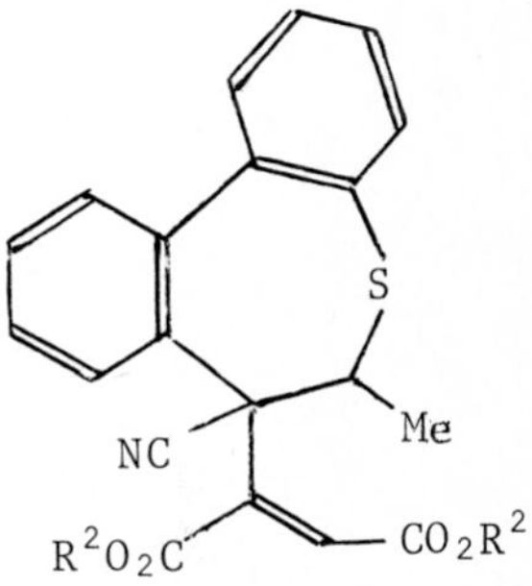
S
Me
NC
R2O2C
CO2R2

R^2 = Me,Et

(15)

5. Bridged ring sulphur compounds and related compounds

(a) Thiabicycloalkanes

Thioaldehydes generated by the fluoride induced β--elimination of stabilized aryl thiolate anions from α--silyldisulphides (1) have been trapped by Diels-Alder cycloaddition with cyclopentadiene to yield 2-thiabicycl-[2.2.1]hept-5-enes (2) (G.A. Krafft and P.T. Meinke, Tetrahedron Letters, 1985, 26, 1947).

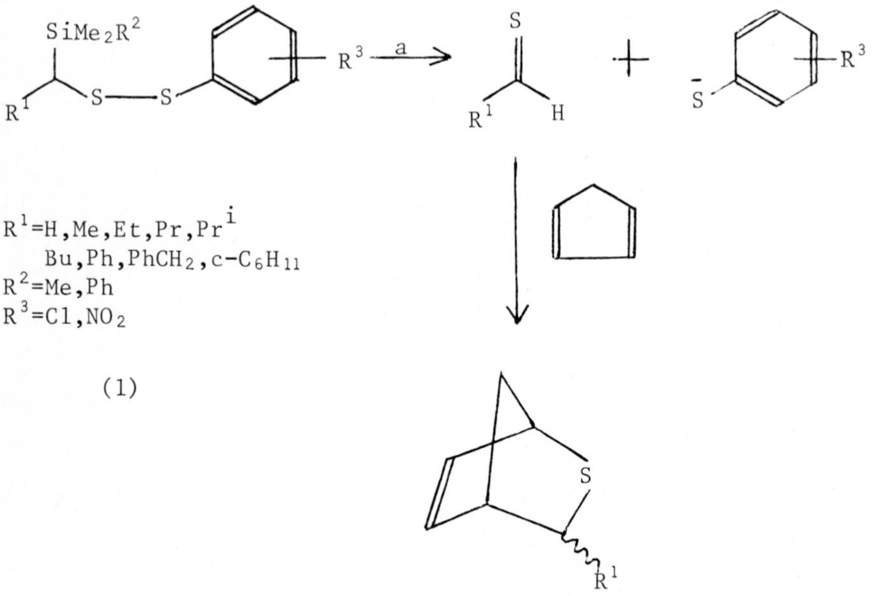
$SiMe_2R^2$
R^1
S
S
R^3
a
S
R^1
H
S
R^3
R^1=H,Me,Et,Pr,Pr^i
Bu,Ph,$PhCH_2$,c-C_6H_{11}
R^2=Me,Ph
R^3=Cl,NO_2
S
R^1

(1)

(2)

Reagents:- (a) C_5F or KF, 18-crowned-6, R.T., or Bu_4^tNF,THF, 0° to -78°

Endo-thioketones(3) selectively rearrange, *in situ*, below 0°, into 4,4a,5,7a-tetrahydrocyclopenta[b]thiopyran-6-enes (4). The same rearrangement is observed with thioketone (5) at slightly above 50° (P. Beslin, D. Lagain and J. Vialle, J. org. Chem., 1980, **45**, 2517).

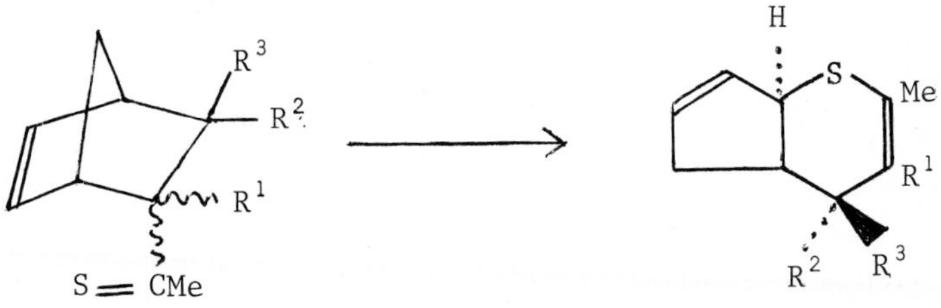

$R^1=R^2=R^3=H$
$R^1=R^2=H,R^3=Me$
$R^1=Me,R^2=R^3=H$

(3)

(4)

(5) $R^1=H,R^2=R^3=Me$

2,7-Methanothia[9]annulene (6) has been synthesized from 1,6-diiodocyclohepta-1,3,5-triene and variable-temperature ^{13}C-nmr spectroscopy and electronic spectroscopy indicate that it is in equilibrium with its valence isomer 9-thia-tricyclo[4.3.1.0^{1,6}]deca-2,4,6-triene (7) (R. Okazaki, T. Hasegawa and Y. Shishido, J. Amer. chem. Soc., 1984, **106**, 5271).

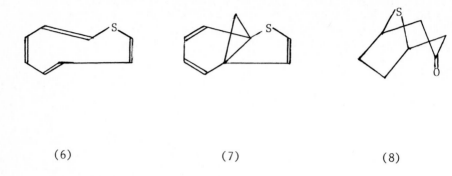

(6)   (7)   (8)

7-Thiabicyclo[3.2.1]octan-3-one (8) is obtained by treating cyclohepta-2,6-dienone with sodium sulphide. Its chemical reactivity has been compared with that of 9-thiabicyclo[3.3.1]nonan-3-one [T. Sasaki, S. Eguchi and T. Hioki, Heterocycles, 1979, 13 (Spec. Issue), 293].

9-Thiabicyclo[3.3.1]nonane-2,6-dione (9) on irradiation in methanol for 2h affords methyl tetrahydrothiopyran-3--one-2-propionate (10) (31%), and after 16h a mixture of compounds. Photolysis of its sulphoxide yields cycloocta--1,5-dione (15%) also obtained along with bicyclo[3.3.0]-octa-2,6-dione (8 and 10%, respectively) on irradiating its sulphone in methanol for 4h (P.H. McCabe and C.R. Nelson, Tetrahedron Letters, 1978, 2819).

$$ \text{(9)} \quad \xrightarrow[\text{MeOH}]{h\nu} \quad \text{(10)} $$

(9)   (10)

The methylsulphonium salts of 2-methylene-8-thiabicyclo-
[3.2.1]octane (11) and 8-thiabicyclo[3.2.1]oct-2-enes
undergo sulphide extrusion by methyllithium to yield
dimethyl sulphide and the hydrogen shift products,
cycloheptadienes and/or bicycloheptenes (T. Uyehara,
M. Takahashi and T. Kato, *ibid.*, 1984, <u>25</u>, 3999).

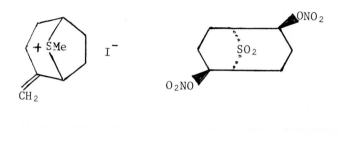

(11)

(12)

X-ray diffraction analysis shows that 2,6-dinitrato-9-
-thiabicyclo[3.3.1]nonane 9,9-dioxide (12) possesses a twin-
-chair conformation with equatorial nitrato groups (McCabe
and G.A. Sim, J. chem. Soc., Perkin II, 1982, 819).

Ylide (13) have been prepared by treating either *cis*- or
trans- 1-thioniabicyclo[4.4.0]decane bromide with sodium
hydride in boiling THF. It reacts with carbonyl compounds
and active methylene compounds to yield the corresponding
epoxide and monosubstituted active methylene compound,
respectively, for example, epoxide (14) and dicyano
derivative (15) (K. Tokuno *et al.*, Yakugaku Zasshi, 1978,
<u>98</u>, 1005).

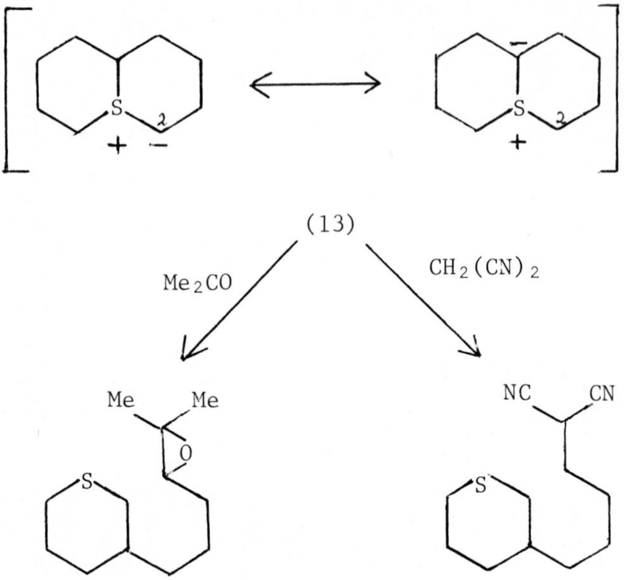
S
S
(13)
Me_2CO
$CH_2(CN)_2$
Me
Me
O
S
NC
CN
S

(14) (15)

Eastman's sulphonium salt has been prepared as an equimolar mixture of the *cis*- (16) and *trans*- (17) 1-thioniabicyclo[4.4.0]decane tetrafluoroborate, each of which has been isolated in a pure form. The 6-carbomethoxy derivative (18) has also been prepared (D.M. Roush *et al.*, J. Amer. chem. Soc., 1979, 101, 1971).

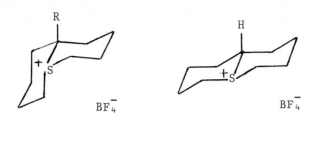

(16) R = H   (17)
(18) R = CO$_2$Me

Thermal cycloaddition reactions between adamantanethione (19) and conjugated dienes, such as, cyclopentadiene, 2,3- -dimethylbuta-1,3-diene, piperylene, a derivative of isoindole, and isobenzofuran occur smoothly at 20-110° to give the [4+2]cycloadducts (20), (21), (22), (23) and (24), respectively, in good yields (T. Katada, Eguchi and Sasaki, J. org. Chem., 1986, 51, 314).

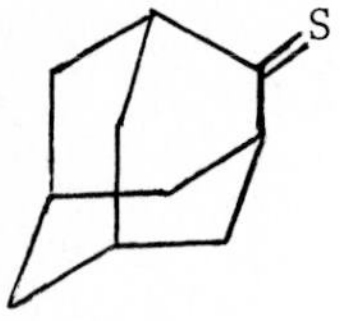

(19)

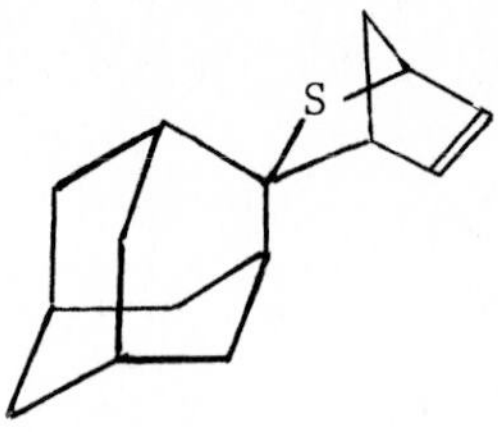

(20)

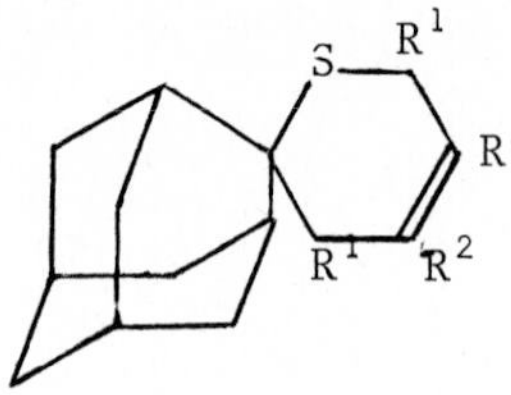

(21) $R^1=H,R^2=Me$
(22) $R^1=R^2=H$

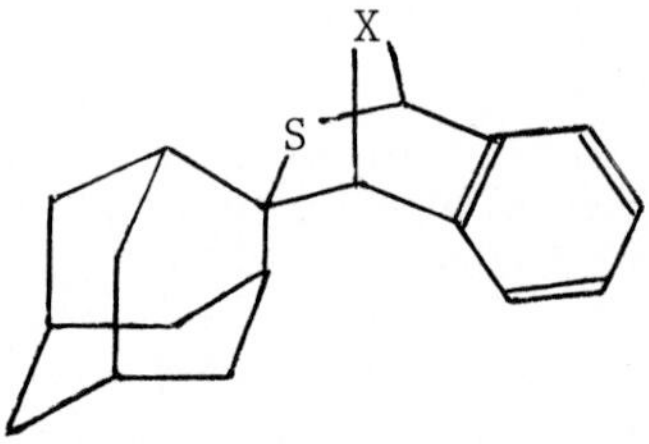

(23) X $=NCO_2Bu^t$
(24) X = 0

Lithium tetradeuterioaluminate reduction of *anti*-4,8-
-dichloro-2-thiaadamantane(25) and *anti*-2,6-dihalogeno-9-
-thiabicyclo[3.3.1]nonanes (26) in ether and THF occurs
with complete retention of configuration. The most likely
explanation of this behaviour is the participation by
sulphur to form intermediate thiopyranium ions. Reactions
of the thiabicyclononanes with LiEt$_3$BD involves mainly
elimination (R.C. Fort Jr, M.H. Stahl and A.F. Sky, *ibid.*,
1987, **52**, 2396).

R=Cl,Br

R^1=Cl,R^2=H
R^1=H,R^2=Cl

(25)   (26)   (27)

7-Thiaprotoadamantane, 7-thiatricyclo[4.3.1.0^{3,8}]decane
(27) and related derivatives have been synthesized *via* a
regiospecific and stereoselective intramolecular Friedel-
-Crafts reaction (Sasaki *et al.*, J. chem. Soc., Perkin I,
1982, 1953).

(b) Perinaphthothiapyrans, thiaphenalenes

Radical bromination of 1H,3H-naphtho[1,8-cd]thiopyran 2,2-
-dioxide (1) yields the monobromo dioxide (2) (48%), dibromo
dioxide (3) (43%; *cis*:*trans* = 64:36) and tribromo dioxide
(5%). The Ramberg-Bäcklund reaction of 1,3-dibromo-1H,3H-
-naphtho[1,8-cd]thiopyran (3) has been investigated under a
variety of conditions with expectation of the formation of
the thiirene dioxide (4) and hence the generation of
acenaphthyne (5) both thermally and photochemically. It
has been found that the use of triethylamine as base
produces 1-bromoacenaphthylene (6) (39%) and debrominated
products; that sodium methoxide gives 1-bromoacenaphthylene
(6) (75%), acenaphthylene (7) (9%) and surprisingly in
trace amounts decacyclene(8); and potassium *tert*-butoxide
affords products (6) (36%), (7) (27%), and (8) (5%)
(J. Nakayama *et al.*, J. org. Chem., 1983, **48**, 60).

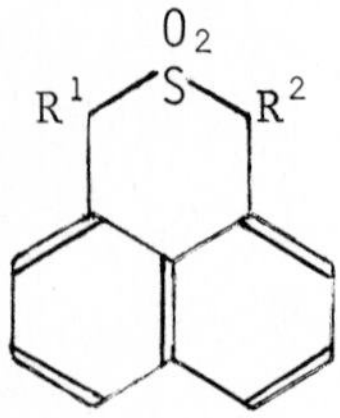

O_2

R^1 S R^2

(1) $R^1=R^2=H$
(2) $R^1=H, R^2=Br$
(3) $R^1=R^2=Br$

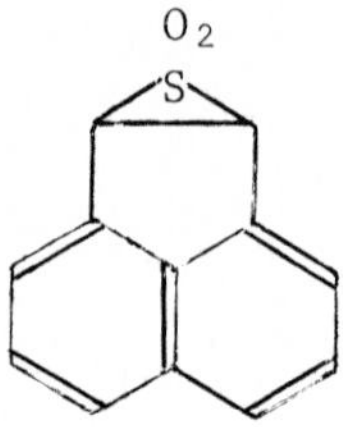

O_2

(4)

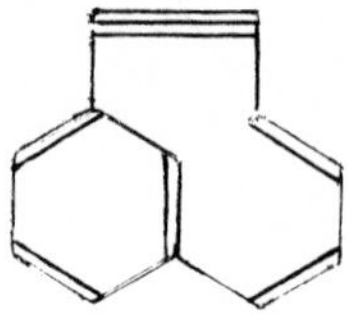

(5)

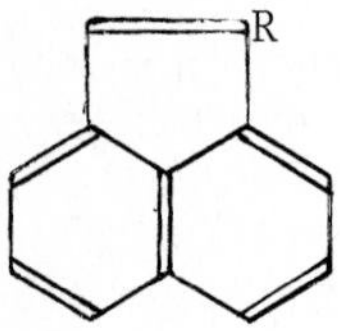

R

(6) R=Br
(7) R=H

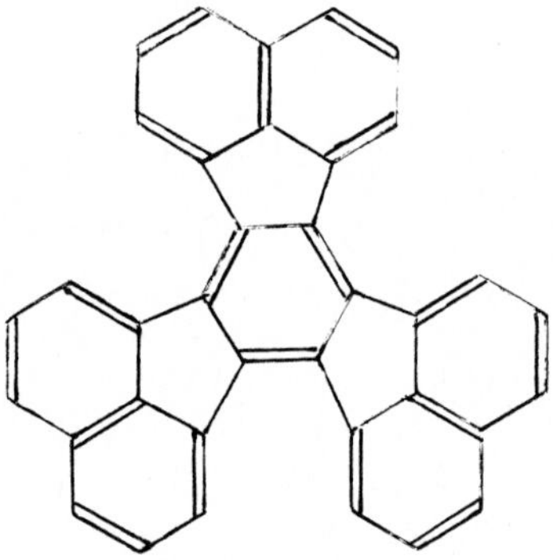

(8)

S   S   S

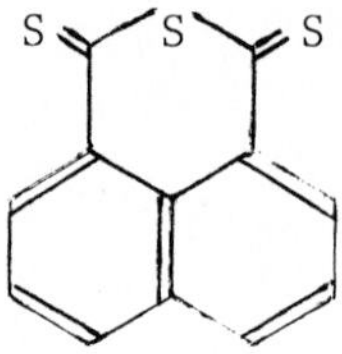

(9)

The first cyclic trithioanhydride, 1,8-naphthalic trithioanhydride, $1H,3H$-naphtho[1,8-cd]thiopyran-1,3--dithione (9) has been synthesized from 1,8-naphthalic anhydride (M.V. Lakshmikantham *et al.*, J. Amer. chem. Soc., 1984, **106**, 6084).

(c) Benzothioxanthene derivatives

5-Di(2-hydroxyethyl)amino- (2) and 5-(2-diethylamino)-
ethylamino- (3) benzo[c]thioxanthone have been prepared by
treating the parent amino derivative (1) with ethylene
oxide and 2-chloroethyldiethylamine, respectively
(A.A. Zayed, T.Y. Lee and J.H. Burckhalter, Pharmazie,
1985, **40**, 318).

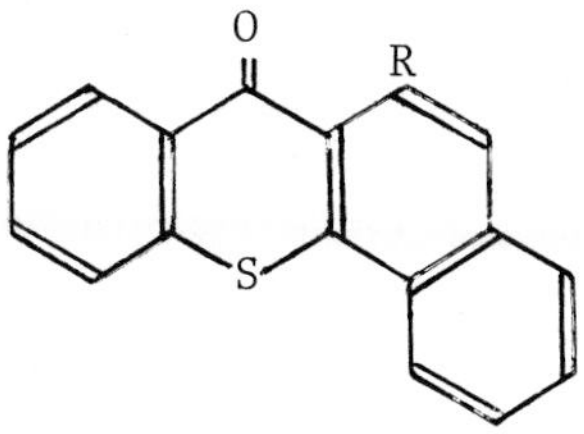

(1) R=NH$_2$
(2) R=N(CH$_2$CH$_2$OH)$_2$
(3) R=NHCH$_2$CH$_2$NEt$_2$

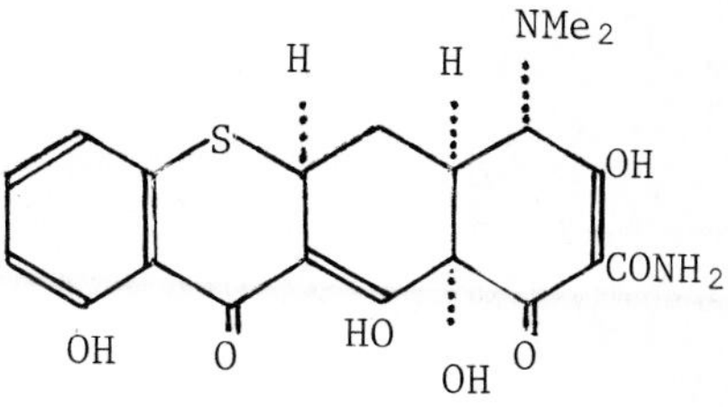

(4)

6-Thiatetracycline (4), an antibacterial agent superior
to all known tetracyclines has been synthesized
(R. Kirchlechner and W. Rogalski, Tetrahedron Letters,
1980, **21**, 247). The structure and conformation of the free
base crystallized from dichloromethane, has been determined
by high resolution X-ray crystallographic analysis at
~120K. Its structure (4) has been compared with those of
related tetracycline antibiotics (R. Prewo and
J.J. Stezowski, *ibid.*, p.251).

6. Selenopyrans and related compounds

(a) Selenopyrans and derivatives

Cyclization of 1,5-diketones, for example, 1,5-diphenyl-pentane-1,5-dione (1) with hydrogen selenide affords 2,6--diphenyl-4H-selenopyran (2). 1,5-Di(4-methoxyphenyl)-pentanedione also yields 2,6-di(4-methoxyphenyl)selenane--2,6-diselenol (3) (B.I. Drevko, Nukleofil'nye Reakts. Karbonil'nykh Soedin., 1982, 78).

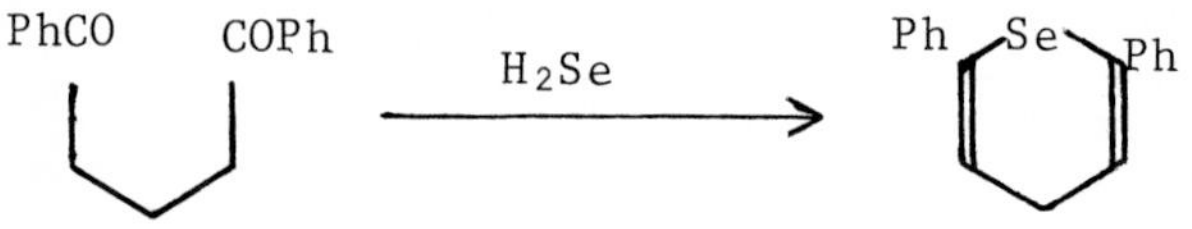

(1)   (2)

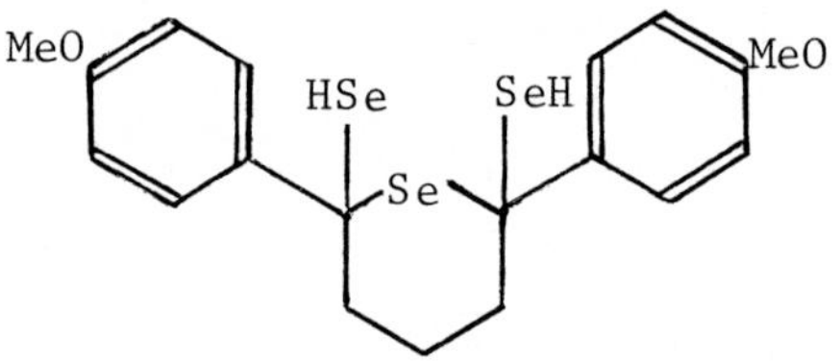

(3)

Vinylaldehydes (4) obtained from the reaction between β-chlorovinylaldehydes, $ClCR^1=CR^2CHO$ and sodium selenide on cyclization with unsaturated aldehydes (5) afford the substituted 2-formylmethylene-2H-selenopyrans (7) *via* intermediate (6) (M. Mueller *et al.*, Z. Chem., 1983, 23, 145).

$$R^1CSeNa \\ \underset{R^2CCHO}{\|} \quad + \quad \underset{CH_2R^3}{\overset{ClC}{\|}} = CR^4CHO \quad \longrightarrow \quad \left[\begin{array}{c} R^4 \\ R^1 \quad Se \quad CHO \\ R^2 \quad CHO \quad CH_2R^3 \end{array} \right]$$

R^1=Ph,Me,R^2=Ph R^3=H,R^4=Me,Ph
R^1=4-BrC$_6$H$_4$,R^2=H R^3R^4=(CH$_2$)$_2$,(CH$_2$)$_3$
R^1=R^2=Me
R^1R^2=(CH$_2$)$_3$,(CH$_2$)$_4$

(4)   (5)

(6)

$$\begin{array}{c} R^4 \\ R^1 \quad Se \quad CHO \\ R^2 \quad R^3 \end{array}$$

(7)

$\Delta^{4,4'}$-2,2',6,6'-Tetraphenyl-4-(selenopyranyl)-4H-seleno-pyran (9), m.p. 297-299°, from acetonitrile, is obtained from 2,6-diphenyl-4H-selenopyran-4-thione (8) with copper powder in boiling toluene under N$_2$ for 16h. $\Delta^{4,4'}$-2,2',-6,6'-Tetra-$tert$-butyl-4-(selenopyranyl)-4H-selenopyran (10), m.p. 260-263°, and the related tetramethyl derivative (11), m.p. 210-212°, are prepared in a similar manner from the corresponding selenopyranthiones. Related sulphur and tellurium compounds have been synthesized and the electrochemistry of all three series have been examined and

compared. The radical-cation states of the tetra-*tert*-
-butyl derivatives have been investigated by esr
spectroscopy (M.R. Detty *et al.*, Tetrahedron, 1985, **41**,
4853). The redox potentials of derivatives (9) and (11)
have been determined (S. Es-Seddiki *et al.*, Bull. Soc.
chim. Fr., 1984, 241).

R=Ph,Me,Bu^t

(9) R=Ph
(10) R=Bu^t
(11) R=Me

(8)

The synthesis and reactions of a selenabenzene have been
described (J. Stackhouse, Diss. Abs. Int. B, 1976, **37**,
2258).

A study of the dependence of uv-visible absorption-band
widths, vinylene shifts, and oscillator strengths of
pyrylocyanine dyes and their N-, S-, and Se-containing
analogues (12) of different electronic asymmetry on n has
been made by quantum-chemical analysis of quadratic
variations in bond orders on excitation (G.G. Dyadyusha *et
al.*, Dopov. Akad. Nauk Ukr. RSR, Ser. B: Geol., Khim.
Biol. Nauki, 1981, 55). Hypsochromic shifts and band
broadening have been observed for (13) on changing solvent
from dichloromethane to nitromethane (*idem*, Ukr. Khim. Zh.,
1980, **46**, 1186).

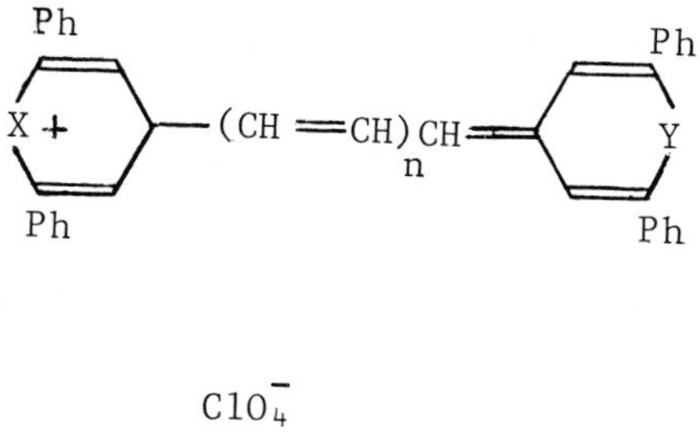

ClO_4^-

(12) X=S, Y=Se
    X=O, Y=S,Se
    X=MeN, Y=O,S,Se
    n=0,1,2
(13) X,Y=MeN,O,S,Se, n=0,1,2

The application of pyrylium, thiopyrylium, and seleno-pyrylium salts to organic syntheses and their reactions with various nucleophiles, including the cyanide ion, active methylene compounds, and organometallic compounds, and their reduction, oxidation, and photochemical reactions have been reviewed (T. Sugimoto, Yuki Gosei Kagaku Kyokaishi, 1981, 39, 1).

5,6-Dihydro-2,6-diphenyl-2H-selenopyran (15) is prepared by mild dehydromesylation of 2,6-diphenyl-4-mesyltetrahydro-selenopyran (14) using commercial grade Woelm W-200 (activity, Super 1) neutral alumina in dichloromethane (C.H. Chen *et al.*, J. heterocyclic Chem., 1978, 15, 289).

(14)   (15)   (16)

The preparation of selenane, *tetrahydroselenopyran*,
selenocyclohexane (16), is achieved by the reaction of
selenium dioxide with n-pentane in the presence of a Zeokar
2 catalyst (zeolite-containing rare earth elements) at 450-
560° (E.Sh. Mamedov *et al.*, U.S.S.R. Pat., 657,031
1979).
The chemical shift differences, δ, between the axial and
the equatorial protons at the 3-position (β position) in
selenane and related heterocycles have been measured from
the [1]H-nmr spectrum of the α,β-deuteriated derivatives
at a temperature below the slow exchange limit for ring
reversal. In virtually every case, the shielding
contribution to $\delta(3)$ [$\delta(\beta)$] from the C-heteroatom bond has
the opposite sign from that of $\delta(2)$ [$\delta(\alpha)$] (J.B. Lambert
and J.E. Goldstein, J. Amer. chem. Soc., 1977, <u>99</u>, 5689).
The second-order rate constants for the acetylation of
epimeric pairs of 3-substituted 2,6-diaryl-4-hydroxy-
selenanes (17) and 6-aryl-2,2-dimethyl-4-hydroxyselenanes
(18) with acetic anhydride in pyridine have been reported
and the conformations of these compounds have been
determined from the kinetic data. The [1]H- and ^{13}C-nmr
spectra have also been recorded and analyzed. The ^{13}C-nmr
chemical shift data suggest that *trans*-6-(4-chlorophenyl)-
-4-hydroxy-2,2-dimethylselenane (18; R = Cl) exists in a
nonchair conformation (P. Nanjappan *et al.*, J. org. Chem.,
1981, <u>46</u>, 2542).

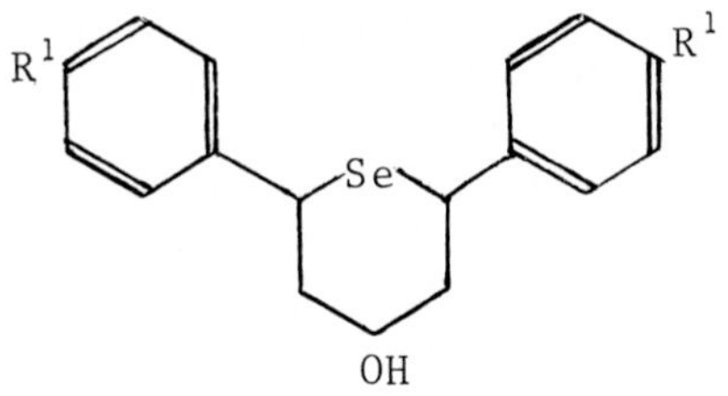

R^1=H,Me,OMe
R^2=H,Me,Et

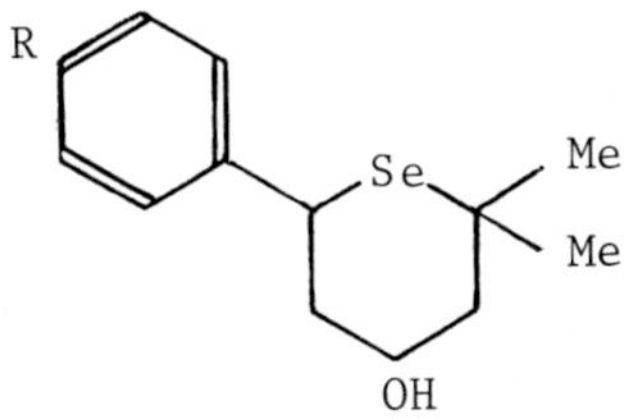

R=H,Cl

(17)   (18)

Several substituted selenan-4-ones and 4-hydroxyselananes have been prepared and their configuration and conformation assigned on the basis of ir, ^{1}H-nmr and ^{13}C-nmr spectral data. The selenane ring has been shown to be predominantly in the chair form. The dissociation constants have been measured for the cyanohydins from some substituted selenan--4-ones in 80% dioxane at 25°. The results have been analyzed in terms of steric parameters in the flattened ring system (*idem*, *ibid*., 1980, **45**, 4622).

(b) Selenopyrans containing fused rings

(i) Benzoselenopyrans and derivatives
4-Methoxybenzo[b]selenopyrylium salt (1) on heating with 2-methylene-3-methylbenzothiazoline in acetic acid at 100°, followed by treatment with perchloric acid yields salt (2) (A.I. Tolmachev, M.A. Kudinova, and L.M. Shulezhko, Khim. Geterotsikl. Soedin, 1977, 178).

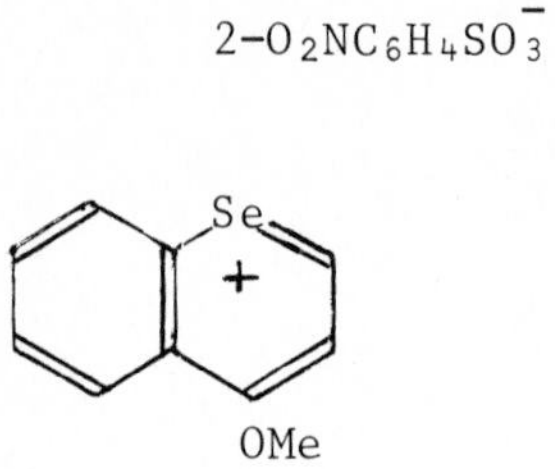

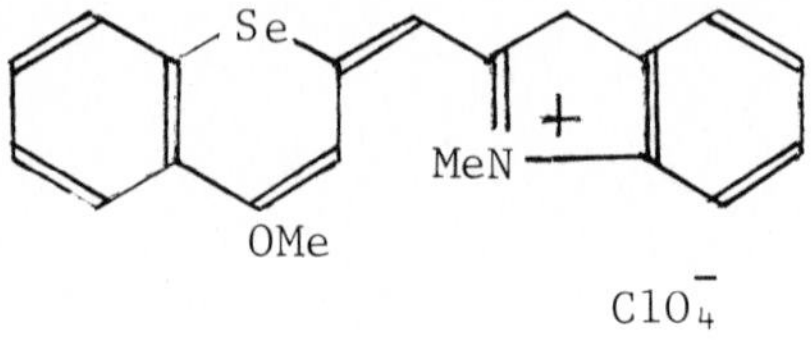

558

$2-O_2NC_6H_4SO_3^-$

(1)

(2)

2-Methylselenochromone, 2-methyl-4H-benzo[b]selenopyran-
-4-one (3) reacts with active methylene compounds, for
instance, malononitrile, and 1,3-indanedione to give 4-
-dicyanomethylene-2-methyl-4H-benzoselenopyran (4) and 4-
-(1,3-dioxo-2-indanylidene)-2-methyl-4H-benzoselenopyran
(5), respectively. Ketone (3) on treatment with 4,6-
-diphenyl-2-methylthiopyrylium perchlorate affords the
thiopyrylium salt (6) and selenoflavone on reaction with
phenylmagnesium bromide, followed by treatment with
perchloric acid yields 2,4-diphenylbenzoselenopyrylium
perchlorate (7) (R. Neidlein and I. Koerber, Arch. Pharm.,
1978, 311, 170).

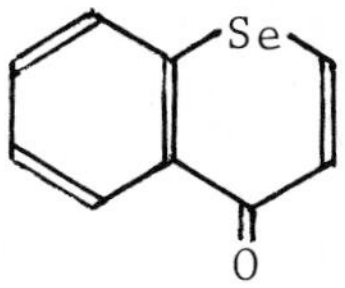

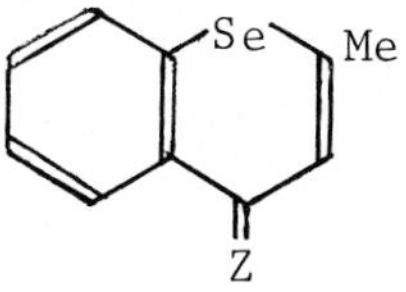

(3)

(4) Z=C(CN)₂

(5) Z=1,3-dioxo-2-indanylidene

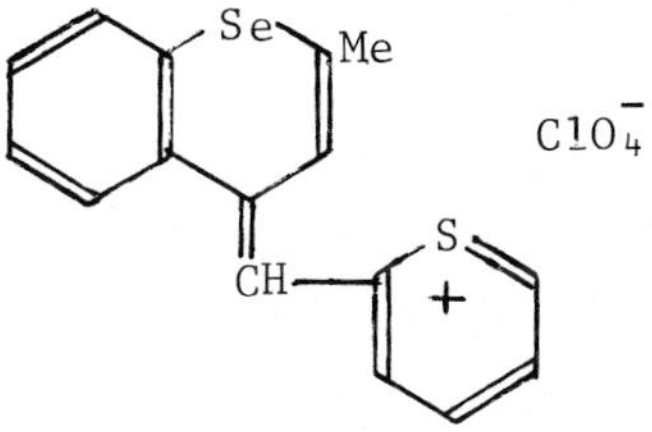

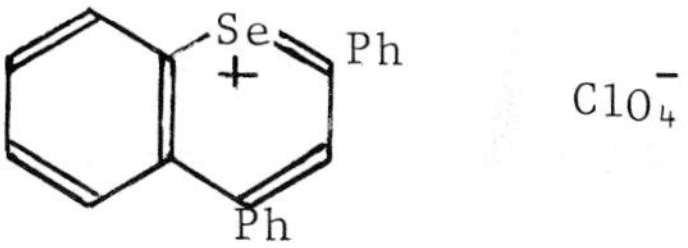

(6)

(7)

Selenoflavones (9) are prepared by treating 2-methyl-selenobenzoyl chlorides (8) with phenylacetylene, followed by cyclization of the products (A.J. Luxen, L.E.E. Christiaens, and M.J. Renson, J. organometal. Chem., 1985, 287, 81). The corresponding oxygen and sulphur heterocycles have also been obtained.

R=H,7-F,6-NO$_2$,6-Me,6-MeO

(8)   (9)

β-(Phenylseleno)cinnamic acid and β-(4-methoxyphenyl-
seleno)cinnamic acid on cyclization with phosphorus
pentoxide/methanesulphonic acid furnishes selenoflavone and
4'-methoxyselenoflavone, respectively (D.H. Wadsworth and
M.R. Detty, J. org. Chem., 1980, <u>45</u>, 4611). The uv-
-absorption spectra in solid films and in solution have been
recorded for selenochromone, selenoxanthene and
selenoxanthone (P.G. Fuochi, G. Giro, and G. Orlandi, Atti
Ist. Veneto Sci. Letters Arti, Cl. Sci. Mat. Nat., 1975,
<u>33</u>, 85). The rapid separation of some selenopyran
derivatives including 4-hydroxyselenochroman,
selenochroman-4-one, and 6*H*-dibenzo[b,d]selenopyran by
thin-layer chromatography has been described (G. Bottura
and M.A. Pavesi, Microchem. J., 1987, <u>35</u>, 223).
   Irradiation of selenochroman-4-one at 3500 Å in methanol
gives selenochromone (5%) (A. Couture, A. Lablache-Combier,
and T.Q. Minh, Tetrahedron Letters, 1977, 2873). The ir,
uv, ^{1}H- and ^{13}C-nmr spectral data and dipole moments of
selenochroman-4-one and the corresponding O, S, and Te
heterocycles have been reported (N. Dereu *et al*., J.
organometal. Chem., 1981, <u>208</u>, 23) and ^{13}C-nmr shielding
effects on C-atoms bearing the heteroatom have been
observed and discussed (L. Laitem, L. Christiaens, and
M. Renson, Org. mag. Reson., 1980, <u>13</u>, 319).
   1*H*-Benzo[c]selenopyran-1-one, 1*H*-2-benzoselenopyran-1-
-one, isoselenocoumarin (10), m.p. 70-80°, is obtained by
electrophilic lactonization of 2-(MeSeCH=CH)C$_6$H$_4$COCl
in dichloromethane in the presence of aluminium chloride,

the temperature being allowed to rise from −80° to room
temperature (A. Luxen, Christiaens, and Renson, J. org.
Chem., 1980, **45**, 3535).

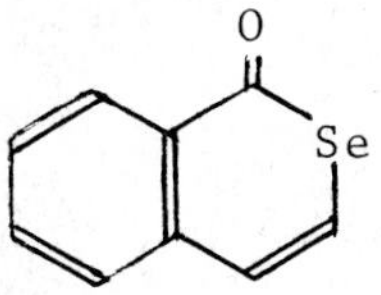

(10)

Isoselenochroman-4-one (12) is obtained in excellent
yield from (2-carboxybenzyl)selenoacetic acid (11). On
methylation it gives 2-methyl-4-oxoisoselenochromanium
tetrafluoroborate (13), m.p. 112-113°, which on
deprotonation with sodium hydride or a sodium alkoxide
affords ylide (14). The ylide (14) gradually decomposes at
room temperature. When generated *in situ* from salt (13) and
an equimolar amount of sodium hydride in boiling aprotic
solvents under N_2, the principal product is *trans*-1,2,3-
-tris {2[(methylseleno)methyl]benzoyl} cyclopropane (15),
with lesser amounts of (*E*)-bis {2[(methylseleno)methyl]-
benzoyl} ethylene (16), m.p. 113-114°, and isoseleno-
chroman-4-one (12) (M. Hori *et al.*, *ibid.*, 1987, **52**, 1397).

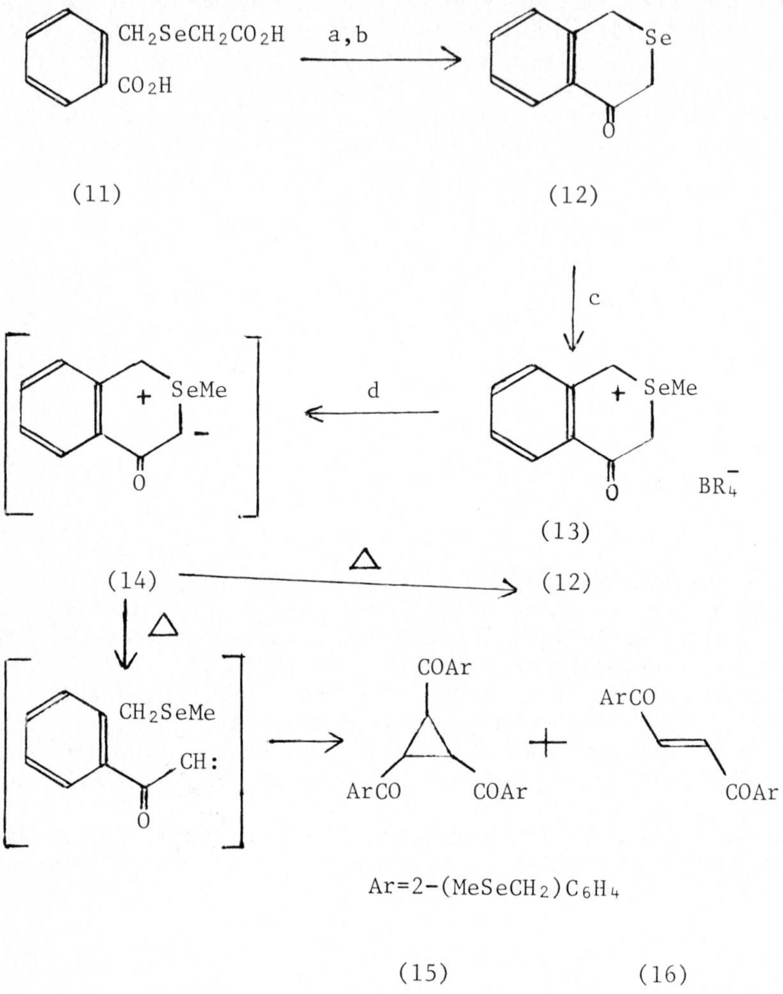

CH2SeCH2CO2H
CO2H
a,b
Se
O
(11)
(12)
c
SeMe
d
BR4
(13)
(14)
(12)
CH2SeMe
CH:
COAr
ArCO
Ar=2-(MeSeCH2)C6H4
(15)
(16)

Reagents:- (a) AcOK, Ac_2O; (b) HCl, H_2O, EtOH; (c) MeI, $AgBF_4$; (d) NaH, solvent one of C_6H_6, THF, CH_2Cl_2, MeCN

(ii) Dibenzoselenopyran, selenoxanthene and derivatives
Irradiation of selenoxanthene azide (1) in benzene gives
a mixture of imine (2) and the hetero dibenzazepine (3) in
a ratio of 9:1. Thermolysis of azide (1) in 1,2-dichloro-
benzene at 190° also results in rearrangement to yield the
same products in a ratio of 2:3 (J.P. Le Roux, P.L. Desbene,
and M. Sequin, Tetrahedron Letters, 1976, 3141).

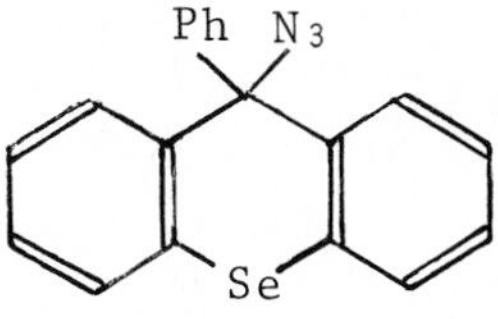

(1) 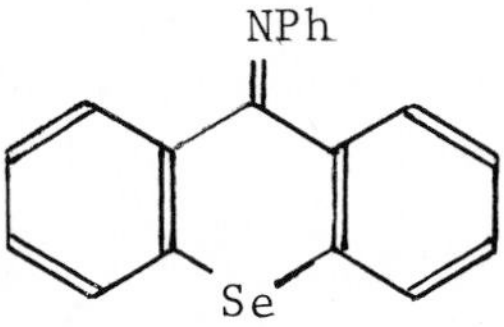   (2) 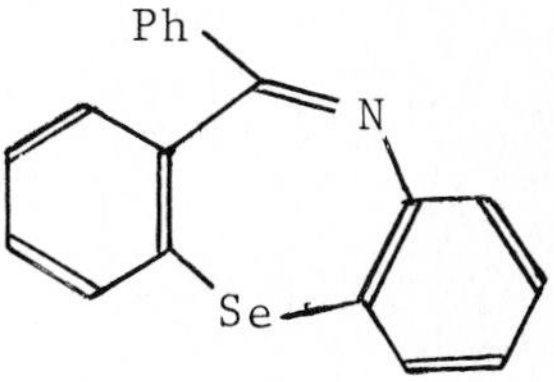   (3)

1,2,5,6,7,8-Hexahydro-3*H*-selenoxanthene-4-carboxaldehyde,
(3,4,5,6,7,8-hexahydro-2H-9-selenaanthracene-1-carboxalde-
hyde)(4) and the related S and Te heterocycles have been
enriched with ^{18}O and a structural study carried out using,
dipole moments, ir, and ^{13}C-nmr spectroscopy. It has been
found that in aldehydes (4; X = Se, Te) there is partial
covalent O...X bonding more important for X = Te than
X = Se, whereas in the sulphur analogue a weak O....S
interaction is involved [R. Close *et al.*, J. chem. Res.
(S), 1978, 4].

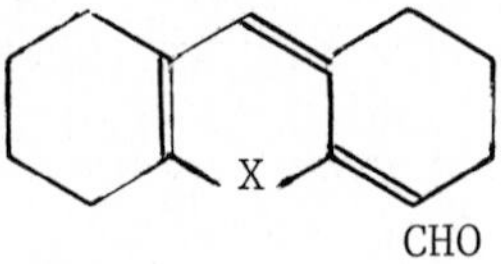

X=Se,Te,S

(4)

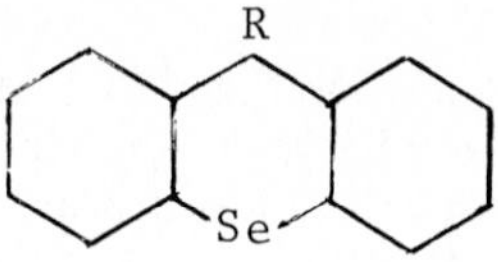

(8)

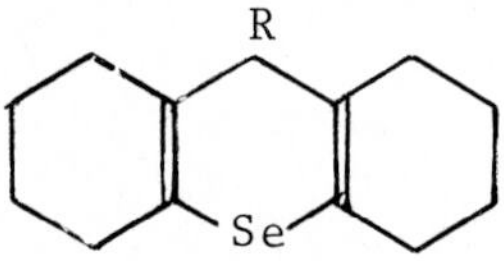

(5) R=H,Ph,4-BrC$_6$H$_4$,2-FC$_6$H$_4$
(6) R=H,Me,Et,Pr,Ph,PhCH$_2$,4-BrC$_6$H$_4$
(7) R=Me,Et,PhCH$_2$,2-FC$_6$H$_4$

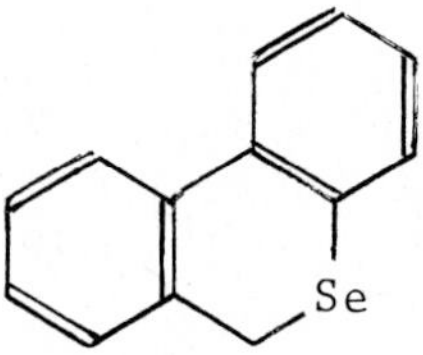

(9)

Cyclization of appropriate substituted methylenedicyclo-hexan-2-ones with hydrogen selenide affords the 9-
-substituted octahydroselenoxanthenes (5). Octahydroseleno-xanthenes (6) are prepared by reduction of the corresponding selenoxanthylium salts with NaBH$_4$, LiAlH$_4$, and Grignard reagents. In the presence of acid derivatives (5) and (6) disproportionate into selenoxanthylium salt and perhydro-selenoxanthenes (A.F. Blinokhvatov and V.G. Kharchenko, Nukleofil'nye Reakts. Karbonil'nykh Soedin., 1982, 71; Blinokhvatov *et al.*, Khim. Geterotsikl. Soedin., 1981, 640).

Ionic hydrogenation of octahydroselenoxanthenes (7) on treatment with triethylsilane and trifluoroacetic acid at 60-70° give perhydroselenoxanthenes (8) (*idem, ibid.,* p.564).

Irradiation of 2-ClC$_6$H$_4$COSePh with uv light yields selenoxanthone (J. Martens, K. Praefcke and H. Simon, Z. Naturforsch., Anorg. Chem., Org. Chem., 1976, **31B**, 1717). Chromatographic method of obtaining small amounts of 6H--dibenzo[b,d]selenopyran (9) have been described (G. Bottura and M.A. Pavesi, Microchem. J., 1987, **35**, 223).

7. Telluropyrans and related compounds

(a) Tellurane, tetrahydrotelluropyran derivatives

1,1-Diiodotellurane (1,1-diiodo-1-telluracyclohexane) on treatment with the necessary silver or potassium salts gives telluranes (1) (T.N. Srivastava, R.C. Srivastava, and H.B. Singh, Indian J. Chem., 1979, **18A**, 71). It also reacts with interhalogen compounds (R[1]I; R[1] = Cl, Br, I) to yield adducts, but no reaction takes place between 1,1--dichloro- or -dibromo-tellurane and R[1]I (Srivastava, Srivastava, and M. Singh, *ibid.*, **17A**, 615). The reliability of the measurement of conformational free energy differences for a series of hydrocarbons and heterocycles with 0, 1, or 2 geminal diMe groups, including 3,3-dimethyltellurane, from the temperature gradient of ^{13}C-nmr chemical shifts, has been investigated (J.B. Lambert, A.R. Vagenas, and S. Somani, J. Amer. chem. Soc., 1981, **103**, 6398).

R=F,Cl,Br,CN,N$_3$,NCO,NCS,
NCSe,AcO,BzO,PhCH$_2$CO$_2$,
Cl$_3$CCO$_2$,Me(CH$_2$)$_{12}$CO$_2$

(1)

(2)

(3)

The reaction of pentane-2,4-dione with tellurium tetrachloride gives 1,1-dichlorotelluran-3,5-dione (2), which can be reduced to telluran-3,5-dione (3). Their structures have been verified by X-ray crystallography (C.L. Raston, R.J. Secomb, and A.H. White, J. chem. Soc., Dalton Trans., 1976, 2307). [1]H- and ^{13}C-nmr spectral data of tellurane-3,5-diones have been reported (J.C. Dewan *et al.*, Org. mag. Res., 1978, **11**, 449).
For $\Delta^{4,4'}$-2,2',6,6'-Tetrasubstituted-4-(telluropyranyl)--4H-telluropyrans see p. 553.

(b) Telluropyrans containing fused rings

(i) Benzotelluropyrans and derivatives

Tellurochromone (1) has been obtained by the intra-molecular cyclization of a β-(2-bromotelluroaroyl)enamine by means of hypophosphorous acid, but an excess of acid gives rise to tellurochroman-4-one (2) (N. Dereu and M. Renson, J. organometal. Chem., 1981, **208**, 11).

(1)   (2)

The ir, uv, [1]H- and ^{13}C-nmr spectral data and dipole moments have been reported for tellurochroman-4-one and the O, S, and Se analogues (Dereu *et al.*, *ibid.*, p.23). For ^{13}C-nmr spectral data see also L. Laitem, L. Christiaens, and M. Renson (Org. mag. Res., 1980, **13**, 319).
Telluroflavones have been prepared from appropriate 2--methyltellurobenzoyl chlorides see A.J. Luxen, Christiaens, and Renson (p. 559).
1*H*-Benzo[c]telluropyran-1-one, 1H-2-benzotelluropyran-1--one, isotellurocoumarin (3), m.p. 83°, is prepared from the same starting material as the seleno analogue (Luxen, Christiaens, and Renson, see p. 560).

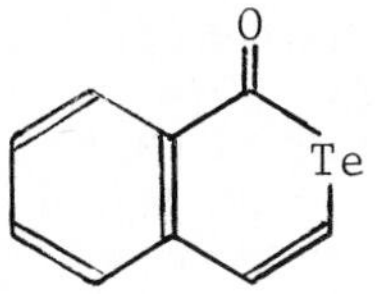

(3) 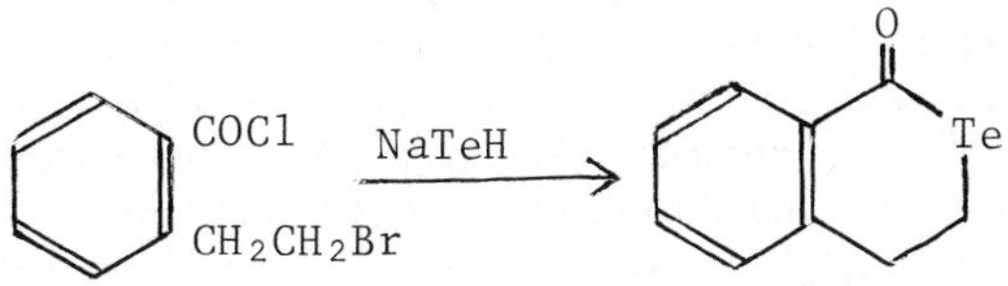   (4)   (5)

Treatment of 2-(2-bromoethyl)benzoyl chloride (4) with NaTeH affords 3,4-dihydroisotellurocoumarin (5), m.p. 52°. Its ir and ^{13}C-nmr data are also reported along with the preparation, by modified methods, of 3,4-dihydroisoseleno-coumarin and the spectral data of related compounds (M. Loth-Compère *et al.*, J. heterocyclic Chem., 1981, 18, 343).

(ii) Dibenzotelluropyran, telluroxanthene and derivatives
    Lithiation of di(2-bromophenyl)methane (1) followed by cyclization with tellurium gives telluroxanthene (2). 2-Methyltelluroxanthene is obtained from (2-bromo-5-methyl-phenyl)(2-bromophenyl)methane (W. Lohner and K. Praefcke, J. organometal. Chem., 1981, 205, 167).

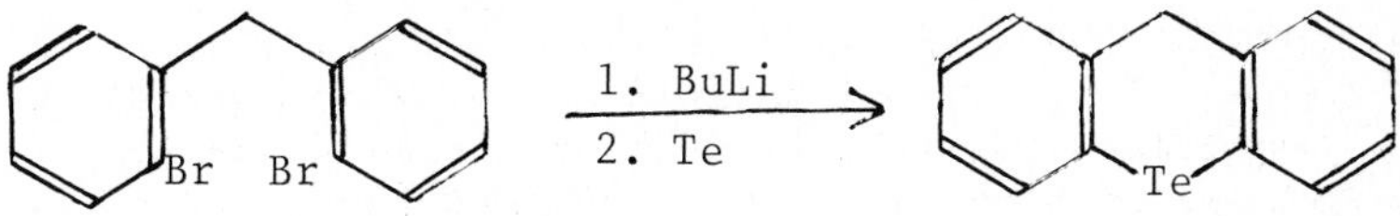

(1)   (2)

Cyclization of $2\text{-}(Cl_3Te)C_6H_4CH_2Ph$ in the presence of aluminium chloride yields 10,10-dichlorotelluroxanthene, which on reduction with $K_2S_2O_5$ affords telluroxanthene (2) (I.D. Sadekov, A.A. Ladatko, and V.K. Minkin, Khim. Geterotsikl. Soedin., 1978, 1567). Treatment of telluroxanthene (2) with halogens furnishes the 10,10--dihalogenotelluroxanthenes (3) and the dichloro derivative (3; R = Cl) with dimedone gives complex (4), which on treatment with acid (HX; X = F, F_3CCO_2) yields (5). Some of the telluroxanthenes have been oxidized to telluroxanthones (*idem, ibid.*, 1980, 1342).

(3) R=Cl,Br,I
(5) R=F,F_3CCO_2

Me$_2$

(4)

Telluroxanthone is obtained in 2% yield by the cyclization of $(2\text{-}N_2^+C_6H_4)_2CO\ 2C\bar{l}$ with Na_2Te (Lohner and Praefcke, Chem. -Ztg., 1979, 103, 265).

Telluroxanthylium perchlorates (7) are prepared by treatment of the 9-hydroxy derivative (6) with perchloric acid. 9-Hydroxy-9-(4-methylphenyl)telluroxanthene (6; R = 4-MeC$_6$H$_4$) is obtained by the Grignard reaction between telluroxanthone and 4-methylphenylmagnesium bromide and the alcohol (6; R = H) by LiAlH$_4$ reduction of telluroxanthone (Sadekov *et al.*, Khim. Geterotsikl. Soedin., 1980, 274).

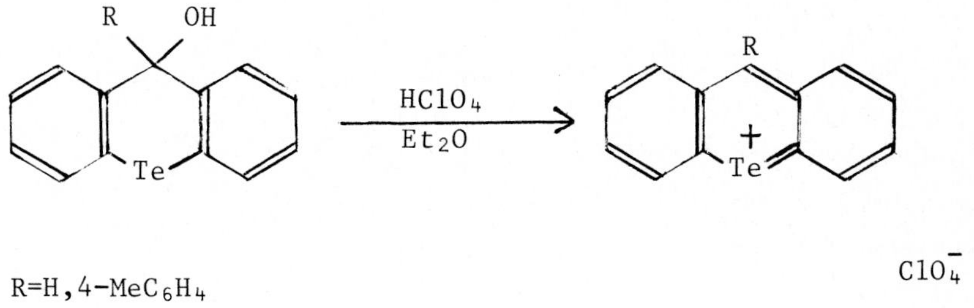

R=H,4-MeC$_6$H$_4$

(6)   (7)

The preparation of the telluroxanthylium perchlorates (7; R = Me, 3-MeC$_6$H$_4$, 4-MeOC$_6$H$_4$, 4-FC$_6$H$_4$, 4-BrC$_6$H$_4$, α-naphthyl, PhCH$_2$) (*idem, ibid.*, 1981, 343). The crystal structure of 4-formyl-1,2,5,6,7,8-hexahydro-3*H*-telluro-xanthene(1-formyl-3,4,5,6,7,8-hexahydro-9-tellura-2*H*- -anthracene) (8) has been determined (J. Lamotte *et al.*, Cryst. Struct. Comm., 1977, **6**, 749).

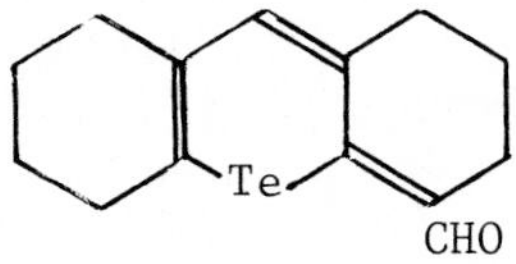

(8)

8. Compounds containing an element from Group 4

(a) Silicon compounds

(i) Silabenzene, silacyclohexenes and their derivatives

Silabenzene and derivatives have been prepared in a number of ways and trapped or obtained as dimers. Dehydrochlorination of 1-chloro-1,4-di-*tert*-butyl-1- -silacyclohexa-2,4-diene produces 1,4-di-*tert*-butyl-1- silabenzene (1), which dimerizes to give (2) and on cycloaddition with buta-1,3-diene and its derivatives affords adduct (3) (G. Märkl and P. Hofmeister, Angew. Chem., 1979, **91**, 863).

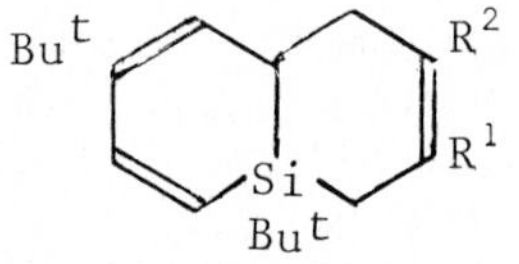

$$R^1 = R^2 = H, Me$$
$$R^1 = H, R^2 = Me$$

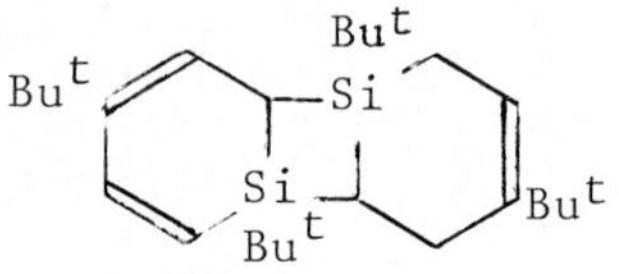

(1) (2) (3)

Pyrolysis of 1-acetoxy- or 1-allyl-1-silacyclohexa-2,4-diene affords silabenzene (4; R = H) (G. Maier, G. Mihm, and H.P. Reisenauer, *ibid.*, 1980, **92**, 58) and similarly 1- -substituted 1-allyl-1-silacyclohexa-2,4-dienes yield 1- -substituted silabenzenes (4) (B. Solouki *et al.*, *ibid.*, p.56).

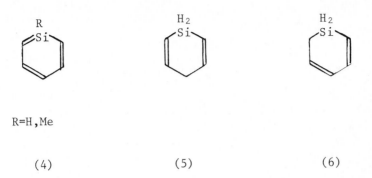

R=H,Me

(4)   (5)   (6)

Flash pyrolysis of 1-silacyclohexa-2,5-diene (5) gives
silabenzene, identified by matrix isolation ir spectroscopy
in argon at 10K. Under identical pyrolysis conditions, 1-
-silacyclohexa-2,4-diene (6) yields cyclopentadiene,
acetylene, other unidentified products, and only a trace of.
silabenzene (Maier, Mihm, and Reisenauer, Ber., 1982, 115,
801). The pyrolysis of 1-silacyclohexa-2,5-diene (5) has
been optimized by photoelectron-spectroscopic, real-time
gas analysis. At 1050K the thermal dehydrogenation affords
exclusively hydrogen and silabenzene (4; R = H) (H. Bock
et al., J. organometal. Chem., 1984, 271, 145).
Silacyclohexane on Pd (110) partially desorbs reversibly
and partially dehydrogenates to give silabenzene
(T.M. Gentle and E.L. Muetterties, J. Amer. chem. Soc.,
1983, 105, 304).
1-Methoxy-1-methyl-4-trimethylsilyl-1-silacyclohexa-2,5-
-diene (7) on flash pyrolysis furnishes 1-methylsilabenzene
(1-silatoluene) (8). The proposed mechanism involves a 1,3-
-sigmatropic migration of Me_3Si followed by β-elimination
of Me_3SiOMe. 1-Methylsilabenzene (8) is trapped by
methanol to afford 1-methoxy-1-methyl-1-silacyclohexa-2,4-
-diene (9) and by acetylene to give 1-methylsilabarralene
(10). When not trapped (8) dimerizes in a Diels-Alder
fashion to give silatoluene dimer (11) (T.J. Barton and
M. Vuper, ibid., 1981, 103, 6789). 1-Silatoluene has also
been obtained by the pyrolysis of 1-allyl-1-methyl-1-sila-
cyclohexa-2,4-diene and trapped in a solid argon matrix,
which allowed both ir and uv spectral observations. The
spectra strongly suggested that silatoluene is aromatic
(C.L. Kreil et al., ibid., 1980, 102, 841).

572

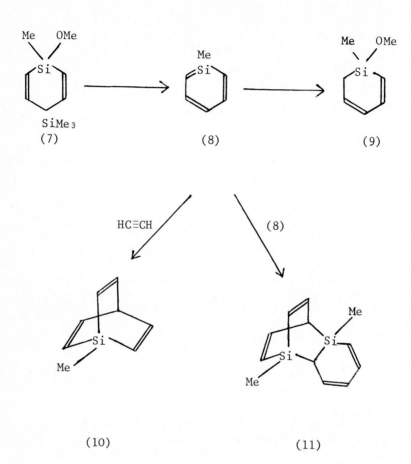

(7)   (8)   (9)

HC≡CH   (8)

(10)   (11)

1,4-Dialkyl-1-silabenzenes (13) have been prepared and
detected in the gas phase, by the pyrolysis of 1-allyl-1-
-silacyclohexa-2,4-dienes (12). Their photoelectron
spectra have been recorded (Märkl *et al.*, Angew. Chem.,
1982, **94**, 211).

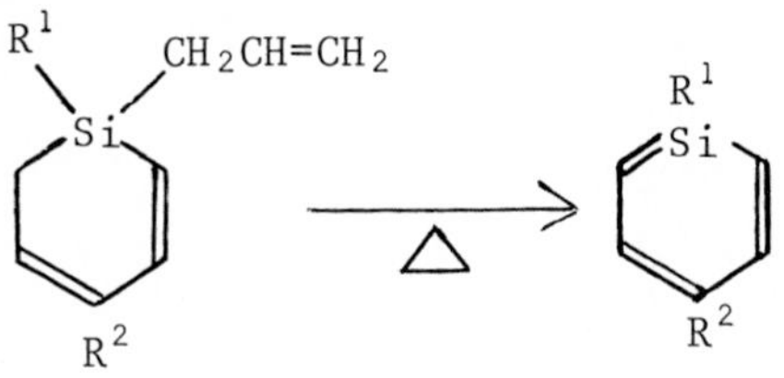

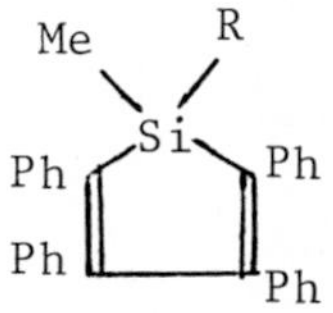

R^1=H,Me
R^2=Me,Et,Pr^i,Bu^t

(12)

(13)

(14) R=CMeN$_2$
(15) R=CH=CH$_2$
(16) R=3-methyl-
        diazirin-3-yl

Thermolysis and photolysis of 1-methyl-2,3,4,5-tetraphenyl-
silacyclopentadienyl-diazomethane and -diazirine affords
silacyclopentadienylmethylene, which rearranges to
silabenzene by ring expansion and to 5-silafulvene by 1,2-
-migration of a methyl group. The former is trapped by
alcohols and 2,3-dimethylbuta-1,3-diene and the latter by
alcohols and diphenyl ketone (W. Ando, H. Tanikawa, and
A. Sekiguchi, Tetrahedron Letters, 1983, 24, 4245).
However, photolysis of 1-methyl-2,3,4,5-tetraphenylsilacyclo-
pentadienyldiazomethylmethane (14) yields normal carbene
products (15) and (16). Photolysis and thermolysis of
diazo silanaphthalene and silaanthracene derivatives have
also been reported (A.H.B. Cheng *et al.*, Organometallics,
1985, 4, 581). Evidence for the generation and trapping of
silabenzene has been discussed (D.S. Banasiak, Diss. Abs.
Int. B, 1978, 38, 5374).

6-Diazo-1,1-dimethyl-2,3,4,5-tetraphenyl-1-silacyclohexa-
-2,4-diene (17) on thermolysis in the presence of copper
(II) sulphate produces 6-silafulvene (18), trapped by
alcohols and carbonyl compounds (Sekiguchi and Ando, J.
Amer. chem. Soc., 1981, 103, 3579).

(17)   (18)

3-Substituted 1(Z),4(Z)-1,5-dilithium-3-methoxypenta-1,4-
-dienes (19) react with dichlorosilanes to yield silacyclo-
hexa-2,5-dienes (20). Cleavage of the ether group with
boron trichloride, or tribromide, or phosphorus trichloride
affords 6-chloro(bromo)-1-silacyclohexa-2,4-dienes or 1,5-
-dichloro-1-silacyclohexa-2,4-dienes, respectively. 4-
-Substituted 4-methoxy-1-silacyclohexa-2,5-dienes on ether
cleavage with sodium produce 4-substituted 1-silacyclohexa-
-dienyl anions, which on hydrolysis yield 1-silacyclohexa-
-2,4-dienes. The phase transfer-catalyzed conversion of 1-
-chlorosilacyclohexadienes into fluoro derivatives and other
substitution reactions of 1-chlorosilacyclohexadienes have
been described (Märkl et al ., J. organometal. Chem., 1979,
173, 125).

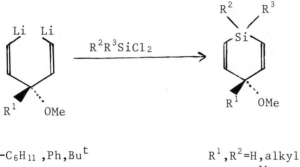

$R^2R^3SiCl_2$

$R^1 = c-C_6H_{11}$,Ph,Bu^t

$R^1, R^2 =$H,alkyl, aryl, alkoxy

(19)   (20)

1,1-Dimethyl-1-silacyclohexa-2,5-dien-4-one has been obtained from [Me$_2$(MeO)Si]$_2$ and cyclopentadiene and its conversion into 4-diazo-1,1-dimethyl-1-silacyclohexa-2,5--diene has been described (Barton and Banasiak, *ibid.*, 1978, 157, 255).

2-Halogeno-1,1,3-trimethyl-1-silacyclohex-2-enes (21) have been prepared and metallated and protonated to give 1,1,3-trimethyl-1-silacyclohex-2-ene (21; R = H) (R. Muenstedt, D. Wrobel, and U. Wannagat, *ibid.*, 1984, 271, 181). Sila-β-ionone (22) (Muenstedt and Wannagat, Monatsh., 1985, 116, 693) and related compounds (23) and (24) have been synthesized and their odours compared with those of natural perfumes (Wannagat, Muenstedt, and U. Harder, Ann., 1985, 950).

Me₂ structures:

$$
\begin{array}{c}
\text{Me}_2 \\
\text{Si} \diagdown \text{R} \\
\diagdown \text{Me}
\end{array}
$$

$$
\begin{array}{c}
\text{Me}_2 \\
\text{Si} \diagdown \text{C} = \text{CHAc} \\
\diagdown \text{Me}
\end{array}
$$

(21) R=Cl,Br
(22) R=CH=CHAc
(23) R=CHO, C≡CAc

(24)

Pyrolysis of 2,3-di(trifluoromethyl)-7-silabicyclo-[2.2.2]octa-2,5-dienes (25) yield $R_2Si=CH_2$, which cyclodimerize to give disilacyclobutanes (26) (Maier, Mihm and Reisenauer, Angew. Chem., 1981, 93, 615).

$$
\begin{array}{ccc}
\text{(structure with SiR}_2\text{, CF}_3\text{, CF}_3\text{)} & \xrightarrow{\triangle} &
\begin{array}{c}
R_2Si \quad \rceil \\
\lfloor \quad SiR_2
\end{array}
\end{array}
$$

R=H,D,Cl

(25)   (26)

(ii) Silacyclohexane and its derivatives

A convenient synthetic route to methylated silacyclohexanes involves the von Braun degradation of substituted piperidines to a variety of substituted primary and secondary 1,5-dibromopentanes. Ring closure with dichlorosilanes and magnesium are carried out using standard procedures at relatively high dilution. The product mixtures which are appreciably enriched in one isomer offer a convenient entry into a variety of derivatives (see Table 1) which can be used for stereo-chemical studies. Assignment of structures based on [1]H-nmr spectroscopy and mass spectral fragmentation patterns has been discussed (B.T. Nguyen and F.K. Cartledge, J. org. Chem., 1986, 51, 2206).

Table 1
Methyl and other derivatives of 1-silacyclohexane

Derivative	% yield[1]	b.p. (mm Hg)	*cis/trans*[2]
1,2-diMe	43	62- 63 (65)	55/45
1,3-diMe	53	55- 56 (71)	50/50
1,4-diMe	60	57- 58 (71)	52/48
1,1,2-triMe	43	64- 65 (40)	
1,1,3-triMe	50	62- 63 (40)	
1,1,4-triMe	55	65- 66 (40)	
1,1,2,6-tetraMe	32	97- 98 (70)	50/50
1,1,3,5-tetraMe	40	80- 81 (65)	50/50
1,1,4,4-tetraMe	48	91- 92 (90)	
1-Cl-1,2-diMe	44	92- 93 (50)	30/70
1-Cl-1,3-diMe	60	94- 95 (50)	70/30
1-Cl-1,4-diMe	60	94- 95 (55)	32/68
1,2-diMe-1-Ph	50	126-127 (5)	58/42
1,3-diMe-1-Ph	64	128-129 (6)	42/58
1,4-diMe-1-Ph	60	100-101 (1)	58/42

[1] Yield in Mg ring closure of appropriate 1,5-
   -dibromopentane and $RMeSiCl_2$ (R = H, Me, Cl, Ph)
[2] Ratio observed using gas chromatograph analysis.

The synthesis, reactivity and stereochemistry of silacyclo-hexanes and their derivatives have been reviewed (Nguyen, Diss. Abs. Int. B, 1986, **47**, 1065).

Some 1-substituted 1-methyl-4-*tert*-butyl-1-silacyclo-hexanes (27) have been prepared and the conformationally stable *cis* and *trans* isomers separated (H. Sakurai and M. Murakami, Bull. chem. Soc., Japan, 1976, **49**, 3185).

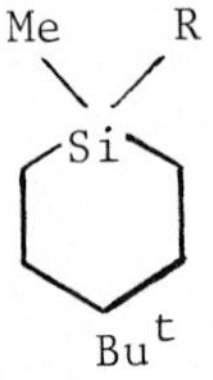

Me   R
\   /
Si

But

(27)

R=H,Cl,OMe,OPr^i,OBu^t,Ph,
4-BrC$_6$H$_4$,4-ClC$_6$H$_4$,SiMe$_2$H,F

Me$_2$
Si

R^2 R^1

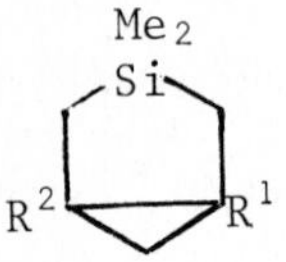

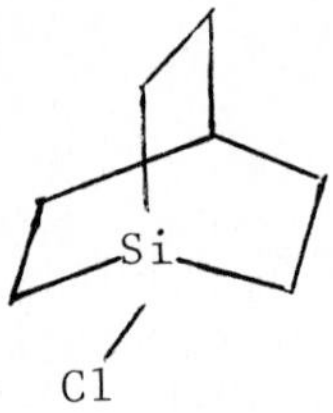

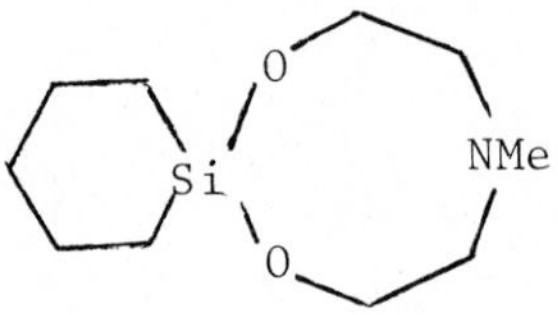

R^1,R^2=H,Me

(28)

Si

Cl

(29)

O

Si   NMe

O

(30)

For the preparation of 1-(4-chlorophenyl)-1-[3-(diethyl-amino)propyl]- and -[2-(diethylamino)ethyl]- 1-silacyclo-hexane and related derivatives see J. Ackermann *et al.*, (Ann., 1979, 1915; Arch. Pharm., 1980, **313**, 129).

The acetolysis of silacyclohexyl tosylates has been studied (S.S. Washburne and R.R. Chawla, J. organometal. Chem., 1977, 133, 7) and conformational (P. Mazerolles *et al.*, J. chim. Phys. Phys.-Chim. Biol., 1980, 77, 329), ir spectral (R. Mathis, A. Faucher, and Mazerolles, Spectrochim. Acta, 1979, 35A, 1311), and [29]Si-nmr and mass spectral (F. Duboudin *et al.*, J. organometal. Chem., 1977, 133, 29) investigations of silacyclohexanes and related compounds have been reported.

For the ^{13}C-nmr spectral data of 1,1,2,6-tetramethyl-1--silacyclohexan-4-one see J.A. Soderquist and A. Hassner (J. org. Chem., 1983, 48, 1801).

Thermolysis of 2-silabicyclo[3.1.0]hexanes (28) affords a mixture of 1-silacyclohexenes and silacyclopentenes (G. Manuel, Faucher and Mazerolles, J. organometal. Chem., 1984, 264, 127).

1-Chloro-1-silabicyclo[2.2.2]octane (29) has been synthesized from 4-chlorotetrahydropyran and on chlorination it gives 1,2-, 1,3-, and 1,4-dichloro-1--silabicyclo[2.2.2]octane (R.F. Cunico and F. Drone *ibid.*, 1978, 150, 179). The [1]H-nmr spectrum of spiro compound (30) and those of related compounds have been discussed (A.D. Naumov *et al.*, Zh. obshch. Khim., 1976, 46, 1808).

(iii) Polycyclic silanes with six-membered rings

1,1-Dimethyl-1-silanaphthalene (33), b.p. 78-79°/3.6 mm, is obtained by ring closure of [3-(2-chlorophenyl)propyl]-dimethylsilane (32), prepared from 1-chloro-2-(3-bromo-propyl)benzene (31). Hydrogenation of (33) affords 1,1--dimethyl-perhydro-1-silanaphthalene (2,2-dimethyl-2-sila-bicyclo[4.4.0]decane) (34), b.p. 71-73°/2.9 mm, in 75% yield [20% (trans) and 80% (cis)] (R.A. Benkeser *et al.*, J. org. Chem., 1979, 44, 1370).

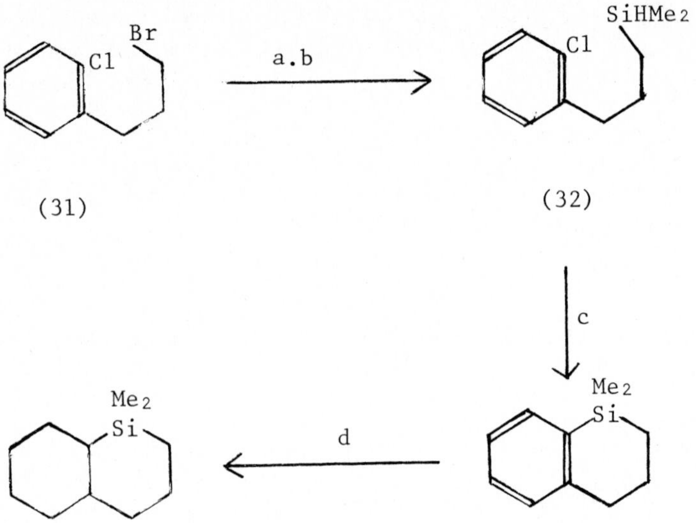

(31)   (32)

(34)   (33)

Reagents:-   (a) $Mg, HSiCl_3$;   (b) $MeMgI$;   (c) $Na, PhMe, \Delta$;
                    (d) H_2, W-2Raney Ni, c-MeC_6H_{11}, 1500 psi,
                    70-150°

    4-Bromo-1-diazo-1,2-dihydro-2,2-diphenyl-2-silanaphthalene,
6-diazo-1,1-dimethyl-2,3,4,5-tetraphenyl-1-silacyclohexa-
-2,4-diene, and 9-diazo-10,10-dimethylsilaxanthene(10-diazo-
-9,10-dihydro-9,9-dimethyl-9-silaanthracene) have been
prepared and their spectroscopic properties determined
(A. Sekiguchi and W. Ando, Bull. chem. Soc., Japan, 1982,
55, 1675).
    The synthesis of 2,2-dimethyl-, -1-trimethylsilyl- and
-1,3-bis(trimethylsilyl)-2,3-dihydro-2-silaphenalene (35)
and (36) and the formation of (2,2-dimethyl-2,3-dihydro-3-
-trimethylsilyl-2-silaphenalene-1,3-diyl)bis[(N,N,N',N'-
-tetramethylethylenediamine)lithium] (37) have been
described (L.M. Engelhardt et al., J. chem. Soc., Dalton,
1984, 311).

$$2(\text{LiL})^+$$

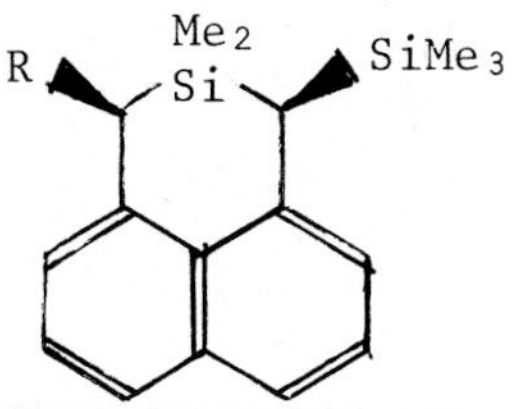

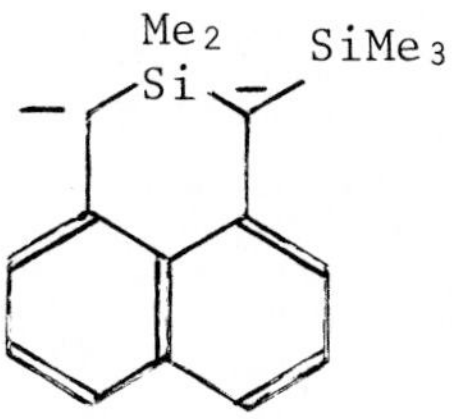

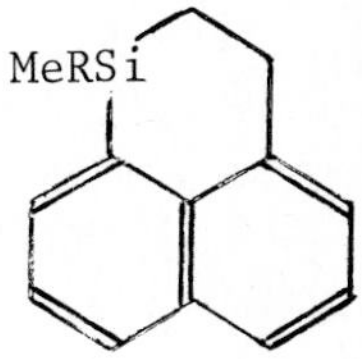

Me$_2$
R   Si   SiMe$_3$

Me$_2$
   Si   SiMe$_3$

MeRSi

L=Me$_2$NCH$_2$CH$_2$NMe$_2$

R=H,Me

(35) R=H
(36) R=SiMe$_3$

(37)

(38)

It has been shown that 1-silaphenalenium ions do not
delocalize the positive charge throughout the entire π
system like the hydrocarbon phenalene. Hydride abstraction
reactions using hydride derivatives of 1-silaphenalene (38)
with Ph$_3$C$^+$SnCl$_6$ does not produce detectable silylenium ions
(R. Sooriyakumaran and P. Boudjouk, J. organometal. Chem.,
1984, 271, 289). The use of 1,8-disubstituted naphthalenes
as intermediates for the synthesis of 1-silaphenalenes and
1-silaacenaphthenes has been reviewed (J.S. Kiely, Diss.
Abs. Int. B, 1979, 40, 1718). 1,1-Dichloro-1-silaphenalene
has been prepared by pyrolysis of 1-naphthylvinyldichloro-
silane (N.G. Komalenkova et al., U.S.S.R. Pat. 730,687,
1980).

582

Condensation of diorganometallic compounds (39) with
trichlorosilane followed by LiAlH$_4$ reduction yields silaxanthene (40; R^1 = R^2 = H). Controlled halogenation of
(40; R^1 = R^2 = H) to 10-chloro- and 10-bromo- silaxanthene
(40; R^1 = H, R^2 = Cl and R^1 = H, R^2 = Br) is accomplished
by 1 molar equivalent of sulphuryl chloride and bromo-
succinimide, respectively, although carbon tetrachloride in
the presence of ClRh(PPh$_3$)$_3$ or PdCl$_2$ results in slow mono-
chlorination. Excess sulphuryl chloride or thionyl chloride
affords 10,10-dichlorosilaxanthene (40; R^1 = R^2 = Cl),
but the former reaction is faster and results in fewer by-
-products. 10-Alkoxy- and 10,10-dialkoxy- silaxanthenes
have been prepared (J.Y. Corey *et al.*, J. organometal.
Chem., 1986, 304, 93).

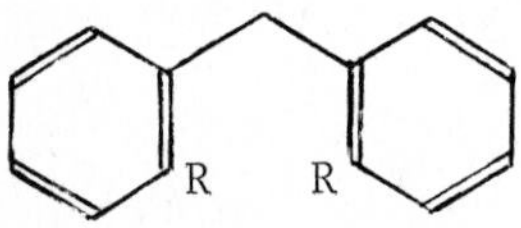

R=Li,MgCl

(39)

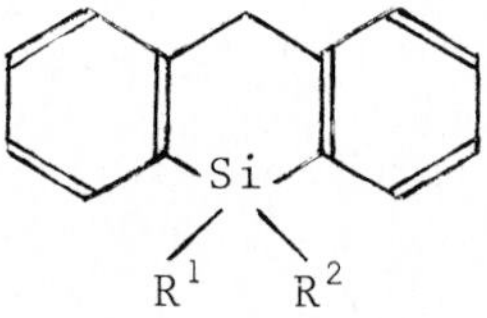

(40)

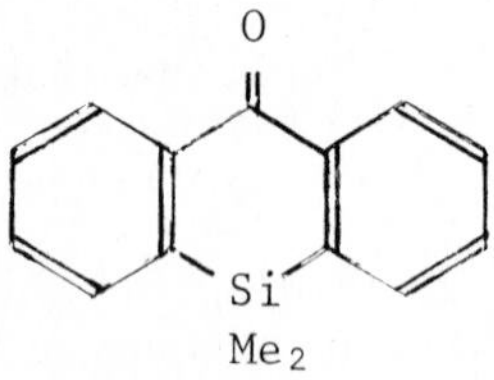

Me$_2$

(41)

R$_2$

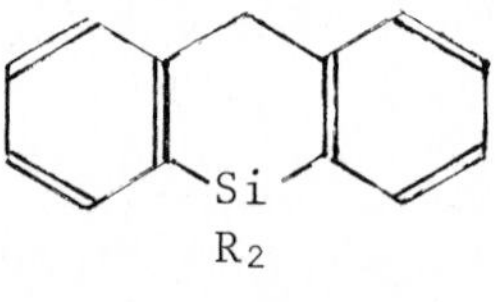

(42) R=H,Me,Cl
(43) R=Cl,F,Et,OMe

The reaction of 10,10-dimethylsilaxanthone (10-oxo-9,10-
-dihydro-9,9-dimethyl-9-silaanthracene) (41) with either
N,N-dimethylaminopropylmagnesium chloride or N-methylpiper-
idinylmagnesium chloride provides alcohols, which can be
dehydrated with thionyl chloride/pyridine, to introduce an
exocyclic double bond in the 9-position of the silaxanthone.
The preparation of other related derivatives have been
reported (*idem, ibid.*, 1978, __153__, 127). The dipole moments
of silaxanthone (41) and silaxanthenes (42) indicate that
the molecules have their central ring in a boat
conformation (G.N. Kartsev *et al.*, Zh. obshch. Khim., 1981,
__51__, 1360). The ring of 9-chloromethylsilaxanthene and
related compounds has been described (Corey *et al.*, J.
organometal. Chem., 1981, __210__, 149).

Photolysis of 9-diazo-10,10-dimethylsilaxanthene gives
10,10-dimethylsilaxanthenylidene. Its triplet nature has
been supported by an esr study (A. Sekiguchi *et al.*,
Tetrahedron Letters, 1982, __23__, 4095). Laser-photolysis
studies have also been carried out on the above compound in
cyclohexane to obtain a series of uv absorptions due to
transient species (T. Sugawara *et al.*, Chem. Letters, 1983,
1257). In aerated solutions, the triplet 10,10-dimethyl-
silaxanthenylidene is quenched with oxygen to give a
characteristic absorption (λ_{max} at 425 nm) due to the
corresponding carbonyl oxide (*idem, ibid.*, p.1261). The
mass spectra of several cyclic silicon compounds, including
silaxanthenes (43) show high-intensity peaks corresponding
to species containing the Si=C group (V.N. Bochkarev *et al.*,
Zh. obshch. Khim., 1980, __50__, 1783). Also the mass spectra
of silicon compounds containing the group $SiCl_2$ and
including 5,5-dichloro-6H-dibenzo[b,d]silapyran (44),
indicate that Cl_2Si: is eliminated from the molecular ions
(*idem, ibid.*, 1981, __51__, 824).

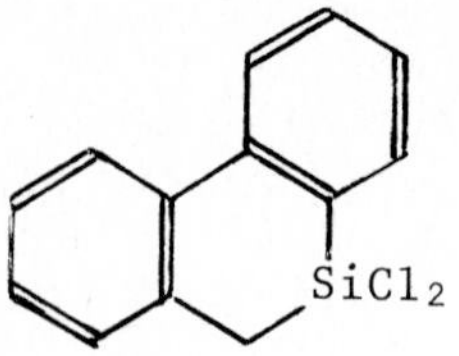

(44)

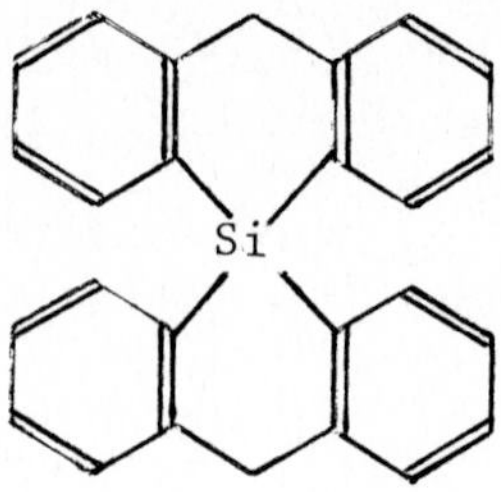

(45)

Treatment of di(2-iodophenyl)methane with butyllithium in ether at $0°$ followed by the addition of silicon tetrachloride at $-70°$ furnishes derivative (45) (F. Bickelhaupt *et al.*, Tetrahedron, 1976, **32**, 1921).

(b) Germanium and tin compounds

(i) Germanium compounds

Dehydrobromination of 1,4-bis(*tert*-butyl)-1-bromo-1- -germacyclohexa-2,4-diene yields the unstable 1,4-bis(*tert*- -butyl)germabenzene (1), which dimerizes to give dimer (2). Germabenzene (1) in the presence of dienes affords Diels- -Alder cycloaddition products, for example, adduct (3) is obtained with 2,3-dimethylbuta-1,3-diene (G. Märkl and D. Rudnick, Tetrahedron Letters, 1980, **21**, 1405).

(1) 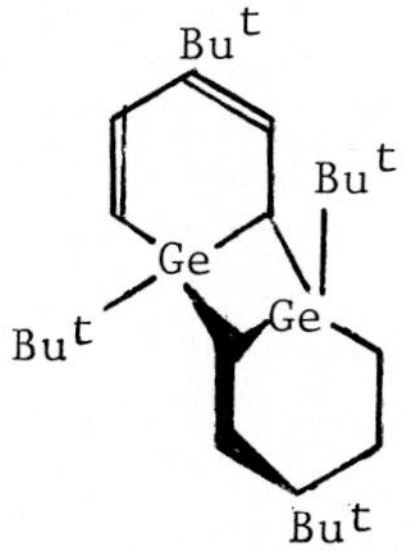   (2) 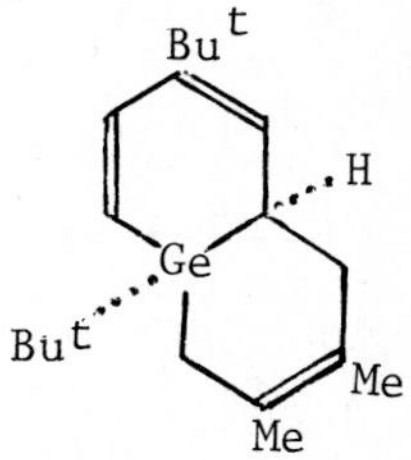   (3)

Pyrolysis of 1,4-disubstituted 1-allyl-1-germacyclohexa-2,4-
-dienes (4) produces 1,4-dialkylgermabenzenes (5), detected
in the gas phase. The photoelectron spectra of
germabenzenes (5) have been recorded (Märkl *et al.*, Angew.
Chem., 1982, **94**, 211).

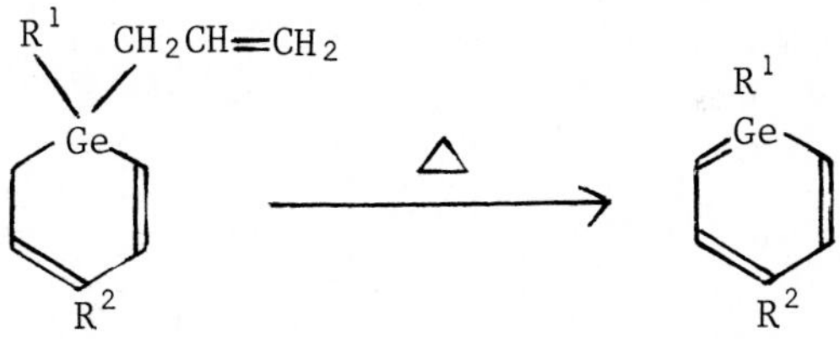

R^1=H,Me
R^2=Me,Et,Pr^i,Bu^t

(4)   (5)

3-Substituted 1(Z),4(Z)-1,5-dilithium-3-methoxypenta-1,4-
-dienes react with dialkyldichlorogermanes and diaryl-
dichlorogermanes to give 4-substituted 1,1-dialkyl- and
1,1-diaryl-4-methoxy- 1-germacyclohexa-2,5-dienes,
respectively. The reduction of 1-chloro-1-ethyl-4-methoxy-
-4-phenyl-1-germacyclohexa-2,5-diene with LiAlH$_4$ affords
1-ethyl-4-phenyl-1H-1-germacyclohexa-2,4-diene (*idem*, J.
organometal. Chem., 1979, **181**, 305).
4-Alkyl(aryl)-1,1-dialkyl(aryl)-4-methoxy-1-germacyclo-
hexa-2,5-dienes undergo ether cleavage with sodium in
pentane or liquid ammonia to give the sodium salts, which
on hydrolysis afford the corresponding 4-alkyl(aryl)-1,1-
-dialkyl(aryl)-1-germacyclohexa-2,4-dienes. 1H-1-Germa-
cyclohexa-2,4-dienes, 1-chloro- and 1-bromo-1-germacyclohexa-
-2,4-dienes, 6-chloro-1-germacyclohexa-2,4-diene, 1-chloro-
-4-methoxy-1-germacyclohexa-2,5-diene, 1,1-diethyl-4-
-methoxy-4-phenyl-1-germacyclohexa-2,5-diene, 4-substituted
1-chloro-4-methoxy-1-phenyl-1-germacyclohexa-2,5-dienes,
4-substituted 1,6-dichloro-1-phenyl-1-germacyclohexa-2,4-
-dienes, and Fe(CO)$_3$ complexes of 1,1-dialkyl(aryl)-1-
-germacyclohexa-2,4-dienes have been synthesized (*idem*,
ibid., 1980, **187**, 175). The mass spectrum of 1,1-dimethyl-
-1-germacyclohex-3-ene has been reported (K. Ujszaszy *et al.*,
Adv. mass Spec. 1978, **7A**, 601).
1,1-Diaryl-1-germacyclohexanes (6) are obtained by the
Grignard reaction between 1,1-dichloro-1-germacyclohexane
and the appropriate aryl bromide or by the cyclization of
BrMg(CH$_2$)$_5$MgBr with R$_2$GeCl$_2$ (I.I. Lapkin *et al.*, Izv.
Vyssh. Üchebn. Zaved., Khim. Khim. Tekhnol., 1985, **28**, 117).

R=2-,4-MeC$_6$H$_4$,2-MeOC$_6$H$_4$,
2,5-Me$_2$C$_6$H$_3$,2,4,6-Me$_3$C$_6$H$_2$,
1-naphthyl

(6)

The ^{73}Ge- and ^{13}C-nmr chemical shifts of a variety of
methyl substituted germacyclohexanes have been determined
(Y. Takeuchi, M. Shimoda, and S. Tomoda, Mag. res. Chem.,
1985, 23, 580).
Treatment of di(2-iodophenyl)methane with butyllithium in
ether at 0°, followed by germanium tetrachloride at -70°
gives the germaxanthene derivative (7) (F. Bickelhaup et al.,
Tetrahedron, 1976, 32, 1921).

(7)

(ii) Tin Compounds
1,1-Dibutyl-1-stannacyclohexa-2,5-diene (8) on
metallation with lithium amides affords the lithium
derivative (9), which on treatment with Me$_3$MCl (M = Si, Ge)
and 1,2-dibromoethane yields (10) and (11) respectively,
and with Me$_3$SnCl the 1-stannacyclohexa-2,4-diene (12).
Other derivatives of stannacyclohexadienes have been
prepared and arsabenzene, 4-trimethylsilyl-, 4-trimethyl-
germyl-, and 4-(2-chloroethyl)-1-arsabenzenes have been
prepared from stannacyclohexadienes and arsenic trichloride
(P. Jutzi and J. Baumgaertner, J. organometal. Chem., 1978,
148, 247).

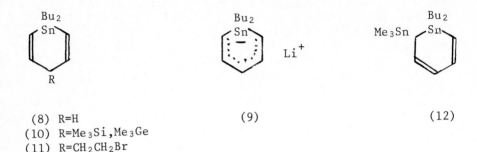

(8) R=H   (9)   (12)
(10) R=Me₃Si,Me₃Ge
(11) R=CH₂CH₂Br

    1,1-Dialkyl-1-stannacyclohexa-2,5-dienes (13) and other
products are synthesized by hydrostannation of 1,4-
-pentadiyne (*idem*, *ibid*., p.257).

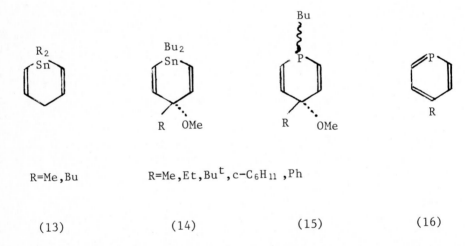

R=Me,Bu   R=Me,Et,Buᵗ,c-C₆H₁₁ ,Ph

(13)   (14)   (15)   (16)

    Cleavage of 4-substituted 1,1-dibutyl-4-methoxy-1-stanna-
cyclohexa-2,5-diene (14) with butyllithium, followed by
cyclization with butyldichlorophosphine produces *cis,trans*-
-dihydrophosphorins (15), which on reduction with LiAlH₄
gives phosphorins (16) (G. Märkl, H. Baier, and R. Liebl,
Ann., 1981, 919).

Cyclization of 1,5-dibromopentane with tin in hexane at 160-180° yields 1,1-dibromo-1-stannacyclohexane (17). The Grignard reaction between (17) and MeCl affords 1,1--dimethyl-1-stannacyclohexane (18) (V.I. Shiryaev *et al.*, Zh. obshch. Khim., 1978, **48**, 2627).

          (17) R=Br
          (18) R=Me

The dilithium derivative obtained from di(2-iodophenyl)-methane on reacting with Me_2SnCl_2 furnishes 10,10-dimethyl-stannaxanthene (19) (Bickelhaupt *et al.*, *loc. cit.*).

(19)

Guide to the Index

This index is constructed in a similar manner to the volume indexes of the first edition of the Chemistry of Carbon Compounds. However, to make the index easier to use, more descriptive entries have been made for the commonly occurring individual, and groups of chemicals.

The indexes cover primarily the chemical compounds mentioned in the text, and also include reactions and techniques, where named, and some sources of chemical compounds such as plant and animal species, oils, etc.

Chemical compounds have been indexed alphabetically under the names used by authors, editing being restricted to ensuring uniformity of entries under the same heading. In view of the alternative nomenclature that can often be used, a limited amount of cross-referencing has been done where it is considered to be helpful, but attention is particularly drawn to Convention 2 below.

For this and the succeeding volumes, the indexing conventions listed below have been adopted.

1. *Alphabetisation*

(a) The following prefixes have not been counted for alphabetising:

n-	*o-*	*as-*	*meso-*	D	C
sec-	*m-*	*sym-*	*cis-*	DL	*O-*
tert-	*p-*	*gem-*	*trans-*	L	*N-*
	vic-				*S-*
		lin-			*Bz-*
					Py-

Some prefixes and numbering have been omitted in the index, where they do not usefully contribute to the reference.

(b) The following prefixes have been alphabetised:

Allo	Epi	Neo
Anti	Hetero	Nor
Cyclo	Homo	Pseudo
	Iso	

(c) A letter by letter alphabetical sequence is followed for entries, firstly for the main entry, followed by the descriptive entry. The only exception to this sequence is the placing of plural entries in front of the corresponding individual entries to prevent these being overlooked by a strict alphabetical sequence which could lead to a considerable separation of plural from individual entries. Thus "butanes" will come before *n*-butane, "butenes" before 1-butene, and 2-butene, etc.

2. *Cross references*

In view of the many alternative trivial and systematic names for chemical compounds, the indexes should be searched under any alternative names which may be indicated in the main body of the text. Only a limited amount of cross-referencing has been carried out, where it is considered that it would be helpful to the user.

3. *Esters*

In the case of lower alcohols esters are indexed only under the acid, e.g. propionic methyl ester, not methyl propionate. Ethyl is normally omitted e.g. acetic ester.

4. *Derivatives*

Simple derivatives are not normally indexed if they follow in the same short section of the text.

5. *Collective and plural entries*

In place of "– derivatives" or "– compounds" the plural entry has normally been used. Plural entries have occasionally been used where compiunds of the same name but differing numbering appear in the same section of the text.

6. *Main entries*

The main entry of the more common individual compounds is indicated by heavy type. Multiple entries, such as headings and sub-headings over several pages are shown by "–", e.g., 67–74, 137–139, etc.

INDEX